River System Analysis and Management

Nayan Sharma
Editor

River System Analysis and Management

Springer

Editor
Nayan Sharma
Department of Water Resources Development
and Management
Indian Institute of Technology Roorkee
Roorkee, Uttarakhand, India

ISBN 978-981-10-1471-0 ISBN 978-981-10-1472-7 (eBook)
DOI 10.1007/978-981-10-1472-7

Library of Congress Control Number: 2016955958

Printed on acid-free paper

This Springer imprint is published by Springer Nature
The registered company is Springer Science+Business Media Singapore Pte Ltd.

Dedicated to
My Late Father
Professor Parameswar Sharma

Preface

River management is a big challenge of the twenty-first century for the scientific community associated with water resources. River basins are complex systems. They are open systems with sometimes ill-defined boundaries. It refers to various aspects essential to achieve a sustainable development of river basins, including water demand and river management. Rivers may have shared delta, watershed limits in flatland areas are either vague or man-made (and alterable), and watershed limits often do not correspond exactly with aquifer limits. On top of this, river basins interact continuously with the atmosphere (precipitation and evaporation, airborne pollution) and the receiving waters (seas and sometimes lakes). Furthermore, the uses made of river basins often transcend river basin boundaries (e.g. interbasin water transfers) as well as political borders of countries.

Rivers are important as they fulfil many important functions, such as water supply for households, industry and agriculture, navigation, fishing, recreation and 'living space'. Economic and social development and even life itself cannot be sustained without sufficient water at the right time and place and of right quality. Integrated river management is about all these things. It is much broader than traditional water management and includes significant parts of land-use planning, agricultural policy and erosion control, environmental management and other policy areas. It covers all human activities that use or affect freshwater systems. In brief, integrated river basin management is the management of water systems as part of the broader natural environment and in relation to their socio-economic environment. The problem has further been compounded by deforestation in catchment areas, siltation of rivers and diversion of water for irrigation and construction of cross structures such as bridges.

Although water is abundant on earth, the amount for mankind is very limited as freshwater is less than 1 % of the world's water which too has to be shared by many competing users. In many areas, freshwater resources are under increasing stress due to escalated multisectoral demand leading to water scarcity.

Growing population is a serious concern in the future as it further dampens the per capita freshwater availability from rivers, lakes, etc. According to the Falkenmark water stress indicators, as the water available per capita falls in the range of 1000–1700 m^3 per year, it indicates water stress, and when it reduces below 1000 m^3 per year, water scarcity starts to take its toll

(Falkenmark 1989). Several areas of the world are in the throes of water stress condition due to sharply rising demand.

Climate change is one of the most important global environmental challenges facing humanity with implications for freshwater supply, natural ecosystems, food production, health, etc. Therefore, before planning and management of any long-term water resources, the assessment of climate change impacts on the hydrological resources is of prime importance.

Intensified erosion, land-water degradation and streamflow pollution call for appropriate river restoration and training measures. This keeps presenting challenges to engineers and social scientists, which subsequently led to a shift from Integrated Water Resources Management (IWRM) to Integrated Land and Water Resources Management (ILWRM).

The successful use of science in formulating public policy for the restoration of rivers relies on accurate communication between researchers and decision makers. Rivers are characterised by multiple scales of length and time. A viable theory of river behaviour must reconcile the various processes that occur at different scales in order to develop a knowledge base by synthesising research and field study results.

The book is intended to augment the knowledge base of behaviour of rivers and analyse the issues related to rivers so as to develop river system management techniques emerging from in-depth scientific analysis as a priority. With the pressure increasing on freshwater reserves and climate change taking its toll, management of rivers system is deemed necessary.

The book comprises 8 sections encompassing 19 chapters and focuses on conduct of analytical treatment preceding sound river management interventions. The inherent analytical issues need to be adequately addressed using a sustained and in-depth study. The multifaceted complex issues related to river system analysis and management call for a multipronged interdisciplinary approach, and this book is an attempt in this direction.

The book is expected to be useful to academics, practitioners, scientists, water managers, environmentalists, administrators, researchers and students who are involved and have stake in water management and river system analysis. The editor avails this opportunity to profusely express umpteen gratitude to the contributors for their devoted effort in producing their chapters. Without the unstinting and dedicated support extended by the local editorial team comprising Dr. Dheeraj Kumar, Dr. Harinarayan Tiwari and Subash Prasad Rai, this book could not have been realised in the present shape. The unceasing secretarial assistance received from Neeraj Kumar is also duly acknowledged here. Last but not the least, the supportive role played by my wife Smrity Sharma and daughters Amrita and Ananya is acknowledged here for their words of inspiration in the course of working on the book manuscript.

Roorkee, Uttarakhand, India Nayan Sharma

Contents

About the Editor

Prof. Dr. Nayan Sharma is with IIT Roorkee, teaching, researching and consulting on water resources including river engineering, hydraulic structures, inland navigation and concrete dams. Earlier, in the Assam Flood Control and Irrigation Department, he was associated with design and implementation of several barrages/dams and river training structures in the Brahmaputra Basin. He has been a supervisor of 170 M.Tech. theses and 21 PhDs in IIT Roorkee and an examiner of 9 PhD theses, with 188 research publications. He is a member of 60 national and international technical committees. He conducted 85 consulting and 9 national and international research projects. He published three books by Springer and INCID. He was a recipient of the Indo-Swiss Bilateral Research Award as visiting professor in the Swiss Federal Institute of Technology Lausanne. He delivered talks in the Imperial College in London, Southampton University and Edinburgh University in the UK, the University of Lausanne in Switzerland, the University of Iowa in the USA, IST Lisbon in Portugal, University of Göttingen in Germany and the Engineers' Conclave 2014 at IISc in Bangalore. He delivered the Dr. K. L. Rao Memorial Lecture in the 26th National Convention of the Institution of Engineers (India). He gainfully researched on the new Piano-Key Weir technology for dam safety and instream storage development as viable adaptation measure for climate change effects and fruitfully implemented his design of RCC Jack Jetty for channelising Ganga reach near Varanasi with encouraging results. Based on his comprehensive model investigation, the new Piano-Key Weir technology was implemented in the Sawra-Kuddu Hydroelectric Project located on Pabbar River in Himachal Pradesh, India. He is recently appointed Honorary Professor of River Science in the School of Geography of The University of Nottingham, UK.

Introduction

1

Nayan Sharma, Subash Prasad Rai, Dheeraj Kumar,
and Harinarayan Tiwari

Abstract

Water is required to ensure food security, feed livestock, and industrial production and to conserve the environment. Human population has always been dependent on the rivers for survival. Management of rivers has been attempted by humans since the ancient times as the civilizations developed in the vicinity of the rivers. As the world's population grows, the demand for water mounts and pressure on finite water resources intensifies. But the importance of population is major factor to consider. However, the brutal challenge of climate change resulting in changes in rainfall regimes, threatening surface as well as groundwater, contributes to making water resource scarcity a reality. Changes in hydrological cycle will certainly alter the precipitation and evapo-transpiration patterns, resulting in significant changes in the discharge regime of rivers. Moreover, it may lead to greater unreliability of dry season flows that possess potentially serious risks to water and energy supplies in the lean season. Therefore, before planning and management of any long-term water resources, the assessment of climate change impacts on the hydrological resources is of prime importance. The two main policy responses to climate change are mitigation and adaptation which is necessary to deal with the impacts of climate change. Adaptation measures may be planned in advance or put in place spontaneously in response to a local pressure. Incidences of waterrelated disasters are showing an upward trend due to climate change impacts thereby increasing the frequency and intensity of extreme weather. These issues intensify the phenomenon of erosion, land–water degradation, and pollution which demand for river restoration and training, as rivers are the essential element for sediment transport from

N. Sharma (✉) • S.P. Rai • D. Kumar • H. Tiwari
Department of Water Resources Development and Management, Indian Institute of Technology Roorkee, Roorkee, Uttarakhand, India
e-mail: nayanfwt@gmail.com

N. Sharma (ed.), *River System Analysis and Management*, DOI 10.1007/978-981-10-1472-7_1

surface land to oceans. The complexity of implementing the Integrated Land and Water Resources Management (ILWRM) increases manifold when the river crosses political border, thus making it a transboundary in nature which are not regulated by ratified international laws. Hence the need for transboundary international water law is overwhelming, constant, and immediate.

1.1 Rising Demand on Freshwater

Glaciers and ice caps (constitute 70 % of the world's freshwater resources) cover about 10 % of the world's landmass mostly concentrated in Greenland and Antarctica. Unfortunately, these resources are not readily accessible for human use as they are majorly located at places far from human habitation. According to the United States Geological Survey (USGS) "96 % of the world's frozen freshwater is at the South and North Poles, with the remaining 4 % spread over 550,000 km^2 of glaciers and mountainous icecaps measuring about 180,000 km^3" (UNEP 1992; Untersteiner 1975). Apart from the poles, the Himalayas is a source of huge freshwater reserves. Numerous glaciers are found in the Himalayas making it apt to be called the third pole of the Earth. Figure 1.1 shows the volume of glaciers and permanent ice caps continent wise.

Management of rivers has been attempted by humans since the ancient times as the civilizations developed in the vicinity of the rivers. Flow control structures (dams) and bank protection works (dyke) were found in the

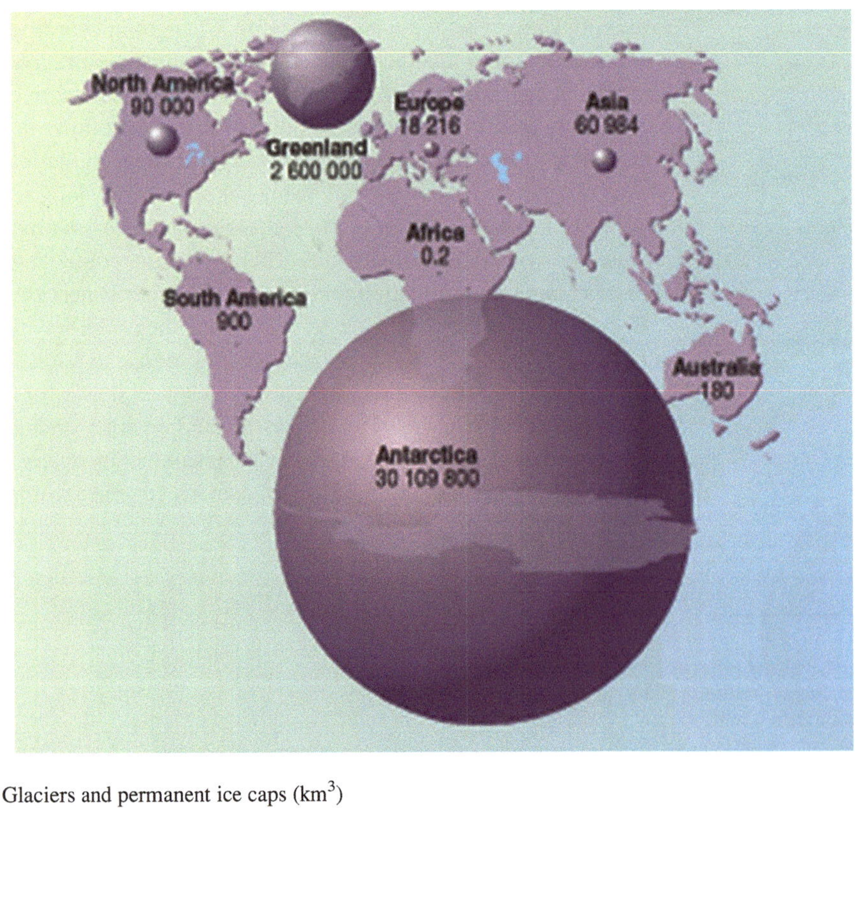

Fig. 1.1 Glaciers and permanent ice caps (km^3)

ancient civilizations of Mesopotamia and Egypt. To increase food supply, they had used indigenous methods to interact with rivers. Over the time, development of science and engineering disciplines had established modern techniques to make river flow regulation easier. But even till date, the knowledge of various intricate aspects of rivers is very limited and case specific. This book is an attempt to support the knowledge base by introducing some latest developments in the field of river management along with novel case studies.

The availability of water and its quality have always played an important role in determining not only where people can live but also their quality of life. Rivers are the most accessible sources of freshwater, and this indicates that for sustainability, river management is a necessity. When river management is discussed in today's scenarios, it would not be wise to ignore the aspects of demographics and climate change. The impacts of ongoing climate change and the rising demographic pressures on the freshwater resources have been widely discussed in literature.

1.2 Demographic Pressures

Human population has always been dependent on the rivers for survival. Water is required to ensure food security, feed livestock, and industrial production and to conserve the environment. As the world's population grows, the demand for water mounts and pressure on finite water resources intensifies. Water though abundant on Earth, the amount for mankind is very limited as freshwater is less than 1 % of the world's water which too has to be shared by many competing users. In many areas freshwater resources are under increasing stress due to escalated demand leading to water scarcity. In the last century, the total water use has grown rapidly; agricultural water use went from about 500–2500 km^3/year, while the total use rose from around 600 to more than 3000 km^3/year. Freshwater finds various uses including but not limited to drinking, bathing, cooking, and washing which comprise the domestic and municipal uses; food and crop production, i.e., agricultural uses; and mining, refinery, power, and energy which constitute the industrial uses. Figure 1.2 shows the pressure

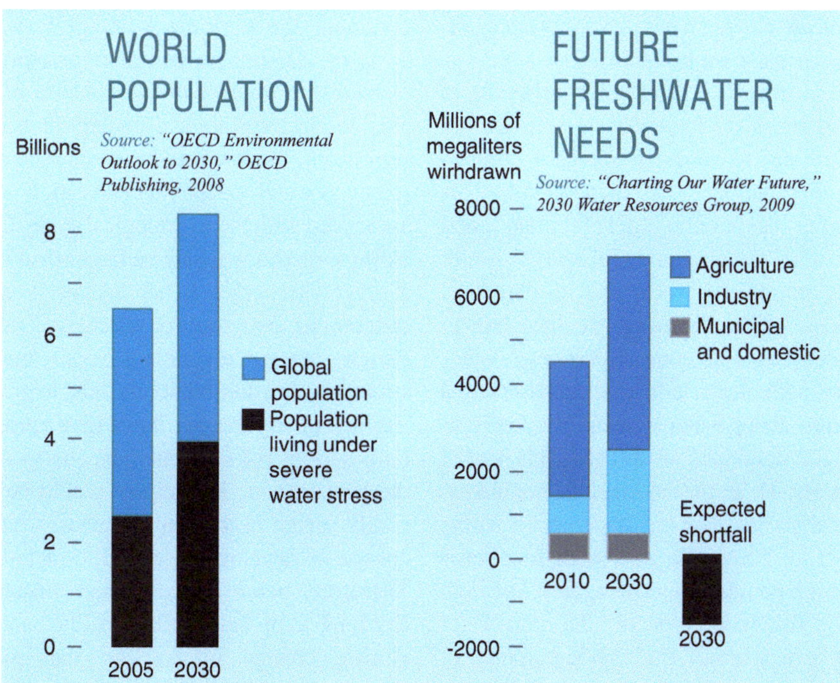

Fig. 1.2 World population and pressure on global freshwater resources

Table 1.1 Per capita water availability in India

Year	Population (Million)	Per capita water availability (m^3/year)
1951	361	5177
1955	395	4732
1991	846	2209
2001	1027	1820
2025	1394	1341
2050	1640	1140

Source: Government of India, 2009

on freshwater resources due to the growing water requirements for agriculture, industry, and municipal and domestic needs.

Growing population is a serious concern in the future as it further dampens the per capita water availability. The Indian example clearly highlights the case shown in Table 1.1, where the per capita water availability has drastically fallen from 5177 m^3 per year in 1951 to 1820 m^3 per year by 2001 which is estimated to reach 1140 m^3 per year in 2050. According to the Falkenmark water stress indicators, as the water available per capita falls in the range of 1000–1700 m^3 per year, it indicates water stress, and when it reduces below 1000 m^3 per year, water scarcity starts to take its toll (Falkenmark 1989). Some of the major water-scarce countries with per habitant freshwater availability have been shown in Table 1.2.

Population growth is a major contributor to the growing stress on freshwater resources and hence contributes to water shortages leading to scarcity. Population explosion—with its accompanying economic development and industrialization—has transformed water ecosystems around the world and resulted in a massive loss of biodiversity. It has led to escalated demand and competition over water for domestic, agriculture, industrial, and municipal uses. Concern about water availability grows as freshwater use continues at unsustainable levels. Approximately 41 % of the world's population lives in river basins that are under water stress (Revenga et al. 2000). Regions with few water resources, high population growth rates, and high population densities are typically the most water stressed or scarce areas. But the importance of population is not the only factor to consider. The brutal challenge of climate change resulting in changes in rainfall regimes, threatening surface as well as groundwater, contributes to making water resource scarcity a reality.

Table 1.2 Top water-scarce areas of the world in 2005 (http://www.fao.org/nr/water/aquastat)

Country	Total actual renewable water resources per habitant (m^3/year)
Kuwait	7.4
UAE	48.2
Qatar	71
Maldives	96
Saudi Arabia	98
Yemen	100
Libya	106
Bahrain	160

1.3 Climate Change and Impacts

Climate change is one of the most important global environmental challenges facing humanity with implications for freshwater supply, natural ecosystems, food production, health, etc. Intergovernmental Panel on Climate Change (IPCC) (2007a) reported that for the Indian subcontinent, the warming would be between 1 and 2 °C by 2030. This will alter precipitation and evapotranspiration patterns because of vigorous changes in hydrological cycle which may have significant impact on the discharge regime of rivers, as water availability and runoff are directly associated with the hydrological cycle. Changes in extreme discharges may be larger than changes in average discharges which makes it an area of concern as extremes of discharge are used to design water regulations and management structures. In addition, it may lead to greater unreliability of dry season flows that possess potentially serious risks to water and energy supplies in the lean season. The impact would be predominantly severe in the tropical areas, which mainly consist of developing countries, including India. Therefore, before planning and management of any long-term water resources, the assessment of climate change impacts on the hydrological resources is of prime importance.

1.3.1 Climate and Hydrology

Climate is defined as the general weather conditions over a certain time span and a certain area (IPCC 2001). In the United Nations Framework Convention on Climate Change (UNFCCC), climate change refers only to the anthropogenic changes over comparable periods. However, in IPCC usage, climate change consists of both natural variability and human-induced change, despite the fact that most of the observed increase in global average temperature since the mid-twentieth century is likely related to anthropogenic activity (IPCC 2007b).

In the hydrological cycle, water moves continually between oceans and the atmosphere through different processes such as precipitation, percolation, and evaporation over various temporal and spatial scales. Under natural conditions, climate variations are considered to be one of the major causes of hydrological change and have crucial social and economic implications for water resources and flood risk (Acreman 2000; Wheater 2002). As anthropogenic climate change affects the energy and mass balance of the fundamental hydrological processes, the water cycle is expected to be intensified (Huntington 2006), and hydrological patterns are very likely to be different under different climate scenarios (Bates et al. 2008). Although, there are distinctions between natural variability and anthropogenic climate abnormality, both human activity and natural climate influence intertwine with current climate events, and the changes in climate are expected to affect the balance of water distribution and living organisms on the Earth (IPCC 2007c).

1.3.2 Effect of Climate Change on Freshwater Reserves and Glaciers

Any alteration in discharges of river, levels of lake, and wetland due to climate change is hinged upon changes in the volume, timing and intensity of precipitation (Chiew and McMahon 2002), snowmelt, and forms of precipitation (i.e., snow, rain, etc.). Evapotranspiration is directly affected by changes in temperature, wind speed, atmospheric humidity, and radiation. Alterations in evapotranspiration can upset spatial and temporal precipitation patterns further exaggerating the effects on surface waters. Lake levels are on a constant decline in many regions, for example, the Laurentian Great Lakes region (Sellinger et al. 2007; White et al. 2008) and western basins (Barnett et al. 2008).

In basins primarily dependent on snowmelt, higher temperatures lead to reduced streamflow and thus decreased water supply in summer (Barnett et al. 2005), particularly evident in the basins of the Hindu Kush Himalaya and the South American Andes, which are sustained by glacier melt during the summer season (Singh and Kumar 1997; Mark and Seltzer 2003; Singh and Bengtsson 2003; Barnett et al. 2005; Coudrain et al. 2005). Higher temperatures, a consequence of global warming, can accelerate melting of glaciers augmenting river flows in the short term, but in the long run, river flows will gradually subside. There is considerable evidence highlighting decline in the volumes of water stored in lakes and rivers due to decrease in long-term average precipitation and runoff as well as increased rates of evaporation. As more than one sixth of the Earth's population relies on freshwater resources obtained from melting of glaciers and seasonal snowpacks for their water uses, the consequence of projected changes for future water availability (Fig. 1.3) predicted with high confidence and already diagnosed in some regions will be adverse and severe.

Issues related to climate change which directly impact human population:

- Increase in incidents of floods and droughts.
- Agriculture is largely affected due to low productivity or damage to crops.
- Significant impact on infrastructural river-based systems like hydropower, inland navigation, etc.
- Ecotourism is hampered.

There are two main policy responses to climate change: mitigation and adaptation. Climate change mitigation consists of actions to limit the magnitude and/or rate of long-term climate

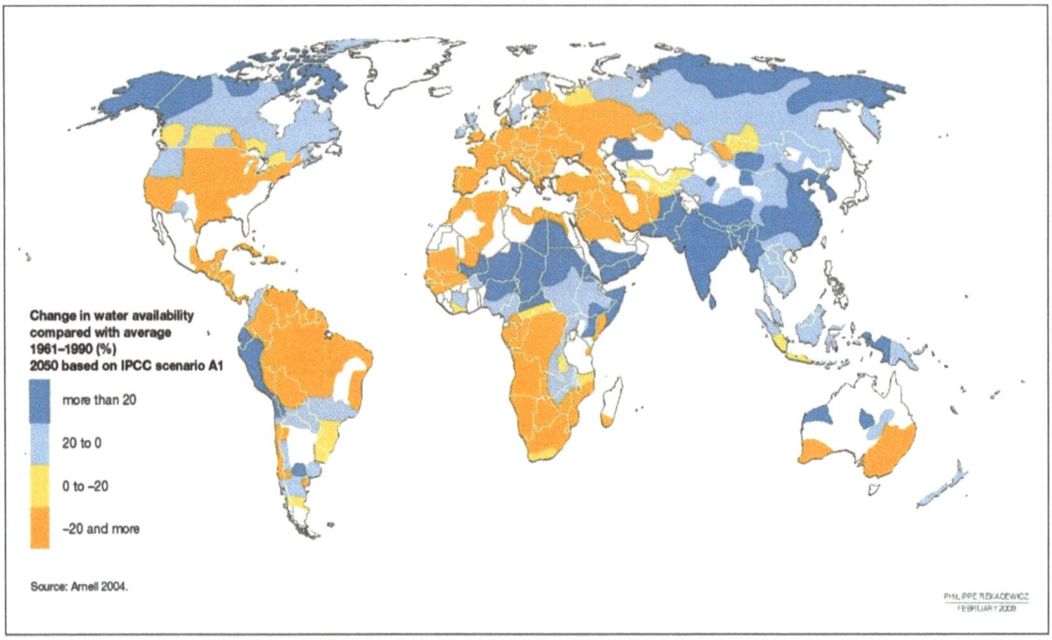

Fig. 1.3 Change in water availability based on IPCC scenario A1

change. It generally involves reductions in human (anthropogenic) emissions of greenhouse gases (GHGs) (IPCC 2007a, b, c). Adaptation means anticipating the adverse effects of climate change and taking appropriate action to prevent or minimize the damage they can cause or taking advantage of opportunities that may arise. Adaptation measures may be planned in advance or put in place spontaneously in response to a local pressure. Adaptation attempts to reduce the risks posed by the consequences of climatic change, while mitigation addresses the root causes (Smit and Pilifosova 2003). Both approaches are indeed necessary to deal with the impacts of climate change.

1.4 River and River System Management

The naturally flowing bodies of waters are named as rivers, mostly having a source from glacier (snowmelt) or spring. Rivers are an essential component of the Earth's water cycle by carrying immense amounts of water from the land to the sea. In general, water accumulates in a river from glacier's melt, rainfall, spring, and subsurface flow from groundwater. Erosion, transport, and deposition can occur in relatively stable conditions and are natural processes in the dynamic stability of rivers (Endreny 2012a). When these processes become extreme, transport becomes degradation, and deposition becomes aggradation. Erosion of streams can have natural, human, or a combined set of triggers. Human modifications to the landscape strongly influence triggers to instability, accelerating the erosion potential and sediment transport regimes of channels, altering their flow by shifts in aggradation and degradation processes (Endreny 2012b).

River mathematical models have been employed to examine diverse scenarios or determine the optimum usage of water resources within control volume. Mathematical models have frequently tuned out the underlying complexity between the various water resource parameters within control volume and generally concentrated on an isolated water resource problem without studying it in the broader river basin context. According to the US National Research, management of the river basin will not be effective unless "scientific analysis is married with public participation, ensuring that decisions based on cultural values are informed decisions with respect to likely consequences and a clear understanding of who benefits and who pays." Integrated hydrologic–ecologic–economic mathematical programming models (also known as holistic models) can be used to surmount the aforesaid issues (Cai et al. 2015).

Precipitation is the most important atmospheric input to land surface hydrological models due to its spatial and temporal variability, which makes rainfall–runoff relationship one of the most complex hydrological phenomena (Kumar et al. 2014). Therefore, accurate precipitation inputs are essential for reliable hydrological prediction. However, in many parts of the developing world, ground-based precipitation measurements are either sparse or nonexistent (Hughes et al. 2006; Su et al. 2008). Precise rainfall data at fine temporal and spatial resolution is essential for many hydrological and water resource management applications and particularly in data-scarce river basins and areas where strong competition for water resources persists. Kumar et al. (2015) concluded that the satellite-derived rainfall estimate (TRMM), with suitable bias correction techniques, can serve as an alternative to the measured rainfall dataset where data are insufficient for runoff prediction (Kumar et al. 2015).

River contamination has been realized as a contributor to a broad range of health problems and disorders in human. The problems of river water quality develop due to the following reasons:

- Poor governance as well as weak policy on the discharges of wastewaters from industries
- Inadequately treated municipal sewage
- Conservative and inadequate land use practices
- Faulty industrial locations
- Uncontrolled agricultural exercises and excessive use of fertilizers
- Lack of integrated watershed management

The consequences of which jeopardize the water resource systems, endanger public health risks, and intensify erosion and sedimentation, leading to land and water resource degradation (Azab 2012).

Intensified erosion, land–water degradation, and pollution demand for river restoration and training. This keeps presenting challenges to engineers and social scientists, which subsequently led to a shift from integrated water resource management (IWRM) to integrated land and water resource management (ILWRM). River restoration science begins with theory, because it is the theory which allows the investigator to identify what to measure and how to construct a conceptual model that connects the measurements together. Like all sciences, geomorphology is in a state of constant change and revision. The successful use of science in formulating public policy for the restoration of rivers relies on accurate communication between researchers and decision makers, but this connection sometimes suffers from failings by both participants (Graf 2008). Rivers are characterized by multiple scales of length and time. A viable theory of river behavior must reconcile the various processes that occur at different scales.

Most natural disasters are caused by water- and climate-related events such as floods, droughts, hurricanes, storm surges, and landslides. Incidences of water-related disasters are showing an upward trend as the impact of climate change increases the frequency and intensity of extreme weather. As a river continues along its course, the surrounding terrain flattens out and the river widens even meandering along their middle course. Due to highly sinuous behavior of rivers, the extensive floodplain may experience severe flooding rather than quickly route water downstream. The core components of fluvial geomorphology and floodplain analysis are assessments of pattern (e.g., meanders, cutoffs), profile (e.g., bed and valley slopes), and dimension (e.g., cross-sectional incision and width) (Endreny 2012a, b). Each of these features impacts the occurrence and conveyance of flood water.

The weather and climate events can cause disruptions in health and social services, scarcities of food and water, and an increase in conflict and migration. People in developing countries are particularly vulnerable due to urbanization, which accelerates the concentration of people in cities and delays the provision of social infrastructure. Natural disasters do more than claim life; they directly impact people's livelihood and aggravate the poverty cycle. Whereas the previous mainstreams of disaster assistance were centered more on structural measures such as construction of dams and levees, there is also a need for compound measures that emphasize nonstructural assistance such as installation of disaster warning systems, creation of hazard maps, and evacuation drills to improve the emergency response of people and society to disasters (JICA).

As most rivers cross national or international borders, it brings transboundary dimension in the management of rivers—hydropolitics. Given the importance of transboundary rivers for societies, economies, and the environment, it is important to understand the conditions under which hydropolitics result in untenable political outcomes, such as militarized disputes, versus more cooperative outcomes, such as international agreements. This importance is underscored by the changing and emerging nature of the issues that affect water management. The challenges of maintaining the supply and health of transboundary rivers and the challenges of deciding how to allocate transboundary water among the countries and sectors can lead to political tensions or hydropolitics. According to Elhance (2000), the complexity of hydropolitics makes it one of the most challenging arenas of international interaction in some regions of the world (Elhance 2000).

1.5 Sediment Yield and Sediment Load

The problem of sediment load and its transport, coming from the watershed, is a major source of river instability. All of the water that reaches a stream and its tributaries carries sediment eroded from the entire area drained by it. Sediment yield is the total amount of erosional debris exported from any drainage basin. Sediment transport from surface land to oceans via rivers is one of the essential processes regulating riverbank stabilization. Due to temporal changes in geological and hydrological activities, sediment loads in rivers are also found to be variable. Syvitski et al. reported that about 65 % of the water and 80 % of the sediment globally come from southern Asia, Oceania, and northeastern South America (Syvitski et al. 2003). Of all the regions, the Asian rivers are most affected by the problem of sediment load. The average sediment load transported by the rivers from different regions of the world has been shown in Fig. 1.4.

The sediment load in a river highly depends on geological (soil character in the watershed), hydrological (rainfall and runoff), as well as social factors (human interference). Human interference (in terms of hydraulic structures or bank protection) can have rigorous consequences (reduction in sediment availability) on agriculture due to fertile soil loss. The construction of Hoover Dam on the Colorado River has decreased the sediment discharge from about 125 million tons per year to 3 million tons per year (Mulder and Syvitski 1996). Climate and land use change and river engineering can drive changes in sediment supply and river floodplain morphology. In India, bulk of the 3×10^9 tons $year^{-1}$ sediment input by the rivers to the world ocean is supplied by the large northern Himalayan river systems (Indus, Ganges, and Brahmaputra). Sediment and morphodynamics influence flood risk through changes in channel dimensions, grain roughness, and form roughness (planform style).

1.6 Transboundary Rivers

As stated earlier in brief, transboundary rivers cross at least one political border, either a border within a nation or an international boundary. Most rivers are transboundary in nature which forms the hurdle in harnessing and managing the rivers. Rivers are shared but in most cases there is no water-sharing mechanism in place. According to the Transboundary Freshwater Dispute

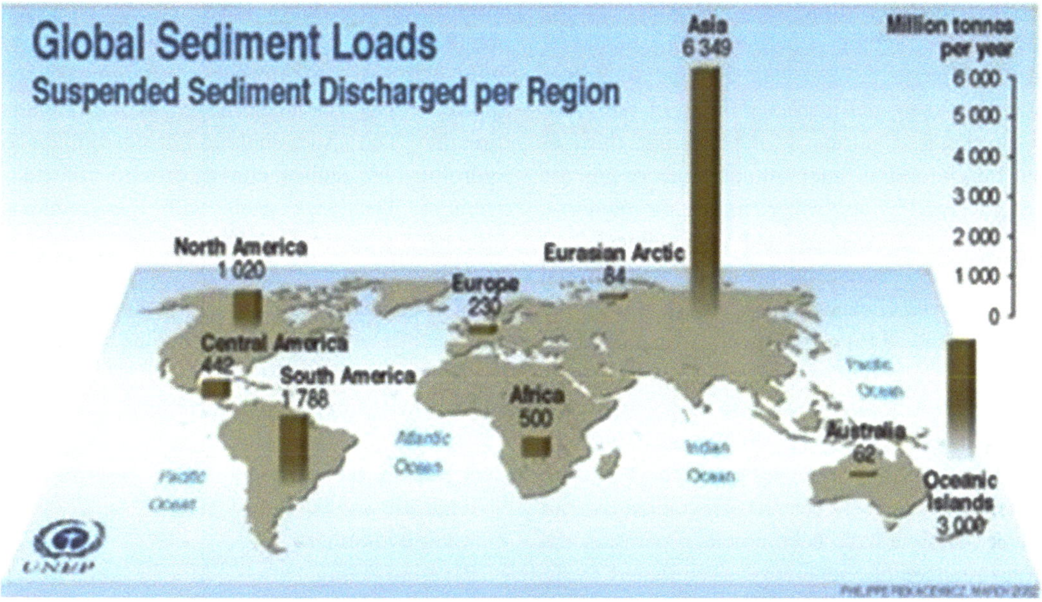

Fig. 1.4 Global sediment loads discharged per region (Source: Peter H. Gleick, Water in Crisis, 1993)

Database (TFDD), the world consists of 276 international transboundary river basins which cover more than 45 % of the land surface on the Earth excluding Antarctica (TFDD 2012). A total of 145 countries which constitute over 75 % of all countries have shared river basins within their boundaries, while 33 countries have more than 95 % of their territorial dominion within international river basins. The internationally shared river basins are home to over 40 % of the world's population and contribute to approximately 60 % of the global river flows (Draper 2002; Giordano and Wolf 2003; Sadoff and Grey 2005; Wolf 1999; Wolf et al. 1999). The total volume of water in the world's rivers is estimated at 2115 km^3 (Groombridge and Jenkins 1998). Even more important is count of the nations that constitute certain individual basins. The Danube basin is shared by 17 countries, the highest in the world. The Zambezi, Rhine, Nile, Niger, and Congo are each shared by nine or more countries, while the La Plata, Neman, Vistula, Ganges–Brahmaputra–Meghna, Tigris–Euphrates, Aral Sea, Volga, Mekong, Jordan, Tarim, Kura–Araks, Amazon, and Lake Chad basins each consist territory of at least five countries (Wolf et al. 1999).

Over the last few decades, there has been a growing speculation about the possibility of an acute conflict or even war over freshwater resources. Scholars increasingly have been pointing out that the twenty-first century might see the battles fought for waters and the prime reason being water scarcity. Water, a scarce resource with no substitute, is perhaps the only scarce resource not regulated by ratified international laws. Hence the need for transboundary international water law is overwhelming, constant, and immediate. Many have predicted that future wars are ought to be fought over water. Due to this, "water" and "war" are two emerging issues being assessed collectively with increasing frequency (Wolf 1998). Rapidly growing population, increasing industrialization, pollution, changing resource, and energy needs bring into focus the situation in which riparians are finding themselves. As the exponential growth of global population continues and as changes in environment threaten the quality and quantity of water resources, the ability of countries to peacefully resolve water conflicts over transboundary international river basins will play a major role in stable and secure international relations.

The need for effective joint water-sharing mechanism among riparians has increased, greatly due to the growing demand for water and increasing detrimental effects of activities in upstream countries. More than one third of the transboundary international river basins are not governed by any water-sharing mechanisms, agreements, or treaties, and only some 30 of them truly have cooperative institutional arrangements (Draper 2007). Hence, there is a significant need for a framework which can provide procedures and guidance to assist the development of institutional water-sharing agreements internationally (Rai et al. 2014). This could provide the base for more efficient and effective water sharing between nations. Scholars of water conflicts have been quick in pointing out that instances of cooperation over shared freshwater rivers outnumber conflicts by a big margin. But the causes of conflict should not be dismissed, and similarly the need for various ways to promote cooperation must be investigated further.

What can be the way forward? Either to continue with the present system which benefits only limited riparians and harms the other riparians, i.e., win–loss (I win, you lose) approach, or move toward cooperative mechanism of river management which promotes harnessing the enormous economic potential of rivers in win–win approach for all. The various stakeholders of water call for joint management of rivers keeping in view the aspect of sustainability.

1.7 Objective of the Book

The objective of the book is to add on to the knowledge base of behavior of rivers and understand the issues related to rivers and to develop river system management as a priority. With the pressure increasing on freshwater reserves and climate change taking its toll, management of rivers system is deemed necessary. The issues highlighted herein seek greater attention from researchers and water managers. These issues need to be adequately addressed using a sustained and in-depth study. The multifaceted complex issues related to river system management call for a multipronged interdisciplinary approach and this book is an attempt in this direction. The issues include but not limited to hydrology, hydraulics, climate change, and management. The book deals with the problems associated with rivers and its management ranging from river hydraulics to river training to water quality and even touching upon transboundary issues. The following aspects of river management have been dealt in this book embellished with novel case studies:

- Sediment transport
- Land use and climate
- River hydraulics
- River modeling
- River training
- Water quality and ecology
- Transboundary river issues
- Disaster management

References

Acreman MC (2000) The hydrology of the UK: a study of change, 1st edn. Routledge, London
Azab AM (2012) Integrating GIS, remote sensing and mathematical modelling for surface water quality management in irrigated watersheds. Delft University of Technology, TU Delft
Barnett TP, Adam JC, Lettenmaier DP (2005) Potential impacts of a warming climate on water availability in snow-dominated regions. Nature 438(7066):303–309
Barnett TP, Pierce DW, Hidalgo HG, Bonfils C, Santer BD, Das T, Dettinger MD (2008) Human-induced changes in the hydrology of the western United States. Science 319(5866):1080–1083
Bates BC, Kundzewicz ZW, Wu S, Palutikof JP (2008) Climate change and water. Technical paper of the intergovernmental panel on climate change. IPCC Secretariat, Geneva
Cai X, Marston L, Ge Y (2015) Decision support for integrated river basin management—scientific research challenges. Sci China Earth Sci 58(1):16–24
Chiew FHS, McMahon TA (2002) Modelling the impacts of climate change on Australian streamflow. Hydrol Process 16:1235–1245
Coundrain A, Francou B, Kundzewicz ZW (2005) Glacier shrinkage in the Andes and consequences for water resources-editorial. Hydrol Sci J 50(6):925–932

Draper SE (2002) Model water sharing agreements for the twenty-first century. ASCE Publications, Reston, p 166
Draper SE (2007) Introduction to transboundary water sharing. J Water Res Plan Manag 133(5):377–381
Elhance AP (2000) Hydropolitics: grounds for despair, reasons for hope. Int Negot 5(2):201–222
Endreny T (2012a) Applying FGM to river conveyance. <http://www.fgmorph.com/fg_2_5.php>. (1/7/2015, 2015)
Endreny T (2012b) River stability & predictability. <http://www.fgmorph.com/fg_2_8.php>. (1/7/2015, 2015)
Falkenmark M (1989) The massive water scarcity now threatening Africa: why isn't it being addressed? Ambio 18:112–118
Giordano MA, Wolf AT (2003) Sharing waters: post-Rio international water management. Nat Res Forum 27(2):163–171
Gleick PH (1993) Water in crisis. Pacific Institute for Studies in Dev., Environment & Security. Stockholm Env. Institute, Oxford University Press. 473p, 9
Graf WL (2008) Sources of uncertainty in river restoration research. River restoration: managing the uncertainty in restoring physical habitat, Wiley, pp 15–19
Groombridge B, Jenkins M (1998) Freshwater biodiversity: a preliminary global assessment
Hughes DA, Demuth S, Gustard A, Planos E, Seatena F, Servat E (2006) An evaluation of the potential use of satellite rainfall data for input to water resource estimation models in southern Africa. Climate variability and change: hydrological impacts, IAHS publication, pp 75–80
Huntington TG (2006) Evidence for intensification of the global water cycle: review and synthesis. J Hydrol 319:83–95
IPCC (2001) Climate change 2001: the scientific basis. In: Houghton JT, Ding Y, Griggs DJ, Noguer M, van der Linden PJ, Dai X, Maskell K, Johnson CA (eds) Contribution of working group I to the third assessment report of the intergovernmental panel on climate change. Cambridge University Press, Cambridge/New York, p 881
IPCC (2007a) Climate change 2007: mitigation: contribution of working group III to the fourth assessment report of the intergovernmental panel on climate change. Cambridge University Press, Cambridge/New York
IPCC (2007b) Impacts, adaptation and vulnerability. Contribution of working group II. In: Parry ML, Canziani OF, Palutikof JP, van der Linden PJ, Hanson CE (eds) Fourth assessment report of the intergovernmental panel on climate change. Cambridge University Press, Cambridge, pp 541–580
IPCC (2007c) The physical science basis. Contribution of working group I to the fourth assessment report of the intergovernmental panel on climate change. Cambridge University Press, Cambridge/New York, p 996
JICA Water Resources and Disaster Management. <http://www.jica.go.jp/english/our_work/thematic_issues/water/overview.html>. (1/7/2015, 2015)
Kumar D, Pandey A, Sharma N, Flügel W-A (2014) Modeling suspended sediment using artificial neural networks and TRMM-3B42 version 7 rainfall dataset. J Hydrol Eng 20(6):C4014007
Kumar D, Pandey A, Sharma N, Flügel W-A (2015) Evaluation of TRMM-precipitation with raingauge observation using hydrological model J2000. Unpublished manuscript
Mark BG, Seltzer GO (2003) Tropical glacier meltwater contribution to stream discharge: a case study in the Cordillera Blanca, Peru. J Glaciol 49(165):271–281
Mulder T, Syvitski J (1996) Climatic and morphologic relationships of rivers: implications of sea-level fluctuations on river loads. J Geol 104:509–523
Rai SP, Sharma N, Lohani A (2014) Risk assessment for transboundary rivers using fuzzy synthetic evaluation technique. J Hydrol 519:1551–1559
Revenga C, Brunner J, Henninger N, Kassem K, Payne R (2000) Pilot analysis of global ecosystems: freshwater systems. World Resources Institute, Washington, DC
Sadoff CW, Grey D (2005) Cooperation on international rivers: a continuum for securing and sharing benefits. Water Int 30(4):420–427
Sellinger CE, Stow CA, Lamon EC, Qian SS (2007) Recent water level declines in the Lake Michigan – Huron System. Environ Sci Technol 42(2):367–373
Singh P, Bengtsson L (2003) Effect of warmer climate on the depletion of snow-covered area in the Satluj basin in the western Himalayan region. Hydrol Sci J 48(3):413–425
Singh P, Kumar N (1997) Impact assessment of climate change on the hydrological response of a snow and glacier melt runoff dominated Himalayan river. J Hydrol 193(1–4):316–350
Smit B, Pilifosova O (2003) Adaptation to climate change in the context of sustainable development and equity. Sustain Dev 8(9):9
Su F, Hong Y, Lettenmaier DP (2008) Evaluation of TRMM Multisatellite Precipitation Analysis (TMPA) and its utility in hydrologic prediction in the La Plata Basin. J Hydrometeorol 9(4):622–640
Syvitski JP, Peckham SD, Hilberman R, Mulder T (2003) Predicting the terrestrial flux of sediment to the global ocean: a planetary perspective. Sediment Geol 162(1):5–24
TFDD (2012) International river basin register. College of Earth, Ocean, and Atmospheric Sciences, Oregon State University, Corvallis, USA
UNEP (1992) Glaciers and the environment. UNEP/GEMS Environment Library No. 9, p.8. UNEP, Nairobi
Untersteiner N (1975) Sea ice and ice sheets and their role in climatic variations. WMO The Phys. Basis of Climate and Climate Modelling, p 206–224(SEE N 76-19675 10-47)
Wheater HS (2002) Progress in and prospects for fluvial flood modelling. Philos Trans Math Phys Eng Sci 360(1796):1409–1431
White MS, Xenopoulos MA, Hogsden K, Metcalfe RA, Dillon PJ (2008) Natural lake level fluctuation

and associated concordance with water quality and aquatic communities within small lakes of the Laurentian Great Lakes region. Hydrobiologia 613(1):21–31

Wolf AT (1998) Conflict and cooperation along international waterways. Water Policy 1(2):251–265

Wolf AT (1999) The transboundary freshwater dispute database project. Water Int 24(2):160–163

Wolf AT, Natharius JA, Danielson JJ, Ward BS, Pender JK (1999) International river basins of the world. Int J Water Res Dev 15(4):387–427

Dheeraj Kumar Department of Water Resources Development and Management, Indian Institute of Technology Roorkee, Roorkee, Uttarakhand, India

Nayan Sharma Department of Water Resources Development and Management, Indian Institute of Technology Roorkee, Roorkee, Uttarakhand, India

Harinarayan Tiwari Department of Water Resources Development and Management, Indian Institute of Technology Roorkee, Roorkee, Uttarakhand, India

Subash Prasad Rai Department of Water Resources Development and Management, Indian Institute of Technology Roorkee, Roorkee, Uttrakhand, India

Part I

Sediment Transport

Forest Impact on Flood Peak Discharge and Sediment Yield in Streamflow

2

James C. Bathurst, Steve J. Birkinshaw, Felipe Cisneros Espinosa, and Andrés Iroumé

Abstract

Two recent studies help to define the extent to which forest cover, compared with a cover of shorter vegetation, can reduce flood peaks and sediment yields at the catchment scale as part of an integrated flood control programme. First, field data analysis and model analysis tested the hypothesis that, as the size of the rainfall event increases, the effect of forest cover on peak discharge becomes less important. Second, a systematic model analysis assessed the relationship between specific sediment yield and catchment area for various land use scenarios. The results show that the change in forest cover must apply to 20–30 % of the catchment area to affect the hydrological response; forest cover can affect the peak discharges for small to moderate floods but has little effect on large floods; increased cultivation in headwater areas can increase sediment yield, but the effect becomes attenuated over an order of magnitude increase in catchment area. In an Indian context, these results suggest that altered land use in the Himalayas has little immediate effect on flood magnitude and sediment yield in Bangladesh. However, forests can have a role in controlling floods and sediment yield in smaller headwater catchments.

J.C. Bathurst (✉) • S.J. Birkinshaw
School of Civil Engineering and Geosciences, Newcastle University, Newcastle upon Tyne NE1 7RU, UK
e-mail: james.bathurst@ncl.ac.uk

F. Cisneros Espinosa
Programa para el Manejo de Agua y Suelo (PROMAS), Department of Water and Soil Resources Engineering, Faculty of Engineering, Cuenca University, Avenida 12 de abril S/N, Cuenca, Ecuador

A. Iroumé
Facultad de Ciencias Forestales y Recursos Naturales, Instituto de Conservación, Biodiversidad y Territorio, Austral University of Chile (UACh), Independencia 631, Valdivia, Chile

2.1 Introduction

River floods are a major cause of misery and destruction in the world and are therefore a major concern for the authorities charged with protecting populations, housing and infrastructure from the impacts of natural hazards. Traditionally protection has been provided through river engineering works, such as flood embankments and river straightening, and through flood control reservoirs. However, land

N. Sharma (ed.), *River System Analysis and Management*, DOI 10.1007/978-981-10-1472-7_2

use in the river catchment is also recognized as a potentially significant control on flood generation, and catchment management may therefore offer an additional means of flood protection, under certain circumstances. Conversely, unsympathetic land management practices may increase flood hazard under certain circumstances. However, the particular circumstances in which land use, especially forest cover, can influence flood development are not generally well understood in the wider community. Thus deforestation and logging are regularly blamed by the public and the media for exacerbating the disastrous effects of floods generated by extreme rainfall, such as hurricanes. Government and development agency water resource policies tend to adhere to this perception; consequently large sums of money are invested in reforesting headwater areas of river catchments, and forest cover controls are imposed on the (typically poor) populations living in these areas (CIFOR and FAO 2005; Calder 2005; Calder and Aylward 2006). China, for example, is afforesting an area the size of Wales in the belief that this will reduce flood devastation. There has similarly been a continuing debate about the effect of land use changes, including deforestation, in the Himalayas on flood magnitudes in India and Bangladesh (e.g. Hamilton 1987; Ives 1987; Hofer and Messerli 2006). However, the impact of forest management on catchment response for extreme rainfall events is an area in which there is considerable scientific uncertainty as well as poorly conceived policy. In particular, while forests may reduce floods for small to moderate storms, there is growing evidence that this effect is increasingly reduced as rainfall increases to more extreme levels (e.g. Beschta et al. 2000; Sikka et al. 2003; Calder 2005; López-Moreno et al. 2006). Less controversially, forests do offer significant protection against soil erosion for a wide range of events. Removal of forests can therefore significantly increase sediment delivery to the channel network, as evidenced by the effects of widespread tree felling to make way for agricultural development by European settlers in the USA (e.g. Trimble 1976) and New Zealand (e.g. Glade 2003), with subsequent impacts including increased flooding (because channel capacity is reduced), reservoir sedimentation and deterioration of aquatic habitat. The extent to which changes in forest cover can alter downstream flood peak discharges or sediment yields, though, depends on the proportion of the catchment affected and on the location of the change in the catchment. Catchment managers concerned to minimize flood and erosion impacts therefore need guidance on those areas of a catchment to which flood response is most sensitive (allowing targeted, and therefore more efficient, mitigation measures). They also need guidance on the minimum area for which a change in land cover will modify flood or sediment yield response, which may help to decide whether carrying out the change is practically feasible or can deliver benefits which outweigh the costs.

This paper considers the potential for catchment land management (specifically for forest cover) to form a complementary approach to river engineering for flood management. It reviews the current understanding of forest impact on catchment flood response and sediment yield, concentrating in particular on the results of two recently completed research studies at Newcastle University, UK, in order to determine the conditions under which that impact may be significant. It concludes by discussing the applicability of the results to the Himalayan issue mentioned above. The paper also shows how computer models can be used to complement field data and provide additional insights to the problems of interest.

2.2 SHETRAN Catchment Modelling System

The two Newcastle University studies involved computer simulations using SHETRAN. This is a physically based, spatially distributed, hydrological and sediment yield modelling system applicable at the river catchment scale (Ewen et al. 2000; http://research.ncl.ac.uk/shetran/). Spatial distribution of catchment properties, rainfall input and hydrological response (including soil erosion and sediment transport) is represented in the horizontal direction through an orthogonal grid network and in the vertical direction by a column of horizontal layers at each grid square.

The hydrological component provides an integrated surface and subsurface representation of water movement through a river catchment. The sediment transport component models soil erosion by raindrop impact, leaf drip impact and overland flow, modified according to the protection afforded by vegetation cover (Bathurst et al. 1995; Wicks and Bathurst 1996; Bathurst 2011). Eroded material is carried to the stream network by overland flow. Within the channel, bank erosion may add to the supply of material, as a function of excess flow shear stress above a threshold value. Material is transported by the channel flow, and the sediment yield is determined from the accumulated total sediment discharge modelled at the catchment outlet. An additional component enables the sediment yield from shallow landslides to be modelled (Burton and Bathurst 1998; Bathurst et al. 2005). Application of the predecessor model (Système Hydrologique Européen, SHE) to Indian catchments is described in Refsgaard et al. (1992) and Jain et al. (1992).

2.3 Forest Impacts: What We Know

Decades of research with paired catchments and process studies have shown that, relative to shorter vegetation, forest cover reduces run-off at the annual scale because trees have higher rainfall interception rates and higher transpiration rates during dry periods (e.g. Bosch and Hewlett 1982; Calder 1990; Stednick 1996; Andréassian 2004; Bruijnzeel 2004). Dry season flows are particularly likely to be affected by forest plantation as the tree roots can abstract soil water from greater depths than can shorter vegetation. For small catchments ($<$1–10 km^2), Bosch and Hewlett (1982) and Stednick (1996) found that the change in forest cover required to provoke a measurable change in annual run-off was about 20 %.

The effects of forests on flood characteristics are less well understood, and there has been relatively little research into quantifying the impacts of forests on response to rainfall events of different frequencies of occurrence. The biophysical controls on floods are complex, typically involve conflicting processes and can be explained only through multidisciplinary considerations (e.g. Hamilton 1987; Forsyth 1998; Calder 2005). Theoretical considerations suggest that interception by forests should also reduce floods by removing a proportion of the storm rainfall and by allowing higher soil moisture deficits, relative to areas without forests. However, these effects are expected to be most significant for small storms and to be increasingly less effective as rainfall amounts increase. Research on small catchments (of around 1 km^2), mainly in the Pacific Northwest of North America (Jones and Grant 1996; Thomas and Megahan 1998; Beschta et al. 2000; Moore and Wondzell 2005) but also elsewhere including southern India (Sikka et al. 2003), tends mostly to support the hypothesis that, as the size of the rainfall event increases, the effect of forest cover becomes less important. From limited data, convergence of response for different vegetation covers appears to occur for flood return periods of 5–10 years. However, while the flood magnitude resulting from a given large rainfall event might then be the same for two catchments with different land covers, the flood frequencies remain different. A given flood magnitude is likely to occur more frequently (with a smaller return period) in a grassland catchment than in a forested but otherwise identical catchment (e.g. Alila et al. 2009; Kuraś et al. 2012).

At the larger catchment scale, the picture is less clear and the pattern may be masked by diluting and attenuating effects. Studies, at scales of 60–600 km^2 in the northwest USA (Jones and Grant 1996; Thomas and Megahan 1998; Beschta et al. 2000), 259 km^2 in central Chile (Pizarro et al. 2006), 12,100 km^2 in Thailand (Wilk et al. 2001) and 175,360 km^2 in South America (Costa et al. 2003), show in some cases a response and in others no response to changes in forest cover. Overall, the evidence in favour of the hypothesis from the past literature is relatively weak.

Forests are perceived to limit catchment sediment yield by protecting the soil against erosion. Thus forest logging generally results in an increase in soil erosion and sediment delivery to the river network (e.g. Amaranthus et al. 1985; Davies and Nelson 1993; Guthrie 2002), although much of this may be attributed to logging technique and forest roads (e.g. Reid and Dunne 1984; Grayson et al. 1993; Croke et al. 1999). (Under normal circumstances of best management practice, the maximum increase in catchment specific sediment yield (in t km^{-2} $year^{-1}$) in the years immediately following logging is typically within one order of magnitude above pre-logging levels, at both the annual and storm event scales (Bathurst and Iroumé 2014).) Forests protect the soil from erosion by raindrop impact but may cause erosion as a result of leaf drip impact (Hall and Calder 1993; Calder 2005). Forests also protect against shallow landslide erosion through their root-binding action and by generating higher soil moisture deficits (i.e. drier soils) which allow reduced soil pore water pressures (e.g. Amaranthus et al. 1985; Bruijnzeel 2004). This protection has been found to remain effective at extreme rainfalls (Phillips et al. 1990), at least for mature forests. Forests do not provide protection against deep-seated landslides, though.

Land use also affects the downstream variation in sediment yield. Specific sediment yield (in t km^{-2} $year^{-1}$) is conventionally considered to decrease as catchment area increases. In the last decade or so, though, a number of studies have indicated that the relationship can be direct as well as inverse. From an analysis of 1872 mountain rivers around the world, Dedkov and Moszherin (1992) conclude that, where hillslope erosion (i.e. sheet and gully erosion) is the main source of sediments, specific sediment yield decreases as catchment area increases, whereas where channel (e.g. bank) erosion is dominant, erosion rates and specific sediment yield increase as catchment area increases. On the basis that hillslope erosion tends to be dominant in areas disturbed by human activity (e.g. agriculture) and that the early sediment yield studies were often dominated by data from the USA (where the land is heavily affected by human activity), the possibility has been raised that the inverse relationship is in fact a reflection of human impact on the fluvial system rather than a basic principle (Walling and Webb 1996). Evidence that this may indeed be the case has been provided recently by Dedkov (2004) who, analysing 352 Eurasian catchments, found that the inverse relationship is characteristic only of intensively cultivated catchments (where hillslope erosion may be presumed to be significant). A direct relationship between sediment yield and catchment area was observed for uncultivated catchments or catchments with limited cultivation (where bank erosion may be presumed to be the principal source of sediment). There remains, though, a lack of systematic study into the separate influences of sediment source and the spatial distribution in land use and rainfall erosivity on the spatial variation in catchment sediment yield.

2.4 Forest Impact on Flood Peak Magnitude

Building on the above knowledge, the effect of forest cover on flood discharge for rain-derived floods was studied within the EPIC FORCE project (Evidence-Based Policy for Integrated Control of Forested River Catchments in Extreme Rainfall and Snowmelt) (http://research.ncl.ac.uk/epicforce), led by Newcastle University with seven other partners and funded by the European Commission (Bathurst et al. 2010a). A summary of the study is given below, and full details are available in Bathurst et al. (2011a, b) and Birkinshaw et al. (2011).

A combination of field data analysis and model application for focus catchments in Costa Rica, Ecuador, Chile and Argentina was used to test the hypothesis that, as the size of the rainfall event increases, the effect of forest cover on the peak discharge becomes less important. This hypothesis may be considered in terms of the relationship between peak discharge and rainfall event frequency (represented by return period) for catchments which are identical except for the level of forest cover. In both cases, the less frequent the rainfall event (i.e. with a larger return period), the greater is the resulting peak discharge. For moderate floods, which are relatively frequent, the forested catchment is assumed to be able to absorb more rainfall into the soil and therefore has lower peak discharges than the non-forested catchment. This is because the greater interception of rainfall by the forest, combined with a higher transpiration by the trees, allows the build-up of greater soil moisture deficits compared with the non-forested case. However, the effect of the deficit is expected to decrease as rainfall amounts increase. The hypothesis thus proposes a convergence of peak discharge response for the more extreme rainfall events (and floods).

The aim of the field data analysis was to use available discharge records to quantify directly the impact of forest cover on catchment response to rainfall. The model applications were intended to extrapolate the data analyses over a wider range of conditions and to provide a systematic analysis of the impact of forest cover on flood peak discharge using a standard approach. For these applications SHETRAN was applied to contrasting land use scenarios for each focus catchment (generally with and without a forest cover) for generated 1000-year precipitation time series. (These series, representative of current conditions, provided a more appropriate statistical basis for defining the flood response for events with return periods of up to 100 years or so than was possible with the limited data record.) The maximum daily discharges for the contrasting scenarios were compared for each day of the 1000-year simulations so as to investigate the degree to which the contrasting responses converge as the size of the flood peak increases.

The focus catchments and the principal results were:

1. *The 131-km^2 Pejibaye catchment in Costa Rica, in which almost all the native forest has been replaced by cattle pasture, coffee plantation and some arable cultivation since the late 1940s.* Data available from 1970 onwards show that the percentage of the catchment covered by native forest fell steadily from about 20 to 3.5 % during 1970–2000 but that this was to some extent compensated for by an increase in the area under coffee cultivation from 9 to 20 % during 1973–2000. Analysis of the discharge record shows that the resulting net change in forest cover had little effect on the return periods of the annual maximum discharges. This agrees with the previous studies concerned with annual run-off, which found that the net percentage change had to exceed around 20 % to produce a measurable response (Bosch and Hewlett 1982; Stednick 1996; Bruijnzeel 2004). The SHETRAN simulations show an overall convergence of the peak discharges for the forested and unforested scenarios as discharge increases. At all discharges the difference between the scenarios decreases through the wet season, as the respective soil moisture conditions develop similar levels of saturation.
2. *Two sets of neighbouring paired catchments in highland Ecuador (0.6–10 km^2), selected for their contrasting vegetation covers.* The data analysis (albeit for a limited period of record) and the SHETRAN simulation results support each other. A forested catchment has lower flood peak discharges than a grassland catchment but the difference between the two decreases as discharge increases; the convergence may be absolute or relative (i.e. the difference between the two is constant but decreases as a percentage of the discharge as discharge increases) (Fig. 2.1). The simulations show that at all discharges, the difference between the forested and unforested cases decreases through the wet season.
3. *A 0.35-km^2 catchment in southern Chile that has undergone clearfelling of its plantation cover of radiata pine.* Comparison of peak flows between the pre- and post-logging conditions shows that the percentage difference for “large” rainfall events (for rainfall

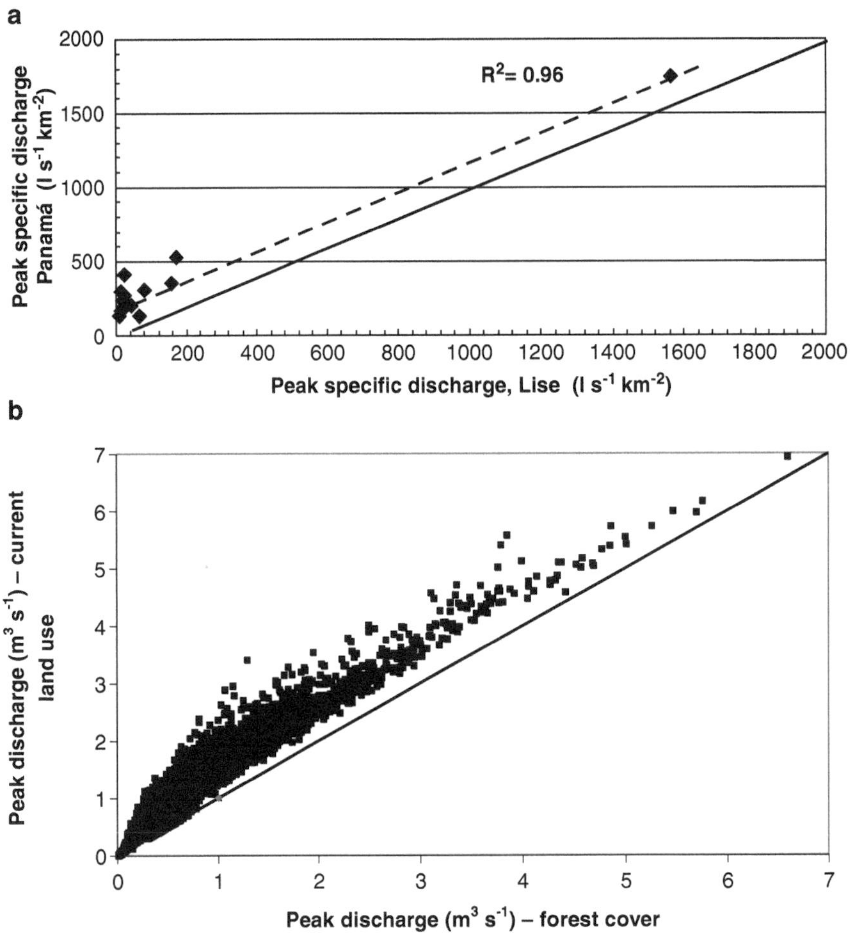

Fig. 2.1 (**a**) Comparison of observed peak specific discharges for the Panamá and Lise catchments (Ecuador) for the same storm events (Reproduced from Bathurst et al. (2011a) by permission of Elsevier). (**b**) Comparison of maximum daily discharges ($m^3\ s^{-1}$) for the current pasture vegetation and a hypothetical forest cover from the 1000-year SHETRAN simulations of the Panamá catchment (Reproduced from Bathurst et al. (2010a) by permission of Taylor & Francis Ltd (http://www.informaworld.com)). *Dashed line* fitted to data; *solid lines* are lines of equality

depths greater than 50 mm) is less than that for "medium" events (rainfall depths of 10–50 mm) and "small" events (rainfall depths of 5–10 mm). The relationships between peak flow and rainfall depth for the pre- and post-logging conditions also tend to converge as rainfall depth increases. On the basis of limited data, the convergence is estimated to occur for flood discharges with a return period of around 10 years. The SHETRAN simulations show relative convergence of the peak discharges for the forested and unforested scenarios as discharge increases but indicate also the moderating effects of season, type of event, soil depth and antecedent soil moisture condition.

4. *Three large catchments (94, 434 and 1545 km²) in central-southern Chile that have undergone significant plantation affecting at least 30 % of the catchment area.* These underwent data analysis only. Comparison of peak discharges for paired events (i.e. of similar rainfall) from the pre-afforestation and post-afforestation periods shows a very consistent pattern: for small floods peak discharges are higher for

the pre-afforestation conditions but the difference decreases as the size of the rainfall event and the peak discharge increase and convergence occurs for flood return periods of greater than 5 years (Fig. 2.2).

5. *A 12.9-km² catchment in Tierra del Fuego, Argentina, which formed a basis for examining the effect of forest cover on snowmelt response*. This was subjected to model analysis only. The simulations show that, in general, removal of the trees increases the peak river discharge but, for certain conditions of snowmelt, there can be a reduction. This result indicates the complicating effect of snowmelt and the difficulty in distinguishing trends concerning land use effect on peak discharges for events incorporating snowmelt.

The combined data and model analyses suggest that the effect of forest cover on flood peak discharge does indeed diminish either relatively or absolutely as rainfall (and discharge) increases and that, at least for small catchments, convergence of response is likely for flood return periods of greater than about 10 years (although this figure requires further confirmation). These results agree with and extend those of the previous studies. For larger catchments (up to 1500 km²), the results from Costa Rica and Chile suggest that there must be a change of at least 20–30 % in forested area for a change in peak discharge to be measurable. However, it is then possible for the same effect of decreasing forest cover impact with increasing discharge to be observed. The flood return period at which convergence occurs is lower (5 years) but it is not clear if this reduction is due to a general variation with catchment size or whether it reflects purely local conditions.

The pattern is nevertheless complicated by a number of factors. The effect of the forest cover depends on more than the size of the rainfall event: the SHETRAN simulations show that factors such as soil depth, antecedent moisture content and season also play a role. Other anthropogenic impacts (e.g. the effects of roads) and climate changes may have greater effects and so drown out the forest signal. The study was also limited in considering the effect of a change in only vegetation cover, i.e. in forest biomass. It did not consider the effect of the practices used in that removal, including logging technique and road building (e.g. La Marche and Lettenmaier 2001; DeWalle 2003). The provision of forest roads, for example, may effectively increase stream network density and contribute to an

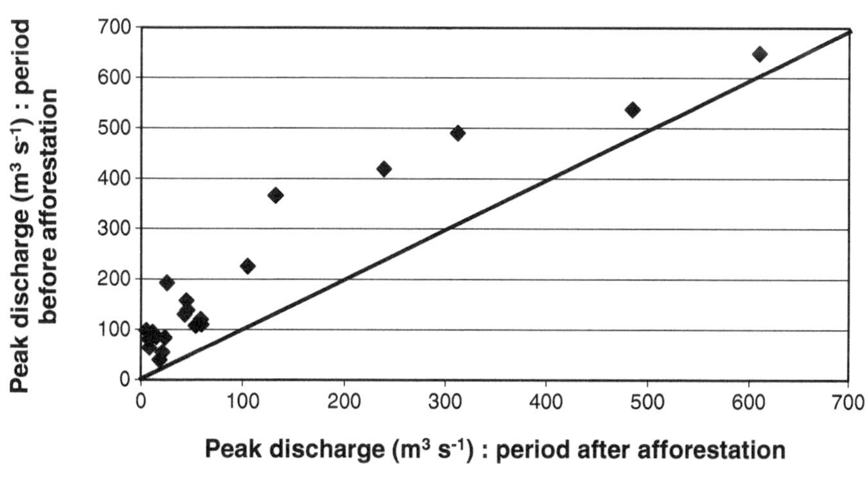

Fig. 2.2 Comparison of paired peak discharges from before and after forest plantation in the Duqueco catchment (1545 km²), Chile; line is line of equality (Reproduced from Bathurst et al. (2010a) by permission of Taylor & Francis Ltd (http://www.informaworld.com))

increase in flood levels. The traditional UK practice (now abandoned) of digging drainage ditches prior to forest plantation could similarly increase flood levels. Roads, along with poor logging techniques, can also increase sediment yield. The study was further limited to the impact of forest cover on flood magnitude and did not consider flood frequency.

2.5 Forest Impact on Sediment Yield

A modelling study of the effect of land use on sediment yield at the catchment scale was carried out by Newcastle University as part of the MEDACTION project (Policies for land use to combat desertification), funded by the EC. A summary of the study is given below and full details are available in Birkinshaw and Bathurst (2006).

Using SHETRAN, a systematic assessment was made of the relationship between specific sediment yield (in t km^{-2} $year^{-1}$) and catchment area for a number of land use and sediment source scenarios. Two contrasting catchments for which SHETRAN had already been validated in earlier studies were selected to provide the test topographies and river networks for the study: the 1532-km^2 Agri catchment in southern Italy (largely upland, rising to 1976 m) and the 701-km^2 Cobres catchment in southern Portugal (moderate, homogeneous topography, 115–376 m). The original model calibrations and applications are described in Bathurst et al. (2002) for the Agri catchment and in Bathurst et al. (1996) for the Cobres catchment. It should be noted, though, that the simulations for the sediment yield study were not representations of the existing real-world catchments. They simply used the catchment topographies and river networks as a framework within which to investigate systematically the relationship between sediment yield and catchment area for a number of specified conditions.

The simulations were designed to indicate the effect of sediment source (comparing hillslope with channel bank), spatial distribution of rainfall and spatial distribution of land use on the sediment yield/catchment area relationship, considering each effect separately and in combination with the others. The input data consisted of measured rainfall and potential evapotranspiration time series for each catchment and the simulation periods were 5–6 years. For each catchment the accumulated sediment yield over the total duration of the simulation was determined at several points along the river network, so as to show the variation with upstream catchment area. The contributing areas so selected were defined by the catchment channel network and ranged from 172 to 1532 km^2 for the Agri catchment and 32–701 km^2 for the Cobres catchment.

The simulation results indicated the following:

1. For uniform land use and rainfall distribution, hillslope erosion supports an inverse or near-constant relationship between specific sediment yield and catchment area (see Fig. 2.3 for uniform wheat and uniform forest covers). Bank erosion supports a direct relationship but with lower specific sediment yields. This agrees with the analyses of Dedkov and Moszherin (1992) and Dedkov (2004). In the model, mean hillslope gradient decreases as catchment size increases, reducing the potential for catchment-scale erosion. Absolute sediment discharge derived from hillslope erosion increases in the downstream direction but at a lesser rate than does catchment area. By contrast, the sediment load derived from bank erosion grows in the downstream direction at a rate greater than the increase in catchment area, and its specific sediment yield therefore increases downstream.
2. Distributing the land use while maintaining uniform rainfall has a significant impact on the sediment yield derived from hillslope erosion. If erosion is reduced in the higher ground relative to the lower (e.g. a pine forest cover on the higher land, wheat on the lower), the inverse relationship can be reversed. The downstream increase in soil erodibility allows

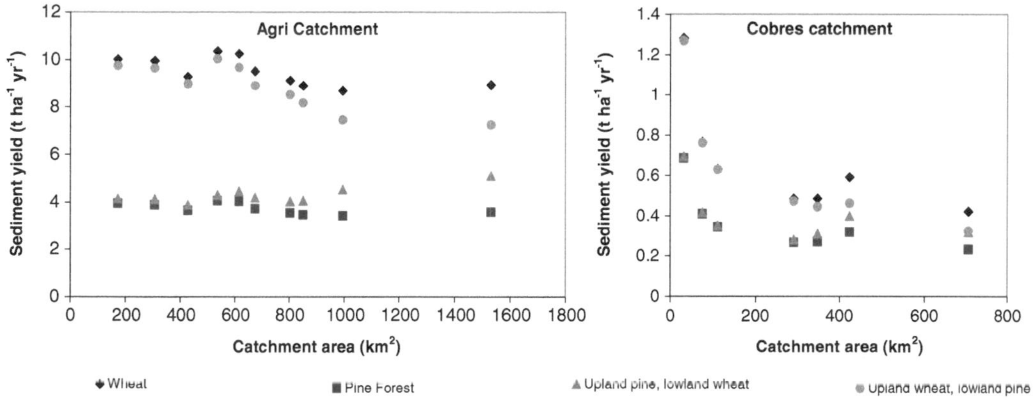

Fig. 2.3 Simulated hillslope erosion sediment yield for different land uses with uniform precipitation (Reproduced from Birkinshaw and Bathurst (2006) by permission of Wiley)

specific sediment yield to increase as catchment area increases (Fig. 2.3). If, on the other hand, the land use pattern is reversed, the greater erosion on the higher ground relative to the lower enhances the inverse relationship. This suggests that cultivation patterns can potentially alter the variation of sediment yield with catchment area, in agreement with the field measurements of Krishnaswamy et al. (2001).

3. Variations in rainfall distribution may dominate the effect of land use change. In agreement with field measurements by Krishnaswamy et al. (2001), the inverse relationship between specific sediment yield and catchment area is enhanced by a more realistic rainfall pattern of heavier rainfall on the higher ground. Similarly, combining this rainfall pattern with the distributed land use of upland pine and lowland wheat (see the preceding point) restores the downstream decrease in sediment yield derived from hillslope erosion and accentuates the same trend for all the other land uses.

The model results are consistent with the recent literature. They show both inverse and direct relationships depending on the principal source of sediment in transport, on rainfall spatial distribution and on land use distribution. If the sediment is supplied solely from hillslope erosion (no channel bank erosion), then, with uniform land use, specific sediment yield either decreases or is nearly constant as area increases. The downstream decrease is accentuated if rainfall (and thence erosion) is higher in the headwaters than at lower elevations. Introducing a nonuniform land use (e.g. forest at higher elevations, wheat at lower elevations) can reverse the trend, so that sediment yield increases downstream. If the sediment is supplied solely from bank erosion (no hillslope erosion), the sediment yield increases downstream for all conditions.

The results generated for this study do not of course automatically apply to all catchment conditions. Different combinations of rainfall and land use distributions (and the effect of processes not included in SHETRAN) could create a different balance between the variations in sediment yield and catchment area. The results are therefore an illustration of the variety of forms which the sediment yield/catchment area relationship may adopt, not a general description for all catchments.

The study also shows the usefulness of physically based, spatially distributed modelling in illustrating the effects of different controls on sediment yield, in exploring the topic in a systematic manner and in avoiding the data limitations of field-based studies. By identifying important controls, it allows field studies to be efficiently targeted and to be designed with minimum data needs.

Forest cover does offer significant protection against soil erosion for a wide range of events; removal of forest can therefore significantly increase sediment delivery to the channel network. However, as with flood impact, the extent to which this affects downstream sediment yield is likely to depend on the proportion of the catchment affected. For both of the above test catchments, additional simulations were therefore carried out first for a uniform vegetation cover of wheat and second for a uniform wheat cover except in the top headwater subcatchment which had a forest cover. In the model the forest cover reduces the sediment yield of the headwater catchment. Comparison of the downstream variation in sediment yield for the two cases shows that sensitivity to the changed land use is high immediately downstream of the headwater subcatchment but decreases downstream as the proportion of the contributing area represented by the headwater subcatchment decreases (Fig. 2.4). Rather approximately the difference

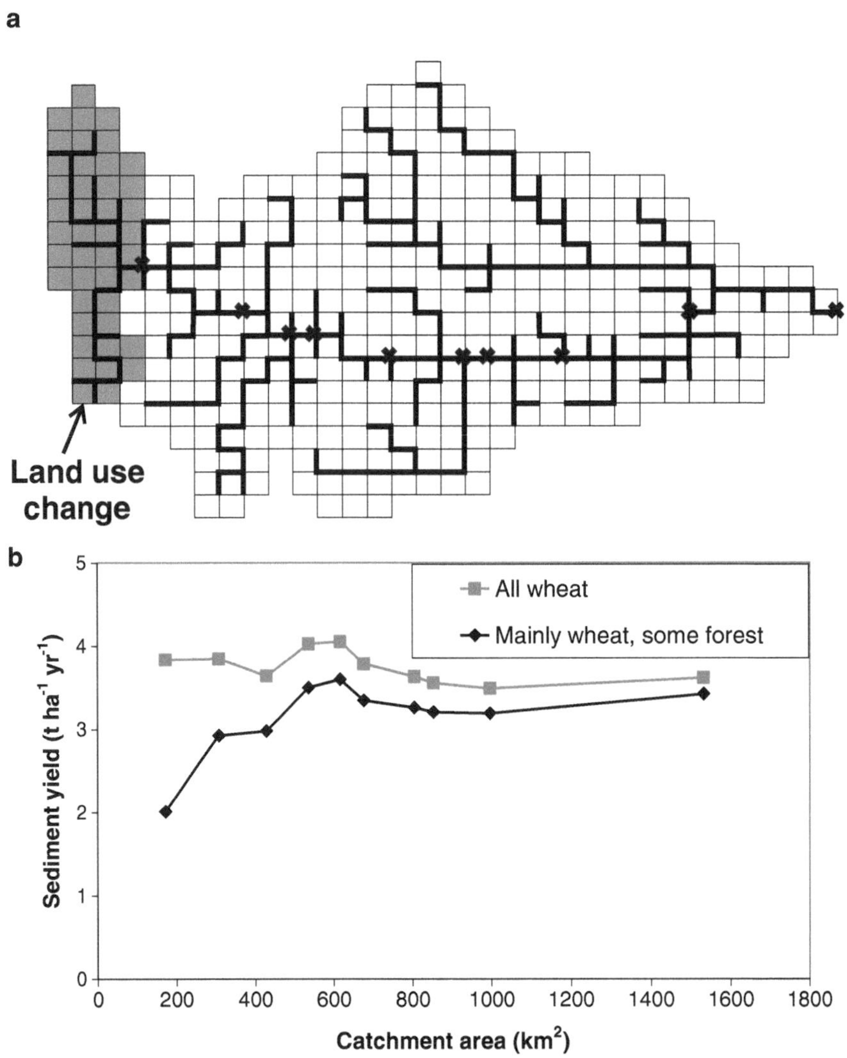

Fig. 2.4 SHETRAN simulation showing the sensitivity of downstream sediment yield to afforesting the headwaters of a catchment otherwise covered in wheat. (**a**) The SHETRAN model grid (2-km resolution) and river network; *crosses* indicate subcatchments for which specific sediment yield was calculated. (**b**) Comparison of variations in specific sediment yield for uniform wheat cover and for wheat cover with an afforested headwater subcatchment

in sediment yield is reduced to 25 % of the initial value for a catchment area three to four times the headwater subcatchment area and to 10 % for an area ratio of 10.

These results show that changes in land use in headwater catchments can indeed alter both the sediment yield and the downstream variation in sediment yield. However, the impact progressively declines in the downstream direction and is unlikely to be significant at an area of more than ten times the area affected by the land use change in the headwater catchment. Given that the land use change is unlikely to be as effective as represented in the model (where it is imposed over the entire headwater catchment area), in reality the impact will be even less noticeable.

2.6 Highland-Lowland Interaction

Over the past two decades, there has been considerable debate over the effect which land use changes in the Himalayas may have had on flooding along the Ganges and Brahmaputra river systems and especially in Bangladesh. Schumm (1977) introduced the concept of the fluvial system, in which the river system is divided into three linked zones. The headwater zone is characterized primarily by erosion and sediment production; this feeds into the central zone which is characterized primarily by sediment transport; and this feeds into the downstream zone which is characterized primarily by deposition. This integrated view of the catchment suggests that activities in the headwaters (e.g. forest logging triggering erosion) will eventually have an impact, such as siltation, in downstream zones. This is a logical connection, borne out, for example, by the effects of forest logging by European settlers in the USA and New Zealand (Trimble 1976; Glade 2003). In a Himalayan context, it provides part of the basis for the so-called Himalayan Environmental Degradation theory: this proposes that a rapidly growing population in the Himalayas causes a loss of forest cover through expansion of fuel gathering and agriculture (the latter also extending under pressure to increasingly steep and thus unstable slopes), thereby triggering greater soil erosion and flood run-off, the results of which are transmitted down the Ganges and Brahmaputra rivers to generate increased flooding and siltation in the delta regions of Bangladesh (see Ives (1987) for an overview). Research over the last two decades has shown that this theory is in fact somewhat simplistic and misleading: it is based on a number of assumptions, not all of which are supported by evidence, it has not been demonstrated that the flood hazard has in fact increased in Bangladesh, there is considerable uncertainty in process understanding and a lack of reliable data, and the links within the fluvial system must be seen in the context of the large physical scale of the system (e.g. Ives 1987; Hamilton 1987; Hofer 1993; Forsyth 1998; Hofer and Messerli 2006). Nevertheless, the apparently logical nature of the theory has ensured its survival, and it is likely that land use policies and allocation of relevant funds have been affected accordingly.

The two studies outlined above may be added to the other evidence in the literature to highlight the simplistic nature of the theory and the consequent dangers of using it as the basis for drawing up policy. In general terms:

1. Forest cover is likely to reduce only small to medium floods and has little effect on the large floods which appear in the news headlines; furthermore, even this effect will decline through the monsoon season as the soil water store fills up. Therefore, if there has been any loss in forest cover (and the extent and form of this in the Himalayas are not well defined), it is unlikely to have caused any significant increase in downstream large-scale flooding in recent decades.
2. For a change in forest cover to have a measurable effect on hydrological response, it must apply to at least 20–30 % of the catchment area. This finding is based on data from catchments of up to 1500 km^2, and it is not clear if it applies to much larger scales where attenuation effects, nonuniformity in rainfall, vegetation and other catchment characteristics, and a greater likelihood that

flood waves from individual subcatchments contribute sequentially rather than simultaneously at the outlet may mask the effects of land use or vegetation cover change. However, if it did apply, noting that the combined catchment area of the Ganges and Brahmaputra is about 1,621,000 km^2, an area of about 400,000 km^2 (roughly equal to the area of California), would have had to have been changed from a fully forested state to a cover of arable crop, grassland or wasteland in the past few decades to cause any significant effect at the delta; it must be questioned whether such a massive change has in fact occurred.
3. Loss of forest cover in individual headwater catchments could increase the sediment yield (by surface erosion and shallow landslides), but, against the naturally high erosion related to monsoon rainfall, steep slopes and deep-seated landslides, the relative effect would be small and would rapidly attenuate downstream. The sediment yields of the Ganges and Brahmaputra are already amongst the highest in the world for large rivers (e.g. a total of 1037 million tonnes carried into Bangladesh every year and a combined specific sediment yield of about 900 t km^{-2} $year^{-1}$ (Islam et al. 1999)). In smaller catchments the effect of enhanced sediment yield would be felt at the outlet relatively quickly; e.g. in Taiwan the additional sediment yield caused by earthquakes may be cleared out (by typhoon floods) and transferred to the sea in a few years (Dadson et al. 2004). However, at the scale of the Ganges and Brahmaputra catchments, the effects of increased erosion in the headwaters will not be felt at the delta for decades, if not longer (Hamilton 1987). Further, these effects will not be significant unless the headwater area affected is a substantial proportion of the total catchment area.

In common with the recent literature, these results suggest that land use change in the Himalayas is unlikely to have significantly increased flood hazard or siltation problems in the downstream reaches of the major rivers. Locally, though, within individual catchments, land use change may have had a significant impact. In such cases, afforestation, or protection of the existing forest cover, may be beneficial in reducing the small to moderate floods and the overall level of soil erosion. Bathurst et al. (2010b) show how in such circumstances computer modelling can help to define the areas of a catchment which are most susceptible to erosion and where targeted afforestation can have proportionately the greatest effect in reducing sediment yield. Flood risk in the downstream rivers will be more efficiently tackled through a mixture of on-site engineering structures (including dams), land use zoning and land management, flood preparedness (e.g. education, flood shelters and flood warning systems) and flood emergency measures (e.g. evacuation plans, provision of emergency accommodation and clean up), all of which admittedly pose major challenges in areas of poverty and high population density and where funding is severely limited.

2.7 Conclusions

Building on two recent studies, this paper has examined the extent to which the management of forest cover at the catchment scale can reduce flood peaks and sediment yield, as part of an integrated flood control programme. The results of the studies show that:

1. For a change in forest cover to affect the hydrological response at a catchment outlet, it must apply to at least 20–30 % of the catchment area.
2. Changes in forest cover have little effect on the peak discharges of large floods, with return periods of around 10 years and larger.
3. Forest cover can reduce flood peaks for moderate floods but in areas with a prolonged wet season, this effect decreases as the soil water store fills up.

4. The above patterns have been observed in catchments of up to 1500 km^2 but it is not clear if they can be extrapolated to larger scales.
5. Increased intensity of cultivation in headwater catchments can increase sediment yield and thus enhance the pattern of a downstream decrease in specific sediment yield. Equally, good erosion protection measures and forest plantation according to best practice would be expected to reduce sediment yield.
6. Spatial variation of rainfall in the headwater catchments may have a greater effect than land use change on sediment yield.
7. Changes in sediment yield caused by land use change may be significant at the local scale but become attenuated to relatively insignificant levels over an order of magnitude increase in area.

Application of the results to the debate about the Himalayan Environmental Degradation theory reinforces the recent research which suggests that altered land use in the Himalayas currently has little effect on flood magnitude and sediment yield in Bangladesh. Therefore planting forests to reduce downstream flooding may show little benefit, especially for the larger floods, while having the unwelcome effect of reducing water availability at the annual scale or at times of low flow (e.g. Sikka et al. 2003). A more appropriate approach is likely to be a combination of downstream engineering works and zoning and land use management to reduce the impact of the flood. Forests also have a relatively moderate effect on sediment yield where erosion is dominated by the occurrence of deep-seated landslides. Nevertheless, forests still offer significant benefits, especially at the local scale: as proposed by López-Moreno et al. (2006), forest cover could be effective during the more frequent, less intensive rainstorms, noticeably reducing the frequency and intensity of more moderate but more frequent floods. Forests also provide a good level of protection against soil erosion and sediment transport associated with shallow landslides, gullying and overland flow. Forests therefore have a role to play in controlling floods and sediment yield in smaller headwater catchments. Computer models can be used, through systematic study, to define this role and help target forest plantation on those parts of a catchment where it can be most effective.

Acknowledgements The MEDACTION and EPIC FORCE projects were funded by the European Commission through its Framework Programme under contract numbers EVK2-CT-2000-00085 and INCO-CT2004-510739, respectively.

References

Alila Y, Kuraś PK, Schnorbus M, Hudson R (2009) Forests and floods: a new paradigm sheds light on age-old controversies. Water Resour Res 45. doi:10.1029/2008WR007207

Amaranthus MP, Rice RM, Barr NR, Ziemer RR (1985) Logging and forest roads related to increased debris slides in Southwestern Oregon. J For 83:229–233

Andréassian V (2004) Waters and forests: from historical controversy to scientific debate. J Hydrol 291:1–27

Bathurst JC (2011) Predicting impacts of land use and climate change on erosion and sediment yield in river basins using SHETRAN. In: Morgan RPC, Nearing MA (eds) Handbook of erosion modelling. Blackwell, Oxford, pp 263–288

Bathurst JC, Iroumé A (2014) Quantitative generalizations for catchment sediment yield following forest logging. Water Resour Res 50(11):8383–8402. doi:10.1002/2014WR015711

Bathurst JC, Wicks JM, O'Connell PE (1995) The SHE/SHESED basin scale water flow and sediment transport modelling system. In: Singh VP (ed) Computer models of watershed hydrology. Water Resources Publications, Highlands Ranch, pp 563–594

Bathurst JC, Kilsby C, White S (1996) Modelling the impacts of climate and land-use change on basin hydrology and soil erosion in Mediterranean Europe. In: Brandt CJ, Thornes JB (eds) Mediterranean desertification and land use. Wiley, Chichester, pp 355–387

Bathurst JC, Sheffield J, Vicente C, White SM, Romano N (2002) Modelling large basin hydrology and sediment yield with sparse data: the Agri Basin, Southern Italy. In: Geeson NA, Brandt CJ, Thornes JB (eds) Mediterranean desertification: a mosaic of processes and responses. Wiley, Chichester, pp 397–415

Bathurst JC, Moretti G, El-Hames A, Moaven-Hashemi A, Burton A (2005) Scenario modelling of basin-scale, shallow landslide sediment yield, Valsassina, Italian Southern Alps. Nat Hazards Earth Syst Sci 5:189–202

Bathurst JC, Amezaga J, Cisneros F, Gaviño Novillo M, Iroumé A, Lenzi MA, Mintegui Aguirre J, Miranda M, Urciuolo A (2010a) Forests and floods in Latin

America: science, management, policy and the EPIC FORCE Project. Water Int 35(2):114–131
Bathurst JC, Bovolo CI, Cisneros F (2010b) Modelling the effect of forest cover on shallow landslides at the river basin scale. Ecol Eng 36:317–327
Bathurst JC, Iroumé A, Cisneros F, Fallas J, Iturraspe R, Gaviño Novillo M, Urciuolo A, de Bièvre B, Guerrero Borges V, Coello C, Cisneros P, Gayoso J, Miranda M, Ramirez M (2011a) Forest impact on floods due to extreme rainfall and snowmelt in four Latin American environments 1: field data analysis. J Hydrol 400:281–291
Bathurst JC, Birkinshaw SJ, Cisneros F, Fallas J, Iroumé A, Iturraspe R, Gaviño Novillo M, Urciuolo A, Alvarado A, Coello C, Huber A, Miranda M, Ramirez M, Sarandón R (2011b) Forest impact on floods due to extreme rainfall and snowmelt in four Latin American environments 2: model analysis. J Hydrol 400:292–304
Beschta RL, Pyles MR, Skaugset AE, Surfleet CG (2000) Peakflow responses to forest practices in the Western Cascades of Oregon, USA. J Hydrol 233:102–120
Birkinshaw SJ, Bathurst JC (2006) Model study of the relationship between sediment yield and river basin area. Earth Surf Process Landf 31:750–761
Birkinshaw SJ, Bathurst JC, Iroumé A, Palacios H (2011) The effect of forest cover on peak flow and sediment discharge – an integrated field and modelling study in central-southern Chile. Hydrol Process 25(8):1284–1297
Bosch JM, Hewlett JD (1982) A review of catchment experiments to determine the effect of vegetation changes on water yield and evapotranspiration. J Hydrol 55:3–23
Bruijnzeel LA (2004) Hydrological functions of tropical trees: not seeing the soil for the trees? Agric Ecosyst Environ 104:185–228
Burton A, Bathurst JC (1998) Physically based modelling of shallow landslide sediment yield at a catchment scale. Environ Geol 35:89–99
Calder IR (1990) Evaporation in the uplands. Wiley, Chichester, 148pp
Calder IR (2005) Blue revolution, integrated land and water resource management, 2nd edn. Earthscan, London, 353pp
Calder IR, Aylward B (2006) Forest and floods: moving to an evidence based approach to watershed and integrated flood management. Water Int 31:87–99
CIFOR, FAO (2005) Forest and floods: drowning in fiction or thriving on facts? http://www.cgiar.org/insightdev/upload/291/145_BCIFOR0501.pdf. Accessed 17 Aug 2015
Costa MH, Botta A, Cardille JA (2003) Effects of large-scale changes in land cover on the discharge of the Tocantins River, Southeastern Amazonia. J Hydrol 283:206–217
Croke J, Hairsine P, Fogarty P (1999) Sediment transport, redistribution and storage on logged forest hillslopes in south-eastern Australia. Hydrol Process 13:2705–2720
Dadson SJ, Hovius N, Chen H, Dade WB, Lin J-C, Hsu M-L, Lin C-W, Horng M-J, Chen T-C, Milliman J, Stark CP (2004) Earthquake-triggered increase in sediment delivery from an active mountain belt. Geology 32(8):733–736
Davies PE, Nelson M (1993) The effect of steep slope logging on fine sediment infiltration into the beds of ephemeral and perennial streams of the Dazzler Range, Tasmania, Australia. J Hydrol 150:481–504
Dedkov A (2004) The relationship between sediment yield and drainage basin area. In: Sediment transfer through the fluvial system. Intl Ass Hydrol Sci, Wallingford, Oxon, UK, Publ 288:197–204
Dedkov AP, Moszherin VI (1992) Erosion and sediment yield in mountain regions of the world. In: Erosion, debris flows and environment in mountain regions, Intl Ass Hydrol Sci, Wallingford, Oxon, UK, Publ 209:29–36
DeWalle DR (2003) Forest hydrology revisited. Hydrol Process 17:1255–1256
Ewen J, Parkin G, O'Connell PE (2000) SHETRAN: distributed river basin flow and transport modeling system. Proc Am Soc Civ Engrs, J Hydrol Engrg 5:250–258
Forsyth T (1998) Mountain myths revisited: integrating natural and social environmental science. Mt Res Dev 18(2):107–116
Glade T (2003) Landslide occurrence as a response to land use change: a review of evidence from New Zealand. Catena 51:297–314
Grayson RB, Haydon SR, Jayasuriya MDA, Finlayson BL (1993) Water quality in mountain ash forests – separating the impacts of roads from those of logging operations. J Hydrol 150:459–480
Guthrie RH (2002) The effects of logging on frequency and distribution of landslides in three watersheds on Vancouver Island, British Columbia. Geomorphology 43:273–292
Hall RL, Calder IR (1993) Drop size modification by forest canopies – measurements using a disdrometer. J Geophys Res 90:465–470
Hamilton LS (1987) What are the impacts of Himalayan deforestation on the Ganges-Brahmaputra lowlands and delta? Assumptions and facts. Mt Res Dev 7(3):256–263
Hofer T (1993) Himalayan deforestation, changing river discharge, and increasing floods: myth or reality? Mt Res Dev 13(3):213–233
Hofer T, Messerli B (2006) Floods in Bangladesh: history, dynamics and rethinking the role of the Himalayas. United Nations University Press, Tokyo
Islam MR, Begum SF, Yamaguchi Y, Ogawa K (1999) The Ganges and Brahmaputra rivers in Bangladesh: basin denudation and sedimentation. Hydrol Process 13:2907–2923
Ives JD (1987) The theory of Himalayan environmental degradation: its validity and application challenged by recent research. Mt Res Dev 7(3):189–199
Jain SK, Storm B, Bathurst JC, Refsgaard JC, Singh RD (1992) Application of the SHE to catchments in India — part 2: field experiments and simulation studies with the SHE on the Kolar subcatchment of the Narmada River. J Hydrol 140:25–47

Jones JA, Grant GE (1996) Peak flow responses to clear-cutting and roads in small and large basins, Western Cascades, Oregon. Water Resour Res 32(4):959–974
Krishnaswamy J, Richter DD, Halpin PN, Hofmockel MS (2001) Spatial patterns of suspended sediment yields in a humid tropical watershed in Costa Rica. Hydrol Process 15:2237–2257
Kuraś PK, Alila Y, Weiler M (2012) Forest harvesting effects on the magnitude and frequency of peak flows can increase with return period. Water Resour Res 48. doi:10.1029/2011WR010705
La Marche JL, Lettenmaier DP (2001) Effects of forest roads on flood flows in the Deschutes River, Washington. Earth Surf Process Landf 26:115–134
López-Moreno JI, Beguería S, García-Ruiz JM (2006) Trends in high flows in the central Spanish Pyrenees: response to climatic factors or to land-use change? Hydrol Sci J 51(6):1039–1050
Moore RD, Wondzell SM (2005) Physical hydrology and the effects of forest harvesting in the Pacific northwest: a review. J Am Wat Res Ass 41(4):763–784
Phillips C, Marden M, Pearce A (1990) Effectiveness of reforestation in prevention and control of landsliding during large cyclonic storms. In: Proceedings of the 19th IUFRO World Congress, Montreal
Pizarro R, Araya S, Jordán C, Farías C, Flores JP, Bro PB (2006) The effects of changes in vegetative cover on river flows in the Purapel River basin of central Chile. J Hydrol 327:249–257
Refsgaard JC, Seth SM, Bathurst JC, Erlich M, Storm B, Jørgensen GH, Chandra S (1992) Application of the SHE to catchments in India — part 1: general results. J Hydrol 140:1–23
Reid LM, Dunne T (1984) Sediment production from forest road surfaces. Water Resour Res 20(11): 1753–1761
Schumm SA (1977) The fluvial system. Wiley, New York, 338pp
Sikka AK, Samra JS, Sharda VN, Samraj P, Lakshmanan V (2003) Low flow and high flow responses to converting natural grassland into bluegum (*Eucalyptus Globulus*) in Nilgiris watersheds of south India. J Hydrol 270:12–26
Stednick JD (1996) Monitoring the effects of timber harvest on annual water yield. J Hydrol 176:79–95
Thomas RB, Megahan WF (1998) Peak flow responses to clear-cutting and roads in small and large basins, Western Cascades, Oregon: a second opinion. Water Resour Res 34(12):3393–3403
Trimble SW (1976) Sedimentation in Coon Creek Valley, Wisconsin. In: Proceedings of the 3rd Federal Inter-Agency Sedimentation Conference, Denver, Symposium 5. Sedimentation Committee, Water Resources Council, pp 5.100–5.112
Walling DE, Webb BW (1996) Erosion and sediment yield: a global overview. In Erosion and sediment yield: global and regional perspectives. Intl Ass Hydrol Sci, Wallingford, Oxon, UK, Publ 236:3–19
Wicks JM, Bathurst JC (1996) SHESED: a physically based, distributed erosion and sediment yield component for the SHE Hydrological Modelling System. J Hydrol 175:213–238
Wilk J, Andersson L, Plermkamon V (2001) Hydrological impacts of forest conversion to agriculture in a large river basin in northeast Thailand. Hydrol Process 15:2729–2748

James C. Bathurst School of Civil Engineering and Geosciences, Newcastle University, Newcastle upon Tyne, UK

Steve J. Birkinshaw School of Civil Engineering and Geosciences, Newcastle University, Newcastle upon Tyne, UK

Felipe Cisneros Espinosa Programa para el Manejo de Agua y Suelo (PROMAS), Department of Water and Soil Resources Engineering, Faculty of Engineering, Cuenca University, Cuenca, Ecuador

Andrés Iroumé Facultad de Ciencias Forestales y Recursos Naturales, Instituto de Conservación, Biodiversidad y Territorio, University of Chile Austral (UACh), Valdivia, Chile

On the Physical and Operational Rationality of Data-Driven Models for Suspended Sediment Prediction in Rivers

3

Nick J. Mount, Robert J. Abrahart, and Christian W. Dawson

Abstract

Suspended sediment remains an important variable for prediction in river studies. Knowledge of suspended sediment concentration or load at different downstream locations within a channel allows the temporal and spatial patterns of catchment sediment yield to be determined, as well as within-channel sediment budgets that provide important insight into the patterns and processes governing channel siltation and scour. However, in many rivers, the comprehensive and long-term downstream monitoring of discharge contrasts with relatively sparse and temporally discontinuous monitoring of suspended sediment. As a result, the generation of suspended sediment data through physical and empirical modelling approaches is commonplace. The emergence of a data-driven modelling paradigm in the last two decades has resulted in the adoption of new methods for suspended sediment modelling. To a large extent, these methods mirror traditional empirical approaches, except that the a priori determination of the form of the response function by the modeller is replaced by machine learning and artificial intelligence algorithms that 'learn' the response function (both its form and associated parameters) directly from data. Data-driven models have been shown to result in improved goodness-of-fit metric scores, but many hydrologists remain critical about the lack of attempts by data-driven modellers to demonstrate the physical rationality and operational validity of their models. In this chapter, we examine this criticism; highlight specific research challenges facing data-driven, suspended sediment modellers; and detail the research directions through which advances may be made.

N.J. Mount (✉) • R.J. Abrahart
School of Geography, University of Nottingham, Nottingham NG7 2RD, UK
e-mail: nick.mount@nottingham.ac.uk

C.W. Dawson
Department of Computer Science, University of Loughborough, Loughborough LE11 3TU, UK

3.1 Introduction

Estimation and prediction of a river's suspended sediment concentration and/or load remains an important operational task for those engaged in

N. Sharma (ed.), *River System Analysis and Management*, DOI 10.1007/978-981-10-1472-7_3

river engineering and management practice. Where instantaneous values are needed, rating curve approaches, which deliver a prediction based upon the modelled relationship between discharge and suspended sediment concentration, remain the de facto standard throughout the majority of the world's rivers (e.g. Toth and Bodis 2015). Where short-term, future predictions are required, forecasting methods commonly develop autoregressive models that define the relationship between temporally lagged and current suspended sediment and discharge observations and their value at one or more time steps into the future (e.g. Hodgkins 1996). The methods by which such models are developed have remained at the forefront of hydrological research for more than 40 years, and in the last decade the potential of data-driven modelling (DDM) techniques has caught the attention of the hydrological modelling community (e.g. Kumar et al. 2015; Singh and Chakrapani 2015; Olyaie et al. 2015; Mehdi 2013; Mount and Abrahart 2011). DDMs are empirically based models that utilise intelligent computing methods to learn their structure and parameters directly from data, requiring little, if any, human input to control or constrain the model development process. Numerous different types of DDM exist, and opportunities for their practical implementation have been demonstrated across many of the world's largest rivers including the Nile (Negm et al. 2007), the Mississippi (e.g. Nourani and Gholamreza 2015), the Yangtze (e.g. Wu and Hu 2009) and the Yellow River (e.g. Ni et al. 2004), as well as a plethora of smaller rivers. Significant praise has been given to their apparent ability to offer better performing representations of the sediment delivery system over standard rating curve and multiple regression approaches (e.g. Kisi 2005). However, the use of a DDM is not without its dangers, and it should be recognised that the ability of such mechanisms 'to find connections between the system state variables (input(s), internal parameters and output(s)) without explicit knowledge of the physical behaviour of the system' (Solomatine et al. 2008: 17) can lead to erroneous model conceptualisation and structure and, potentially, operationally irrational solutions. In particular, examples of models that offer significant improvements in predictive capabilities are emerging unchallenged, despite their limited degree of physical rationality with respect to the predictor variables used and the form/parameterisation of the response function. This can constrain the scope of the operational applications in which such models are useful.

In this chapter we explore the need for legitimacy in DDMs. We define a model as being legitimate if the modeller is able to demonstrate that the *means* by which the model's output (s) is generated is sufficient in terms of its physical, mechanistic and operational rationality (c.f. Mount et al. 2011, 2013). We begin by considering the particular characteristics of DDMs that have sometimes led to inadequate model rationalisation and the importance of physical rationalisation in empirical modelling. We then consider physical, mechanistic and operational rationalisation, highlighting the specific research challenges and providing some examples of the pitfalls that may be encountered by drawing upon previously published studies. Finally, we call for renewed research efforts to address the problem and highlight some research directions through which advances may be made.

3.2 Suspended Sediment Modelling Approaches in Hydrology

Hydrological models for suspended sediment prediction can be classified into three basic types: physically based, conceptual (including lumped conceptual and distributed physically based models) and empirical (Solomatine et al. 2008). The first two types (e.g. Wicks and Bathurst 1996; Kothyari et al. 1997; Fang and Wang 2000; Hardy et al. 2000; Shaw et al. 2006; Xie et al. 2008) involve the use of numerical equations, considered sufficiently representative of sediment delivery and transport processes, to simulate and predict hydrological responses in a physically meaningful way. These types of model require calibration of their parameters in order to ensure their efficacy in a given physiographic/hydrologic setting – commonly achieved by the iterative adjustment of parameter values such that the fit between model outputs and observed data is maximised. The third type (e.g. Colby 1956; Glysson 1987; Crowder et al. 2007; Yang and Stall 1974; Syvitski et al. 2000) attempts to predict suspended sediment responses through extraction and reapplication of the numerical patterns that can be identified in time series data comprising one or more input drivers and outputs. To many physically minded hydrologists, it should be possible to demonstrate a quantitative rationalisation of a model's inputs, structure and parameter values in the context of the hydrological system under study if confidence in its legitimacy is to be maximised. Indeed, those engaged in empirical and data-driven modelling are sometimes criticised for avoiding the requirement to demonstrate the physical rationality of their model structures and parameters (Abbot 2008) – instead relying on the fit that is achieved between model outputs and observed data as their primary measure of confidence in the model. However, physically based, reductionist models do not offer a nirvana to hydrologists. Crucially, they become difficult to apply in situations where the spatial or temporal scale and context of the modelling challenge result in significant uncertainty in the parameter knowledge needed by the model. Equally, where the complexity of the modelling challenge increases, model equifinality (Beven and Freer 2001, Beven 2006) is a common problem which arguably challenges the entire pretext for using a physically based model (i.e. how can two different models, each with different physical representations of the hydrological system, be correct?). Indeed, to discount empirical suspended sediment models too readily is to overlook their continued importance as simple, operationally useful tools that have been successfully applied in thousands of catchments worldwide.

Traditional examples of the empirically based, statistical approach in hydrology include the unit hydrograph method (e.g. Duband et al. 1993), linear regression models (e.g. Jansson 1996, 1997) and autoregressive, integrated moving average (ARIMA) models (e.g. Wei and Zhang 2010). Despite the plethora of different statistical model types available to hydrologists, the general aim of any empirical model is to define the response function that maximises the fit of input to output data. Importantly, traditional statistically based models achieve this optimisation within a framework of assumptions that constrain their form and rationality. For example, a linear regression model is a simple response function that is constrained by a linear structure and assumptions of data normality. These constraints and assumptions can limit the contexts in which the application of such models is appropriate. Non-normally distributed data (inputs and outputs) may need to be transformed in order to model them, or log transformations may be required (Mount and Abrahart 2011), and in the case of highly non-linear input/output responses, it may be impossible to produce useful models. Such constraints can ensure a degree of coarse hydrological rationality in an empirical model's outputs (e.g. a linear regression-based sediment rating curve must always produce a monotonic increasing trend of suspended sediment with discharge). However, the lack of flexibility that they impose may also mean that they are unlikely to fully capture a range of important and complex hydrological processes (e.g. hysteresis, which can often result in high degrees of local non-linearity and unmodelled scatter in sediment rating curves) (Eder et al. 2010). As a consequence new, more flexible empirically based methods have been sought that learn a response function from data that describe a hydrological system, without any reference to explicit physical system knowledge or the use of a priori assumptions to constrain the nature of the modelled solution that is generated. These approaches are collectively termed data-driven modelling.

In DDMs, input/output data are interrogated so that the response function's structure and parameter values are informed by the patterns discovered in the data itself. The resulting

Table 3.1 The contrasting approaches of physical, mathematical, statistical and data-driven modelling

Physically based and conceptual mathematical models	Empirically based statistical models	DDMs
Formalise the fundamental physical relationships in the system and the parameters involved	Conceptualise the hydrological system in terms of state variables (inputs, outputs and internal variables)	Conceptualise the hydrological system in terms of state variables (inputs, outputs and internal variables)
⇩	⇩	⇩
Create mathematical representations of these relationships	Decide the form of the optimal statistical relationship by which inputs and outputs are to be optimally related	Interrogate the datasets to identify optimal endogenous relations that relate inputs to outputs
⇩	⇩	⇩
Integrate the mathematics	Collate data for state variables and, where necessary, preprocess to ensure conformability to statistical assumptions	Formalise the optimal response function numerically
⇩	⇩	⇩
Identify optimum parameters for particular application or scenario	Validate and apply the model	Validate and apply the model
⇩		
Validate and apply the model		

functions used in DDMs need not, therefore, conform to any global characteristics, and a high degree of local non-linearity is common. The contrasts between physically/conceptually based, empirical statistically based and data-driven modelling approaches are summarised in Table 3.1. The most important difference between statistical models and DDMs lies in the way that the form of the response function is determined and the constraints imposed on this form. In statistical modelling, the optimal form is ultimately an a priori decision for the modeller with the decision made on the basis of some pre-existing knowledge of how a given catchment's hydrological system is likely to function and, thus, should be represented. In data-driven modelling, it is 'learned' algorithmically using computational intelligence and machine learning techniques (see Solomatine et al. 2008 for a comprehensive review of approaches) but often with only limited consideration of how the particular catchment to which the techniques are applied operates as a hydrological system. Instead, it adopts a paradigm based on the belief that, where sufficient data exist that provide a comprehensive encoding of the hydrological system function, it should be capable of guiding response function development towards a form that best represents the hydrological system.

A large number of different machine learning algorithms have been used for the data-driven modelling of daily suspended sediment concentration or daily suspended sediment load across a wide range of catchments. These include individual or modular feedforward neural networks (Jain 2001, 2008), individual feedforward neural networks (Ciğizoğlu 2004), generalised regression neural networks (Kisi 2004a; Ciğizoğlu and Alp 2006), radial basis function neural networks (Kisi 2004a; Alp and Ciğizoğlu 2007), fuzzy differential evolution algorithms (Kisi 2004b, 2009a, b), neuro-fuzzy and fuzzy logic algorithms (Kisi 2005), support vector machines (Cimen 2008), genetic programming algorithms (Aytek and Kisi 2008) and neuro-wavelet/neuro-fuzzy-wavelet conjunction modelling algorithms (Partal and Ciğizoğlu 2008; Rajaee et al. 2010).

The wide range of algorithmic techniques adopted in data-driven suspended sediment modelling highlights the particular interest that modellers afford to the algorithm and its specific input-output optimisation capabilities. Indeed, a large proportion of the studies listed are concerned with algorithmic performance benchmarking and comparing the optimisation performance of different algorithms so that the optimal model is identified for a given case study. However, what is considered 'optimal' is often founded on little more than the cross-model comparison of goodness-of-fit statistics, which constitutes model validation. This arguably provides a limited perspective on the process of modelling that can be characterised by a philosophical stance in which the end products of a modelling process (i.e. model outputs) may be a sufficient justification for the process itself. In other words, the extent to which a model and its outputs are generated for the right or wrong reasons is not of fundamental concern. There is, therefore, a danger that a data-driven modeller's views of what constitutes a good model can become overly focussed on a partial validation, to the detriment of other vital questions, including:

1. Can the modeller's conceptualisation of the physical system and its state variables be justified and rationalised in terms of what is known about the physical processes operating in the catchment in which it is to be applied?
2. Are the patterns of response function sensitivities to changes in state variables mechanically and/or physically realistic?
3. Are the solutions produced operationally useful?

An optimal goodness-of-fit solution should not be considered a sufficient justification of a DDM in hydrology without adequate consideration of the model's physical and operational rationality. Yet the frequency and extent to which these crucial questions are addressed remain limited and have led to criticisms about data-driven solutions being little more than implicit, black-box predictive mechanisms that have no physical meaning (Mount and Abrahart 2011).

3.3 Empirical Modelling, Physical and Mechanistic Rationality

The need for the physical and mechanistic rationalisation of models developed within the empirical epistemology (including statistical and data-driven approaches) stems from the underlying precept that knowledge is gained through experience in isolation from any pre-existing or innate knowledge of the subject of interest (Markie 2008). In other words, pure empiricism promotes the view that useful suspended sediment models may be developed by those with little, if any, pre-existing hydrological knowledge, so long as the model outputs replicate the testing data adequately. Indeed, DDMs can be argued to conform quite closely to this extreme empirical notion by the very fact that:

1. They are models that can be generated without detailed physical knowledge of the system that they seek to represent.
2. They rely on the assumption that the encapsulation of the physical system in the data from which they are developed is of sufficient scope and richness to support response function development so that it is representative of the hydrological system under study.
3. They are usually validated on the sole basis of the goodness of fit of their outputs to observed data.

Moreover, their attachment to the pure empirical epistemology can be considered even more marked than statistical models because the latter at least utilise some hydrological system knowledge to constrain the response function that is developed (although it is possible to introduce some constraints in DDMs Kingston et al. 2005, 2006; Abrahart et al. 2012a). This adherence to empirical notions results in dangers for data-driven modellers, which can result in the generation of models which offer limited heuristic value to the hydrological discipline because the way in which they generate their outputs bears little resemblance to functioning of the physical system.

For example, the lack of a formal requirement for the input/output relationship encoded in a DDM to be physically rational can lead to incomplete or incorrect selection of input variables with serious errors such as common causality presenting a real danger. Substantive efforts have been made by data-driven modellers to address concerns about the rationale for input variable selection through methods that evaluate the relative information content of potential inputs (e.g. May et al. 2008). Such methods ensure that the inputs selected can be justified in terms of their relative salience as drivers of the phenomenon being modelled. However, they do not provide a test of conceptual rationality of the variables that they select. Thus, it is possible for coincidental trends existing between potential input and output variables to result in the selection of physically irrational inputs to DDMs. Sober (2001) illustrates the problem through the hypothetical, and clearly nonsensical, analysis of British bread prices and Venetian sea level. Coincidental trends between the two variables mean that it is possible to develop a highly significant, yet irrational, empirical model to predict each from each other (Fig. 3.1), producing excellent goodness-of-fit performance for the model output but delivering a model that is clearly nonsense. Sober's paper (ibid) provides a simple, yet clear, example of what can happen when the primary driver of a model development process is its ability to fit independent observations and the conceptual and/or hypothetical foundations that would otherwise guide the model development in a manner that ensures a basic degree of legitimacy are ignored.

Similarly, the lack of a formal requirement for the numerical mechanisms in a DDM's response

Fig. 3.1 Sober's (2001) hypothetical Venetian sea level rise and British bread price data series

Year	Venetian Sea Level	British Bread Price
1	22	4
2	23	5
3	24	6
4	25	10
5	28	14
6	29	15
7	30	19
8	31	20

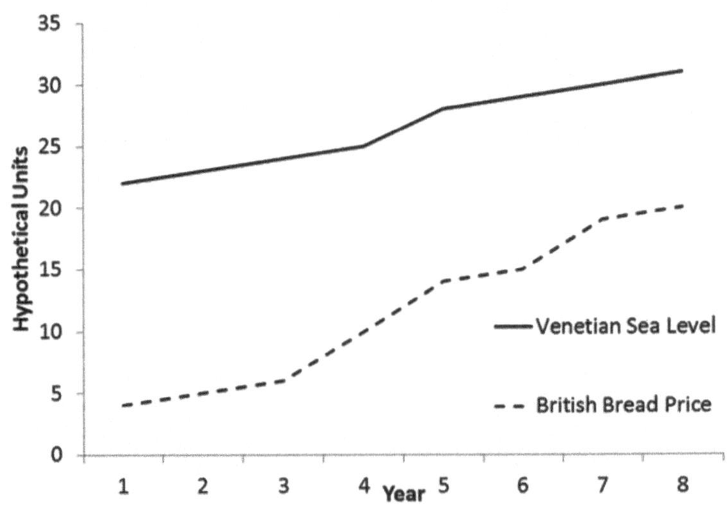

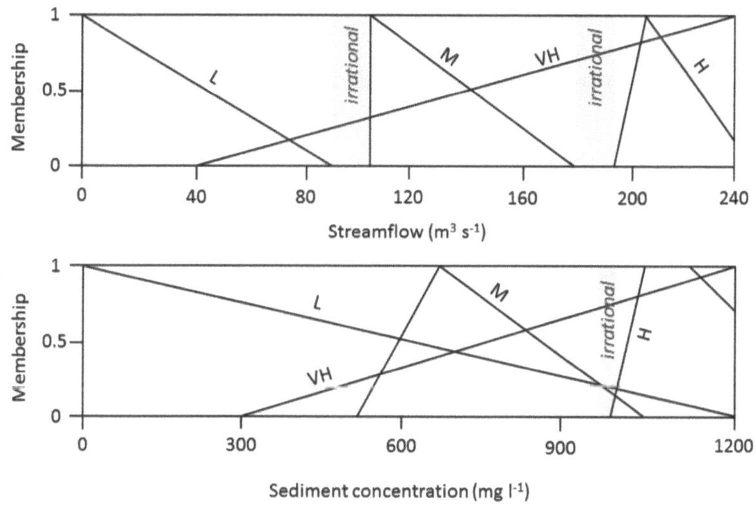

Fig. 3.2 Membership functions from a fuzzy differential evolution model for the prediction of sediment concentration from stream flow on the Quebrada Blanca, Puerto Rico (Kisi 2004b), showing regions of the model with irrational modelling mechanisms. Each plot shows the relative membership of any observed flow or sediment concentration measurement to each of low (L), medium (M), high (H) and very high (VH) membership functions

function to reflect the basic hydrological patterns that one might expect to exist (as long as the model output has adequate goodness of fit) can lead to irrational mechanisms that fail to reflect hydrological axioms of cause and effect. Where DDMs are to be used for anything other than abstract, curve-fitting tasks, this risks compromising its legitimacy. For example, DDMs based on fuzzy logic (e.g. Abrahart and See 2002) have become popular, arguably because they provide a transparent means for capturing and documenting the numerical mechanisms that constitute the response function. These models use codifications of the real world that are expressed in membership diagrams that relate the degree to which specific values in an observed record (e.g. of streamflow or suspended sediment concentration) belong to one or more categories (e.g. high/medium/low). The degree of belonging is then expressed as a series of membership functions. In model development, learning algorithms are used to adjust the membership functions that are applied to the data so as to deliver an optimised input/output, rule-based mapping of the observed data values. However, as is characteristic of DDMs, understanding of the physical system is not used to constrain the adjustment. This can lead to models that produce excellent goodness-of-fit performance, but whose membership diagrams indicate the existence of numerical representations that lack physical and/or logical rationality.

An example of this problem is presented in Fig. 3.2. Here, moderate streamflow observations of ~100 m^3 s^{-1} have partial membership of the very high category but no membership of the medium or high categories that would logically be expected. Similarly, extremely high suspended sediment concentration observations of ~950 mg l^{-1} have a greater membership of the low category than of the medium or high categories. Indeed, such physically irrational examples raise valuable questions about whether DDMs in hydrology are true 'models' at all or simply advance curve-fitting solutions (Mount and Abrahart 2011). In particular, the need to demonstrate conformance between knowledge of physical processes and the employed modelling mechanisms and the solutions that are being produced by a model is often cited as a requirement for hydrological models that can be legitimately used as agents of heuristic knowledge (Thomas 1988). This necessitates a greater degree of rationalism than is commonly encompassed in many DDMs.

3.4 Addressing Physical and Mechanistic Rationality Through Methodological Protocols

One way of increasing the rationalism of DDMs is to develop and adopt model development protocols which can assess the physical and/or mechanistic legitimacy of alternative model realisations in combination with goodness of fit. The purpose of such protocols is to provide a context for model performance assessment that explicitly incorporates the physical processes and geomorphological/hydrological conditions which the data represent and interrogates the response function to determine whether its numerical mechanisms adequately reflect hypotheses about the form and function of hydrological processes. In this way, an increased degree of rationalism may be introduced into the model development process as different models are evaluated, not only against global goodness-of-fit statistics but against a combined data-/hypothesis-led evaluation of how different components of the sediment-discharge relationship are encapsulated and represented and how well this conforms to expectations for a given catchment.

It is, perhaps, surprising that the development of such protocols has not been a primary focus for the data-driven modelling community (Abrahart et al. 2012a) despite the established use of such protocols to direct those engaged in statistically based, empirical modelling. For example, Glysson (1987) developed an important and widely used procedure for assessing the validity of sediment rating curves that highlights the importance of understanding the hydrological context of the data and the identification of structures in the data that need to be represented in any solution. Only in the last few years have generalised protocol for DDMs emerged (e.g. data-driven appraisal modelling protocol (DAMP), Abrahart et al. (2011a)) which identifies the following requirements:

1. A sound understanding of the physiographic and hydrologic context of the data and the likely spatio-temporal processes by which sediment is generated and transported within the catchment
2. Consideration of the quality of the dataset in terms of how well it captures these processes, along with the identification of random and structured errors that may be present
3. Identification of the immediately observable structures present in the dataset and their respective origins (both physical processes and measurement artefacts) with which they are most likely associated and which, therefore, need to be either modelled or rectified
4. Determination of the relative importance of these structures in the context of the operational purpose of the model

It is clear from this list that Points 1 and 2 respond directly to the need for greater understanding of a dataset's complexity and that Points 3 and 4 respond to the value of data exploration and visualisation as a key element in DDM appraisal. However, it could be argued that one important limitation of DAMP is that it focuses too greatly on the data whilst ignoring the need to explicitly interrogate the specific internal numerical mechanisms of the DDM's response function in order to ascertain that it is physically and mechanistically sound. Therefore, an additional element is required that exposes and interrogates the model's response function behaviour so that its conformance to hypothetical expectations can be evaluated.

3.5 Addressing Physical and Mechanistic Rationality Through Sensitivity Analysis

Sensitivity can be defined as the 'rate of change of one factor with respect to change in another' (McCuen 1973). Importantly, a model's sensitivity can be explored in the context of either its global (e.g. Muleta and Nicklow 2005; Saltelli et al. 2008) or local (e.g. Turanayi and Rabitz 2000; Spruill et al. 2000; Holvoet et al. 2005; Hill and Tiedman 2007) responses. If the dominant response function behaviour is relatively consistent across the model's output range, global sensitivity analysis approaches will offer a sufficient basis for exploring the model's physical and mechanistic rationality. If the response function behaviour is more variable across the model's output range, local sensitivity explorations may be more appropriate.

The principal considerations that should be applied in a sensitivity analysis operation are:

- That the response to representative variation of input data and parameter values results in realistic behaviour in the model, at least in a theoretical sense (i.e. does the response function suitably capture gross trends, local patterns of variability and/or the effects of phenomena such as hysteresis?)
- That the model is sufficiently sensitive to representative variations in the hydrological system it is attempting to model, such that it produces realistic behaviour across the whole data range
- That the model is sufficiently sensitive to each of its inputs and parameters that their inclusion is warranted

Two basic computational approaches for sensitivity analysis are commonly employed as a means of examining the extent to which these principles are met. The first is a simulation-based approach in which the model is executed repeatedly for different combinations of input variable or parameter values and the impact on the model output is observed. This can involve exploring the impact of a limited number of fixed adjustment increments (e.g. Sudheer and Jain 2004) or more comprehensive approaches (such as Monte Carlo estimation methods) that seek to elicit probability density functions that describe how adjustments to model inputs or parameters affect its output response across the model's operational range. The second involves the computation of partial derivatives for the specific mathematical response function that comprises the model. This enables the relative influence (expressed as rate of change) of each input/output pairing to be evaluated for any or all defined points in the model's input/output range.

The two approaches can be thought of as different sides of the same coin: simulation focuses on elucidating the general sensitivities for the model's response function (e.g. for specific quantiles), whilst calculated partial derivatives provides a greater degree of granularity – enabling the specific response function sensitivities to be isolated and characterised at any point in the model's operational space. Computing partial derivatives for an empirical model is particularly useful for assisting the physical interpretation and/or justification of a model's internal behaviour because it reveals patterns of response function sensitivity associated with *real* combinations of input and/or parameter values, enabling their physical and/or mechanistic rationality to be explored. Indeed, examination of global sensitivity has limited potential for informing the physical and/or mechanistic justification of a model as there is no guarantee that the complex nature of the interactions between the different input variables and their internal influences on the response function will be adequately reflected. This is particularly true for DDMs, which commonly utilise large numbers of input variables with differing degrees of independence and for which the level of internal complexity of the response function that is generated through the training process is relatively unconstrained.

Partial derivative sensitivity analysis is an established method for assessing statistical models of suspended sediment, and this is largely due to the fact that statistical model response functions are both explicit and constrained. This limits the complexity of their numerical mechanisms and facilitates the algebraic derivation of partial derivatives. As an example, we can consider the following standard multiple linear regression (Eq. 3.1):

$$y = \beta_0 + \beta_1 x_1 + \beta_2 x_2 + \in \tag{3.1}$$

which can be converted to its partial derivatives:

$$\frac{\delta y}{\delta x_1} = \beta_1 \text{ and } \frac{\delta y}{\delta x_2} = \beta_2 \tag{3.2}$$

However, the same cannot be said for many DDMs whose response functions are seldom explicit (i.e. the formulae that define them are not immediately accessible to the modeller) and are characterised by highly complex internal numerical mechanisms. However, progress has been made in determining algebraic methods for supporting sensitivity analysis of certain DDMs (e.g. Yeung et al. 2010). A case in point is the artificial neural network (ANN) model, perhaps the most established tool for developing a DDM (for a recent review, see Maier et al. (2010)). The partial derivatives for ANN models can be computed through the application of a backward chaining partial differentiation rule (Hashem 1992) so that for an ANN with sigmoid activation functions, one hidden layer, i input nodes, j hidden nodes and one output node (O), the partial derivatives can be computed thus (Abrahart et al. 2012b; Mount et al. 2013; Dawson et al. 2014):

$$\frac{\partial O}{\partial I_i} = \sum_{j=1}^{n} w_{ij} w_{jO} h_j (1 - h_j) O (1 - O) \quad (3.3)$$

in which w_{ij} is the weight from input node i to hidden node j, w_{jO} is the weight from hidden node j to the output node O, h_j is the output of hidden node j and O is the output from the network.

Such equations are not generally available for other types of DDM, meaning that the use of partial derivative sensitivity analysis in data-driven hydrological modelling in general and suspended sediment modelling in particular is infrequently applied in published studies. However, where they have been used, partial derivatives of a DDM enabled detailed and specific arguments for the mechanistic and physical rationality of the model to be advanced through characterisation of local and/or global sensitivity patterns. Figure 3.3 provides an illustrative example of the partial derivative-based relative sensitivity analysis using the 'brute force' method (see Dawson et al. 2014) applied to the case of a simple single-input-single-output discharge/SSC rating curve model. Figure 3.3a shows the pattern of relative sensitivity for a

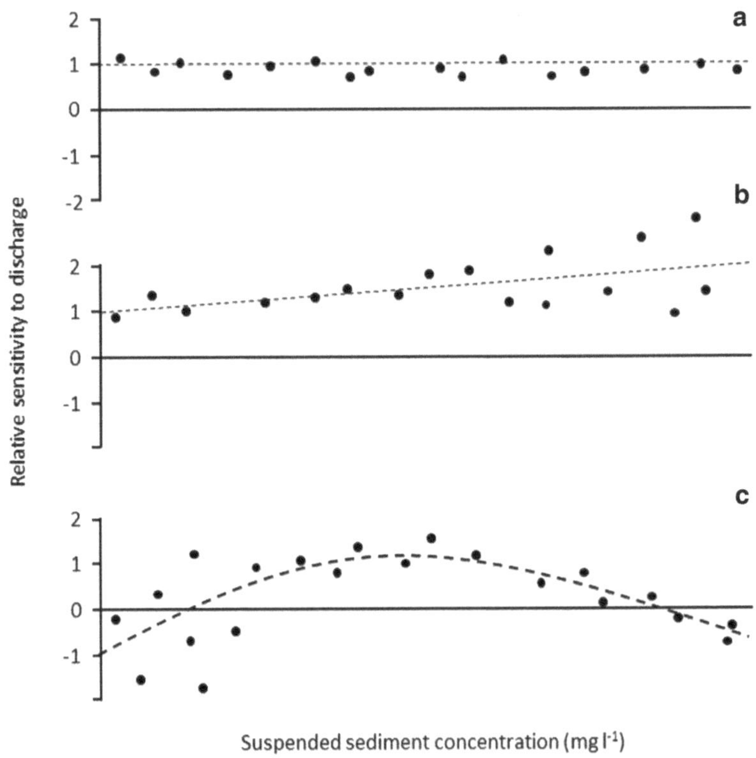

Fig. 3.3 Three characteristically different relative sensitivity plots for a simple rating curve model

model whose response function behaves in a positive, linear, 1:1 manner. There is little variability about the dominant global pattern of sensitivity as demonstrated by the low degree of scatter, which is consistently low throughout the prediction range. This characterises a response function with non-complex internal mechanisms (i.e. approximating linear regression). However, it arguably fails to encapsulate the curvilinear nature of the underlying suspended sediment transport processes that might reasonably be expected nor the inherent heteroscedasticity that is common in suspended sediment-discharge relationships and that one might expect to see reflected in the modelling mechanism. The response function that is characterised by the relative sensitivity plot shown in Fig. 3.3b is arguably more physically realistic than that shown in Fig. 3.3a. The overall trend of positive, increasing relative sensitivity across the model range is consistent with a positive, curvilinear relationship between SSC and discharge. Moreover, the increased scatter at greater levels of concentration implies a modelling mechanism whereby the response function is modelling enhanced levels of local variability (heteroscedasticity) in the relationship at higher discharges. It is difficult to develop either a physical or mechanistic rationale for the relative sensitivity plot presented in Fig. 3.3c. Negative relative sensitivity values at low and high model output ranges suggest an illogical physical relationship in which increasing discharge results in decreasing SSC. Importantly, at two specific points, the relative sensitivity is zero, meaning that SSC is invariant to change in discharge. Should this occur more generally in a more complex model, it would indicate a redundant model component. The global trend is also physically irrational as it implies a similar, yet directly opposing relationship at low to moderate and moderate to high model output ranges. In addition, the high degree of scatter at low output values and the low degree of scatter at high output values are in direct contradiction to hypothetical, real-world settings where the relationship between SSC and discharge at low ranges is generally consistent.

A more detailed discussion on the protocols, methods and interpretation approaches that can be used with relative sensitivity analysis of DDMs can be found in Mount et al. (2013) and Dawson et al. (2014). Such analyses should be an important and urgent research goal of data-driven modellers (Abrahart et al. 2012a). It should also be noted that, even in the absence of partial derivative-based relative sensitivity analysis, basic simulation-based sensitivity analysis can still be applied to DDM outputs (for an example, see Sudheer 2005). This can deliver valuable information about the overall pattern of sensitivity of the model output generated across the range of the input variables and from which the physical rationality of the output can be inferred, albeit without full examination of the internal numerical mechanisms used to generate it.

3.6 Operational Rationality

Alongside concerns about the need for greater physical and mechanistic rationality in DDMs of suspended sediment, concerns exist that not all DDMs of suspended sediment have operational rationality (e.g. Mount et al. 2011). The problem occurs when models that are able to generate excellent goodness-of-fit statistics are developed from input predictor variable sets that make it impossible for the model to be operationalised in a meaningful, predictive manner. Usually, this occurs in DDMs that are developed for predicting time series in tasks allied to forecasting of river, where the aim is to deliver a short step-ahead, suspended sediment prediction. This issue is of particular relevance since one of the often-cited advantages of DDMs over traditional statistical rating curve models is their flexibility with respect to the number and type of predictors that can be used as inputs to the model and the fact that a very high degree of complexity between inputs and outputs can be handled. Moreover, the lack of formal assumptions governing DDMs means that important issues such as the statistical independence of predictors are afforded far less attention in DDMs compared to their statistical counterparts.

In suspended sediment modelling, the issue of operational irrationality is perhaps best exemplified by DDMs that utilise lagged suspended sediment values (i.e. at time t_{-1}, t_{-2}, t_{-3}, etc.) to predict suspended sediment at time t (Aytek and Kisi 2008; Kisi 2009a, b). The use of lagged sediment as a predictor is supported by claims that DDMs perform better when such inputs are used (e.g. Kisi 2005; Aytek and Kisi 2008), and at first glance, this may seem a reasonable justification for their inclusion as measures of model fit do improve when lagged sediment data are included (Table 3.2). However, it is important to recognise that the use of past sediment records as an input to the modelling process not only reduces the problem to something approaching a near-linear modelling operation (in which case the need for using a DDM becomes questionable anyway), but it also makes little operational sense. This is due to the fact that, in order to run the model, a full set of observed measurements of suspended sediment is needed as an input to the model for the period being modelled minus the lag. Consequently, there would be no need to model the suspended sediment record as it would already, by definition, be known at least up until the period of the lag. Similarly, extrapolation of suspended sediment outputs beyond the period for which suspended sediment values are known is not supported. Indeed, even if one makes the case that the model's value lies in its ability to predict sediment response within just the period of the lag, thereby allowing high-frequency, short-term prediction of sediment response, problems remain. The standard inclusion of current discharge input means that the related sediment response would have happened before it could be modelled – negating the value of the prediction.

Finally, model solutions cannot be transferred to similar rivers or to different periods, if no observed suspended sediment records are available for use as inputs to the sediment prediction model. The supposed advantages of DDMs are thus negated on several counts since the resultant models become restricted to the original stations for which sediment data must already have been collected and the transferability of solutions to different reaches or catchments is operationally invalid.

In the data-driven modelling paradigm, the utility of the final model is not a constraint on what sort of model is developed. A high degree of flexibility in the model development process, coupled with the freedom to use and manipulate data in an unconstrained manner, makes it possible to develop models that are operationally irrational even though their goodness of fit may be excellent. Once again, this highlights the dangers associated with DDM's over-reliance on the use of goodness-of-fit statistics as the main model justification and selection criterion. Moreover, in the case of recursive model structures, where a model output is subsequently used as one or more inputs (e.g. Coulibaly 2010; Abrahart

Table 3.2 Goodness-of-fit statistics (RMSE and R^2) obtained by Aytek and Kisi's (2008) for a suite of genetic programming models of suspended sediment utilising a range of lagged discharge (Q) and sediment (S) inputs as predictor variables at an upstream and downstream gauging station of the Tongue River, USA

	Upstream station		Downstream station	
Input combinations	RMSE (ton day^{-1})	R^2	RMSE (ton day^{-1})	R^2
Q_t	421	0.767	621	0.878
S_{t-1}	333	0.851	343	0.937
Q_t, S_{t-1}	331	0.848	381	0.930
S_{t-1}, S_{t-2}	333	0.842	357	0.937
Q_t, S_{t-1}, S_{t-2}	348	0.851	398	0.932
Q_t, Q_{t-1}, S_{t-1}	231	0.941	331	0.934
Q_t, Q_{t-1}, S_{t-1}, S_{t-2}	262	0.903	338	0.933

Improved fit statistics can be observed for models which include suspended sediment at t−1 and t−2 as inputs, yet inclusion of past sediment as an input variable renders the model operationally invalid et al. 2011b), the degree of independence between input and output reduces sharply, and the modelling task becomes increasingly simplistic – resulting in improved goodness-of-fit statistics achieved as the DDM correctly replicates the increasingly simple relationships that map input to output. However, by asking questions about whether a DDM's structure can actually be operationalised, such pitfalls should easily be identified and avoided.

3.7 Data-Driven Modelling of Suspended Sediment: Research Challenges

Data-driven modelling of suspended sediment is no longer a new, immature research activity. Indeed, more than 50 research papers have now been published on the use of ANN models alone (Mount and Abrahart 2011). However, the thrust of published work is case study based, with relatively little consideration of fundamental questions about the rationality of the modelling inputs, methods and mechanisms used and the model outputs generated. However, as this area of hydroinformatics matures, questions about how data-driven modellers conceptualise the hydrological system in their models and the ability of different DDMs to represent physical, hydrological processes in a valid manner must be addressed. It is only by demonstrating a combination of physical, mechanistic and operational rationality as well as enhanced strong model fit that data-driven hydrological modelling will reach its potential. It is clear that, at present, data-driven modelling of suspended sediment lacks the rationality of its statistical, conceptual and physical modelling counterparts. Arguably, this restricts its usefulness and potential to empirically based curve-fitting tasks.

Moving the research agenda forward requires a number of developments (c.f. Abrahart et al. 2012a). Crucially, methodological protocols are urgently required that mirror those existing for empirically based, statistical modellers and that can direct the choice and evaluation of DDMs. This can be achieved via an explicit contextual appraisal of the specific modelling needs and challenges associated with a given dataset, the physical processes responsible for them, the specific data structures that will result and the potential of the data-driven technique to capture and replicate these structures. At a most basic level, this will require modellers to engage more fully in the hydro-physiographical context of the modelling problem through the development of a physical and/or mechanistic rationale that, at a minimum, is based upon a qualitative appraisal of the data within the problem domain. However, one goal must be to move beyond qualitative approaches, for example, through the inclusion of sensitivity analysis techniques. In this way, the physical and/or mechanistic rationality of both the model output and the internal numerical mechanisms of the model's response function may be ensured, and any constraints to the adopted data-driven techniques that are necessary to improve the physical rationality of their outputs may be identified, developed and tested. Initial steps towards this exist in the form of both simulation-based and partial derivative analysis approaches. The ongoing development of techniques to enhance our ability to appraise the rationality of DDMs is likely to be of fundamental importance in the next few years (c.f. Mount et al. 2016). Finally, data-driven modellers should remain mindful of the operational value of their models and be careful to ensure that their choice of input variables does not negate the usefulness of the model at the expense of enhanced goodness-of-fit statistics.

References

Abbot MB (2008) Some future prospects in hydroinformatics. In: Abrahart RJ, Solomatine D, See LM (eds) Practical hydroinformatics: computations intelligence and technological developments in water applications. Springer, Berlin, pp 3–16

Abrahart RJ, See LM (2002) Multi-model data fusion for river flow forecasting: an evaluation of six alternative methods based on two contrasting catchments. Hydrol Earth Syst Sci 6(4):655–670

Abrahart RJ, Mount NJ, Ab Ghani N, Clifford NJ, Dawson CW (2011a) DAMP: a protocol for contextualising goodness-of-fit statistics in sediment-discharge data-driven modelling. J Hydrol 409(3–4): 596–611

Abrahart RJ, Mount NJ, Shamseldin AY (2011b) Discussion of 'Reservoir computing approach to Great Lakes water level forecasting' by P. Coulibaly [J. Hydrol. 381(2010) 76–88]. J Hydrol 422:76–80

Abrahart RJ, Anctil F, Coulibaly P, Dawson CW, Mount NJ, See LM, Shamseldin AY, Solomatine DP, Toth E, Wilby RL (2012a) Two decades of anarchy? Emerging themes and outstanding challenges for neural network river forecasting. Prog Phys Geogr 36(4):480–513

Abrahart RJ, Dawson CW, Mount NJ (2012b) Partial derivative sensitivity analysis applied to autoregressive neural network river forecasting. In: Hinkelmann R, Nasermoaddeli MH, Liong SY, Savic D, Fröhle P, Daemrich KF (eds) Proceedings of the Tenth International Conference on Hydroinformatics, 14–18 July 2012, Hamburg

Alp M, Ciğizoğlu HK (2007) Suspended sediment load simulation by two artificial neural network methods using hydrometeorological data. Environ Model Softw 22:2–13

Aytek A, Kisi Ö (2008) A genetic programming approach to suspended sediment modelling. J Hydrol 351:288–298

Beven KJ (2006) A manifesto for the equifinality thesis. J Hydrol 320:18–36

Beven KJ, Freer J (2001) Equifinality, data assimilation and uncertainty estimation in mechanistic modelling of complex environmental systems. J Hydrol 249:11–29

Ciğizoğlu HK (2004) Estimation and forecasting of daily suspended sediment data by multi-layer perceptrons. Adv Water Resour 27:185–195

Ciğizoğlu HK, Alp M (2006) Generalized regression neural network in modelling river sediment yield. Adv Eng Softw 37:63–68

Çimen M (2008) Estimation of daily suspended sediments using support vector machines. Hydrol Sci J 53:656–666

Colby BR (1956) Relation of sediment discharge to streamflow. US Geological Survey Open File Report, p 170

Coulibaly P (2010) Reservoir computing approach to Great Lakes water level forecasting. J Hydrol 381(1):76–88

Crowder DW, Demisse M, Markus M (2007) The accuracy of sediment loads when log-transformation produces nonlinear sediment load discharge relationships. J Hydrol 336:250–268

Dawson CW, Mount NJ, Abrahart RJ, Louis J (2014) Sensitivity analysis for comparison and physical legitimacy of neural network-based hydrological models. J Hydroinf 16(2):1–18

Duband D, Obled C, Rodriguez JY (1993) Unit-hydrograph revisited – an alternate iterative approach to UH and effective precipitation identification. J Hydrol 150:115–149

Eder A, Strauss P, Kreuger T, Quinton JN (2010) Comparative calculation of suspended sediment loads with respect to hysteresis effects. J Hydrol 389:168–176

Fang H-W, Wang G-Q (2000) Three-dimensional mathematical model of suspended sediment transport. J Hydraul Eng 126:578–592

Glysson GD (1987) Sediment transport curves. US Geological Survey Open File Report, 87–218, p 47

Hardy RJ, Bates PD, Anderson MG (2000) Modelling suspended sediment deposition on a fluvial floodplain using a two-dimensional dynamic finite element model. J Hydrol 229:202–218

Hashem S (1992) Sensitivity analysis for feedforward artificial networks with differentiable activation functions. In: Proceedings 1992 international joint conference on neural networks, Baltimore, MD, USA, 7–11 June 1992, vol 1, pp 419–424, IEEE, N.J.

Hill MC, Tiedeman CR (2007) Effective groundwater model calibration with analysis of sensitivities, predictions and uncertainty. Wiley, New York

Hodgkins R (1996) Controls on suspended sediment transfer at a high-Arctic glacier, determined from statistical modelling. Earth Surf Process Landf 24(1):1–12

Holvoet K, van Griensven A, Seuntjens P, Vanrolleghem PA (2005) Sensitivity analysis for hydrology and pesticide supply towards the river in SWAT. Phys Chem Earth 30:518–526

Jain SK (2001) Development of integrated sediment rating curves using ANNs. J Hydraul Eng 127:30–37

Jain SK (2008) Development of integrated discharge and sediment rating relation using a compound neural network. J Hydrol Eng 13:124–131

Jansson MB (1996) Estimating a sediment rating curve of the Reventazón river at Palomo using logged mean loads within discharge classes. J Hydrol 183:227–241

Jansson MB (1997) Comparison of sediment rating curves developed on load and concentration. Nord Hydrol 28:189–200

Kingston GB, Maier HR, Lambert MF (2005) Calibration and validation of neural networks to ensure physically-plausible hydrological modelling. J Hydrol 314:158–176

Kingston GB, Maier HR, Lambert MF (2006) A probabilistic method to assist knowledge extraction from artificial neural networks used for hydrological prediction. Math Comput Model 44:499–512

Kisi Ö (2004a) Multi-layer perceptrons with Levenberg-Marquardt training algorithm for suspended sediment concentration prediction and estimation. Hydrol Sci J 49:1025–1040

Kisi Ö (2004b) Daily suspended sediment modelling using a fuzzy differential evolution approach. Hydrol Sci J 49:183–197

Kisi Ö (2005) Suspended sediment estimation using neuro-fuzzy and neural network approaches. Hydrol Sci J 50:683–696

Kisi Ö (2009a) Evolutionary fuzzy models for river suspended sediment concentration estimation. J Hydrol 372:68–79

Kisi Ö (2009b) Daily suspended sediment estimation using neuro-wavelet models. Int J Earth Sci [online first]. doi:10.1007/s00531-009-0460-2

Kothyari UC, Tiwari AK, Singh R (1997) Estimation of temporal variation of sediment yield from small catchments through the kinematic method. J Hydrol 203:39–57

Kumar D, Pandey A, Sharma N, Flugel WA (2015) Modeling suspended sediment using artificial neural networks and TRMM-3B42 Version 7 Rainfall Dataset. J Hydrol Eng ASCE 20(6):C4014007

Maier HR, Jain A, Dandy GC, Sudheer KP (2010) Methods used for the development of neural networks for the prediction of water resource variables in river systems: current status and future directions. Environ Model Softw 25(8):891–909

Markie P (2008) Rationalism vs. empiricism, The Stanford Encyclopedia of Philosophy (Fall 2008 Edition). http://plato.stanford.edu/archives/fall2008/entries/rationalism-empiricism/

May RJ, Maier HR, Dandy GC, Fernando TMK (2008) Non-linear variable selection for artificial neural networks using partial mutual information. Environ Model Softw 23(10–11):1312–1326

McCuen RH (1973) The role of sensitivity analysis in hydrologic modelling. J Hydrol 18(1):37–53

Mehdi V (2013) Comparison of cokriging and adaptive neuro-fuzzy inference system models for suspended sediment load forecasting. Arab J Geosci 6(8):3003–3018

Mount NJ, Abrahart RJ (2011) Load or concentration, logged or unlogged? Addressing ten years of uncertainty in neural network suspended sediment prediction. Hydrol Process 25(20):3144–3157

Mount NJ, Abrahart RJ, Dawson CW, Ab Ghani N (2011) The need for operational reasoning in data-driven rating curve prediction of suspended sediment. Hydrol Process 26(26):3982–4000

Mount NJ, Dawson CW, Abrahart RJ (2013) Legitimising data-driven models: exemplification of a new data-driven mechanistic modelling framework. Hydrol Earth Syst Sci 17:2827–2843

Mount NJ, Maier HR, Toth E, Elshorbagy A, Soolomatine D, Chang F-J, Abrahart RJ (2016) Data-driven modelling approaches for socio-hydrology: opportunities and challenges within the Panta Rhei Science Plan. Hydrol Sci J, 61(5-8):1192–1208

Muleta MK, Nicklow JW (2005) Sensitivity and uncertainty analysis coupled with automatic calibration for a distributed watershed model. J Hydrol 306:127–145

Negm AM, Elfiky MM, Owais TM, Nassar MH (2007) Modelling of suspended sediment in Nile River using ANN. In: Helfert FJ, Shishkov B (eds) Proceedings of the Second International Conference on Software and Data Technologies, 22–25 Jul 2007, Barcelona, Spain, ISDM/WSEHST/DC, pp 209–214

Ni JR, Zhang HW, Xue A, Wieprecht S, Borthwick AGL (2004) Modeling of hyperconcentrated sediment-laden floods in Lower Yellow River. J Hydraul Eng ASCE 130(10):1025–1032

Nourani V, Gholamreza A (2015) Daily and monthly suspended sediment load predictions using wavelet-based artificial intelligence approaches. J Mt Sci 12(1):85–100

Olyaie E, Banejad H, Chau KW, Melesse AM (2015) A comparison of various artificial intelligence approaches performance for estimating suspended sediment load of river systems: a case study in United States. Environ Monit Assess 187(4):189

Partal T, Ciğizoğlu HK (2008) Estimation and forecasting of daily suspended sediment data using wavelet-neural networks. J Hydrol 358:317–331

Rajaee T, Mirbagheri SA, Nourani V, Alikhani A (2010) Prediction of daily suspended sediment load using wavelet and neuro-fuzzy combined model. Int J Environ Sci Technol 7:93–110

Saltelli A, Ratto M, Andres T, Campolongo F, Cariboni J, Gatelli D, Saisana M, Tarantola S (2008) Glboal sensitivity analysis. The primer. Wiley, Chichester, p 304

Shaw SB, Walter MT, Steenhuis TS (2006) A physical model of particulate wash-off from rough impervious surfaces. J Hydrol 327:618–626

Singh N, Chakrapani GJ (2015) ANN modelling of sediment concentration in the dynamic glacial environment of Gangotri in Himalaya. Environ Monit Assess 187(8):494

Sober E (2001) Venetian sea levels, British bread prices and the principle of the common cause. Br J Philos Sci 52(2):331–346

Solomatine D, See LM, Abrahart RJ (2008) Data-driven modelling: concepts approaches and experiences. In: Abrahart RJ, Solomatine D, See LM (eds) Practical hydroinformatics: computations intelligence and technological developments in water applications. Springer, Berlin, pp 17–30

Spruill CA, Workman SR, Taraba JL (2000) Simulation of daily and monthly stream discharge from small watersheds using the SWAT model. J Am Soc Civ Eng 43:1431–1439

Sudheer KP (2005) Knowledge extraction from trained neural network river flow models. ASCE J Hydrol Eng 10(4):264–269

Sudheer KP, Jain A (2004) Explaining the internal behaviour of artificial neural network river flow models. Hydrol Process 18:833–844

Syvitski JP, Morehead MD, Bahr DB, Mulder T (2000) Estimating fluvial sediment transport: the rating parameters. Water Resour Res 36:2747–2760

Thomas RB (1988) Monitoring baseline suspended sediment in forested basins: the effects of sampling on suspended sediment rating curves. Hydrol Sci J 33:499–514

Toth B, Bodis E (2015) Estimation of suspended loads in the Danube River and God (1668 river km), Hungary. J Hydrol 523:139–146

Turanayi T, Rabitz H (2000) Local methods. In: Saltelli A, Chan K, Scott EM (eds) Sensitivity analysis, Wiley Series in Probability and Statistics. Wiley, Chichester, pp 81–99

Wei XH, Zhang MF (2010) Quantifying stream flow change caused by forest disturbance at a large spatial scale: a single watershed study. Water Resour Res 46: WR00925010

Wicks JM, Bathurst JC (1996) SHESED: a physically-based distributed erosion and sediment yield component for the SHE hydrological modelling system. J Hydrol 175:213–238

Wu DA, Hu GD (2009) Interpolation calculation methods for suspended sediment concentration in the Yangtze estuary. In: Chen W, Li SZ, Wang YL (eds) Proceedings IEEE international conference on intelligent computing and intelligent systems location, 20–22 Nov, Shanghai, China, vol 1, pp 634–638

Xie Z-T, Hou W-G, Ren H (2008) 2D horizontal modelling for the movement of flow and sediment from Yichang to Yangjianao reach at the Gezhouba downstream. Adv Water Sci 3, pp 309–316

Yang CT, Stall JB (1974) Unit stream power for sediment transport in natural rivers, Research Report Number 88. Water Research Centre, University of Illinois at Urbana, Champaign

Yeung DS, Cloete I, Shi D, Ng WWY (2010) Sensitivity analysis for neural networks. Springer, New York, USA, 86 pp

Nick J. Mount School of Geography, University of Nottingham, Nottingham, UK

Robert J. Abrahart School of Geography, University of Nottingham, Nottingham, UK

Christian W. Dawson Department of Computer Science, University of Loughborough, Loughborough, UK

Sediment Dynamics in a Large Alluvial River: Characterization of Materials and Processes and Management Challenges

4

Chandan Mahanta and Lalit Saikia

Abstract

Sediments of a large alluvial river often hold the key to its geoenvironmental behavior. Several physical and chemical behaviours of high sediment carrying river including water quality, bank line shifting, sediment budget and bank erosion are influenced by sediments. The current study carries out a partial evaluation on some of the key factors. Considering severity of bank erosion problem in the last few years, six locations on the river Brahmaputra were considered for this study. Role of individual bank material properties in bank erosion were studied with binary logistic regression using SPSS.

Young lithology, seismicity, unconsolidated sedimentary rocks of the Himalayas, steep slope of the Brahmaputra in the Himalayas, heavy rainfall in monsoon, deforestation and Jhum cultivation were the causes of sediment generation in the river, whereas decrease of slope in Assam plains was the main cause of sediment deposition. Steepness of the Brahmaputra River is high compared to most of the large rivers of the world including Amazon, Congo, Yangtze, Volga, Mississippi, Ganges and Indus. Slope of the Brahmaputra River created by using data of 175 points along the channel taken from GoogleEarth demonstrated a much clear picture with marked differences compared to the earlier figure based on a few points proposed by other researcher.

Erosion was more severe in the south bank whereas more deposition is taking place in the north bank of the Brahmaputra, particularly in upstream of Dibrugarh. Erosion during 1990–2008 in north bank and south bank of the Brahmaputra (within Assam) were 544 km^2 and 920 km^2 respectively, whereas corresponding deposition were 145 km^2 and 68 km^2 respectively.

An attempt was made to construct a sediment budget for the Brahmaputra River in Assam using a mass balance equation. Major

C. Mahanta (✉) • L. Saikia
Department of Civil Engineering, Indian Institute of Technology, Guwahati, India
e-mail: mahanta_iit@yahoo.com

N. Sharma (ed.), *River System Analysis and Management*, DOI 10.1007/978-981-10-1472-7_4

tributaries were found to contribute 326 × 106 t suspended sediment in a year to the Brahmaputra. Considering 30 % of riverine sediments trapped in the river beds and flood-plains, 228 × 106 t sediments were considered in suspension at downstream, whereas 98 × 106 t was estimated to be deposited. From scouring and deposition data, mass of deposited sediment on river bed has been 69 × 106 t in a year. Area of land lost due to bank erosion in a year was found to be 81 × 106 m^2. Total sediment load in the river at downstream was estimated 869 × 106 t/year. Considering 10 % of sediment load of the Brahmaputra as bed load, suspended sediment load at downstream was 782 ×3 106 t/year. Tributaries, bank erosion and scouring of river bed were found to contribute 26 %, 54 % and 20 % to suspended load of the Brahmaputra at downstream. The findings differed somewhat from previous estimates possibly due to relatively dependable volume of currently measured data.

Among the analysed parameters pH, OC, CC, CEC, ESP, d10, d50 and d90, were more fluctuating in bank materials of erosion sites than that of non-erosion sites. Low (<2 %) organic content, dominance of silt and sand sized particles, particularly in lower layer of sediment profile were the major geochemical properties of bank materials contributing to bank erosion in sites like A, B, E and F. Decrease of SAR, ESP and CEC towards deeper depth showed similarity to results of other erodible areas elsewhere. Presence of clay pocket and high organic content at lower layer of sediment profile had no role in stability of bank profile in C area due to overall dominance of silt particles in the bank.

The present study has the potential scope of study of fate and transport of sediments of a large alluvial river using more data from tributaries as well as the main stem. Inclusion of hydrological parameters can led to reliable erosion prediction in a particular area. Consideration of bank material properties and erosion behavior can be effective in site specific protection measures of erosion problem.

4.1 Introduction

4.1.1 Sediments: Erosion, Transportation, and Deposition

Sediment is typically a naturally occurring substance, originally derived from underlying bedrocks, broken down by weathering and erosion and subsequently transported by the action of fluids such as wind, water, or ice and/or by the force of gravity acting on the particle. Coarse materials are generally transported as bed load along the bed of a river through rolling, sliding, or saltation. Finer materials are carried aloft, suspended above the channel bed by turbulent eddies, and is transported downstream as suspended load and wash load by the process of advection and turbulent diffusion. Hjulström diagram (Fig. 4.1) proposed by F. Hjulström (1935) shows whether, for a given velocity, particles will be transported, deposited, or eroded.

Erosion, transportation, and deposition are the three major processes in a fluvial system (Schumm 1977), and these are influenced by supply of sediment at the upstream end and sediment that is locally eroded from bed and banks (Charlton 2008). Bank erosion is a major contributor of sediment load to many streams and rivers (Simon and Darby 1999; Sekely et al. 2002; Evans et al. 2006; Wilson et al. 2008). Physicochemical properties of the soil and the chemistry of the pore and eroding fluids determine the erosion

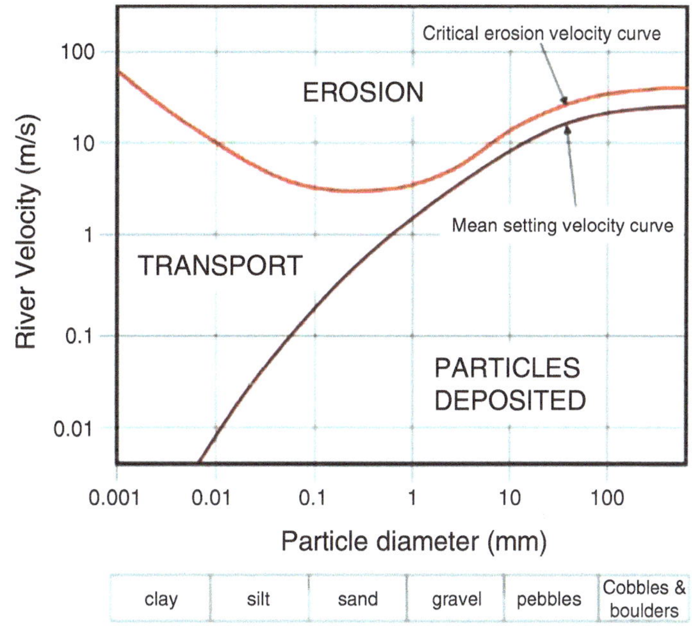

Fig. 4.1 Hjulström diagram

resistance of cohesive banks (Arulanandan et al. 1980). Studies on riverbank stability analysis have accounted the effect of multiple horizontal layers of soil with different physical properties (Takaldany 2003; EPA 2003). Soil pH is the single most important chemical property of the soil. For a high-silt soil, increased pH increases erodibility if the structure is very fine or fine granular. Sodium adsorption ratio is the relative proportion of monovalent (Na^+) to divalent (Ca^{2+} and Mg^{2+}) cation concentrations in the soil or sediment. The clay minerals absorb more water at high SAR, and this expansion and dispersion of the minerals result in a soil with high porosity, low permeability, and high erodibility (Rowell 1994; Brady and Weil 2002). Organic content has long been recognized as a critical factor determining the erodibility of soils (Morgan 2005). Wischmeier and Mannering (1969) postulated that organic matter content ranked next to particle size distribution as an indicator of erodibility. Robinson and Phillips (2001) showed that a high concentration of organic matter can facilitate aggregate stabilization in the topmost horizons. Soil with less than 2 % organic content is considered erodible, and erodibility is negatively correlated with organic content from 0 to 10 % (Brady and Weil 2002; Morgan 2005). Soil particle size and, particularly, the silt–clay content of the soil have been recognized to influence fluvial erosion and mass failures (e.g., Wolman 1959; Schumm 1960; Walker et al. 1987; Ashworth and Lewin 2012). Dade et al. (1992) demonstrated that erosion threshold is positively correlated to particle size. While it is convenient to describe sediment by a mean particle size, most natural cohesive sediment is composed of a range of particle sizes, and the relative proportions of these different sized particles can substantially affect sediment erodibility. Adding clay to a sand bed makes it more resistant to erosion, up to a maximum erosion threshold at 30–50 % clay (Grabowski et al. 2011).

4.1.2 Physical and Chemical Characteristics of Sediments

Investigation of sediment quality has been seen as a logical extension to studies of sediment dynamics and yields and has the potential to further understanding of sediment behavior in rivers (Web et al. 1995). During transportation process, sediment size is reduced on its way by wear and sorting, and the size distribution narrows (Raudkivi 1999). Coarse-grained sediments generally have a limited transportation history and are often sourced from a limited continental area. By contrast, fine-grained river sediments consist of large amounts of particles with history of long distance transportation. Examination of mineralogical, mineral-magnetic, chemical, organic, radiometric, isotopic, and physical characteristics of sediments allows tracing of original source of sediments by the technique of sediment fingerprinting (Collins et al. 1997).

Understanding of physicochemical characteristics of river sediment is important as sediments act both as source and sink of pollutants. Due to smaller particles' high ratio of surface area to volume, clay particles can account for a large amount of adsorbed and/or partitioned mass of the pollutant in solution. During flood and bank erosion, soil and sediments wash into rivers, and the particles aggregate and form larger particles that will settle in calm (quiescent) waters. When these particles have accumulated pollutants, the settling of these particles to the sediments can act as a removal mechanism for pollutants. These sediments may be released to the overlying water under suitable conditions of turbulence, change in pH, etc., and hence, deposited contaminated sediment was termed as "chemical timebomb" by Stigliani (1991). Again, the pollutants accumulated in the riverbed sediment may affect the bio-community through food chain for a long period of time (Yujun et al. 2008).

4.1.3 Role of Sediments in River Hydrology

The evolution of alluvial rivers in nature, such as riverbed deformation, sandbar deposition, and river broadening accompanied with bank collapse, is triggered by the interaction of flow and riverbed or bank in the form of sediment transport (Zhang 1989). Thus, the morphology of an alluvial river channel is the consequence of sediment transport and sedimentation in the river (Church 2006). Sediment discharge is a dependent river variable for short-term period and independent variable for long-term equilibrium (Schumm 1971; Chang 1988).

The rivers of the world annually discharge about 35,000 km^3 of freshwater and 20–22 $\times$ 10 ^{9} tons of solid and dissolved sediment to the ocean (Milliman and Meade 1983; Milliman and Syvitski 1992). Asian rivers contribute about 50 % of sediment flux to world oceans (Milliman and Meade 1983). Sediment load of the Himalayan rivers is disproportionally (~4 times) higher compared to the fraction of areal coverage of the rivers to the global drainage area, underscoring the important role of physical erosion in these river basins on the global sedimentary budget (Tripathy et al. 2014). Thirty to fifty percent river-derived sediments of the Tibetan rivers are trapped in the river's low reaches and contribute to extensive floodplain and delta plain development in Eastern and Southern Asia, while the rest is discharged to the sea (Liu et al. 2009).

4.1.4 Present Study

Tibetan plateau, the "water tower" of Asia, is the source of almost all of Asia's major rivers: the Yellow River, the Yangtze, the Mekong, the Salween, the Ganges, the Indus, the Irrawaddy, and the Yarlung Tsangpo (which become the Brahmaputra downstream). Compared to other large rivers like the Amazon, the Congo, and the Mississippi, where the channel patterns are developed in basins with gentle slopes with fairly homogenous and stable underlying geological structure with little or no control on river morphology (Twidale 2004); Tibetan drainage systems are complex and distinct with different sub-basins. Tibetan drainage systems have been developed in a setting where the underlying geological structure is heterogeneous and active and is the major factor controlling the course of rivers and the landscapes they carve out (Tandon and Sinha 2007). Among the Tibetan rivers, the Brahmaputra is unique due to its peculiar drainage pattern, diverse geological setting, high sediment load, and critical erosion problem. The Brahmaputra is one among the large alluvial rivers (Coleman 1969; Bristow 1987) with plain widths of up to 20 km, individual channel widths of several kilometers, and maximum scour depths of up to 50 km. The

Brahmaputra supplies an estimated 1,000 million tons of suspended material to the Bay of Bengal (Milliman and Meade 1983; Milliman and Syvitski 1992; Hay 1998). Sediment yield of the Brahmaputra (852 t/km^2/year) is the highest in the world (Latrubesse 2008). There is additional strong indication that erosion is a major factor of river instability in the Brahmaputra due to the very large amount of sediment intrusion from bank erosion itself. This sediment causes further instability downstream, triggers more bank erosion, and, apart from loss of land and flood protection, hampers navigation (ADB 2007).

The role of sediments including bank material properties in overall process of erosion in a large alluvial river like the Brahmaputra has not yet been explored adequately. In this study, an attempt has been made to explore sediments of the Brahmaputra including influence of sediments on geo-environments of the Brahmaputra floodplains. The main objectives were:

(i) To study the genesis and deposition of sediments in a large alluvial river Brahmaputra
(ii) To carry out a sediment characterization of the Brahmaputra as an example of large alluvial river
(iii) To examine how different sediment-related parameters influence major geo-environmental behavior of a large alluvial river Brahmaputra

4.2 Materials and Methods

4.2.1 Study Area: The Brahmaputra River System in Assam

Originated as the Yarlung Tsangpo ("Tsangpo" means "purifier") from a glacier mass near Manasarovar Lake in the Kailash range in southern Tibet, the Brahmaputra [in Bengali, Jamuna; Tibetan, Tsangpo; Chinese (Pinyin), Yarlung Zangbo Jiang or Ya-lu-tsang-pu Chiang] is an international river flowing to the Bay of Bengal through China (Tibet), India, and Bangladesh. The source of Brahmaputra was previously believed to be either on the Chema-yungdung glacier, as proposed by Indian geographer Swami Pranavananda in the 1930s, or on the Kobei glacier, as determined by Swedish explorer Sven Hedin in 1907. Analysis and field investigations by Chinese Academy of Sciences showed that the Brahmaputra River originates on the Angsi Glacier, located on the northern side of the Himalayas in Burang County of China's Tibet Autonomous Region (http://news.xinhuanet.com). According to the new findings, the Brahmaputra is 3848 km long, and its drainage area is 7,12,035 km^2. The Brahmaputra River system in India part is shown in Fig. 4.2.

The Brahmaputra is an extremely dynamic, predominantly braided river with scours even 40–45 m deep (Coleman 1969). Uniqueness of the Brahmaputra can be viewed from different aspects:

(i) It provides the only example in the world where the drainage pattern runs in a diametrically opposite direction. In Tibet, it flows from the west to the east. But then it takes a U-turn in the south of Tibet and flows from the east to the west in Assam (Hussain 1994). However the other large river systems like the Amazon, Congo, and Mississippi have relatively simple dendritic tributary networks that resemble a branching tree.
(ii) The river drains diverse environments as the cold dry Tibetan Plateau, the steep rain-drenched Himalayan slopes, alluvial Assam Valley, and fluvio-deltaic Bangladesh Plain (Sarma 2005).
(iii) The Himalayas, where the origin of the Brahmaputra lies, is considered to be younger in age than the Brahmaputra River (Ojha and Singh 2004).
(iv) Besides flood hazards, in no other rivers of the world, the bank erosion hazard is so critical as in the Brahmaputra Valley.

The Brahmaputra and its tributaries and sub-tributaries cause major problems during the monsoon period every year in the shape of flood, bank erosion and drainage congestion. The Brahmaputra River has destroyed nearly

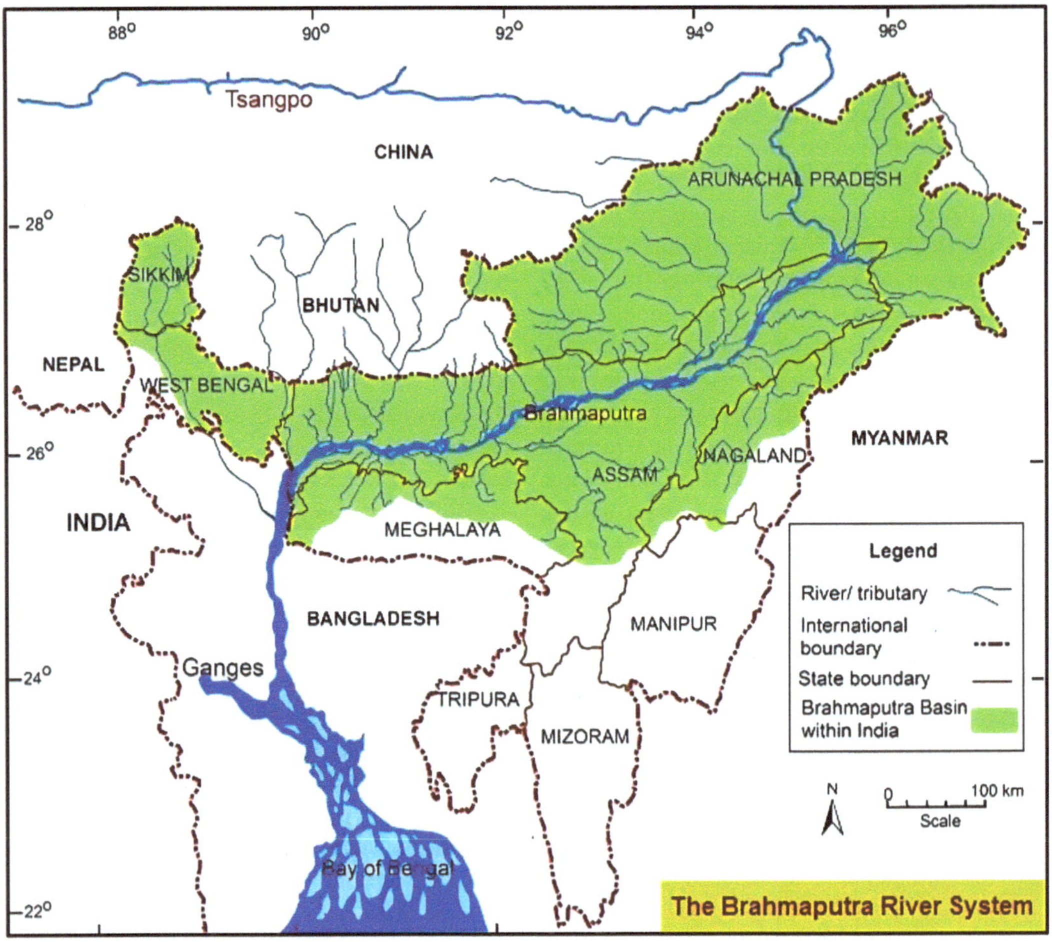

Fig. 4.2 Map showing the Brahmaputra River system within India

4000 km^2 since the last five decades at a rate of 80 km^2 per year and erosion also wiped out more than 2500 villages affecting nearly 5,00,000 people. The stretch falling within Assam, India, has already lost about 7.4 % of its total land due to bank erosion and channel migration (Kakati and Changkakati 2013). The recurrent erosion has caused irreparable damage to many important places along the river bank in addition to other cultivable and homestead lands. The flood in Brahmaputra valley is a recurring phenomenon and has been causing large-scale damages every year. Assam accounts for 9.4 % total flood-prone areas in India. But in very high flood years, the picture is quite different, e.g., damage in Assam during high flood years, 1954, 1959, 1962, 1966, 1972, 1987, 1988, and 1993, were much higher in comparison to that of the total areas affected in the country. Adverse physiography of the region, inadequate capacity of the river channel due to braided nature, heavy rainfall, excessive sedimentation, hill/landslides, reduction of forest areas, and encroachment of riverine areas are some of the major factors causing floods in the region.

4.2.1.1 Factors Responsible for Erosion in the Brahmaputra

Highly erosive characteristics of the Brahmaputra River have been studied by various authors (Coleman 1969; Goswami 1985; Sarma 2005; Pahuja and Goswami 2006; Sarkar and Thorne 2006; Wiebe 2006). The dynamically linked

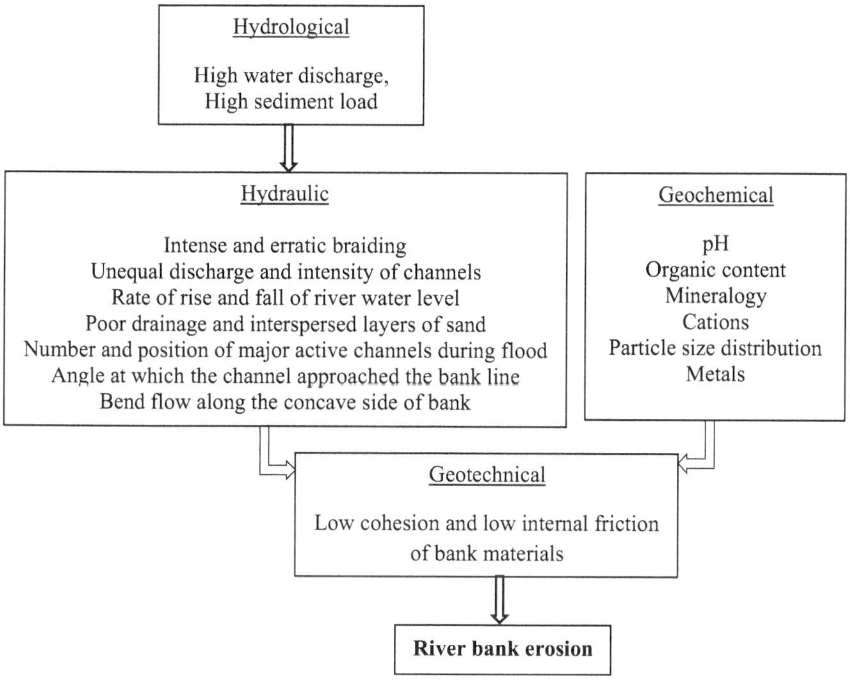

Fig. 4.3 Linked hydrological, hydraulic, geochemical, and geotechnical factors responsible for bank erosion

hydrological, hydraulic, geochemical, and geotechnical factors responsible for bank erosion in the Brahmaputra River are outlined in Fig. 4.3.

Although hydrologic, hydraulic, and geotechnical factors are studied for erosion in the Brahmaputra River, there has been little study on the impact of geochemical properties on erodibility of bank materials. Erodibility is a measure of these resistive forces, which can be determined by evaluating the biogeochemical properties of the sediment (Winterwerp and van Kesteren 2004; Sanford 2008). Precise evaluation of physical and geochemical characteristics can lead to long-term solution options for mitigating the erosion of the Brahmaputra River.

4.2.2 Sources of Secondary Data

There were a few limitations of the secondary data related to the Brahmaputra River:

(i) Nonavailability of enough hydrological data for the Brahmaputra and its tributaries, particularly in Assam plains.

(ii) Most of the available data is discrete, i.e., not continuous and confined to a particular site/location. Several organizations/departments such as Brahmaputra Board, Water Resources Department, Assam PWD, Central Water and Power Commission, and Joint River Commission, are dealing with different hydrological aspects of the Brahmaputra, with the result that data remained dispersed in different offices and is difficult to locate and access.

Besides published articles and reports, following volumes/publications were considered in the present study:

(i) Morphological studies of the Brahmaputra (Water and Power Consultancy Services (India) Limited, 1993)
(ii) Morphological studies of River Brahmaputra at Nagaghuli-Maijan-Oakland area upstream of Dibrugarh town, Assam (Central Water and Power Research Station, Technical Report no: 3590, March 1999)

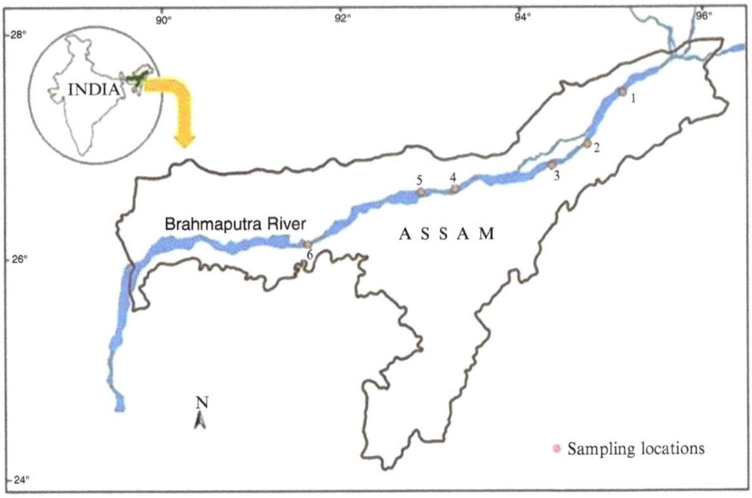

Fig. 4.4 Map showing sampling locations of suspended sediment and bed sediments (*1*) *Dibrugarh*, (*2*) *Disangmukh*, (*3*) *Nematighat*, (*4*) *Biswanathghat*, (*5*) *Tezpur*, (*6*) *Pandu*]

(iii) Additional volume for the master plan of Brahmaputra Basin (Brahmaputra Board, 1995)
(iv) Hydrometeorology of the Brahmaputra Basin (Brahmaputra Board, 1986)
(v) Task force for flood management/erosion control report (Ministry of Water Resources (Govt. of India), 2004)
(vi) Master plan of Brahmaputra Basin, Part I, mainstem (Brahmaputra Board, 1986)

4.2.3 Collection of Sediments

Suspended sediment and bed sediment samples of the Brahmaputra River were collected from six different locations of Assam in India (Fig. 4.4). For suspended sediment samples, four water samples were collected from each sampling location during pre-monsoon (January–March, 2011), monsoon (June, 2011), and post-monsoon (November, 2011) seasons. Wide-mouthed high-grade plastic bottles of 1 l volume were used, which were dipped into the river water, and after being filled, they were capped inside the river water itself. Bed sediments were collected by scooping of freshly deposited materials in the months of December 2010 and January 2011. All samples were brought to laboratory and preserved till analysis. Suspended sediments were obtained from those water samples by filtration using micro-pore filters.

4.2.4 Collection of Bank Materials

Considering severity of riverbank erosion in the last few years, six locations were selected for the study of geochemical evaluation of bank materials. The locations were (A) *Rohmoria*, (*B*) *Dibrugarh*, (*C*) *Nematighat*, (*D*) *Majuli*, (*E*) *Gamerighat*, and (F) *Palashbari* (Fig. 4.5). Soil samples were collected at 10 cm depth and then at 30 cm intervals from vertical profiles of exposed steep banks of erosion sites and nearby non-erosion sites. Multiple samples were collected from each location. All samples were brought to laboratory and preserved till analysis.

4.2.5 Evaluation of Physicochemical Properties of Sediments/Bank Materials

pH, organic matter, carbonate content, sodium absorption ratio, exchangeable sodium percentage, cation exchange capacity, mineralogy, and particle size were estimated using methodologies/instruments as shown in Table 4.1.

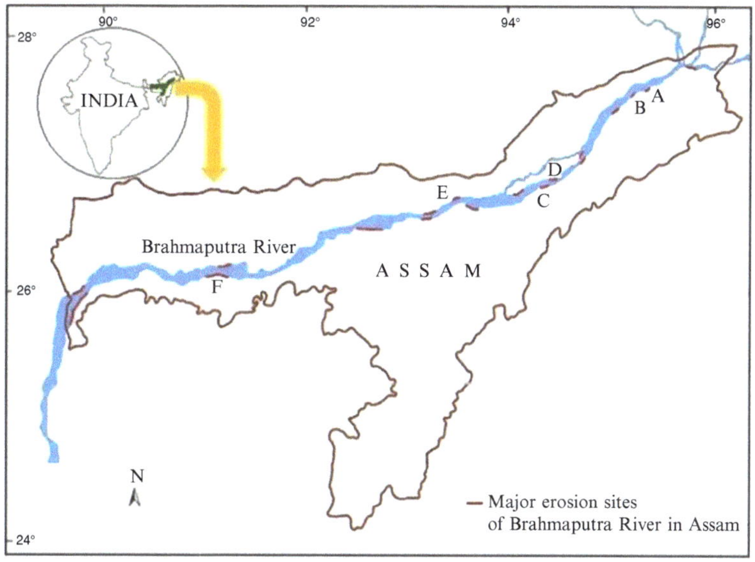

Fig. 4.5 Map showing major erosion sites and sampling locations

Table 4.1 Instruments/methods of estimation of different parameters

Parameter	Instrument/method
pH	pH meter
Organic matter	Loss on ignition method
Heavy metal	USEPA 3050B method
Geochemical fractionation of metals	Modified BCR sequential extraction method (Rauret et al. 1999)
SAR, ESP, CEC	Standard method, flame photometer
Elemental composition	Scanning electron microscope
Mineral composition	X-ray diffraction
Particle size	Laser particle size analyzer

4.2.6 Analysis of Data

The role of individual bank material properties was studied with binary logistic regression using SPSS. The goal of logistic regression is to find the best model to describe the relationship between a dependent variable and multiple independent variables (Lee 2005; Ohlmacher and Davis 2003). The general form of logistic regression is

$$y = a + b_1x_1 + b_2x_2 + b_3x_3 + \ldots + b_mx_m \tag{4.1}$$

$$y = \log_e[P/(1-P)] = \text{logit}(P) \tag{4.2}$$

$$P = e^y/(1+e^y) \tag{4.3}$$

where $x_1, x_2, \ldots, x_m$ are explanatory variables and y is a linear combination function of the explanatory variables representing a linear relationship. The parameters $b_1, b_2, \ldots, b_m$ are the regression coefficients to be estimated. P refers to the probability of the occurrence of an event.

4.3 Results and Discussions

4.3.1 Genesis, Transport and Deposition of Sediments in Brahmaputra

Young lithology, seismicity, unconsolidated sedimentary rocks of the Himalayas, steep slope of Brahmaputra in the Himalayas, heavy rainfall in monsoon, and deforestation and *Jhum* cultivation were the causes of sediment generation in the river, whereas decrease of slope in Assam plains was the main cause of sediment deposition.

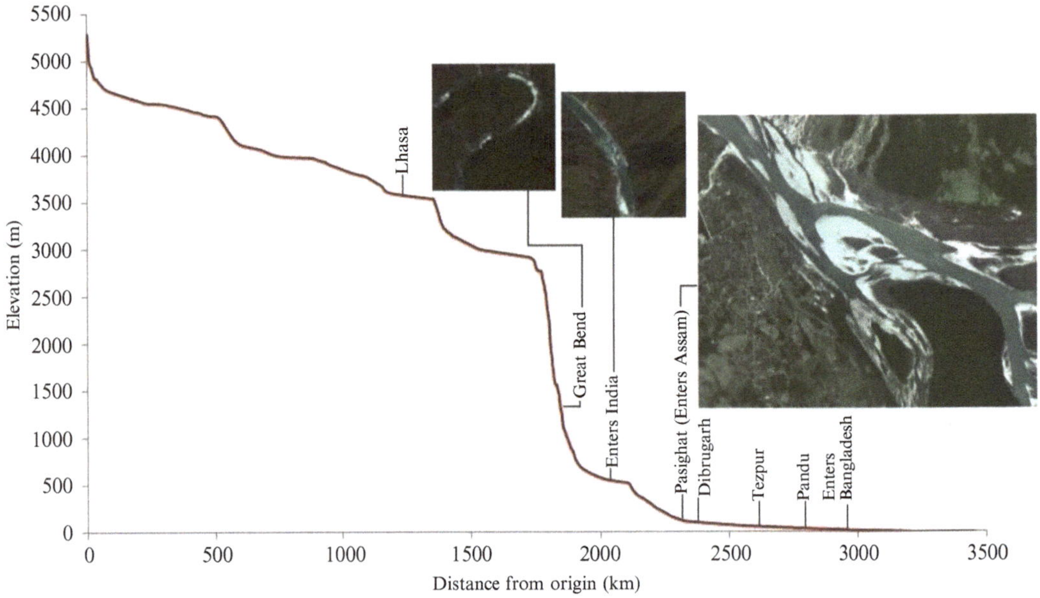

Fig. 4.6 Slope of the Brahmaputra. (Source of data and images: 2013 DigitalGlobe [Google Earth], Eye altitude of images: 17 km)

The steepness of Brahmaputra is high compared to most of the large rivers of the world including the Amazon, Congo, Yangtze, Volga, Mississippi, Ganges, and Indus. The slope of Brahmaputra created by using data of 175 points along the channel taken from Google Earth (Fig. 4.6) demonstrated a much clear picture with marked differences compared to the earlier figure based on a few points proposed by other researcher, e.g., slope near *Pasighat* and *Guwahati* were 1.52 m/km and 0.13 m/km, respectively, whereas the corresponding values in the earlier finding were 0.62 m/km and 0.11 m/km, respectively.

Average gradient of Brahmaputra was 1.52 m km^{-1}. Steep slopes of the river (slope of Brahmaputra in the reach between PE and India border is 8.27 m/km) and tributaries in the mountainous reaches led to high amount of sediment generation and transportation, whereas a sudden decrease in slope (slope in the reach between India border and *Pasighat* was 1.52 m/km) resulted in a large amount of sediment deposition, developing a prominent braided pattern. There was considerable aggradation in the head reaches (C/S 51–61) of Brahmaputra at the rate of 16.85 cm per year during the period 1957–1989. Slope was further decreased during the course of the river in Assam plains (from 0.81 m km^{-1} between *Pasighat* and *Dibrugarh* at upstream to 0.09 m km^{-1} between *Guwahati* and Bangladesh border at downstream), causing more prominent braided pattern and river instability in terms of lateral shifting and erosion of non-cohesive banks. Average scour of 1.92 cm per year was noticed in the entire reaches with maximum scouring of 4.96 cm per year in the middle reaches (C/S 31–41). Slope variation in different cross sections was the maximum in 1971. Decrease in average bed slope at the second reach (C/S 42–52) and then increase in the third and fourth reaches (C/S 32–42, 22–32) favored more erosion in C/S 42–52 and deposition in other locations.

Aggradation and degradation in the riverbed of the Brahmaputra were studied considering data from six reaches as mentioned in Table 4.2 (Fig. 4.7).

The amount of total deposition and scour in the six reaches during 1957–1971, 1971–1977, 1977–1981, and 1981–1989 is shown in Fig. 4.8.

During 1957–1971, scouring was highest in the middle reaches and upstream of the river (cross sections 21–31, 41–51, and 51–61) with consequent deposition in downstream reaches (cross sections 2–11 and 11–21) including cross section 41–51. During 1971–1977 and

Table 4.2 Average width of the river, length and area of reaches

Sl no	Cross section to cross section	Average width of river (km)	Length of reach (km)	Area (km^2)
1	2–11	11.46	83.64	958.85
2	11–21	12.35	88.23	1088.41
3	21–31	9.62	83.14	799.47
4	31–41	11.22	117.31	1316.22
5	41–51	10.40	109.14	1134.73
6	51–61	13.04	103.02	1343.38
Total	61–2	11.25	584.38	6574.28

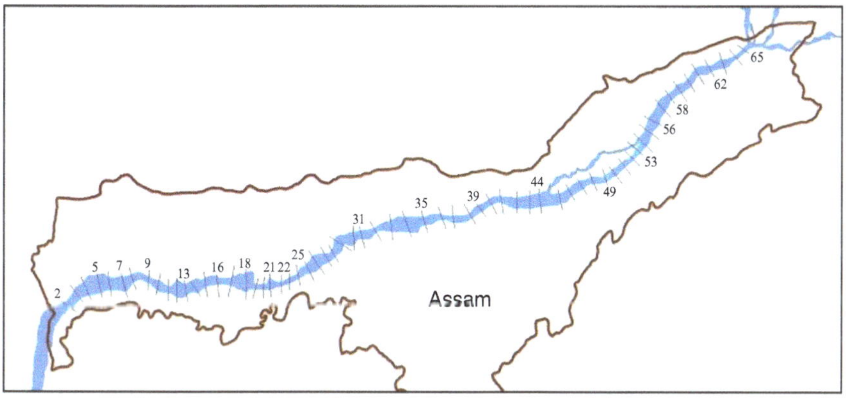

Fig. 4.7 Map showing different cross sections of the Brahmaputra River

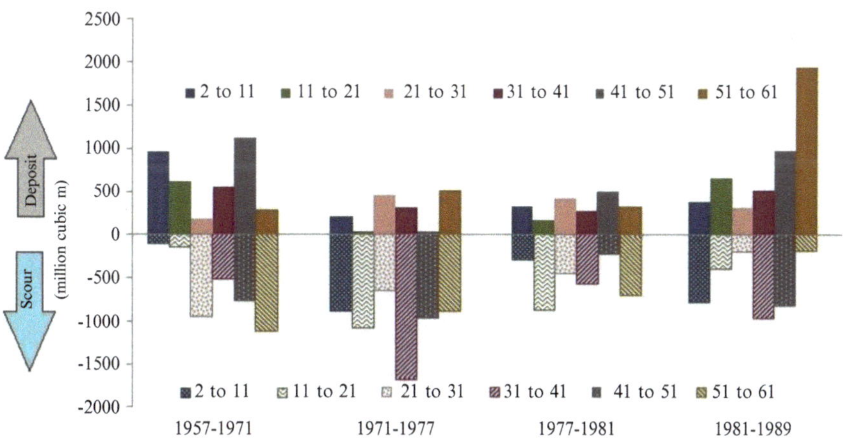

Fig. 4.8 Scour and deposition in the Brahmaputra

1977–1981, scouring was dominant in all the reaches (exception cross sections 41–51) with maximum scouring in the cross sections 31–41 during 1971–1977.

Comparison of net deposition/scour in different reaches from 1957 to 1971 shows relatively lesser deposition or scour along the river, although the scouring and deposition at different reaches were not uniform. During 1957–1971, aggradation of about 6 cm per year and a scour of 7 cm per year appears to have taken place (Fig. 4.9). Beyond 1971, however, scour was more pronounced and

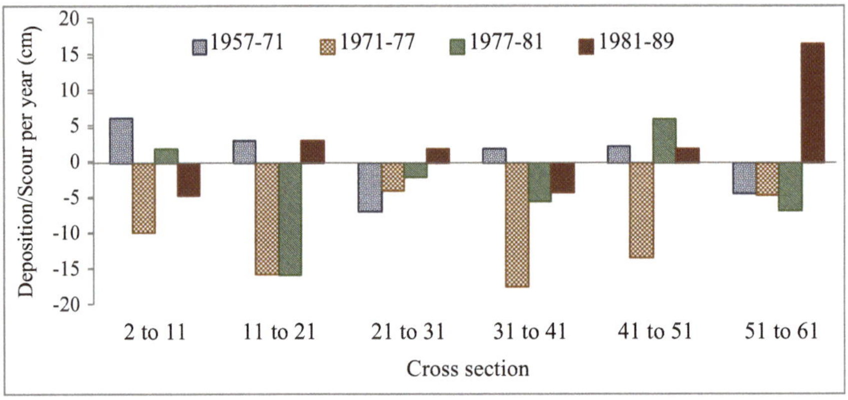

Fig. 4.9 Deposition/scour in different reaches in different periods

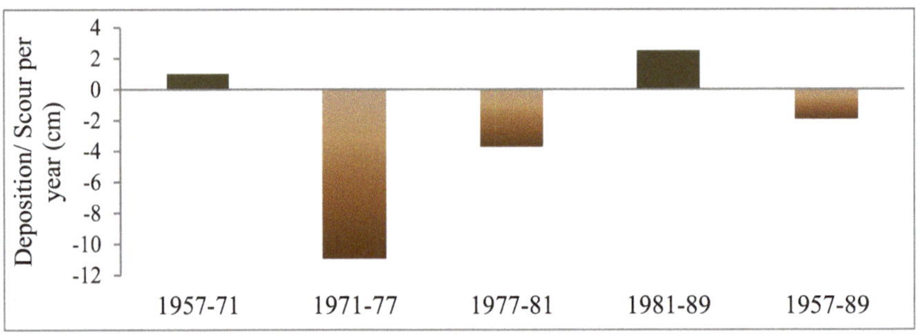

Fig. 4.10 Net deposition/scour in all the reaches in different periods

approximately 11 cm per year was observed from 1971 to 1977. During 1977–1981, scour of about 4 cm per year occurred. During 1981–1989, aggradation of 2.5 cm per year was noticed. Considerably higher aggradation in the head reaches upstream at the rate of 17 cm per year was recorded. Considering the period of 1957–1989 for which cross sections are available, average scour of 2 cm per year was noticed with maximum scouring of 5 cm per year in the middle reaches (cross sections 31–41) except for aggradation of 0.5 cm per year in the upper most reaches (Cross sections 51–61) (Figs. 4.10 and 4.11).

Deposition of sediments results in bar development island formation. Braid bars are highly unstable. Their size, shape, and position change radically both seasonally and annually (Coleman 1969; Bristow 1987), but the islands are relatively stable. Large-scale bedforms (bars and islands) and microscale bedforms (ripples and dunnes) are probably the most important riverbed features inducing resistance to flow and thus influences bed shear stress. Ripples and dunes also exist in abundance in the Brahmaputra. Sand dunes, with heights $\leq$6 m, wavelengths $\leq$330 m, and migration rates $\leq$17 m/h (Garzanti et al. 2010) exist through the entire course of the monsoon in the Brahmaputra. Dunes remain as major driving forces for friction and turbulence to the flow. The distribution of flow is the most complicated in the Brahmaputra due to the presence of braidbars, ripples, and dunes. Dunes are the main driver of bedload transports and responsible for the formation of local bed morphology.

Erosion was more severe in the south bank, whereas more deposition was taking place in the north bank of Brahmaputra, particularly in upstream of *Dibrugarh*. Erosion during 1990–2008 in north bank and south bank of Brahmaputra (within Assam) were 544 km^2 and 920 km^2 respectively, whereas corresponding deposition were 145 km^2 and 68 km^2 respectively.

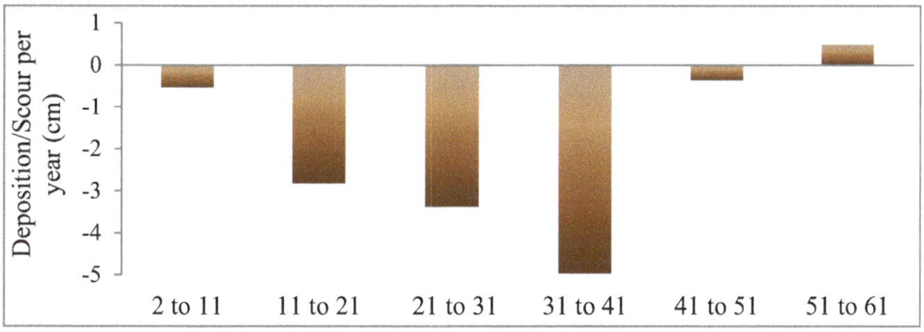

Fig. 4.11 Net deposition/scour in different reaches during 1957–1989

4.3.2 Discharge and Suspended Sediment Characteristics of the Brahmaputra

Barman (2013) observed increase of suspended sediment concentration (C) with discharge (Q), and figure eight form of hysteresis exhibited by C-Q relations at *Pandu* location indicated contribution of sediments from both nearby (e.g., flowage failure of river banks: Coleman 1969; bed sediment resuspension and bank scour: Datta et al. 1999) and distant sources (e.g., Singh and France Lanord 2002). Similar pattern of C–Q data was exhibited at Bahadurabad (Bangladesh) Coleman's (1969) data, indicating possibility of such sediment discharge relationship to be true at other locations of the river as well. The suspended sediment transport in the Brahmaputra River was found to be controlled by flow hydraulics at lower discharge and by supply at higher discharge. This explains the formation of figure eight form of hysteresis in C–Q relationship.

4.3.3 A Sediment Budget for Brahmaputra in Assam

A sediment budget for Brahmaputra in Assam was attempted using a mass balance equation:

Total sediments at downstream = Sediment contribution from main stream and tributaries + sediment contribution from river bank erosion + sediment contribution from scouring − sediment deposition in river bed/banks/floodplains

In the calculation process:

1. Annual average suspended load of eighteen numbers of tributaries along with *Dihang*, *Dibang*, and *Lohit* were considered.
2. It was estimated that 70 % suspended sediments of the tributaries contribute to sediments of the Brahmaputra (after Goodbred and Kuehl 1998, 1999; Liu et al. 2009).
3. Contribution from smaller tributaries was not taken into account due to unavailability of data. However, sediment load of smaller tributaries are negligible compared to bigger tributaries.
4. Sediment input from bank erosion and sediment sink due to deposition in bank were calculated from erosion data for the period 1990–2008.
5. Sediment inputs from scouring and deposition in riverbed were calculated from aggradation and degradation data of the Brahmaputra for the period 1957–1989.

A schematic sediment budget for the Brahmaputra within Assam is shown in Fig. 4.12.

Major tributaries were found to contribute 326×10^6 t suspended sediment in a year to the river. Considering 30 % of riverine sediments trapped in the riverbed and floodplains, 228×10^6 t sediments were considered in suspension at downstream, whereas 98×10^6 t was estimated to be deposited. From scouring and deposition data, mass of deposited sediment on river bed has been 69×10^6 t in a year. Land area lost due to bank erosion in a year was found to be 81×10^6 m^2. Assuming depth of erosion as 4.7 m (average difference of yearly observed highest and the lowest water levels of

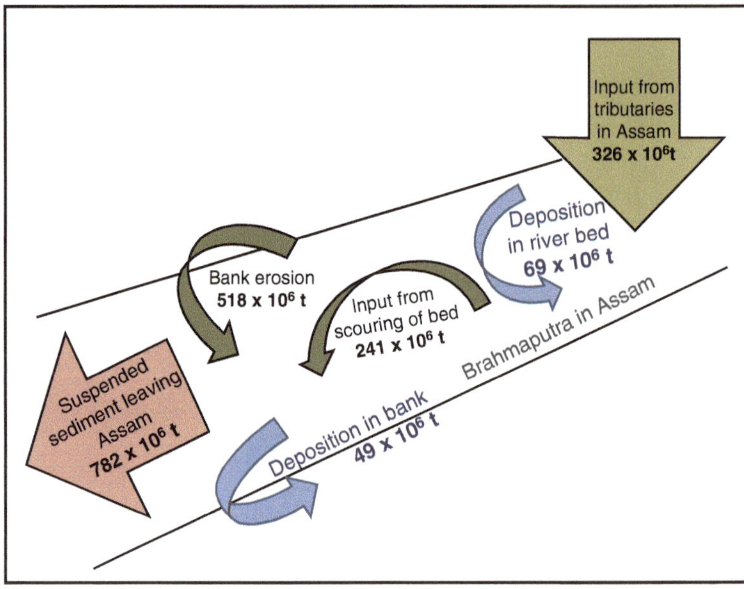

Fig. 4.12 A sediment budget for the Brahmaputra River in Assam

Brahmaputra for the period 1914–1990), the mass of eroded soil was found to be 518 × 10^6 t. Assuming 3 m average uniform height of sand bars, mass of deposited sediments on 12 × 10^6 m ^{2} area was 49 × 10^6 t. Total sediment load at downstream was estimated 869 × 10^6 t/year. Considering 10 % of sediment load of Brahmaputra as bed load, suspended sediment load at downstream was 782 × 10^6 t/year. Tributaries, bank erosion, and scouring of river-bed were found to contribute 26 %, 54 %, and 20 %, respectively, to sediment load of Brahmaputra at downstream. Thus, riverbank erosion was the major contributor of sediment in Brahmaputra. More than half of total sediment of Brahmaputra at downstream came from riverbank erosion. Calculated sediment load of 782 × 10^6 t year^{-1} at downstream of the Brahmaputra (India–Bangladesh border) is comparable with findings of earlier studies (Table 4.3).

4.3.4 Physicochemical Properties of Sediments of Brahmaputra

Main observations on sediments of the Brahmaputra River obtained from the available data for different locations (*Bechamara*, *Bhurbandha*, and *Pandu*) and major tributaries, namely, *Burhi Dihing*, *Disang*, *Dhansiri*, *Jia-Bhareli*, and *Manas*, are as follows (NEC 1993):

1. The annual sediment concentration at Pandu varies from a maximum of 0.068 % (1958) to a minimum of 0.004 % (1974).
2. From observation of sediment data of *Pandu* for the period 1955 to 1983, it was observed that sediment concentration was more and almost constant up to 1958. It followed a linearly decreasing trend from 1958 to 1974 and it followed an increasing trend up to 1983.
3. The maximum annual suspended load at *Pandu* has been 36,609 ha m in 1958 and the minimum 1487 ha m in 1973.
4. The composition of suspended load at *Pandu* (downstream) is 5.3 % coarse, 26.1 % medium and 68.7 % fine grade.
5. The suspended load brought down by the north bank tributaries comprises 14.5 % coarse, 29.3 % medium, and 56.3 % fine grade against 8.2 % coarse, 17.0 % medium, and 74.8 % fine grade carried by the south bank tributaries.

4.3.4.1 Particle Size and Mineralogy

Particle size distribution of suspended sediments revealed dominance of silt (3.9–62.5 μm), very

Table 4.3 Sediment load of Brahmaputra from different studies

Suspended sediment load (10^6 t $year^{-1}$)	Reference	Gauging stations/sampling locations	Period of measurement
617	Coleman (1969)	Bahadurabad, Bangladesh	1958–1962
541	BWDB (1972)	Bahadurabad, Bangladesh	1967–1969
1157	Milliman and Meade (1983)	Bahadurabad, Bangladesh	1966–1967
402	Goswami (1985)	Pandu, Assam	1955–1979
650	Hossain (1992)	Bahadurabad, Bangladesh	1982–1988
721	Islam et al. (1999)	Bahadurabad, Bangladesh	1989–1994
782	**Present study**	Based on secondary data compiled from multiple sources for the river in Indian part	Different period for different data

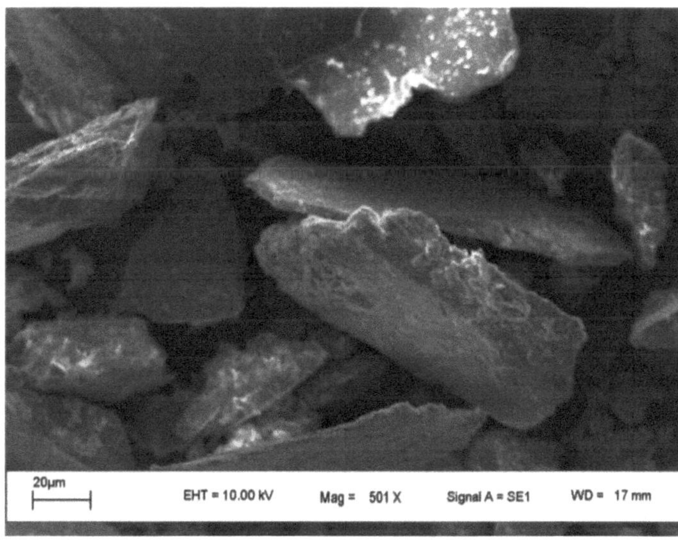

Fig. 4.13 SEM image of suspended sediment of location 5

fine sand (62.5–125 μm), and fine sand (125–250 μm) fractions. XRD analysis revealed dominance of silt-sized quartz, kaolinite, illite, and albite. Dominance of silt-sized quartz, kaolinite, and montmorrillonite in suspended sediments suggests less adsorbing capacity of metals. Prolate, oblate, and bladed shape of suspended sediments (Fig. 4.13) suggested immature sediments coming from bank erosion or scouring with short transportation history. The presence of montmorillonite in most of the sediment samples suggested chemical transformation of aluminosilicates during sediment transport (Fayed 1970; Singh et al. 2005; Campos et al. 2008) and/or contribution from volcanic rocks present in upper drainage basin of Brahmaputra. High values of Q:F (quartz: feldspar) and Q:M (quartz: mica) indicated loss of feldspar and mica due to chemical weathering (Sarin et al. 1989).

4.3.4.2 Metal Content in Sediments

Concentrations of Fe, Mn, Cu, Zn, and Ni in pre-monsoon season were higher compared to that of Monsoon and post-monsoon seasons. All metal concentrations except Mn were in higher concentrations in Brahmaputra River (Fe, 6.5 %; Cu, 41.3 μg/g; Pb, 23.3 μg/g; Zn, 94 μg/g; Mn, 503.2 μg/g; N, 40.3 μg/g) compared to that of

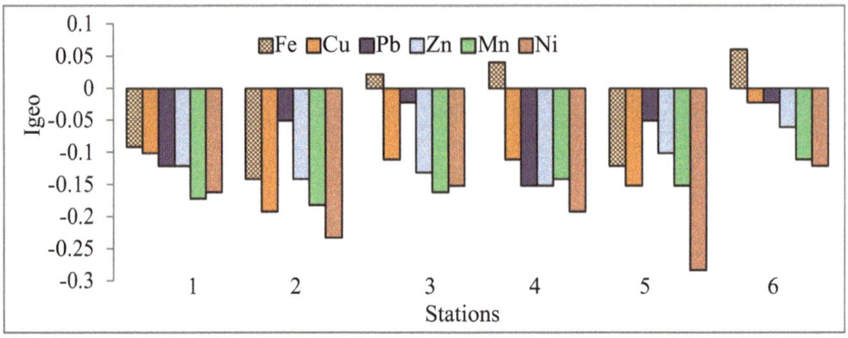

Fig. 4.14 Geo-accumulation indices of different metals at different locations

Indian average river (Fe, 2.9 %; Cu, 28 μg/g; Pb, 15 μg/g; Zn, 16 μg/g; Mn, 605 μg/g; Ni, 37 μg/g) and Bay of Bengal (Fe, 3.9 %; Cu, 26 μg/g; Mn, 529 μg/g; Ni, 64 μg/g).

Geo-accumulation indices showed uncontaminated sediment quality for all the metals except moderately Fe contamination in three locations (Fig. 4.14). Fe enrichment in the sediments points to the presence of primary ferromagnesian silicates or Fe-rich secondary phases in the basin (Pokrovosky et al. 2005). Enrichment of Fe in the Brahmaputra sediments may be attributed to the iron-rich soils (Sabri et al. 1993). Dominance of residual fraction and less amount of bioavailable portion of heavy metals may be attributed to low organic matter, dominant coarser fraction in sediments coupled with low pollution in the river from less industrial activities in the region.

The comparison of studied metal concentration (average) in suspended sediments of Brahmaputra River with that of the world average (from two data sets), Indian River sediment, Bay of Bengal, World surface rock, and average shale, is presented in Table 4.4. Except Fe, concentrations of other heavy metals are lower in the Brahmaputra River compared to the world's average suspended sediment as presented by Martin and Meybeck (1979) and Viers et al. (2009). All metal concentrations except Mn are in higher concentrations in the Brahmaputra River compared to that of Indian average river and Bay of Bengal.

4.3.5 Geochemical Evaluation of Bank Materials

Different geochemical properties of bank materials collected from six locations, i.e., A, B, C, D, E, and F, were evaluated. Among the analyzed parameters, pH, OC, CC, CEC, ESP, d_{10}, d_{50}, and d_{90} were more fluctuating in erosion sites than that of non-erosion sites (Tables 4.5 and 4.6).

Low (<2 %) organic content, dominance of silt, and sand-sized particles, particularly in lower layer of sediment profile, were the major geochemical properties of bank materials contributing to bank erosion in sites like A, B, E, and F (Fig. 4.15). Decrease of SAR, ESP, and CEC toward deeper depth has similarity with findings of Santis et al. (2010) for eroded profiles of Aliano area (Southern Italy). The presence of clay pocket and high organic content at the lower layer of sediment profile had no role in stability of bank profile in C area due to overall dominance of silt particles in the bank.

Low amount of exchangeable Na, K, Ca, and Mg was observed in all locations including non-erosion site of D, which resulted in very low value of CEC in all locations. Low value of SAR in erosion sites could be due to low available Na in soils. High variation of SAR with depth along with high clay content and less variation of sand particles had been observed as significant geochemical properties contributing to non-erosion in the locations of D. Homogenous structure of bank profile (Fig. 4.16) with high organic content

Table 4.4 Metal concentration (average) in suspended sediments (Fe in %, others in μg/g)

	Suspended sediment of the Brahmaputra (present study)			Word's average suspended sediment		Indian average river sediment (Subramanian et al. 1985)	Bay of Bengal (Sarin et al. 1979)	World surface rock (Martin and Meybeck 1979)	Average shale (Turkian and Wedephol 1961)
	Pre-monsoon	Monsoon	Post-monsoon	(Martin and Meybeck 1979)	(Viers et al. 2009)				
Fe	6.5 %	5.9 %	6.2 %	4.8 %	5.81 %	2.9 %	3.90 %	3.59 %	4.72 %
Cu	41.3	39.3	39.5	100	75.9	28	26	32	45
Pb	23.3	23	21.7	150	61.1	15	–	16	20
Zn	94	90.7	90.2	350	208	16	–	129	95
Mn	503.2	469	485	1050	1679	605	529	720	850
Ni	40.3	34.3	36.8	90	74.5	37	64	49	50

Table 4.5 Descriptive statistics of parameters of soil samples from erosion sites ($N = 36$)

	Range	Minimum	Maximum	Mean	SD
pH	3.17	4.93	8.10	6.81	.93
OC	6.48	.12	6.60	1.74	1.75
CC	5.45	.22	5.67	1.30	1.15
SAR	3.90	.26	4.16	1.12	.765
CEC	3.97	.24	4.22	1.32	.96
ESP	14.41	1.26	15.67	4.15	3.15
d_{10}	343.10	1.90	345.00	35.22	75.35
d_{50}	605.00	15.00	620.00	105.98	133.65
d_{90}	1340.90	59.10	1400.00	343.21	346.81

Table 4.6 Descriptive statistics of parameters of samples from non-erosion sites ($N = 34$)

	Range	Minimum	Maximum	Mean	SD
pH	2.13	6.67	8.80	7.34	0.38
OC	5.50	1.10	6.60	2.49	1.13
CC	2.32	.68	3.00	1.82	.69
SAR	6.85	.75	7.60	2.61	1.62
CEC	3.54	0.64	4.19	1.48	.71
ESP	4.42	1.47	5.89	3.22	1.25
d_{10}	32.10	2.10	34.20	7.25	6.23
d_{50}	84.08	10.72	94.80	36.78	17.43
d_{90}	272.62	34.38	307.00	110.17	54.96

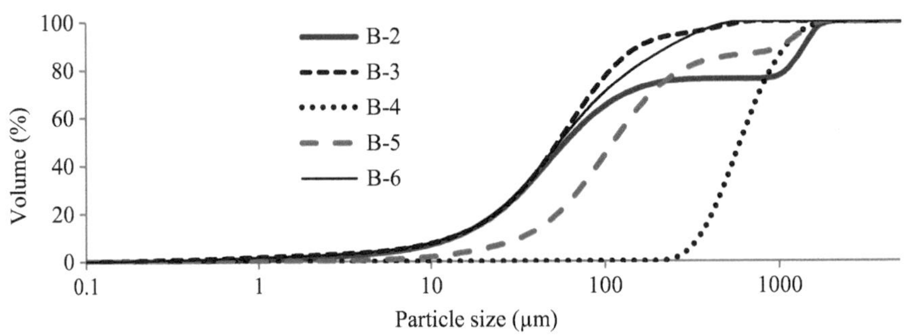

Fig. 4.15 Particle size distribution of bank materials of erosion site B. (B-2, 3, 4, 5, and 6 represent samples at depth of 40 cm, 70 cm, 100 cm, 130 cm, and 160 cm, respectively)

are likely to be responsible for non-erodibility of those two locations.

Erosion site A had comparatively high sand content and the lowest value of minimum clay content (0.3 %) among the study areas. Erosion sites of B, D, E, and F also had higher sand content (52 %, 44 %, 36 %, and 31 %, respectively) with very low amount of clay-sized particles (3–8 %). These two factors including low organic content are likely to be considered as significant geochemical factors for low cohesion in the studied locations as revealed from direct shear test results (Fig. 4.17).

High angle of internal friction of samples in spite of low cohesion could be explained by aggregation of soil particles (Lebert and Horn 1991). The aggregates were strong enough to resist the stresses and the soil acted like very dense sand with high angle of internal friction (Lebert and Horn 1991). Lohnes and Handy

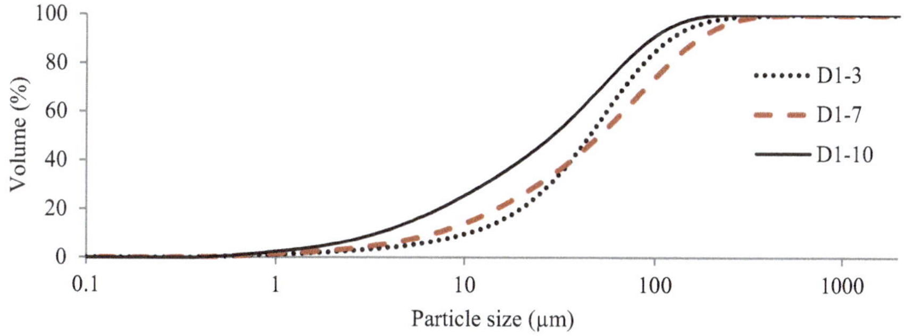

Fig. 4.16 Particle size distribution of bank materials of non-eroded site of D1. (D1-3, D1-7, and D1-10 represent samples at depth of 70 cm, 190 cm, and 280 cm, respectively)

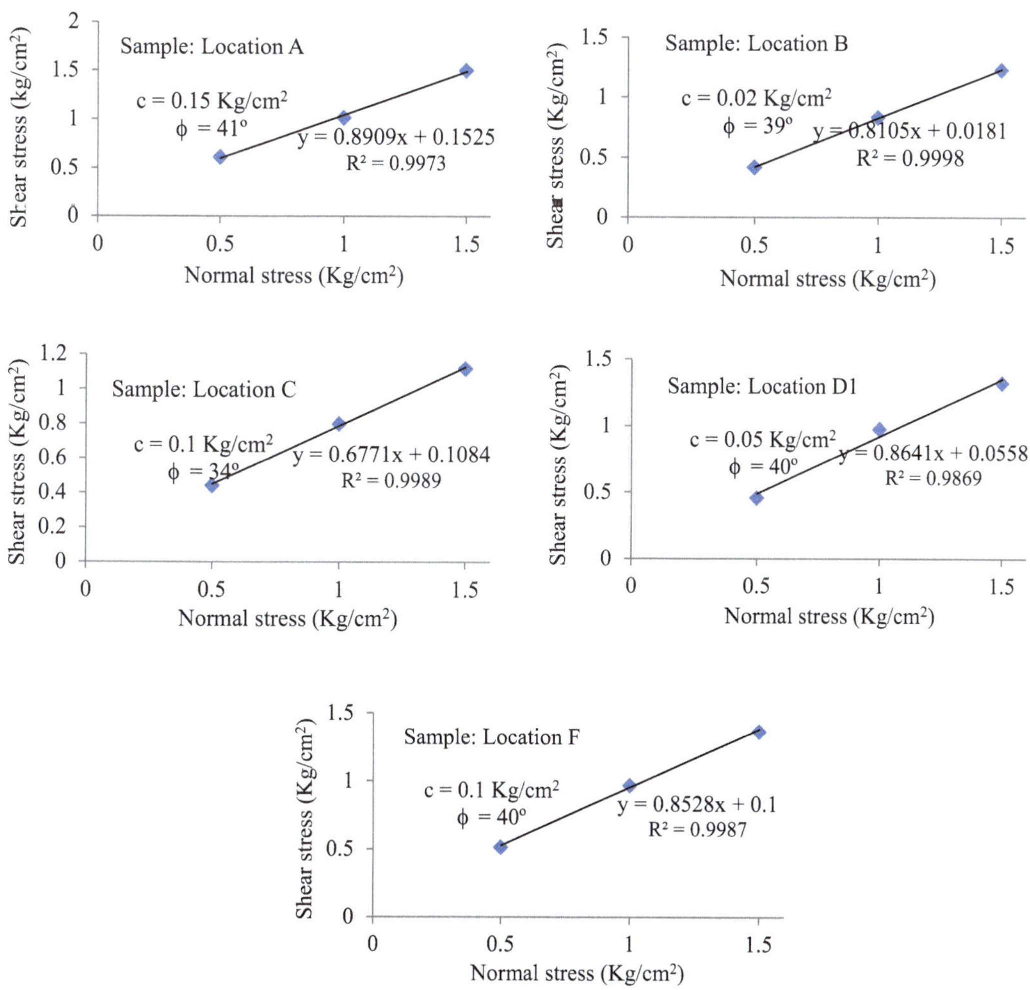

Fig. 4.17 Direct shear test results of bank materials

Table 4.7 Particle size and corresponding velocities for threshold movement of sediments

Location **B**				Location **F**			
Pre-monsoon discharge = 15,000 m^3/s				Pre-monsoon discharge = 15,000 m^3/s			
Monsoon peak discharge = 64,619 m^3/s				Monsoon peak discharge = 76,303 m^3/s			
Pre-monsoon velocity = **0.82 m/s**				Pre-monsoon velocity = **0.80 m/s**			
Monsoon velocity = **1.01 m/s**				Monsoon velocity = **0.86 m/s**			
d_{50}	TV	d_{90}	TV	d_{50}	TV	d_{90}	TV
0.06	0.07	0.16	0.40	0.01	0.009	0.03	0.03
0.06	0.07	0.24	0.50	0.02	0.02	0.07	0.09
0.06	0.07	1.40	3	0.02	0.02	0.10	0.35
0.11	0.35	1.10	2.7	0.02	0.02	0.06	0.07
0.62	0.80	1.10	2.7	0.02	0.02	0.12	0.38

TV velocities for threshold movement of sediments

Table 4.8 Correlation matrices of different parameters of all samples

	Erosion	pH	OC	CC	SAR	CEC	ESP	d10	d50	d90
Erosion	1									
pH	$-.35^{**}$	1								
OC	$-.25^{*}$	.07	1							
CC	$-.27^{*}$	.05	$.51^{**}$	1						
SAR	$-.52^{**}$	$.27^{*}$	.21	$.29^{*}$	1					
CEC	−.09	.17	$.34^{**}$	$.47^{**}$	.13	1				
ESP	.19	$-.48^{**}$	$-.24^{*}$	$-.38^{**}$	−.10	$-.56^{**}$	1			
d_{10}	.25	.21	$-.27^{*}$	−.22	−.15	−.15	.13	1		
d_{50}	$.34^{**}$	.07	−.22	$-.26^{**}$	−.21	−.20	.22	$.93^{**}$	1	
d_{90}	$.43^{**}$	.04	−.21	−.13	−.22	.01	.10	$.63^{**}$	$.68^{**}$	1

**Correlation is significant at 0.01 level
*Correlation is significant at 0.05 level

(1968) suggested that unconsolidated sediments with low cohesion can stand temporarily at steep angles. Steep river banks in erosion sites of the Brahmaputra might be due to high angle of internal friction resulting from aggregation of cohesion less soil particles.

Brahmaputra at erosion sites A, B, and F is extremely braided. So, section average velocity at A, B, and F is comparatively low than that of single channel in other locations. However these average velocities corresponding to a pre-monsoon discharge and peak monsoon discharge are two- to hundredfolds higher than that required for threshold movement of sediment, as observed from Hjulström diagram (Table 4.7).

4.3.5.1 Role of Geochemical Properties of Bank Materials in Erosion

Potential correlation between predictor variables (different geochemical properties of bank materials) and the outcome, i.e., erosion, were observed. Correlation matrices (Table 4.8) showed that pH, organic content, carbonate content, and SAR had negative correlation with erosion, and particle size (d_{50} and d_{90}) had positive correlation with erosion.

Negative correlation (0.48) was observed between pH and ESP. Increasing pH of soil will lead to decrease in ESP, and decreasing ESP will lead to increase in CEC (Sumner 1993; Rengasamy and Churchman 1999; Quirk 2001).

Table 4.9 Variables in the equation

	B	S.E.	Wald	Sig.	Exp(B)
OC	−.073	.199	.135	.713	.929
SAR	−1.321	.485	7.405	.007	.267
d_{90}	.010	.004	5.888	.015	1.010
Constant	.796	.951	.701	.402	2.217

Table 4.10 Model summary

−2 Log likelihood	Cox & Snell R square	Nagelkerke R square
56.520[a]	.439	.586

[a]Estimation terminated at iteration number 6 because parameter estimates changed by less than .001

OC had significant positive correlation with carbonate content (0.51) and CEC (0.34) and negative correlation with particle size, i.e., d_{50} and d_{90}. SAR had positive correlation with carbonate content (0.29). d_{10}, d_{50}, and d_{90} are strongly correlated to each other. d_{10} had significant positive correlation with d_{50} (0.93) and d_{90} (0.63). With increase in finer fraction of soil particles, there was increase of coarse particles in the study area. The presence of coarse particles was dominant in erosion sites although a few sites had clay pockets.

pH, OC, carbonate content, and SAR had negative correlation with erosion, and particle size (d_{50} and d_{90}) had positive correlation with erosion. To evaluate contribution of selected geochemical properties to erosion event, binary logistic regression in SPSS was used with "erosion" ("yes" or "no") as dependent variable and OC, SAR, and particle size (d_{50} and d_{90}) as predictor variables (outputs are shown in Table 4.9).

Nagelkerke's R^2 of .586 (Table 4.10) indicated a moderate relationship between prediction and grouping. Thus values of OC, SAR, and d_{90} of bank materials could be used for rapid assessment of probability (P) of erosion in Brahmaputra using the equation (following Eq. 4.3):

$$P = \frac{e\{0.796 + (-.073)\mathrm{OC} + (-1.321)\mathrm{SAR} + (.010)\mathrm{d}90\}}{1 + e\{0.796 + (-.073)\mathrm{OC} + (-1.321)\mathrm{SAR} + (.010)\mathrm{d}90\}}$$

From the above equation, the odd ratio was .267 for an additional unit in SAR, which suggested that for one unit increase of SAR, the odds of erosion was lowered by 73 % (= .267 × 100–100). Brady and Weil (2002) suggested that SAR and erodibility were positively correlated for clay soils. The anomaly in the present study might be due to abundance of sand content as clay size particles were responsible for expansion and dispersion at high SAR.

The odd ratio was 1.010 for an additional unit in d_{90} values, which suggested that for one unit increase of d_{90}, the odds of erosion was increased by 1 %. The resistance of a bank to fluvial erosion and mass failures tends to increase with increasing silt–clay content (Thorne and Tovey 1981; Osman and Thorne 1988), i.e., decrease in particle size. Thus, with increase in particle size, susceptibility of erosion increases. Role of particle size and OC in erosion was already documented by different researchers.

4.4 Conclusions

Based on primary as well as secondary data, this study addresses sediments of a large alluvial river Brahmaputra and major processes including bank line shifting and river bank erosion.

Particle size distribution of suspended sediments revealed dominance of silt, very fine sand, and fine sand fractions. Dominance of silt-sized quartz, kaolinite, and montmorrillonite in suspended sediments suggest less adsorbing capacity of metals. Prolate, oblate, and bladed shape of suspended sediments suggested immature sediments coming from bank erosion or scouring with short transportation history.

Concentrations of Fe, Mn, Cu, Zn, and Ni in pre-monsoon season were higher compared to that of monsoon and post-monsoon seasons. All metal concentrations except Mn were in higher concentrations in Brahmaputra compared to that of Indian average river and Bay of Bengal. Geo-accumulation indices showed uncontaminated sediment quality for all the metals except moderately Fe contamination in a few locations. Fe enrichment in the sediments points to the presence of primary ferromagnesian silicates or Fe-rich secondary phases in the basin. Dominance of residual fraction and less amount of bioavailable portion of heavy metals may be attributed to low organic matter, dominant coarser fraction in sediments coupled with low pollution in the river from less industrial activities in the region.

Key factors in causing the Brahmaputra unstable at many vulnerable reaches are decreased slope of the Brahmaputra bed and that of its tributaries in Assam plains, aggradation of the riverbed and intense braiding resulting from high sediment load (Lewin and Ashworth 2014). Slope of the river is abruptly decreased during its course in Assam plains (from 0.81 m km^{-1} between *Pasighat* and *Dibrugarh* at upstream to 0.09 m km^{-1} between *Guwahati* and Bangladesh border at downstream), causing more prominent braided pattern and river instability in terms of lateral shifting and erosion of non-cohesive banks.

In an attempt to construct a sediment budget for the Brahmaputra in Assam plains, tributaries, bank erosion, and scouring of riverbed were found to contribute 26 %, 54 %, and 20 %, respectively, to sediment at downstream. Thus, riverbank erosion was the major contributor of sediment in Brahmaputra.

High sand content and low clay-sized particles along with low organic content are likely to be considered as significant geochemical factors for low cohesion in the studied locations. High angle of internal friction of bank materials in spite of low cohesion could be explained by aggregation of soil particles as suggested by different researchers. Average velocities of Brahmaputra corresponding to a pre-monsoon discharge and peak monsoon discharge were two to hundred folds higher than that required for threshold movement of sediment.

This study has demonstrated how different geochemical properties of sediments (bank materials) can be used for assessment of vulnerability of a particular location to bank erosion problem. Inclusion of hydrological parameters like discharge, depth, and velocity along with sediment data as input parameters can give better picture of erosion prediction in a particular area.

The sustainable solutions for sediment management and mitigation of riverbank erosion in the Brahmaputra will involve integrated river management planning and design on holistic approach with understanding of the river processes and adequate quality data along with understanding of processes and pertaining data as highlighted in the present study.

References

ADB (2007) Technical assistance consultant's report, India: Northeastern integrated flood and riverbank erosion management project (Assam), TA (Phase 1) Workshop Materials

Arulanandan K, Gillogley E, Tully R (1980) Development of a quantitative method to predict critical shear stress and rate of erosion of natural undisturbed cohesive soils. Report GL-80-5, U.S. Army Engineers, Waterways Experiment Station, Vicksburg, Mississippi

Ashworth PJ, Lewin J (2012) How do big rivers come to be different? Earth Sci Rev 114(1–2):84–107

Barman B (2013) Evaluation of suspended sediment flux along a cross-section of the Brahmaputra River. PhD thesis submitted to IIT Guwahati

Brady NC, Weil RR (2002) The nature and properties of soils. Prentice Hall, Upper Saddle River, p 960

Bristow CS (1987) Brahmaputra river: channel migration and deposition. In: Ethridce FG, Flores RM, Harvey MD (eds) Recent developments in fluvial sedimentology. Society of economic paleontologists and mineralogists special publications, SEPM Society for Sedimentary Geology, USA. 39, pp 63–74

BWDB (1972) Sediment investigations in main rivers of Bangladesh, 1968 & 1969, BWDB water supply paper No. 359. Bangladesh Water Development Board, Bangladesh

Campos EJD, Mukherjee S, Piola AR, de Carvalho FMS (2008) A note on a mineralogical analysis of the sediments associated with the Plata River and Patos Lagoon outflows. Cont Shelf Res 28:1687–1691

Chang HH (1988) Fluvial processes in river engineering. Wiley, New York, p 5

Charlton R (2008) Process of erosion, transport and deposition: fundamentals of fluvial geomorphology. Routledge, London, p 97

Church M (2006) Bed material transport and the morphology of alluvial river channels. Annu Rev Earth Planet Sci 34:325–354

Coleman JM (1969) Brahmaputra river: channel processes and sedimentation. Sedimentol Geol 3(2–3): 129–239

Collins AL, Walling DE, Leeks GJL (1997) Finger printing suspended sediment sources in larger river basins: combining assessment of spatial provenance and source type. Geogr Ann 79A:239–254
Dade WB, Nowell ARM, Jumars PA (1992) Predicting erosion resistance of muds. Mar Geol 105(1–4):285–297
Datta DK, Gupta LP, Subramanian V (1999) Distribution of C N and P in the sediments of the Ganges – Brahmaputra – Meghna river system in the Bengal basin. Org Geochem 30:75–82
EPA (2003) River morphology, low land river bank stability and erosion assessment. pp 6–52
Evans DJ, Gibson CE, Rossell RS (2006) Sediment loads and sources in heavily modified Irish catchments: a move towards informed management strategies. Geomorphology 79:93–113
Fayed LA (1970) Clay minerals of some of the river Nile sediments near Cairo. Int J Rock Mech Miner Sci 7:249–252
Garzanti E, Andò S, France-Lanord C, Vezzoli G, Galy V, Najman Y (2010) Mineralogical and chemical variability of fluvial sediments 1. Bedload sand (Ganga – Brahmaputra, Bangladesh). Earth Planet Sci Lett 299(3–4):368–381
Goodbred SL Jr, Kuehl SA (1998) Floodplain processes in the Bengal Basin and the storage of Ganges–Brahmaputra river sediment: an accretion study using Cs-137 and Pb-210 geochronology. Sediment Geol 121(3):239–258
Goodbred SL Jr, Kuehl SA (1999) Holocene and modern sediment budgets for the Ganges-Brahmaputra River: evidence for high stand dispersal to floodplain shelf and deep-sea depocenters. Geology 27:559–562
Goswami D (1985) Brahmaputra river, Assam, India: physiography, basin denudation and channel aggradation. Water Resour Res 21(7):959–978
Grabowski RC, Droppo IG, Geraldene W (2011) Erodibility of cohesive sediment: the importance of sediment properties. Earth Sci Rev 105:101–120
Hay R (1998) Sense of place in developmental context. J Environ Psychol 18(1):5–29
Hjulstrom F (1935) Studies of the morphological activity of rivers as illustrated by the river Fyris. Bull Geol Inst Univ Upps 25:221–257
Hossain MM (1992) Total sediment load in the lower Ganges and jamuna. J Inst Eng Bangladesh 20(1–2):1–8
Islam MR, Begum SF, Yamaguchi Y, Ogawa K (1999) The Ganges and Brahmaputra rivers in Bangladesh: basin denudation and sedimentation. Hydrol Process 13(17):2907–2923
Kakati H, Changkakati PP (eds) (2013) Proceedings of the Assam Water Conference 2013. Water Resources Department, Government of Assam, Dispur, p 41
Latrubesse E (2008) Patterns of anabranching channels: the ultimate end-member adjustment of mega rivers. Geomorphology 101(1):130–145
Lebert M, Horn R (1991) A method to predict the mechanical strength of agricultural soils. Soil Tillage Res 19(2–3):275–286
Lee S (2005) Application of logistic regression model and its validation for landslide susceptibility mapping using GIS and remote sensing data. Int J Remote Sens 26(7):1477–1491
Lewin J, Ashworth PJ (2014) Defining large river channel patterns: alluvial exchange and plurality. Geomorphology 215:83–98
Liu JP, Xue Z, Ross K, Wang HZ, Yang ZS, Li AC, Gao S (2009) Fate of sediments delivered to the sea by Asian large rivers: long-distance transport and formation of remote alongshore clinothems. Sediment Rec 7(4):4–9
Lohnes RA, Handy RL (1968) Slope angles in friable loess. J Geol 76(3):247–258
Martin JM, Meybeck M (1979) Elemental mass-balance of material carried by major world rivers. Mar Chem 7(3):173–206
Milliman JD, Meade RH (1983) World-wide delivery of river sediment to the oceans. J Geol 91(1):1–21
Milliman JD, Syvitski PM (1992) Geomorphic/tectonic control of sediment discharge to the ocean: the importance of small mountainous rivers. J Geol 100:525–544
Morgan RPC (2005) Soil erosion and conservation. Blackwell, Oxford, p 304
NEC (1993) Chapter VIII, morphological studies of river Brahmaputra, WAPCOS. pp VIII–2
Ohlmacher CG, Davis CJ (2003) Using multiple regression and GIS technology to predict landslide hazard in northeast Kansas, USA. Eng Geol 69(3–4):331–343
Ojha CSP, Singh VP (2004) Introduction. The Brahmaputra basin water resources. V.P Singh et al. (eds). p 2
Osman AM, Thorne CR (1988) Riverbank stability analysis I: theory. J Hydraul Eng ASCE 114:134–150
Pahuja S, Goswami D (2006) A fluvial geomorphology perspective on the knowledge base Brahmaputra. Draft paper No. 3, Study of natural resources, water and the environmental nexus for development and growth in Northeast India, World Bank and Gauhati University
Pokrovsky OS, Schott J, Kudryavtzev DI, Dupre B (2005) Basalt weathering in central Siberia under permafrost conditions. Geochim Cosmochim Acta 69(24):5659–5680
Quirk JP (2001) The significance of the threshold and turbidity concentrations in relation to sodicity and microstructure. Aust J Soil Res 39(6):1185–1217
Raudkivi AJ (1999) Loose boundary hydraulics-grey zones, river sedimentation. In: Jayawardena L, Wang (eds) Balkema, Rotterdam, p 5
Rauret G, Lopez-Sanchez JF, Sahuquillo A, Rubio R, Davidson C, Ure A, Quevauvillerc P (1999) Improvement of the BCR three step sequential extraction procedure prior to the certification of new sediment and soil reference materials. J Environ Monit 1(1):57–61
Rengasamy P, Churchman GJ (1999) Cation exchange capacity, exchangeable cations and sodicity. In: Peverill KI, Sparrow LA, Reuter DJ (eds) Soil

analysis: an interpretation manual. CSIRO Publishing, Collingwood
Robinson DA, Phillips CP (2001) Crust development in relation to vegetation and agricultural practice on erosion susceptible, dispersive clay soils from central and southern Italy. Soil Tillage Res 60(1):1–9
Rowell DL (1994) Soil science: methods and applications. Longman, London, p 350
Sabri AW, Rasheed KA, Kassim TI (1993) Heavy metals in the water, suspended solids and sediment of the River Tigris impoundment at Samarra. Water Res 27(6):1099–1103
Sanford LP (2008) Modeling a dynamically varying mixed sediment bed with erosion, deposition, bioturbation, consolidation, and armoring. Comput Geosci 34(10):1263–1283
Santis FD, Giannossi ML, Medici L, Summa V, Tateo F (2010) Impact of physico-chemical soil properties on erosion features in the Aliano area (Southern Italy). Catena 81(2):172–181
Sarin MM, Borole DV, Krishnaswami S (1979) Geochemistry and geochronology of sediments from the Bay of Bengal and the equatorial Indian Ocean. Proc Indian Acad Sci Sect A, Phys Sci 88 (pt. 2, 2):131–154
Sarin MM, Krishnaswami S, Dilli K, Somayajulu BLK, Moore WS (1989) Major ion chemistry of the Ganga–Brahmaputra river system: weathering processes and fluxes to the Bay of Bengal. Geochim Cosmochim Acta 53(5):997–1009
Sarkar MH, Thorne CR (2006). Morphological response of the Brahmaputra-Padma-Lower Meghna river system to the Assam Earthqake of 1950. Braided Rivers, Special publication no. 36, International Association of Sedimentologists, University of Nottingham, UK
Sarma JN (2005) Fluvial processes and morphology of the Brahmaputra River in Assam, India. Geomorphology 70(3–4):226–256
Schumm SA (1960) The shape of alluvial channels in relation to sediment type. US Geological Survey Professional Paper No. 352-B
Schumm SA (1971) Fluvial geomorphology: historical perspective, river mechanics. In: Shen HW (ed) P.O. Box 606, Ft. Collins, Colorado, FAO, USA
Schumm SA (1977) The fluvial system. Wiley, New York, p 388
Sekely AC, Mulla DJ, Bauer DW (2002) Streambank slumping and its contribution to the phosphorus and suspended sediment loads of the Blue Earth River, Minnesota. J Soil Water Conserv 57:243–250
Simon A, Darby SE (1999) The nature and significance of incised river channels. In: Darby SE, Simon A (eds) Incised river channels. Wiley, Chichester, pp 3–18
Singh SK, France-Lanord C (2002) Tracing the distribution of erosion in the Brahmaputra watershed from isotopic compositions of stream sediments. Earth Planet Sci Lett 202(3–4):645–662
Singh M, Sharma M, Tobschall HJ (2005) Weathering of the Ganga alluvial plain, northern India: implications from fluvial geochemistry of the Gomati River. Appl Geochem 20:1–21
Stigliani WM (1991) Chemical time bombs: definition, concepts and examples, IIASA executive report 16. International Institute for Applied Systems Analysis, Austria
Subramanian V, Richey JE, Abbas N (1985) Geochemistry of river basins of India, Pt II: Preliminary studies on the particulate C and N in the Ganges-Brahmaputra river system. In: Degens, ET, Kempe S, Herrera R (Eds) Transport of carbon and minerals in major world rivers, Pt.3, Mitt. Geol.- Paläont. Inst. Univ. Hamburg, SCOPE/UNEP Sonderbd Heft 58, pp. 513–518
Sumner ME (1993) Sodic soils: new perspectives. Aust J Soil Res 31:683–750
Takaldany AE (2003) Bank Stability analysis for predicting reach scale land loss and sediment yield. J Am Water Resour Assoc 39(4):897–909
Tandon SK, Sinha R (2007) Geology of large river systems. In: Gupta A (ed) Chapter 2: large rivers. Wiley, UK
Thorne CR, Tovey NK (1981) Stability of composite river banks. Earth Surf Process Landf 6:469–484
Tripathy GR, Singh SK, Ramaswamy V (2014) Major and trace element geochemistry of Bay of Bengal sediments: implications to provenances and their controlling factors. Palaeogeogr Palaeoclimatol Palaeoecol 397:20–30
Turkian KK, Wedephol KH (1961) Distribution of the elements in some major units of the earth crust. Geol Soc Am Bull 72(2):175–192
Twidale CR (2004) River patterns and their meaning. Earth Sci Rev 67(3–4):159–218
Viers J, Dupre B, Gaillardet J (2009) Chemical composition of suspended sediments in world rivers: new insights from a new database. Sci Total Environ 407(2):853–868
Walker J, Arnborg L, Peippo J (1987) Riverbank erosion in the Colville Delta, Alaska. Geogr Ann 69(1):61–70
Web BW, Foster IDL, Gurnell AM (1995) Hydrology, water quality and sediment behaviour. In: Foster IDL, Gurnell AM, Webb BW (eds) Sediment and water quality in river catchments. Wiley, England
Wiebe H (2006) Brahmaputra flooding and erosion in Northeast India, Draft paper No. 4, study of natural resources, water and the environmental nexus for development and growth in Northeast India, World Bank and Gauhati University
Wilson CG, Kuhnle RA, Bosch DD, Steiner JL, Starks PJ, Tomer MD, Wilson GV (2008) Quantifying relative contributions from sediment sources in conservation effects assessment project watersheds. J Soil Water Conserv 63:523–531
Winterwerp JC, Kesteren WGM (2004) Introduction to the physics of cohesive sediment in the marine environment. Elsevier, Amsterdam, p 576
Wischmeier WH, Mannering JV (1969) Relation of soil properties to its erodibility. Soil Sci Soc Am Proc 33:131–137

Wolman MG (1959) Factors influencing erosion of a cohesive river bank. Am J Sci 257(3):204–216
Yujun YI, Zhaoyin WANG, Zhang K, Guoan YU, Xuehua DUAN (2008) Sediment pollution and its effect on fish through food chain in the Yangtze River. Int J Sediment Res 23(4):338–347
Zhang RJ (1989) River dynamics. Water Power Press, Beijing

Web References

http://news.xinhuanet.com/english2010/china/2011-08/22/c_131067137.htm. Assessed on 12 Dec 2012
http://www.britannica.com/EBchecked/topic/77154/Brahmaputra-River. Assessed on 18 Feb 2014

Chandan Mahanta Department of Civil Engineering, Indian Institute of Technology, Guwahati, India

Lalit Saikia Department of Civil Engineering, Indian Institute of Technology, Guwahati, Guwahati, India

Sediment Runoff Modelling Using ANNs in an Eastern Himalayan Basin, India

5

Archana Sarkar, Nayan Sharma, and R.D. Singh

Abstract

The amount of sediments transported by headwater rivers plays a crucial role in planning of water resources. Most widely used methods of estimating the sediment in rivers are the empirical methods, and in India, the rating curve technique is most popular. The present study is focused on the application of artificial neural network (ANN) technique for sediment-discharge modelling of a headwater river. For ANN development, daily discharge and suspended sediment concentration data of Subansiri River (an eastern Himalayan river) in India have been used. Rating curves have also been developed with similar data, and comparison of the two techniques has been carried out. It has been observed that the estimates of suspended sediment concentration obtained by ANNs compared to the rating curve technique were much closer to the observed values.

5.1 Introduction

The rainfall-run-off process in a catchment causes the detachment of soil material which is then transported to the river system and also deposited at various locations. The amount of sediment carried by the river system forms a crucial information for various water resource planning, management, and operation projects, for example, design of dams and reservoirs; hydroelectric power generation and water supply; water quality estimation of lakes, rivers and estuaries; watershed management; as well as environmental impact studies. The soil erosion and sediment yield is one of the major problems in Himalayan region. Water resources planners and environmentalists are concerned about the Himalayan region more because of its fragile ecosystem. Steep slopes and increasing deforestation along with seismic vulnerability are the major factors in the Himalayas responsible for soil erosion and subsequent sedimentation in

A. Sarkar (✉) • R.D. Singh
National Institute of Hydrology (NIH), Roorkee, Uttarakhand, India
e-mail: archana@nih.ernet.in

N. Sharma
Department of Water Resources Development and Management, Indian Institute of Technology Roorkee, Roorkee, Uttarakhand, India

N. Sharma (ed.), *River System Analysis and Management*, DOI 10.1007/978-981-10-1472-7_5

Himalayan-fed rivers (Varshney et al. 1986). The rivers emerging out from the Himalayan region transport the sediment at a very high rate. Twenty five percent of the dissolved sediment load in the world oceans is supplied by the Tibetan and Himalayan region alone which forms only about 5 % of the land surface on Earth (Raymo and Ruddiman 1992). In the Himalayan Mountains, as a consequence of loss of forest cover coupled with the influence of the monsoon pattern of rainfall, the fragile catchments have become prone to low water retention and high soil loss associated with run-off (Rawat and Rawat 1994). Keeping this in view, Subansiri River, which flows through the eastern Himalayan region of India, has been selected for this study.

For estimating the sediment load/yield, there exist various models and techniques, such as sediment rating curves, erosion modelling, etc. The models vary from a simple regression relationship to complex simulation models. As the sediment-discharge relationship is not linear, conventional statistical tools like regression and curve fitting methods are unable to model the non-linearity in the relationship. However, physically distributed models for sediment simulation are able to produce more realistic simulations, but their applications are limited due the huge amounts of spatio-temporal data required and the model complexities. Therefore, some simple yet efficient method is required which can be applied with the kind of data easily available and still able to model the inherent non-linearity of the sediment-discharge relationship. One possible solution to this complex sediment simulation problem is neurocomputing. In recent years, artificial neural networks (ANNs) which represent the working of human brain in simplified mathematical models have been widely used for run-off and sediment yield modelling. It has been shown by many authors that three-layer feed-forward ANNs are capable of mapping the input-output hydrological processes and have been widely used in water resource problems (ASCE Task Committee 2000).

Application of ANN methodology for sediment-discharge modelling started recently with highly encouraging results. Rosenbaum (2000) used the ANN methodology for prediction of sediment distribution in Swedish harbours. Baruah et al. (2001) developed ANN models for assessment of lake water quality of the Kasumigaura lake in Japan utilizing Landsat TM imagery to estimate the amounts of surface chlorophyll and sediment in the lake. The authors reported that three-layer back-propagation ANN models produced much better results than conventional regression techniques for both the variables. Jain (2001) integrated ANN models to simulate the stage-discharge-sediment concentration relation for two sites on the Mississippi River and reported the superiority of ANN results over the conventional rating curve methods. Nagy et al. (2002) also developed ANN models for natural sediment-discharge (sediment concentration) estimation in rivers and addressed the importance of using field data for ANN training and selection of a suitable ANN architecture. Sarkar (2005) and Sarkar et al. (2008) applied ANN technique for sediment-discharge modelling in Sutlej River of western Himalaya in India and Kosi River of Bihar in India, respectively, and found that sediment load estimations in the rivers estimated by ANNs were much closer to the observed values compared to the ones estimated through the sediment rating curve.

In the present study, two techniques, namely, sediment rating curve (SRC) and artificial neural networks (ANNs), have been applied for modelling the sediment-discharge relationship of Subansiri River in eastern Himalayan region of India, and a comparison of these techniques has been made.

5.2 Sediment Rating Curves

Sediment rating curves (SRCs) are widely used to calculate the sediment concentration/load being transported by a river. A sediment rating curve is a mathematical relation between the sediment concentration/load and discharge. Graphically, sediment rating curves may be plotted showing sediment concentration or load as a function of discharge on varying time scales, e.g., daily, ten daily, monthly etc. The underlying assumption in the development of rating curves is a stable relationship between sediment concentration/load and discharge. It is assumed that mean sediment yield can be calculated with historic data of discharge based on the rating curve which shows certain amount of scatter. However, the major problem of the rating curve technique is the high degree of scatter which may be reduced but not eliminated. Sediment concentration will always not increase as a function of discharge (Ferugson 1986).

A sediment rating curve is mathematically constructed as the best line of fit in the linear least square regression using log-transformed data of sediment concentration/load and discharge. The mathematical relationship is of the form:

$$\log S = \log a + b \log(Q) \tag{5.1}$$

And the log-transformed form of the above relationship is of the following form which will plot a straight line on log-log paper:

$$S = aQ^b \tag{5.2}$$

where

S = sediment concentration (or load)
Q = discharge,
a and b are regression constants.

5.3 Artificial Neural Networks (ANNs)

The development of artificial neural networks was inspired by a desire to understand the functioning of human brain that processes the patterns with high efficiency and speed. ANN is a simplified computing model that tries to mimic the nervous system and the human brain in a very primitive way to emulate the capabilities of the human being in a very limited sense. ANNs have been developed as a generalization of mathematical models of human cognition or neural biology (Fig. 5.1).

ANN development is based on the following rules:

- The processing of information takes place at many single elements called neurons, also referred to as nodes, cells or units.
- Connection links pass the signals between nodes.
- A weight is associated with each connection link which signifies its connection strength.
- Each node receives certain net input which undergoes a non-linear transformation in the form of activation function to determine its output signal.

Development of ANN models generally involves four discrete steps. The very first step requires identification of input and output data and transformation/scaling of the same. In the next step, it is required to set the architecture of the ANN model in the form of number of hidden layers to be used, number of neurons each layer will have and the interneuron connectivity. The third step requires identification of an appropriate learning algorithm for training of the ANN model using historical data of input and output which is similar to calibration of a mathematical model. The fourth and the final step requires testing of the trained ANN model based upon already identified statistical performance evaluation criteria. This last step is analogous to validation of a mathematical model. The detailed theory of ANN modelling can be found in dedicated text books (Haykin 1994).

5.4 Study Area, Data Availability and Selection of Input/Output Variables

Study area chosen for the present study is the Subansiri River basin (Fig. 5.2). Subansiri River is one of the main tributaries of Brahmaputra River originating in the north from Himalayas. The origin of Subansiri River takes place at an elevation of about 5340 m from a point which is further ahead of the Great Himalayan range. The Subansiri basin comprises of four distinct parts

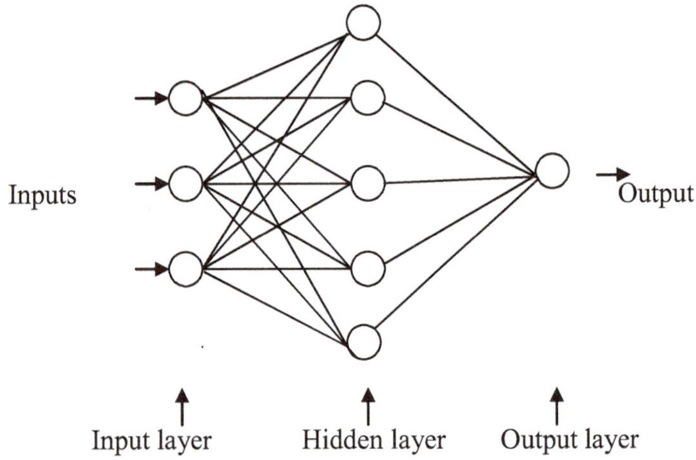

Fig. 5.1 Structure of a multilayer feed-forward artificial neural network model

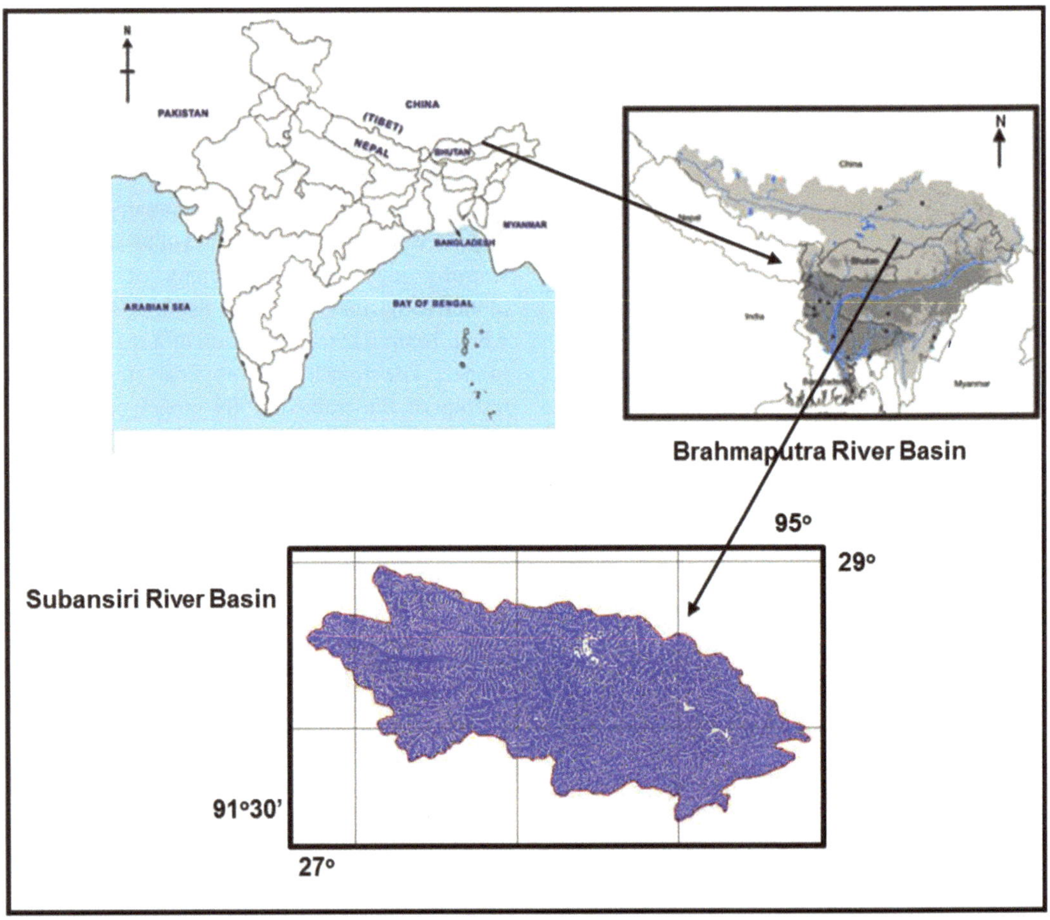

Fig. 5.2 Study area

having different climatic conditions. The first (Tibetan) and second (Indian) parts lying in the distant Tibetan mountains up to Indo-Tibetan boundary and the river reaching up to Miri Hills after the international boundary in Arunachal Pradesh, respectively, belong to the Great Himalayan range (central Himalaya). The third (Indian) part from Miri Hills in Arunachal Pradesh up to the state boundary of Assam belongs to the sub-Himalayan range. The fourth (Indian) part up to the joining of Subansiri River with the main stem of Brahmaputra River lying in Assam belongs to the fertile flood plains. In its upper reaches, Subansiri River flows through the mountainous terrain for a distance of about 208 km after which it flows in the plains for 70 km and falls into the Kherkutia Suti and flows further for a distance of 60 km until it joins Brahmaputra River. In its entire flow regime, the riverbed elevation varies from 4206 m in the upper reaches to 80 m in the foothills near Dulangmukh. In its entire journey from the Great Himalayan range to the foothills of Arunachal Pradesh, the Subansiri River is joined by many small and big mountainous tributaries. The total length of tributaries joining the Subansiri River is about 1960 km. The discharge contribution of Subansiri River to the Brahmaputra River measured at Pandu near Guwahati is about 10.66 % of the total discharge of Brahmaputra River. The total basin area of Subansiri is about 37,000 km^2. The basin area lying in Tibet is about 38 % of the total area of Subansiri basin.

The sub-Himalayan range of Subansiri generally consists of soft sandstones and weathered rocks. Land formation of the catchment having steep slope and shallow braided channels, therefore, carries heavy charge of coarse sediment particles of sizes 0.2 mm diameter up to 2 mm diameter in suspension. Due to granulometric characteristics of the soils present in this river basin, their density of deposition and coverage is guided by the energy potentialities of the flood volume based on the run-off. As the precipitation is intensive during the period May to October and the grade of flowing surface is a steep one, sediment deposits at areas nearer to and along the foot hills are heavier materials and debris such as boulders and shingles, and towards the lower portions of the catchment they are of porous detritus materials such as fine sands and silt.

The daily data of suspended sediment concentration and discharge were available at the Chaulduaghat gauging site for 10 years (1997–2006) constituting a total of 3652 patterns. Out of this, 4 years (1997–2002) consisting of 2191 patterns were used for training, 2 years (2003–2004) consisting of 731 patterns for testing and 2 years (2005–2006) consisting of 730 patterns for validation.

As stated in the earlier sections, the first step of ANN model development is the identification of input and output variables. In the present study of sediment-discharge modelling, the suspended sediment concentration/load at time step t; S_t forms the output variable. Discharge at a time step t; Q_t forms the input variable. However, the lagged values of data variables when included in the input enhance the simulation/forecasting accuracy of the ANN model as they cater for the routing component of the hydrological modelling on a catchment scale. Therefore, in the present study, additional input variables such as Q_{t-1}, Q_{t-2}, S_{t-1} and S_{t-2} have also been considered besides Q_t.

5.5 Rating Curve Analysis

Based on the sediment rating curve (SRC) technique given by Eq. (5.1), the sediment rating equation between suspended sediment concentration and discharge for Subansiri River at Chaulduaghat gauging site for the training period is

$$S = 7\text{E-}06\ Q^{1.2923} \quad (5.3)$$

where

S = suspended sediment concentration in the Subansiri River at Chaulduaghat in g/l at time t

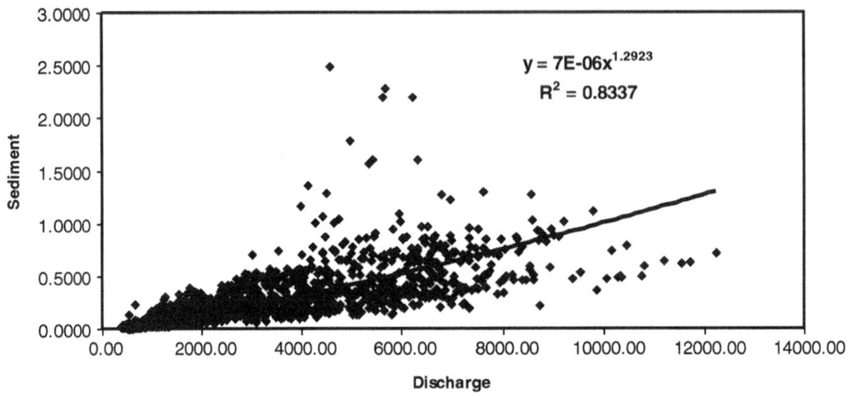

Fig. 5.3 Sediment-discharge relationship using rating curve technique

Q = discharge in the Subansiri River at Choulduaghat in cumec at time t

Figure 5.3 is a graphical representation of the SRC obtained based on Eq. (5.3) above.

5.6 Design and Training of ANN Models

Table 5.1 provides details of various ANN models developed using different combinations of input data variables in increasing number of input variables. However, the input-output variables of ANN-1 have been used for the conventional SRC analysis.

A back-propagation ANN with the generalized delta rule as the training algorithm has been employed in this study (Rumelhart et al. 1986). An ANN software package, namely, Neural Power (2003), has been used for ANN development and training for the present case study. This package is freely available on the internet. The structure for all simulation models is three layers BPANN which utilizes a non-linear sigmoid activation function uniformly between the layers. Nodes in the input layer are equal to the number of input variables, nodes in hidden layer are varied from the default value by the NP package for various numbers of input nodes above to approximately double of input nodes (Zhu et al. 1994) and the nodes in the output layer are one as the models provide single output. According to Hsu et al. (1995), "three-layer feed forward ANNs can be used to model real-world functional relationships that may be of unknown or poorly defined form and complexity". Therefore, only one hidden layer was used in the ANN models developed in this study.

To begin with the ANN modelling, all the input and output data were normalized (transformed) in the range of 0–1 with respect to the maximum value of respective data series. Data rescaling prevents the saturation effect inherent in the application of sigmoid functions. Values were assigned randomly to all interconnection weights (links between nodes of successive layers). Learning rate α and momentum β were assigned values of 0.15 and 0.5, respectively, based on hit and trial technique. Training of all the ANN models was accomplished using quick propagation (QP) learning algorithm. QP is similar to the standard back-propagation learning algorithm but very fast due to some heuristic modification. The updating of interconnection weights was carried out after presenting each pattern from the training data set, instead of updating once per iteration. The criteria of generalization of ANN through cross-validation (Haykin 1994) were adopted to prevent overtraining of the ANN model. For this reason, the entire data were segregated into three sets, viz., training, testing and validation data sets. Training data (2191 patterns) were used for assessment of interconnection weights of the ANN model, and testing data (731 patterns)

Table 5.1 Various ANN run-off-sediment models

ANN model	Output variable	Input variables
ANN-1	S_t	Q_t
ANN-2	S_t	Q_t, Q_{t-1}, S_{t-1}
ANN-3	S_t	$Q_t, Q_{t-1}, Q_{t-2}, S_{t-1}, S_{t-2}$
ANN-4	S_t	$Q_t, Q_{t-1}, Q_{t-2}, Q_{t-3}, S_{t-1}, S_{t-2}, S_{t-3}$
ANN-5	S_t	$Q_t, Q_{t-1}, Q_{t-2}, Q_{t-3}, Q_{t-4}, S_{t-1}, S_{t-2}, S_{t-3}, S_{t-4}$

where *S* represents suspended sediment concentration at Choulduaghat, *Q* represents discharge at Choulduaghat and *t* represents the time step

Table 5.2 Statistical performance evaluation of trained ANN models and rating curve

	Training			Testing			Validation		
ANN model (nodes)	RMSE (abs)	*r*	Nash CE	RMSE (abs)	*r*	Nash CE	RMSE (abs)	*r*	Nash CE
ANN-1 (1-2-1)	0.1788	0.878	0.771	0.1519	0.417	0.468	0.1479	0.996	0.628
ANN-2 (3-3-1)	0.0378	0.995	0.989	0.0375	0.996	0.991	0.0369	0.770	0.990
ANN-3 (5-4-1)	**0.0348**	**0.996**	**0.991**	**0.0344**	**0.995**	**0.990**	**0.0341**	**0.996**	**0.993**
ANN-4 (7-6-1)	0.0379	0.995	0.989	0.0335	0.994	0.988	0.0283	0.996	0.990
ANN-5 (9-7-1)	0.0347	0.996	0.991	0.0336	0.992	0.983	0.0266	0.995	0.989
Sediment rating curve (SRC)	0.1991	0.834	0.457	0.2663	0.512	−7.55	0.2182	0.434	−21.3

were used for performance evaluation of the trained ANN models during training. Training was stopped at a point where errors for the testing data set started increasing. This procedure of using the training and testing data sets for performance evaluation of all the candidate ANN model structures allowed to choose the best performing model during training of the ANN models. The chosen ANN model with best performance was then used to estimate the generalized performance of the trained network using validation data set (730 patterns) which was not used before for ANN training. Three statistical performance evaluation criteria, namely, root mean square error (RMSE), correlation coefficient (*r*) and Nash and Sutcliffe coefficient of efficiency (CE) (Nash and Sutcliffe 1970), have been used in the present study.

5.7 Results and Discussion

Table 5.2 displays the statistical performance of the different trained ANN models as well as rating curves developed. The comparison of model performance has been done in terms of RMSE, *r* and Nash CE.

It can be noticed from Table 5.2 that the RMSE values are generally low (less than 0.15) for all the ANN models except ANN-1 model, during all the three phases, i.e. training, testing as well as validation. There are two candidate models in RMSE criteria, ANN-5 and ANN-3, with very close RMSE values. It can also be noticed from the Table 5.2 that the RMSE values are highest for the rating curve model. The RMSE values for the SRC are 0.1991, 0.2663 and 0.2182 during training, testing and validation, respectively, which are even more than the worst ANN model and more than five times the best ANN model.

Further, it can be noticed from Table 5.2 that the correlation (*r*) values are quite high (more than 0.99) for all the ANN models except ANN-1 model, during all the three phases. The performance of ANN-3 model is the best in r statistic (0.996, 0.995 and 0.996 during the three phases of training, testing and validation, respectively). Moreover, almost equal values of r during all the three phases indicate good generalization capability of the trained ANN model. Statistical performance of the SRC is poorest among all models. The CC values of SRC model are 0.834, 0.512 and 0.434 during training, testing and validation phases, respectively. These values are even lower than the worst ANN model, i.e. ANN-1 model.

In the Nash efficiency (CE) statistic, all the ANN models except ANN-1 perform well. The CE values also are very high (more than 0.98) for ANN-2 to ANN-5 during all the three phases. In CE statistic also, ANN-3 model performs the best with 0.991, 0.990 and 0.993 CE values during the three phases. The performance of SRC in CE statistic has gone down drastically with CE values as low as 0.457, −7.55 and −21.3 during training, testing and validation, respectively. These values are much lower than the worst ANN model also.

Therefore, ANN-3 model is the best performing model in all the three statistical and

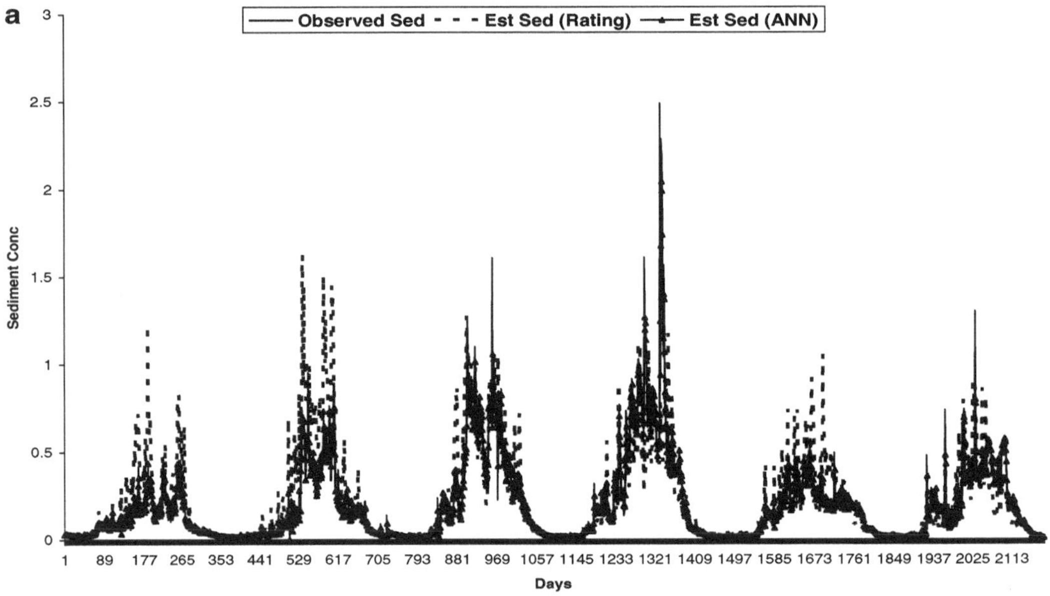

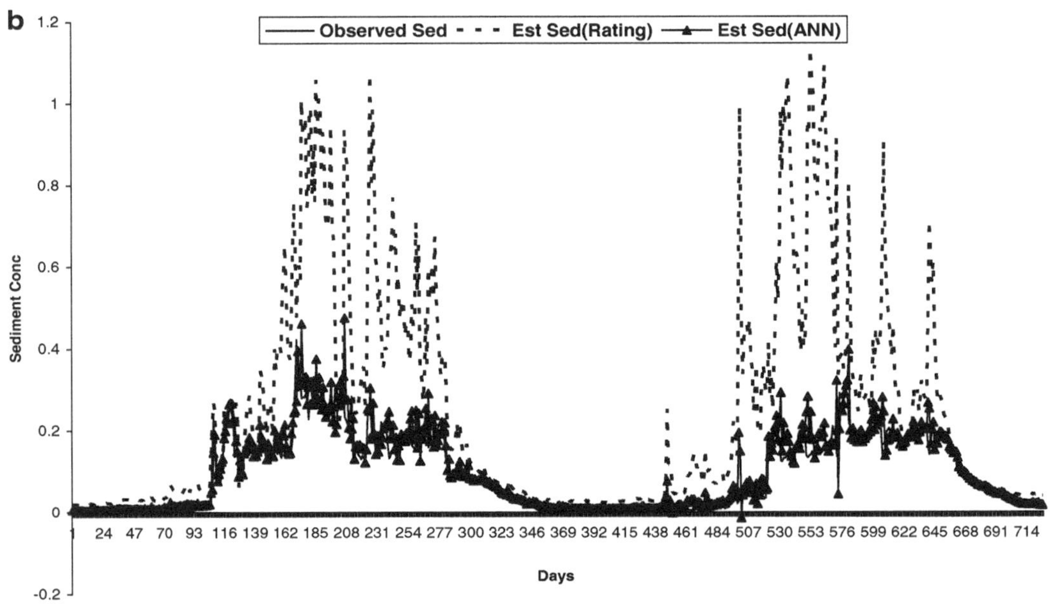

Fig. 5.4 Comparative performance of observed, estimated (ANN) and estimated (rating curve) sediment concentration series. (**a**) Training. (**b**) Testing. (**c**) Validation

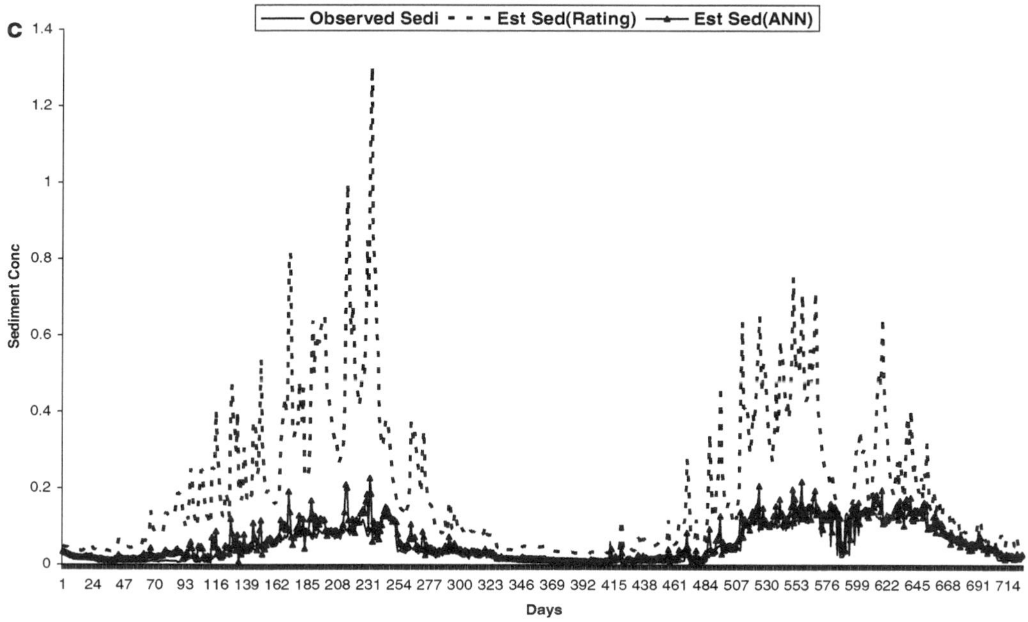

Fig. 5.4 (continued)

hydrological criteria. SRC performance is found to be average in the r criteria but drastically poor in other two criteria. It is because the estimated sediment series (from SRC) follows a good general trend as that of the observed sediment series which gives high r values, but there is a considerable difference in the numeric values of estimated and observed sediment concentration/load due to which the RMSE and CE values are very poor. The accuracy of the corresponding ANN model with only discharge as input, i.e. ANN-1, is also better as compared to the SRC.

Figure 5.4 depicts the comparison of daily suspended sediment concentration time series between SRC and ANN model (ANN-3) with respect to the observed suspended sediment concentration for the training, testing as well as validation period. It can be observed from the plotted time series that the ANN estimated suspended sediment concentration series follows the observed suspended sediment concentration series very closely, whereas the SRC-based suspended sediment concentration series depicts considerable mismatch with the observed curve, especially during the validation phase which conforms to the low Nash efficiency of the rating curve technique.

5.8 Conclusions

In the presented study, ANN methodology has been applied for modelling of the highly non-linear process of sediment-discharge in a river. Primary aim of this research work was to demonstrate the potential of ANN application in the modelling of sediment load in rivers. To achieve the objectives, a case study has been done for Subansiri River in India utilizing the observed data of daily suspended sediment concentration and river discharge at a gauging site, namely, Choulduaghat. The ANN models have been compared with conventional SRC for the Choulduaghat gauging site. Performance of all the models has been evaluated using root mean square error, coefficient of correlation and Nash coefficient of efficiency. ANN models were found to be significantly superior to the conventional SRC which are being frequently used by field engineers. This research work suggests that the ANN modelling approach is able to capture the inherent non-linearity in the river sediment-discharge rating relationships.

The other advantage of using the ANN approach over rating technique is that ANNs can be trained with multiple source input variables that affect the underlying process without any simplification or exclusion which is normally done in the conventional approaches. In the present case study, the ANN model has been developed and trained with the help of river data observed in the field (river gauging site) which does not have any boundary conditions for application. However, a major limitation of the ANN application lies in the fact that the ANN model is unable to simulate/forecast accurately beyond the range of training data. This issue can be handled by considering the extremes of observed data in the training data set. The application of ANN technique is so simple that even the engineers working in the field without any theoretical background of sediment transport can also estimate the sediment load from an ANN model provided they have a fair understanding of the bounds of the data used for ANN model development.

References

ASCE. Task Committee on Application of Artificial Neural Networks in Hydrology (2000) Artificial neural networks in hydrology. II: hydrologic applications. J Hydrol Eng ASCE 5(2):124–137

Baruah PJ, Tamura M, Oki K, Nishimura H (2001) Neural network modelling of lake surface chlorophyll and sediment content from LandsatTM imagery. In: Proceedings of the 22nd Asian conference on remote sensing, 5–9 November 2001, Singapore

Ferugson RI (1986) River loads underestimated by rating curves. Water Resour Res 22(1):74–76

Haykin S (1994) Neural networks – a comprehensive foundation. Macmillan Publishers, New York

Hsu K, Gupta HV, Sorooshian S (1995) Artificial neural network modelling of the rainfall-runoff process. Water Resour Res 31(10):2517–2530

Jain SK (2001) Development of integrated sediment rating curves using ANNs. J Hydraul Eng ASCE 127(1):30–37

Nagy HM, Watanabe B, Hirano M (2002) Prediction of sediment load concentration in rivers using artificial neural network model. J Hydraul Eng ASCE 128(6):588–595

Nash JE, Sutcliffe JV (1970) River flow forecasting through conceptual models. J Hydrol 10:282–290

Neural Power (2003) Neural networks professional version 2.0. CPC-X Software, Copyright: 1997–2003. A demo version downloaded from the Internet

Rawat JS, Rawat MS (1994) Accelerated erosion and denudation in the nana kosi watershed, Central Himalaya, India, Part I: sediment load. J Mt Res Dev 14(1):25–38

Raymo ME, Ruddiman WF (1992) Tectonic forcing of late Cainozoic climate. Nature 359:117–122

Rosenbaum M (2000) Harbours- silting and environmental sedimentology (H-SENSE. Final report, Department of Civil and Structural Engineering, The Nottingham Trent University, Nottingham, UK. http://hjs.geol.uib.no/HSense/

Rumelhart DE, Hinton GE, Williams RJ (1986) Learning representation by back propagating errors. Nature 323(9):533–536

Sarkar A (2005) Sediment runoff modelling using ANNs in a Western Himalayan Basin, India. In: Proceedings of the international conference on headwater control VI: hydrology, ecology and water resources in headwaters, June 20–23, 2005, Bergen, Norway

Sarkar A, Kumar R, Jain SK, Singh RD (2008) Artificial neural network models for estimation of sediment loads in an alluvial river in India. J Environ Hydrol 16:1–16

Varshney RS, Prakash S, Sharma CP (1986) Variation of sediment rate of reservoirs in Himalayan region with catchment. In: Proceedings of the of 53rd R & D Session, CBIP Bhubaneswar, India, Technical session IX paper

Zhu M, Fujita M, Hashimoto N (1994) Application of neural networks to runoff prediction. In: Hipel KW et al (eds) Stochastic and statistical method in hydrology and environmental engineering, vol 3. Kluwer Publishers, Dordrecht, pp 205–216

Archana Sarkar National Institute of Hydrology, Roorkee, Uttarakhand, India

Nayan Sharma Department of Water Resources Development and Management, Indian Institute of Technology Roorkee, Roorkee, Uttarakhand, India

R.D. Singh National Institute of Hydrology (NIH), Roorkee, Uttarakhand, India

Part II

Land Use and Climate

River Basin Impact Assessment of Changing Land Use and Climate by Applying the ILWRM Approach in Africa and Asia

6

Wolfgang-Albert Flügel

Abstract

Methodological aspects of scale-related impact assessment from changing land use/land cover (LULC) management and climate on river basin water resources and their management are discussed. Both control the interactive hydrological process dynamics that transfer precipitation input on the landscape to the different surface and subsurface water resources components and ultimately to river runoff draining the river basin. As the integrated water resources management (IWRM) concept does not sufficiently account for the landscape-related process dynamics associated with LULC management, it is enhanced to the integrated land and water resources management (ILWRM) approach. The latter requires, *firstly*, a consistent methodological concept and, *secondly*, a toolset for its implementation. The DPSIR (D = *Drivers*, P = *Pressures*, S = *State*, I = *Impacts*, R = *Responses*) approach is a suitable analysis concept in this regard and is enhanced by a Decision Information Knowledge System (DIKS). Both are implemented by means of the integrated land management system (ILMS) toolset developed at the University of Jena, Germany, and tested in numerous research projects in Africa, Asia, Australia, Europe and South America. The majority of river catchment studies focus on a particular scale. Upscaling and downscaling of the hydrological knowledge they generate requires the separation of the generic knowledge components from their modifying local specifications. The interdisciplinary ILWRM applications presented in this paper from two projects in South Africa and SE Asia address this challenge by

W.-A. Flügel (✉)
Department of Geoinformatics, Hydrology and Modelling, Friedrich-Schiller-University Jena, Jena, Germany
e-mail: wa.fluegel@gmail.com

N. Sharma (ed.), *River System Analysis and Management*, DOI 10.1007/978-981-10-1472-7_6

applying a multi-scale nested catchment approach (NCA) and respective upscaling and downscaling techniques to regionalize hydrological knowledge between scales.

Abbreviations

CC	Climate change
CLM	Community land model
DEM	Digital elevation model
DIKS	Decision Information Knowledge System
DPSIR	Driver, pressure, state, impact, response
EC	European Commission
ECMWF	European Centre for Medium-Range Weather Forecast
EEA	European Environment Agency
ERA-40	ECMWF reanalysis of the global atmosphere and surface conditions for 45 years
ESF	Ecosystem functions
ESS	Ecosystem services
GCM	General circulation model
GIS	Geographic information system
GLOF	Glacier lake outburst flood
GWP	Global Water Partnership
HD	Human dimension
HRU	Hydrological response units
ISA	Integrated system analysis
ILMS	Integrated land management system
IPCC	Intergovernmental Panel on Climate Change
IPM	Integrated process modelling
IWRM	Integrated water resources management
ILWRM	Integrated land and water resources management
JAMS	Jena Adaptable Modelling System
LULC	Land use/land cover
MFS	Modelling framework system
MODIS	Moderate-resolution imaging spectroradiometer
NCA	Nested catchment approach
NE	Natural environment
RBIS	River basin information system
RCM	Regional climate model
RS	Remote sensing
SRES	Special report on emission scenarios
UBRB	Upper Brahmaputra River basin
UDRB	Upper Danube river basin
USGS	United States Geological Survey
WAM	Water allocation model
WL	Wetlands
WRRU	Water resources response units

6.1 Introduction

Worldwide water resources managers and land use planners are facing the challenge of continuously changing LULC management and climate in their river basin management domains. They both alter the interactive process dynamics within the basin's natural environment (NE) and its human dimension (HD), affect the basin's hydrological cycle and trigger the regeneration dynamics of its water balance components together with associated land erosion (Heathcote 1998; Jain and Singh 2003; Fink et al. 2007a, b). River runoff, interflow, groundwater recharge and evapotranspiration are adapting to such changes in terms of quantity, i.e. seasonal distribution, frequency and magnitudes of extreme events, and quality, i.e. nutrient and sediment transport. To understand and quantify such impacts, a holistic and scale-related approach towards hydrological river basin analysis is required (Flügel 2000, 2010, 2011a). The latter assesses the basins' multi-process driven hydrological cycle at relevant process scales and analyses them with respect to processes active at higher and lower system scales, hence developing knowledge about the river basins' integrated hydrological process dynamics.

To implement such a holistic system's approach, the European Environment Agency (EEA) proposed the DPSIR (*driver*, *pressure*, *state*, *impact*, *response*) framework (EEA 2005,

2008b) for integrated water resources management (IWRM). Chalder (2005) proposed to enhance the latter by the LULC management component to integrated land and water resources management (ILWRM). The integrated land management system (ILMS) developed by Kralisch et al. (2012) puts the DPSIR approach enhanced by a DIKS into practice for ILWRM. Together they were validated applying a multi-scale nested catchment approach (NCA) in river basins in Africa (Helmschrot 2006a; Kralisch et al. 2013; Steudel et al. 2013), Asia (Flügel 2011b; Sharma and Flügel 2015), Australia (Bende-Michl et al. 2007) and Europe (Krause et al. 2006; Krause and Hanisch 2009).

6.2 Scale-Related Impact Assessment and Analysis

Changing LULC management and climate both relate to different scales. Changing LULC management is mainly relevant on the micro- and mesoscale and rarely affects the macro-scale; meanwhile climate change (CC) as a global phenomenon is relevant for macro-scale river basins. Consequently, the analysis of their impacts on river basin water resources and their controlling hydrological regeneration processes need to be scale related as well. The challenge for applied ILWRM is to implement an efficient upscaling and downscaling procedure to transfer knowledge between scales.

Figure 6.1 schematically is showing the respective scale framework to consider in this regard. It comprises the:

- *Spatial scale* ranging from micro-catchments to macro-scale river basins like that of the Amazon, Brahmaputra or Mississippi river
- *Degree of landscape heterogeneity* with respect to landscape physiographic properties (topography, soils, geology) and its LULC management-related socio-economy (forestry, agriculture, settlements, recreation, industry)
- *Impact response time* of the river basin system when adapting to changing NE and HD conditions

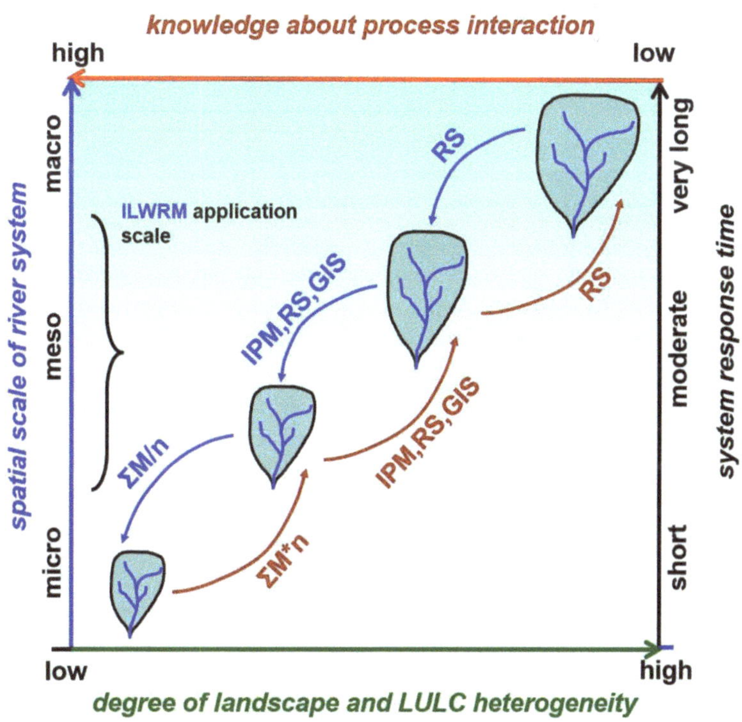

Fig. 6.1 Transfer of process knowledge and understanding of process interaction between scales and the range of scales for applied ILWRM decision-making (*M* measurements, *IPM* integrated process modelling, *RS* remote sensing)

- *Knowledge about process interactions* within and between basin scales required for upscaling and downscaling

The majority of river basin research studies focus on a specific scale. Regionalization of process knowledge they generate by means of upscaling and downscaling to other basins and scales requires a sophisticated understanding of process interactions to separate the generic process components from their modifying local characteristics. Figure 6.1 in a schematic way is grouping hydrological knowledge elaborated by such studies in the scale framework described above and reveals:

(i) Micro-scale catchments ($A < 100$ km^2) have a low degree of landscape and LULC management heterogeneity and a short impact response time to system changes. Therefore, the majority of hydrological process and modelling studies focus on this scale. Measured time series and detailed information about the NE and its HD are either readily available or can be generated within the runtime of the project. Upscaling and downscaling between the micro-scale and the lower mesoscale apply measured time series (M) and field observations. A high level of knowledge about interactive hydrological process understanding for regionalization is available on that scale.

(ii) Mesoscale river systems ($A > 100$–10,000 km^2) have a moderate degree of landscape heterogeneity and LULC management but in turn have longer response times when adapting to system changes. Often measured time series in a sufficient spatial distribution and duration are available to capture the system's dynamics and to cover its impact response time. The same applies for the information about its NE and HD features. Integrated process modelling (IPM) combined with remote sensing (RS) and geographic information system (GIS) methodologies are tools applied for ILWRM analysis, upscaling and downscaling. During the past decades, a moderate level of understanding about the interactive hydrological process dynamics of river basins of this scale has been established.

(iii) Macro-scale river basins ($A > 10{,}000$ km^2), like the Brahmaputra River basin, have a high degree of landscape heterogeneity and LULC management. The system's response time when adapting to changing environment conditions is very long and in most cases are not covered by measured time series data from sparse hydrometeorological station infrastructures. Due to this lack and constraining trans-boundary governance regulations, their NE and HD components are often unknown in detail. ILWRM applies RS and hydrological models (Alcamo and Henrichs 2002) with a simplified process representation to analyse impacts of climate and LULC management heterogeneity. Downscaling of such simplified models is not meaningful but RS techniques offer methodological potential for upscaling and downscaling between the macro-scale and the upper mesoscale. Insufficient understanding of the complex process interaction in river basins of this scale is common as long-term integrated process studies in basins of this scale are rare.

The challenging conclusion from Fig. 6.1 is that ILWRM system analysis of river basin process interactions must *firstly* comprise the micro-, meso- and macro-scale and *secondly* implement efficient techniques of regionalization by means of upscaling and downscaling between them. The nested catchment approach (NCA) applied together with the concept of process-based hydrological response units (HRU) in the multi-scale projects discussed below addresses this challenge and produce enhanced integrated system understanding.

6.2.1 Impacts from Changing LULC Management

Changing LULC management refers to changing economic landscape management, i.e. by agriculture, forestry, ranching, settlements or industry. Numerous literature is presenting the impacts

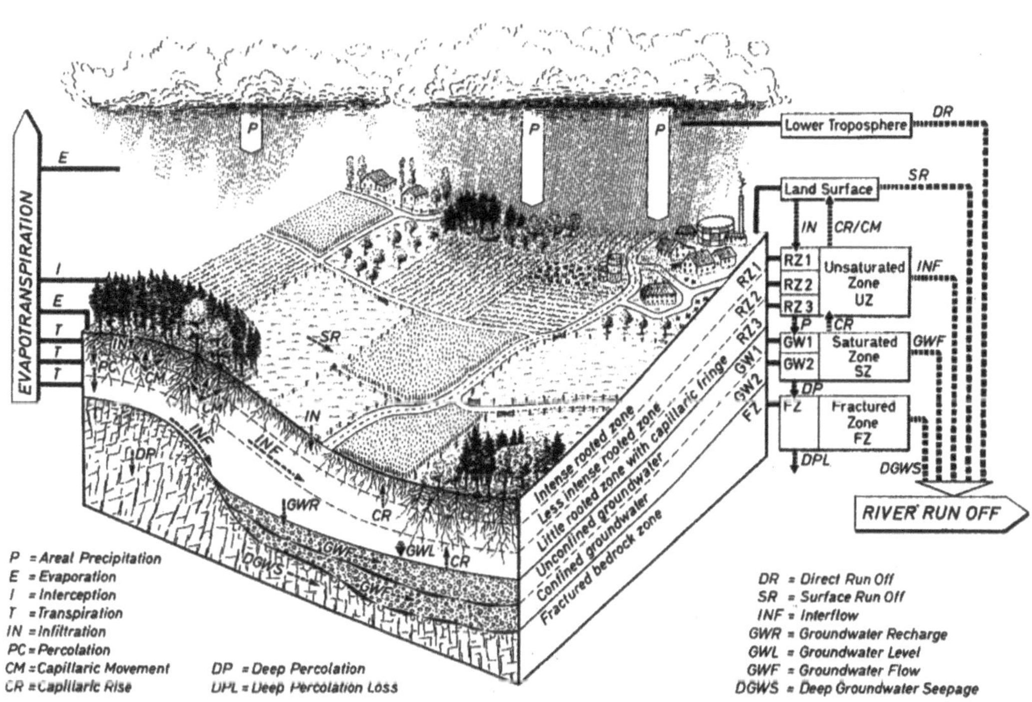

Fig. 6.2 Distributed LULC management in the river basin landscape controls the interactive process dynamics of water balance components and affects the transfer of rainfall input into regenerating surface and subsurface water resources (Flügel 1995a)

of such changes on the hydrological dynamics at different scales as discussed in a comprehensive way, e.g. by DeFries and Eshleman (2004).

As schematically shown for a basin segment in Fig. 6.2, LULC management controls the regeneration of water balance components within the hydrological cycle. Depending on the type of vegetation, location in the topography, soil type and geology, a complex interactive process dynamics is transferring the precipitation input in different quantities to evapotranspiration and river runoff output, thereby generating surface runoff, soil water storage, interflow and groundwater recharge. Exploiting on process knowledge shown in Fig. 6.2, Flügel (1995a) developed the regionalisation concept of process-related HRU for distributed river basin modelling (Flügel 1996; Krause 2002; Krause and Flügel 2005; Fink et al. 2007a) enhanced by Pfennig et al. (2009) for semiautomated delineation within a geographic information system (GIS).

Changing LULC management affects the hydrological relevant vegetation properties, such as leaf area, rooting depth, water consumption and physical soil properties, i.e. infiltration capacity, permeability, porosity and water storage capacity. Extending settlement areas increases the impermeable area and promotes surface runoff at the expense of infiltration and groundwater recharge. From Fig. 6.2 it is obvious that LULC changes of this kind affect the magnitude and seasonal dynamics of water resources and modify their contribution to river runoff that ultimately drains the basin landscape.

Impacts of changing LULC management on the hydrological dynamics have been studied on various scales in agriculture (Thanapakpawin et al. 2006; van Roosmalen et al. 2009; Nejadhashemi et al. 2011), rangeland (Flügel and Märker 2003), forestry (Helmschrot and Flügel 2002; Helmschrot 2006b; Mustafa et al. 2005) and for settlements (Wijesekaraa et al. 2010). Associated with the hydrological dynamics, they affect the interactive processes controlling erosion and nutrient transport (Fink et al. 2007a, b). In historical times, large-scale

LULC management change has caused river salinisation by dryland salinity in Africa, Australia, and North and South America (Flügel 1995b) and impacted erosion, nutrient and sediment transport in Europe (Hoffmann et al. 2011). Changing irrigation management, i.e. from flooding irrigation to drip irrigation, is reducing the soil salinity and the amount of saline irrigation return flow discharging into the river (Flügel 1993).

6.2.2 Impacts from Climate Change (CC)

Since the publication of the 4th IPCC Assessment Report (IPCC 2007a, b, c), climate change (CC) is seen as an ongoing global threat that impacts river basin water resources in various ways (Alcamo and Henrichs 2002; Fischer et al. 2002; Flügel 2010). River basin water managers increasingly become aware that changing LULC and climate likely will alter a wide spectrum of hydrological processes driving river basins' water resources dynamics and supporting its socio-economic development (IUCN 2003).

Many semiarid river basins already suffer from natural and man-made water stress due to ongoing population growth that triggers land degradation by erosion and aggravates floods and droughts (Wallace and Gregory 2002). These river basins are highly vulnerable to impacts from climate change (Fischer et al. 2002; Querner 2002), and it is likely that their scarce water resources will come under even more stress in the forthcoming years (IPCC 2007a, b, c). Adaptation strategies for ILWRM have to be developed to cope with such impacts on both the basins' NE and its HD in an integrated way (IUCN 2003; GWP-TEC 2004).

General circulation models (GCM) provide input for global water balance models (Alcamo and Henrichs 2002). The challenge indicated in Fig. 6.1 is to apply a consistent downscaling approach to regionalise the GCM results to lower basin scales that are relevant for ILWRM (Bossa et al. 2015). At present this is done by means of (i) regional climate models (RCM) based on climate processes (Ahrens 2003; Klein et al. 2005; Kumar et al. 2006; Dobler et al. 2011) or (ii) geostatistics using climate time series and land surface parameters (Schmidli and Frei 2005; Wilby et al. 2004) complemented by satellite-based analysis (Seidel and Martinec 2002). Both of the downscaling procedures generate huge amounts of modelled hydrometeorological data time series that require sophisticated software tools as described by Flügel and Busch (2011) to extract the spatial information for a multi-scale ILWRM analysis.

6.3 Integrated Land and Water Resources Management (ILWRM)

The Global Water Partnership (GWP) defines IWRM as 'a process which promotes the co-ordinated development and management of water, land and related resources, in order to maximize the resultant economic and social welfare in an equitable manner without compromising the sustainability of vital ecosystems' (GWP-TAC 2000, p.22).

Accepted internationally the IWRM approach is an appropriate concept for river basin management (Flügel 2010). It has been implemented to a certain extent for European river basins by the EU Water Framework Directive (WFD) (EC 2000) which according to Rahmann et al. (2004) is not fully compatible with the IWRM demands in Asia.

As pointed out by Calder (2005), the IWRM concept does not adequately account for the LULC management aspects, which control the recharging of the land-based water cycle components, and therefore needs to be enhanced towards the ILWRM approach. The latter accounts for the landscape-related interactive hydrological process network shown in Fig. 6.2 that generates the quantity and quality of water resources subject to sustainable management in the river basin. With reference to Fig. 6.2, ILWRM is defined in this paper as an 'approach towards the integrated assessment and analysis of impacts from changing LULC management and

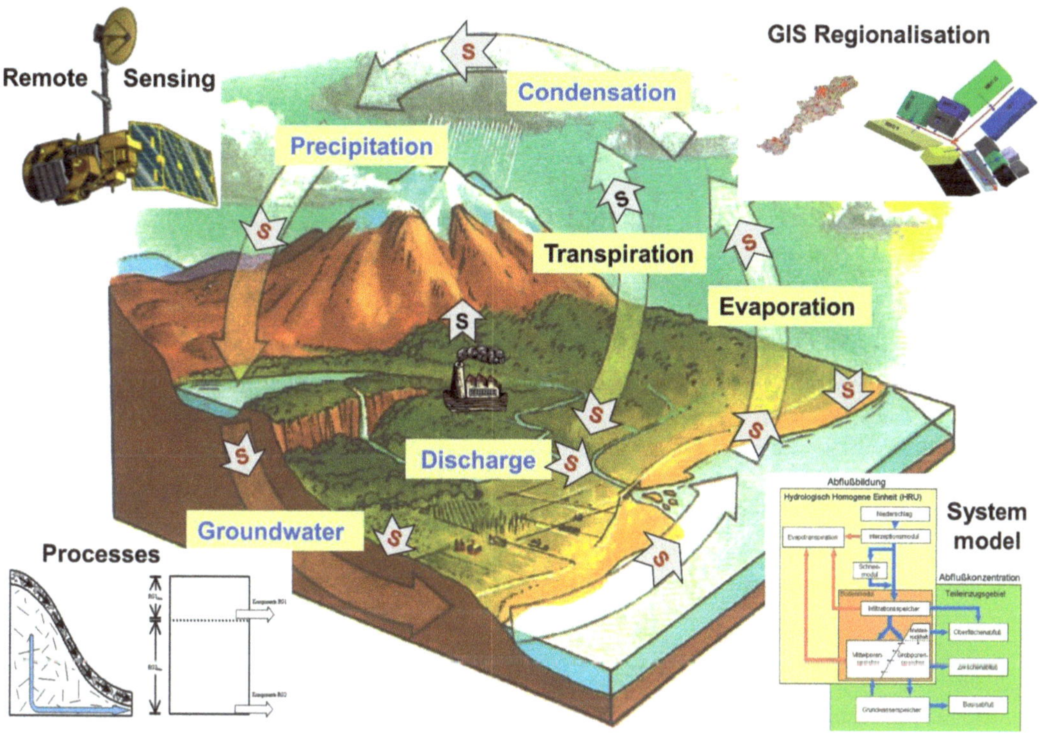

Fig. 6.3 Conceptual implementation of the ILWRM assessment and analysis approach by means of innovative geoinformatics methods, tools and technologies

climate on river basins water resources and their regeneration dynamics for the development and implementation of adaptive land and water resources management strategies' (Fig. 6.3).

With respect to this definition, Flügel (2011a) described the role of geoinformatics methodologies and techniques to implement multi-scale ILWRM analysis in a river basin as follows:

(i) LULC distribution is classified from multi-temporal satellite image analysis applying raster-based or object-oriented classification methods complemented by validating field campaigns.

(ii) Classified LULC maps are input for change analysis in a GIS, and together with other thematic GIS data layers, they are used to delineate process-based HRU as distributed landscape model entities (Flügel 1995a, 1996). HRU are networked by means of a topological model derived from a digital elevation model (DEM) in the GIS (Pfennig et al. 2009) allowing for flow routing in the J2000 river basin model (Krause 2002; Krause et al. 2006; Bende et al. 2007).

(iii) Integrated system analysis (ISA) elaborates on hydrometeorological time series complemented by validating field campaigns. It *firstly* provides insight into the interactive hydrological basin dynamics on different scales, *secondly* reveals trends of changing LULC management and climate and *thirdly* identifies likely impacts on the basins' water resources dynamics (Flügel 2000, 2010).

(iv) Distributed hydrological modelling applies the networked HRU model entities and their parameterisation derived from the LULC analysis. The modelled time series are input for impact analysis of changing LULC management and climate (Bende-Michl et al. 2007; Fink et al. 2007, 2013; Krause 2002; Krause et al. 2006; Nepal et al. 2014).

6.3.1 DPSIR Framework Approach

The DPSIR approach is a holistic concept to assess and analyse impacts from changing LULC management and climate on river basin water resources by linking aspects of LULC management with the hydrological water balance dynamics in an integrated way. For this purpose, the European Environment Agency (EEA 1999, 2008a, b) proposes indicators to quantify process components within a Decision Information Knowledge System (DIKS) as one of the core elements of the ILWRM approach.

The DPSIR approach developed by the EEA (2005) enhanced by the river basin information system, RBIS (Zander et al. 2013), as a DIKS is shown in Fig. 6.4, and the workflow for a multi-scale river basin assessment and analysis application is as follows:

(i) *Drivers* (*D*), *pressures* (*P*) and present *state* (*S*) are elaborated by a thorough ISA comprising (1) remote sensing satellite image and GIS analysis, (2) process studies elaborating on measured data time series complemented by field campaigns, and (3) socio-economic, institutional and governance surveys.

(ii) *State* (*S*) of the system's hydrological dynamics is analysed by modelling, e.g. water balance, runoff generation, erosion and nutrient transport by means of process-based, distributed rainfall-runoff models like the J2000 basin model (Bende-Michl et al. 2007; Fink et al. 2007a, 2013; Krause 2002; Krause et al. 2007; Nepal et al. 2014). The validated model becomes part of the DIKS as an important ILWRM planning tool.

(iii) *Impacts* (*I*) from changing climate, LULC management and socio-economic development on the quantity and quality of water resources are identified from the analysis of hydrological time series provided by the validated river basin model in the DIKS.

(iv) *Indicators* quantify drivers, pressure, state and impacts (EEA 1999; Giannini and Giupponi 2011). They indicate trends of relevant ILWRM processes for the compilation of 'what-if?' management scenarios,

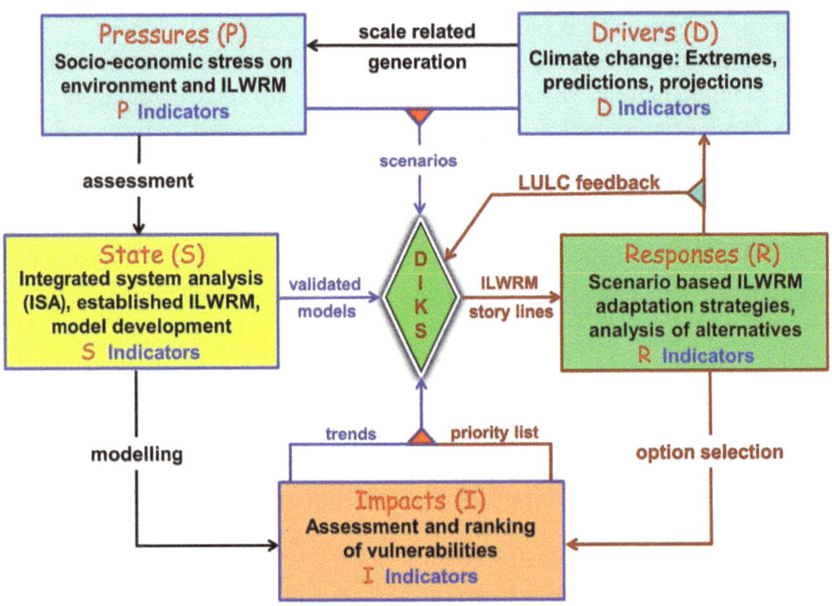

Fig. 6.4 DPSIR framework approach enhanced by a DIKS as a conceptual analysis framework structure to implement ILWRM

which in turn are input to the knowledge base in the DIKS.

(v) 'What-if?' scenario story lines quantify impacts of changing LULC management and climate (IPCC 2000; Giannini et al. 2011) and assimilate the results obtained from (i) till (iv). They feed the knowledge base hold available in the DIKS to develop respective ILWRM adaptation strategy options.

(vi) The ILWRM options are evaluated and priority ranked according to economic-, institutional- and governance-related criteria by means of the Delphi method (Gordon and Peace 2006) applied as part of the stakeholder process (Giannini et al. 2015).

(vii) The priority ranked ILWRM response options and best management practices are parameterized in the validated hydrological model to analyse their implementation potential within the river basin. Mapping updated LULC management distribution of the river basin by means of a GIS provides further thematic input to the knowledge base of the DIKS.

6.3.2 Integrated Land Management System (ILMS)

To support ILWRM applications, Kralisch et al. (2012) developed the comprehensive ILMS toolset, which implements the enhanced DPSIR concept shown in Fig. 6.4. The system was tested extensively in projects carried out in Africa, Asia, Australia, Europe and South America (Bende-Michl et al. 2007; Fink et al. 2007a, 2013; Kralisch et al. 2003, 2005, 2007, 2012; Kralisch and Fischer 2012; Kralisch and Krause 2007; Krause and Flügel 2005; Krause et al. 2006; Künne et al. 2012; Nepal et al. 2014; Steudel et al. 2013; Zander et al. 2013).

ILMS components shown in Fig. 6.5 provide the following methodological means to develop adaptive ILWRM strategies in river basins to cope with impacts from changing LULC management and climate:

(i) **ILMS*info*** as a comprehensive DIKS comprises database management and GIS functionalities for the administration and analysis of data time series, maps and knowledge documents concerning the basin's NE and its HD.

(ii) **ILMS*image*** as a satellite image processing tool for object-oriented LULC classification and LULC management change detection.

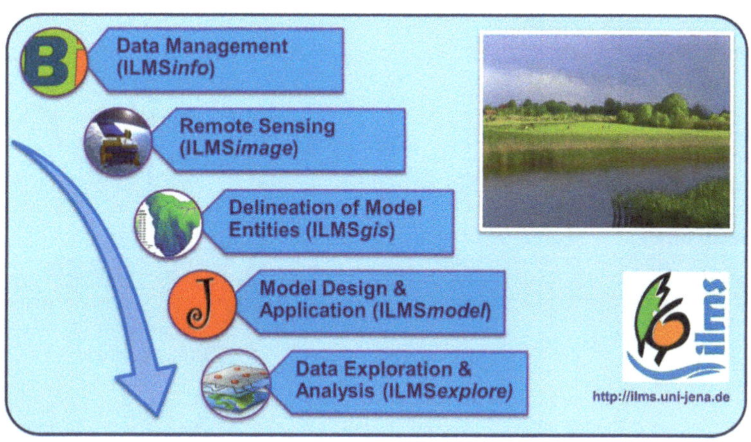

Fig. 6.5 ILMS toolset components providing a consistent processing chain for multi-scale river basin assessment and system analysis for the development of adaptive ILWRM strategies

(iii) **ILMS*gis*** as a GIS system for the delineation of HRU using LULC maps from ILMS*image* and complement thematic GIS data layers quantifying the NE and HD of the river basin.
(iv) **ILMS*model*** comprising the Jena Adaptable Modelling System (JAMS), a model framework system (MFS) for modular model design such as the JAMS/J2000 basin model suite (Kralisch et al. 2007; Kralisch and Fischer 2012). The latter applies the HRU model entities for rainfall-runoff and solute transport modelling (Fink et al. 2007a, 2013; Krause et al. 2006; Nepal et al. 2014).
(v) **ILMS*explore*** as a powerful tool for exploring the modelling results by means of statistical analysis and 2D/3D visualisation.

6.4 Multi-scale ILWRM Applications in South Africa and SE Asia

In the following paragraphs, results from ILWRM and ILMS applications carried out in South Africa (Helmschrot and Flügel 2002; Helmschrot 2006a, b) and SE Asia (Flügel 2011a; Nepal et al. 2014; Sharma and Flügel 2015) are presented and discussed. Applying a multi-scale NCA approach, they analysed and quantified impacts from changing LULC management and climate on the water balance dynamics and related ecosystem functions (ESF) and ecosystem services (ESS) at the micro-, meso- and macro-scale.

6.4.1 Changing LULC Management Impact Analysis for ILWRM in South Africa

The multi-scale ILWRM research study was carried out in cooperation with the University of KwaZulu-Natal and the company Mondi Forest in the headwater zone of the macro-scale Umzimvubu River basin, Eastern Cape Province, Republic of South Africa (Fig. 6.6).

This region has a unique landscape structured by sandstone escarpments and valleys embedded into underlying strata of mudstones and siltstones. Traditional LULC management is cattle ranching complemented by agriculture on patches of fertile soils. LULC management changed in the 1990s when forest plantations began to replace rangeland, and today more than 35,000 ha of former rangeland has turned over into pine and eucalypt industrial forest plantations (Fig. 6.7).

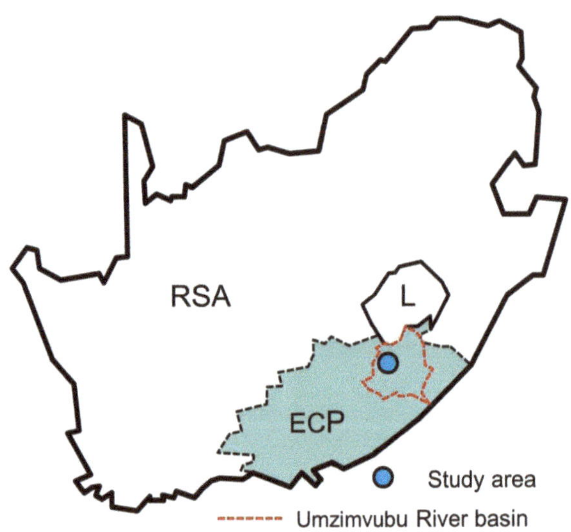

Fig. 6.6 Location of the study area comprising the micro-scale Weatherly Creek in the mesoscale Mooi River headwater catchment of the macro-scale Umzimvubu River basin in the Eastern Cape Province, Republic of South Africa

Fig. 6.7 Changing LULC management from ranching rangeland to industrial pine and eucalypt forestry in the Mooi River catchment

The objectives of the project addressed concerns that the changed LULC management will:

- *Firstly* affect the hydrological dynamics of wetlands in the former rangeland landscape
- *Secondly* reduce the baseflow of the meso-scale headwater catchment of the Mooi River during the dry summer
- *Thirdly* impact the seasonal discharge dynamics of the macro-scale Umzimvubu River (Helmschrot and Flügel 2002)

The research project implemented the following multi-scale NCA:

- The *micro-scale* Weatherly Creek ($A = 1.4\ \text{km}^2$) catchment was selected for detailed process studies, located in the *meso-scale* Mooi River ($A = 307\ \text{km}^2$) headwater catchment modelled by means of the J2000 rainfall-runoff model to analyse the headwater tributary runoff discharging into the *macro-scale* Umzimvubu River (19,845 km^2) basin subject for ILWRM assessment and changing LULC management impact evaluation.

A comprehensive ISA was carried out in the Weatherly Creek and Mooi River catchments and the objectives described above were elaborated by means of the NCA as follows:

(i) Classification of wetlands according to location in the landscape and driving hydrological process dynamics. HRU delineation by means of RS and GIS
(ii) J2000 modelling of the hydrological water balance and process dynamics of the classified wetlands and their contribution to the runoff of the Mooi River catchment
(iii) Modelling of the two 'what-if?' scenarios: (1) no afforestation and (2) afforestation to the full permitted extent of about 60,000 ha applying the validated J2000 model
(iv) Evaluation of impacts from LULC management change in the macro-scale Umzimvubu basin by regionalizing the model results from the Mooi River headwater catchment

6.4.1.1 Wetland Classification and HRU Delineation

Detailed process studies and mapping done in the micro-scale catchment of the Weatherly Creek (Dahlke et al. 2005; Helmschrot and Flügel 2002) were validated in the mesoscale Mooi River catchment during numerous field campaigns (Helmschrot 2006a, b). In result three different types of wetlands (WL) shown in Fig. 6.8 were classified. Each of them has a specific topographic location and composition of geology, soils and vegetation, which together control their climate-driven seasonal hydrological dynamics. Evapotranspiration from all wetland types comprises interception and plant water consumption from the root zone during the growing season. With reference to Fig. 6.8, the hydrological dynamics of the classified wetlands can be described as follows:

- *Plateau WL* developed at the rim of sandstone plateaus. They depend on groundwater seeping out of the unconfined sand aquifer developed in the weathered sandstone that forms the plateaus and their escarpments. The thickness of the sand cover increases with distance from the escarpment but never was found to exceed 10 m in thickness. Recharge of the aquifer is during the rainy season from infiltrating rainfall. Surface runoff is negligible because of the high infiltration capacity of the coarse sand. Saturation overland flow does not exist as no clayey horizon that can store infiltrating rainfall has developed in the soil. Groundwater seepage reaches its maximum in the rainy season and is receding continuously towards the end of the dry season. Consequently, their largest extend is in the rainy season, and they are shrinking to small strips along the rim of the escarpment towards the end of the dry season.
- *Slope WL* are located on the pediment below the sandstone escarpment and receive discharge out of the plateau WL spilling over the rim of the escarpment. In some places, they also receive groundwater flowing through fractions of the overlying sandstone and along the less permeable underlying siltstone onto the pediment. Rainfall is the second recharge source for all slope WL. Rain that infiltrates in the less permeable silty and hydromorphic soils generates interflow and the portion that exceeds the infiltration rate generates overland flow.

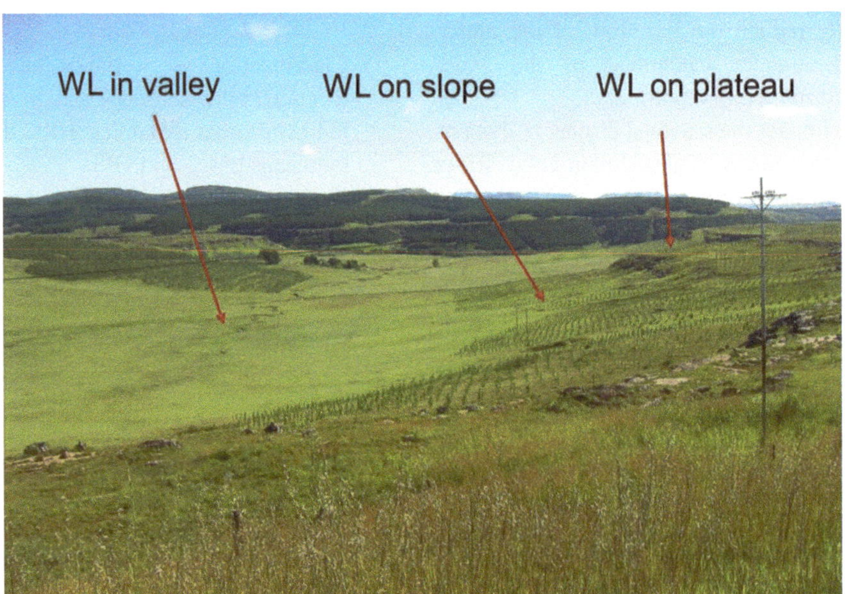

Fig. 6.8 Classified wetlands (WL) and their topographic location in the micro-scale Weatherly Creek catchment

- *Valley floor WL* represent the majority of the total wetland area. They developed in broad and flat valley floors established in mudstones underlying the siltstone pediment and have clayey and hydromorphic gley soils. Their recharge comprises rainfall and overland flow, interflow and groundwater seepage from the adjacent slopes. They are saturated during the rainy season and slowly drain to their perennial creeks in the dry season.

6.4.1.2 Distributed Hydrological Modelling Applying the J2000 Model

The HRU concept is well described by Flügel (1995a), Krause et al. (2006) and Pfennig et al. (2009), and HRU were applied as J2000 process model entities. Their delineation applies GIS analysis of thematic data layers comprising a DEM, LULC classification derived from Landsat satellite image analysis, soils and geology from South African surveys and the wetland classification described above. Linked with each other by means of a topological model (Pfennig et al. 2009), flow paths within the landscape are represented. Their parameterization applies the indicators obtained from the ISA and the process knowledge elaborated from the field studies in the catchments of the Weatherly Creek and the Mooi River, respectively.

The J2000 rainfall-runoff model (Krause 2002; Krause and Flügel 2005; Krause et al. 2006) component of ILMS was applied in the Weatherly Creek catchment for the period 1999 until 2001 and for the Mooi River catchment for the period 1980 until 2003, respectively. Because of data shortage in the Weatherly Creek catchment, only the Mooi River model could be validated.

The calibrated and validated model was then applied for the two 'what-if?' LULC management scenarios: (1) no afforestation of rangeland and (2) afforestation of rangeland to the full permitted extent of about 60,000 ha. Figure 6.9 is showing the results obtained and reveals that the three wetland types and their hydrological contribution to river runoff will be affected in different magnitudes from the full-scale afforestation:

- *Plateau WL* will reduce their runoff contribution by about 34 % in total, as the forest on the sandstone plateaus increases

Valley floor WL	Slope WL	Plateau WL
ΔQ_{Surfr} = - 5,4 %	ΔQ_{Surfr} = - 14,5 %	ΔQ_{Surfr} = - 83,6 %
ΔQ_{Subsr} = - 7,5 %	ΔQ_{Subsr} = - 22,5 %	ΔQ_{Subsr} = - 66,0 %
ΔQ_{Sum} = - 6,9 %	ΔQ_{Sum} = - 14,9 %	ΔQ_{Sum} = - 34,1 %

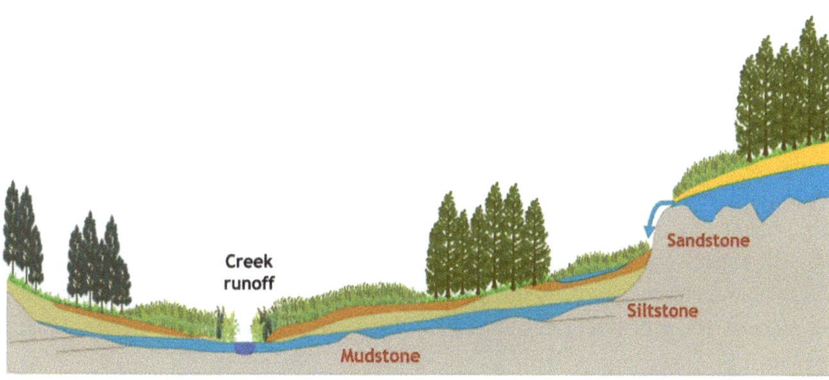

Fig. 6.9 Modelling results quantifying hydrological impacts of afforestation on the classified wetlands (WL) (Modified from Helmschrot 2006a)

evapotranspiration due to higher interception and root water consumption. Litter deposited under the canopy will decrease the already negligible surface runoff to zero, and tree roots will tap the sandy groundwater aquifer on top of the plateaus reducing groundwater seepage by about 66 %.

- *Slope WL* suffer less by losing in average about 15 % of their discharge contribution due to less groundwater seeping out of the plateau aquifer onto the pediment. Hence, they will generate less interflow and surface runoff, but higher evapotranspiration from the trees will tap their wet soil and groundwater, thus reducing their discharge contribution by about 25 %.
- *Valley floor WL* will suffer the least in the afforested scenario and will contribute only 7 % less to the runoff of the Mooi River if compared to the non-afforested rangeland scenario. The reasons are twofold: *Firstly*, only small patches of drier soils can be afforested in the wet valley floors and *secondly* the aquifers in the broad valleys are mainly recharged from rainfall and hence they depend less on inflow from the WL on the slopes and plateaus, respectively.

6.4.1.3 Recommendation for ILWRM

Based on the project results and the modelling exercises, the following ILWRM conclusions were compiled:

1. The present and projected full-scale afforestation in the Mooi River catchment will have no significant impact on the hydrological functioning of the valley floor WL, which provides a substantial part of the baseflow runoff during the dry season.
2. Slope WL will experience a moderate impact by the LULC management change in the landscape, but plateau WL will likely diminish to a large extent.
3. Impacts on the biodiversity in the Mooi River catchment not covered by the ISA might outweigh the hydrological impacts by far. It is therefore recommended to investigate likely ecological impacts by a separate hydro-ecological wetland study.
4. No significant hydrological impacts were identified for the mesoscale Mooi River discharge dynamics, and therefore impacts on the baseflow distribution in the macro-scale Umzimvubu River are unlikely as well. Climate change, i.e. temperature rise and reduced precipitation, might alter this statement at least for the mesoscale Mooi River catchment, but still the impact of afforestation is unlikely to be of significance for the runoff dynamics in the Umzimvubu River.
5. In case afforestation should exceed the projected 60,000 ha and extending further into other Umzimvubu River headwater catchments, the impact assessment and ILWRM conclusions needs to be revised.

6.4.2 Climate Change ILWRM Impact in the Brahmaputra River Basin

The Brahmaputra River basin ($A = 938{,}000\ \mathrm{km}^2$; $L = 2880$ km) is one of the major river systems in Southeast Asia and subject to impacts from climate change (CC) that causes increasing glacier melt and higher risks of glacier lake outburst floods (GLOFs) (Subba 2001). Changing LULC management, i.e. deforestation and drainage of wetlands for extending agriculture and settlements, is another impact factor in the North Eastern Region (NER) of India. Deforestation increases surface runoff, interflow and soil erosion in the Himalaya, which in turn contribute to floods and severe bank erosion in the Brahmaputra floodplain of Assam. Wetland drainage decreases the flood buffering capacity within the floodplain of

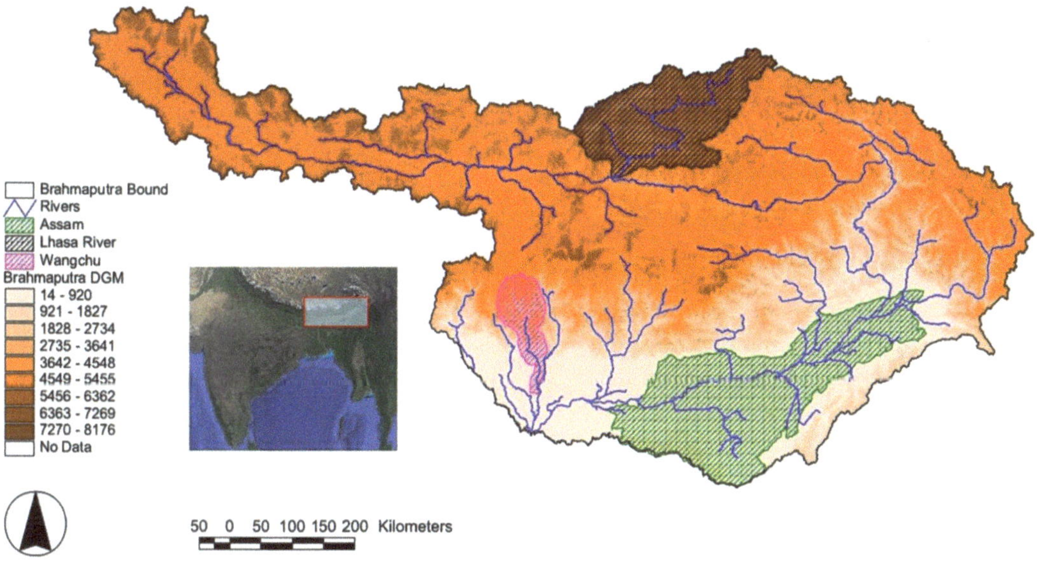

Fig. 6.10 Schematic topographic presentation of the UBRB with the test regions of the Wang Chu and Lhasa River basin in Bhutan and Tibet/China and the Brahmaputra floodplain in Assam, India

the Brahmaputra River and has further socio-economic impacts to be considered.

The European Commission (EC) within the 'Twinning Basin' programme of the 6th Framework Programme (FP6) funded the interdisciplinary BRAHMATWINN research project to study the impacts of climate change in the upper Danube river basin (UDRB) and the upper Brahmaputra River basin (UBRB). Originating as Yarlung Tsangpo in its Tibetan headwaters located in the lee of the Nepal Himalayan ridge, the river flows about 950 km east before it breaches in a sharp U-turn through the mountain ridge reaching as Dihang River the Indian state of Arunachal Pradesh. The UBRB was defined with reference to the gauge at the town of Guwahati in Assam, India, and covers an area of 514,717 km^2 (Flügel 2011b). It is shared by Tibet/China (282,950 km^2), Bhutan (43,546 km^2) and NE India (188,111 km^2). In Assam, the river has formed a broad floodplain in front of the Himalaya with a braided river channel and severe bank erosion.

BRAHMATWINN was carried out between 2006 and 2009 by an international partner team of 18 research institutions and small-medium enterprises (SME) from Bhutan, China, Europe, India and Nepal. Results from this applied ILWRM pilot project were published in detail by Flügel (2011b) and Sharma and Flügel (2015), respectively, and are available for download from http://www.adv-sci-res.net/7/index.html.

BRAHMATWINN applied the following multi-scale NCA in the trans-boundary UBRB (see Fig. 6.10):

- Socio-economic and governance components were assessed and analysed on the *micro-scale* community level in the floodplain of Assam and upscaled to this Indian test region.
- The ISA, wetland analysis, modelling of permafrost and the IPM of melting glaciers and snow storages concentrated on the *mesoscale* Wang Chu catchment ($A = 3550$ km^2) in Bhutan and the lower *macro-scale* Lhasa River basin ($A = 26{,}235$ km^2) in Tibet, respectively.
- The IPM complemented by RS analysis was done for the *macro-scale* UBRB ($A = 514{,}717$ km^2) and to this level the permafrost analysis was upscaled as well.

All deliverables of the project, i.e. time series data, maps and reports, were input to the project DIKS for ILWRM strategy development and evaluation.

6.4.2.1 Prototype Component Structure for ILWRM Implementation

BRAHMATWINN addressed the following important ILWRM components applying methods and technologies from natural and socio-economic sciences and integrating them by means of geoinformatics (Flügel 2010, 2011a):

1. Development of a prototype ILWRM project structure as input to the Web-based DIKS comprising ISA, RS and GIS mapping, rainfall-runoff modelling, socio-economic and governance assessment, stakeholder processes and negotiation structures with actors from the UBRB
2. Assessment and modelling of the basins' atmospheric climate system by downscaling of GCM climate change scenarios to a regional hydrological basin scale by means of a RCM. Evaluation of the basin adherence to present climate conditions and development of appropriate procedures for modelling likely climate parameter scenarios
3. Comprehensive ISA of the NE comprising runoff, groundwater, glaciers, permafrost, topography, LULC and eco-hydrological research to derive the interactive dynamics of the system's components
4. Assessment of the system's HD accounting for water-related issues of socio-economic vulnerability to environmental stress (water quality, water demand and allocation), gender equity in respect to water allocation and labour division, and considering the political structures and policies pertaining to the basins
5. Analysis of present ILWRM practices in place regarding water quantity and quality, user demands, user conflicts, resource allocation and its potential for adaptation to cope with likely impacts of changing flow regimes
6. Transformation of the results obtained from components 2–5 into a set of integrated indicators suitable to quantify the present system status and to evaluate the outcome of 'what-if?' scenario modelling
7. Rainfall-runoff and snowmelt modelling in the monsoonal UBRB to reproduce observed periods of floods and low flow that require adaptive ILWRM strategies
8. Development of likely 'what-if?' scenarios based on the IPCC Special Report on Emission Scenarios (SRES) assumptions (IPCC 2000) and evaluation of identified integrated indicators to quantify system changes
9. Development and implementation of an ILWRM toolset to carry out the ISA and the rainfall-runoff modelling and to organise the upgrade of the knowledge base in the DIKS shown in Fig. 6.4
10. Application of the ILWRM toolset to evaluate adaptive ILWRM options related to the impacts of likely 'what-if?' scenarios defined in BRAHMATWINN
11. Capacity building with ILWRM practitioners, institutions and communities as part of the stakeholder processes and dissemination of BRAHMATWINN's results and best practices at national and international level

These objectives were elaborated in respective work packages (WP) linked with each other as shown in Fig. 6.11. The set-up was provided to project stakeholders as a prototype ILWRM component set-up for the development of adaptive ILWRM strategies and their implementation.

6.4.2.2 Integrated System Analysis (ISA)

The ISA comprising the components WP2, WP3 and WP4 (Fig. 6.11) revealed the hydrological dynamics of the UBRB during the last three decades and established the required knowledge for the development of integrated indicators and their quantification in WP5.

In particular, it provided the following information as input to the modelling exercise and the quantification of the DPSIR components drivers, pressures and state of the system:

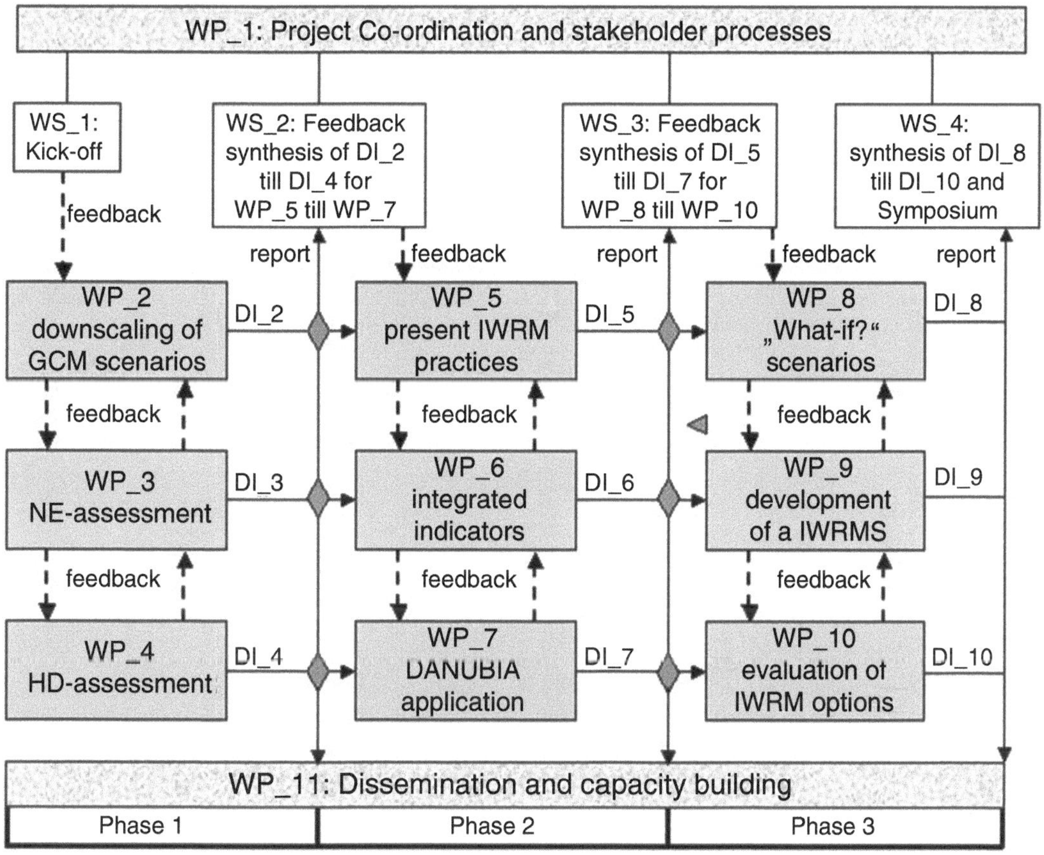

Fig. 6.11 BRAHMATWINN project management structure as a prototype concept for ILWRM implementation in the UBRB

(i) Public domain data sources and national data suppliers provided numerous station data time series. They were quality checked by a thorough data control analysis, which reduced the number of reliable climate stations considerably (Fig. 6.12).

(ii) Temperature and precipitation trend analysis were done for those stations in Fig. 6.12 that have consistent time series from 1974 until 2006, and selected stations from the western, the middle and the eastern part of the UBRB are shown in Fig. 6.13. Together they indicate the following differences between the three parts of the UBRB:

- In the *western part*, precipitation and temperature show opposite trends. Temperature has an increasing and precipitation a decreasing trend (station Shiquanhe in Fig. 6.13). This indicates that the already semiarid headwater region located in the lee of the Nepal Himalayan ridge is getting dryer and warmer. For some decades, this will increase glacier melt during summer but on the other hand will reduce the already small runoff contribution from rainfall even further.
- In the *middle part* of the UBRB, temperature shows an increasing trend meanwhile precipitation has no trend at all (station Dengqen in Fig. 6.13). In result, the warmer climate will drive ongoing glacier melt but evapotranspiration will increase as well from rising temperatures and will likely

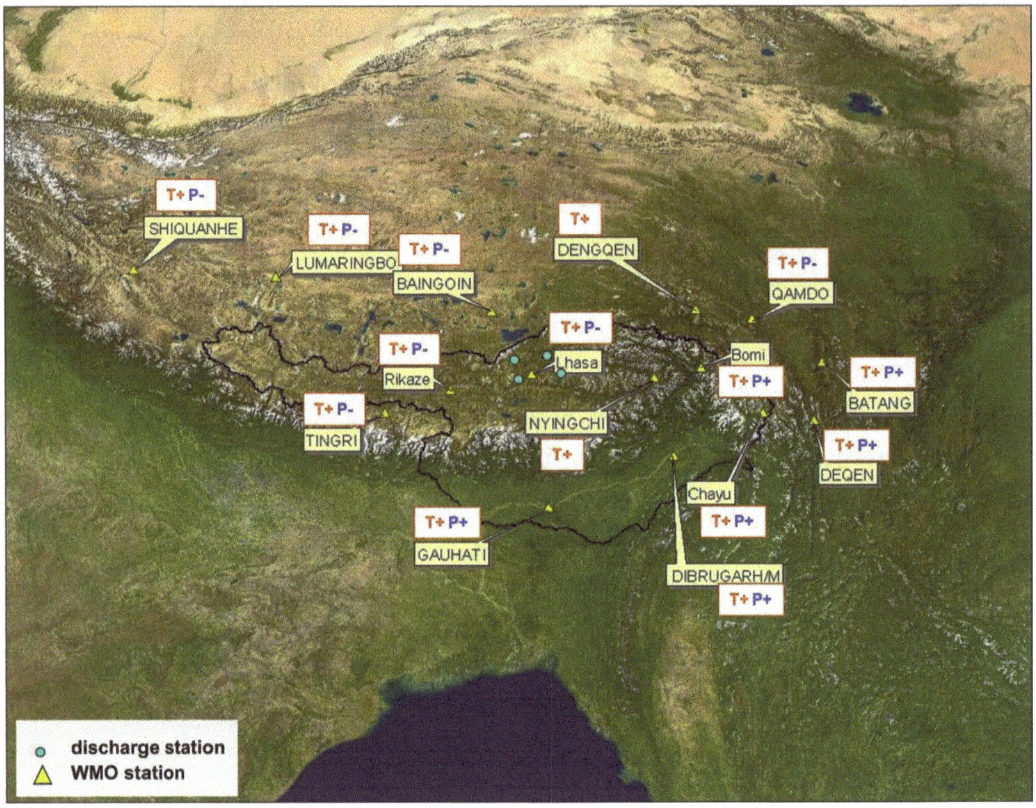

Fig. 6.12 Distribution of temperature and precipitation trends in the western, middle and eastern parts of the UBRB

reduce the runoff contribution from rainfall.

- In the *eastern part* of the UBRB, both temperature and precipitation show a positive trend (station Dibrugarh in Fig. 6.13). Consequently, evapotranspiration will increase, and depending on the rainfall pattern, discharge can increase as well during heavy monsoon rainfall storms.

(iii) Dobler et al. (2011) elaborated on the downscaling of GCM climate projections applying a RCM with about 50 × 50 km grid cell resolution. The model results confirmed the observed positive temperature trend, but did not reflect the observed trends in precipitation. The data time series provided by the RCM were further downscaled to the 1 × 1 km scale of the model entities applying a tool developed by Marke (2008).

(iv) Lang et al. (2011) analysed the NE of the three test regions with respect to LULC, permafrost distribution and wetlands. Thapa et al. (2015) classified changing LULC management of the three test regions by means of satellite image analysis. Kääb et al. (2015) calculated a glacier balance and developed a model to estimate the permafrost distribution for present and projected climate conditions. Exler et al. (2015) classified wetlands with respect to their hydrological dynamics and their eco-hydrological functioning within the basin.

(v) Discharge of the Lhasa River in Tibet/China and the Wang Chu in Bhutan was analysed. Both basins have glaciers in their headwater regions but as shown in Fig. 6.14, they show different annual discharge trends and monthly flow dynamics:

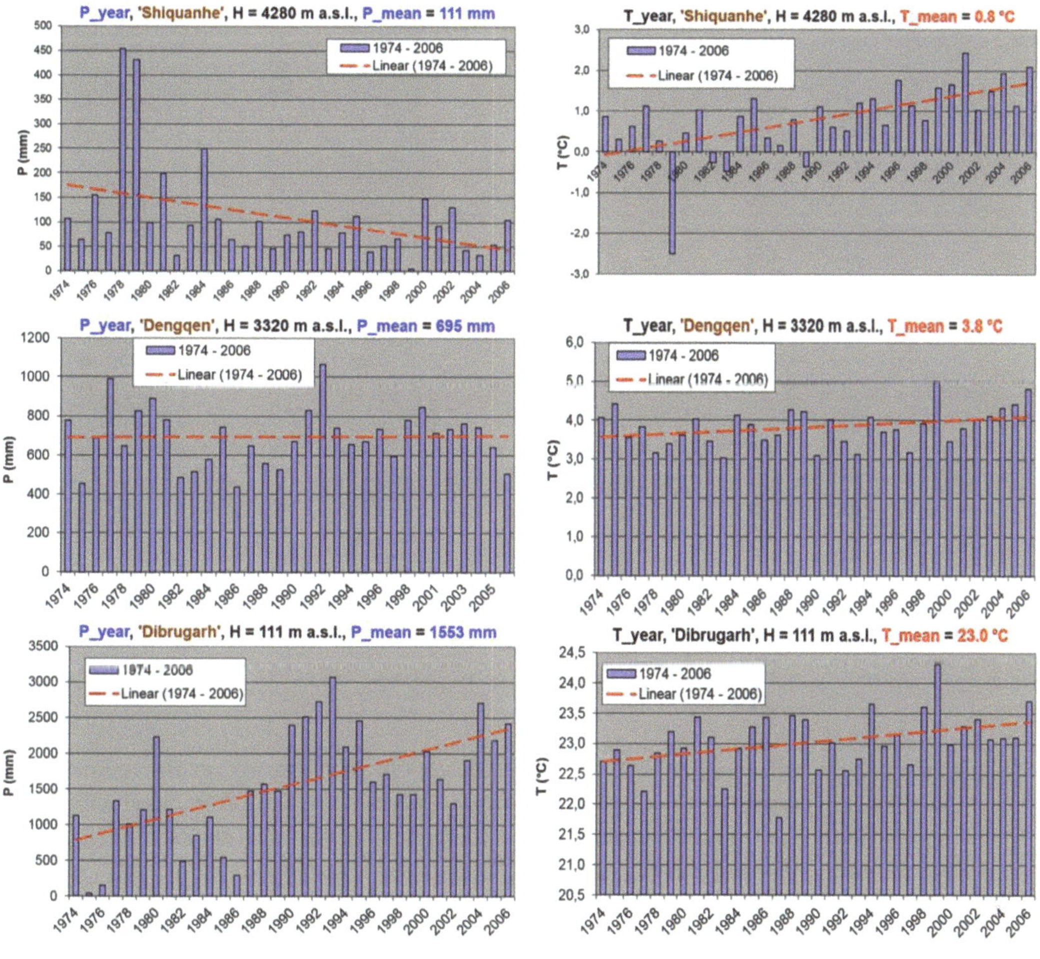

Fig. 6.13 Trends of annual temperature (°C) and precipitation (mm) 1976 until 2006 from selected stations from the western, middle and eastern parts of the UBRB

- Measured river discharge in the Lhasa River located in the lee of the Himalayan mountain ridge has a positive annual discharge trend, which is obvious since 1990. The mean monthly hydrograph also differs in this period as discharge increases more rapidly from May onwards and forming a broader hydrograph peak during the summer months that indicates the increasing contribution of glacier and snowfield meltwater in the headwater region (Prasch 2010).
- The Wang Chu basin located in the windward direction of the Himalayan ridge in Bhutan receives high summer monsoon rainfall. Figure 6.14 shows no annual discharge trend between 1976 and 2006, and the mean monthly flow distribution is not changing in the two 15-year periods as well. This indicates that higher rainfall trends are compensated by higher evaporation rates thus leaving the discharge dynamics of the Wang Chu unchanged.

(vi) Because of its high population density and obvious flood vulnerability (Kienberger et al. 2015), the floodplain of the Brahmaputra River in the state of Assam, North Eastern Region (NER) of India, was selected for the socio-economic studies of the HD (Hutton et al. 2011). They applied

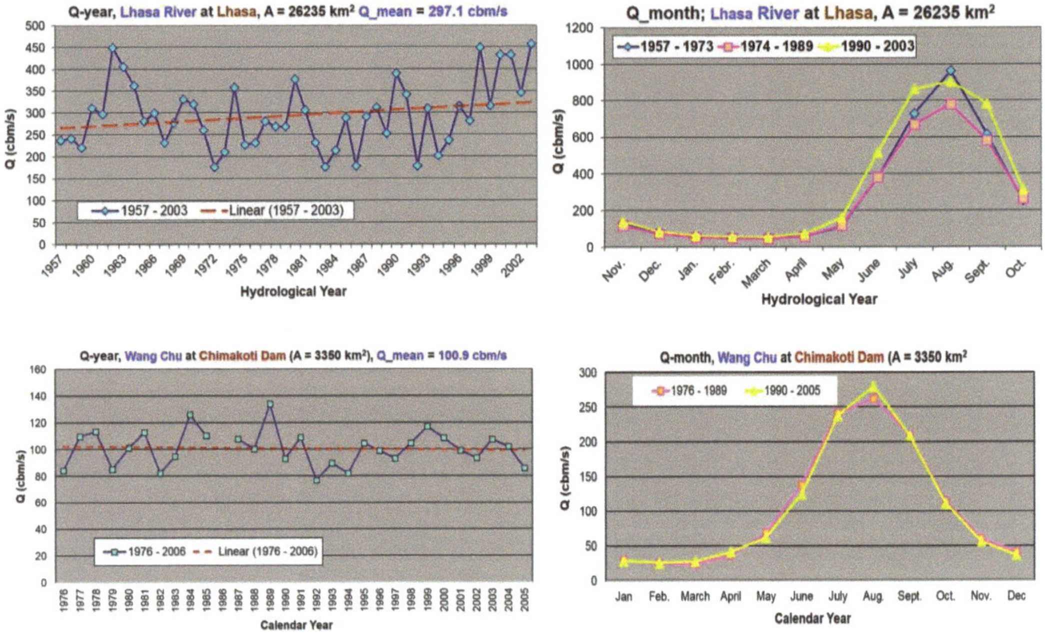

Fig. 6.14 Trend analysis of annual discharge and mean monthly discharge distribution for discharge stations in the Lhasa River in Tibet/China and the Wang Chu in Bhutan

remote sensing methods complemented by validating field campaigns and evaluated results together with stakeholder representatives. Climate change is affecting vulnerabilities to flooding in Assam, but cities with better flood protection measures are less vulnerable than rural areas lacking flood protection infrastructures.

(vii) ILWRM analysis done by Flügel and Bartosch (2011) revealed that neither on the interstate level in the NER nor internationally, any ILWRM strategy is in place. Therefore, the findings of the BRAHMATWINN project are of specific interest for further ILWRM developments in the UBRB.

(viii) Giannini et al. (2011) supported by project experts developed integrated indicators and validated them in their stakeholder processes. The priority ranked indicators were applied later in the project to quantify 'what-if?' scenarios (Giannini and Giupponi 2011).

6.4.2.3 Climate Impact on River Runoff

According to Fig. 6.4, process-based, distributed hydrological modelling is a core methodology for upscaling and downscaling knowledge of the interactive hydrological process dynamics to the relevant ILWRM scales. For the UBRB, Prasch et al. (2011) applied the DANUBIA modelling system for historical time series from 1970 until 2000 processing data from the Community Land Model (CLM) (Oleson et al. 2010) and the ERA-40 reanalysis (Upsala et al. 2005) as hydrometeorological drivers. For the projected SRES scenarios (IPPC 2000), they used the time series provided by the project RCM model between 2010 and 2080 (Dobler et al. 2011). The UBRB model applies 1 × 1 km grid cells as distributed landscape modelling entities parameterized by means of public domain data, i.e. the digital elevation model from the United States Geological Survey (USGS) and LULC from MODIS satellite products. The CLM and ERA-40 data have been downscaled to the model cell size of 1 × 1 km applying a method developed by Marke (2008).

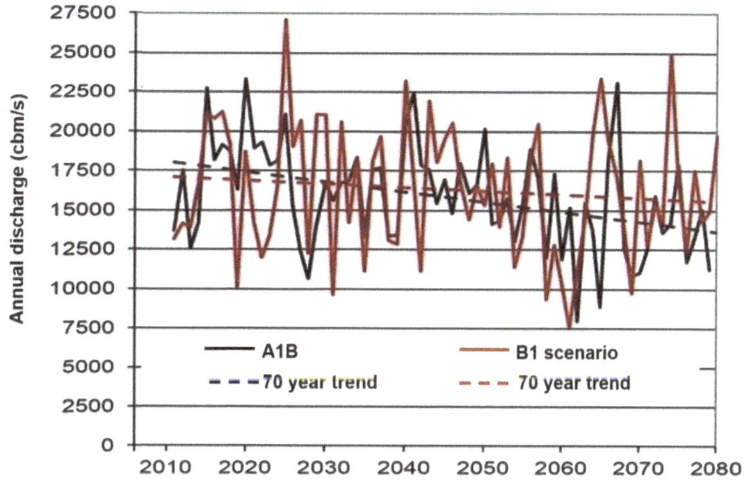

Fig. 6.15 Modelled Brahmaputra River discharge at the gauge in Guwahati, Assam, India, for the projected RCM model period between 2010 and 2080 (Modified from Prasch et al. 2015)

The model was calibrated and validated by means of measured data provided by the ISA and information obtained from the literature (Prasch et al. 2011). The validated model later was applied to the SRES projections B1 and A1B (IPCC 2000; Dobler et al. 2011) employing the modelled meteorological input time series from the project RCM as hydrometeorological drivers. Model results in Fig. 6.15 reveal that both of the modelled SRES climate change scenarios indicate a negative discharge trend for the Brahmaputra River that is responding to increasing evapotranspiration caused by the temperature rise during that period. Tributaries like the Lhasa River with significant meltwater contributions from glaciers and permanent snowfields will be impacted by a shift of hydrograph peaks and changing runoff component contributions until a new input-output balance of the glacier and snowfield systems is established (Prasch 2010).

The modelling exercises produced numerous daily time series of hydrological variables for each of the more than 500,000 grid cells covering the historical period between 1970 and 2000 and the two SERS climate projections from 2010 until 2080. The huge terabyte volume of data could not be processed by available commercial software. For this task, Flügel and Busch (2011) developed an innovative software package that accesses selected pixels out of the data stack and generates daily time series for their modelled hydrometeorological variables for further analysis.

6.4.2.4 Water Resources Response Units (WRRU)

The DANUBIA model also provided mean annual time series (1970–2000) of surface runoff, interflow and groundwater recharge for each of the half a million model grid cells (Prasch et al. 2015). In a stepwise GIS processing, Flügel et al. (2015a, b) regionalised these data for the UBRB by merging the grid model results with the response unit (RU) approach (Flügel 1995a, 1996) as described below:

(i) The modelled mean annual runoff components surface runoff, interflow and groundwater flow of each DANUBIA model grid cell were assigned to the HRU they are located in.
(ii) The summed up flow components were averaged for each HRU, and their flood contribution by surface runoff and interflow was statistically analysed for the UBRB.
(iii) Water Resources Response Units (WRRU) were delineated by reclassifying the HRU applying the flow classes derived from (ii) and mapped as thematic GIS data layer for input to the DIKS.

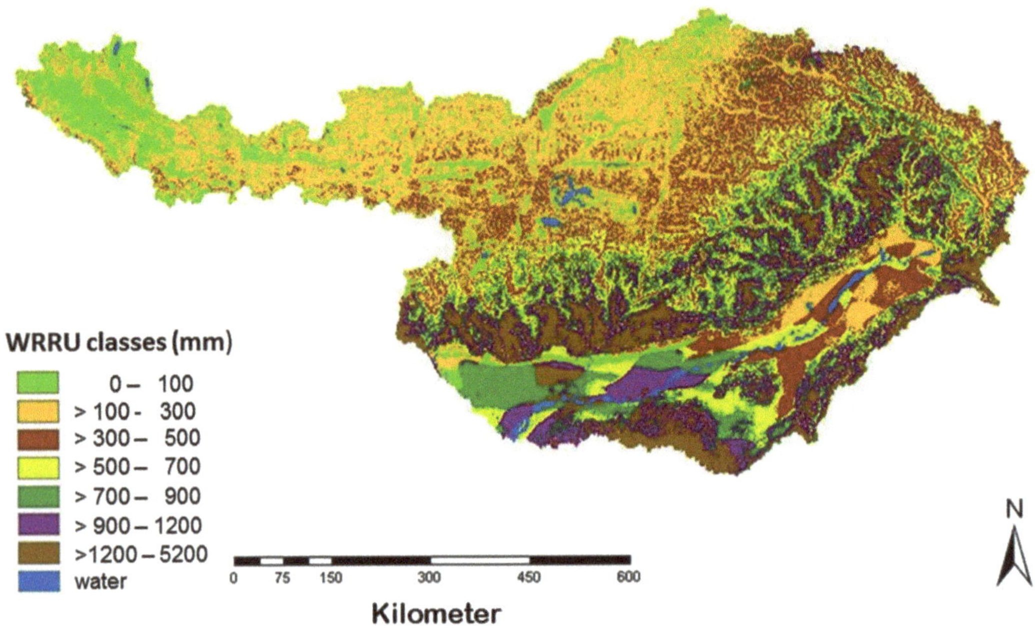

Fig. 6.16 Water resources response units (WRRU) quantifying the flood-related runoff generation by surface runoff and interflow provided by the DANUBIA model (Flügel and Bartosch 2011, CC by 3.0 license)

WRRU integrate the hydrological runoff components provided from the grid cells of the UBRB hydrological model with the landscape-related HRU attributes LULC, topography and soil. As shown in Fig. 6.16, they provide the required information for the ILWRM flood analysis of the UBRB and reveal:

- The Himalayan mountain ridge acts as a barrier for the monsoon rainfall, and therefore the WRRU distribution reflects the luv and lee rainfall distribution of the mountain barrier.
- The *western part* of the UBRB with a semi-arid cold temperate climate and sparse vegetation cover has only little annual runoff contribution mostly not exceeding 200 mm (6.34 l s^{-1} km^{-2}) per year.
- The *middle part* of the UBRB with moderate rainfall, alpine grassland, wetlands and agriculture on the valley floors has a moderate annual runoff generation ranging between 300 and 500 mm (9.51 l s^{-1} km^{-2} till 15.85 l s^{-1} km^{-2}) in the lower mountain range. In isolated spots of the alpine ridge, runoff contribution from rainfall can reach up to 900 mm (28.54 l s^{-1} km^{-2}).
- Towards the *eastern part* of the UBRB covered with forest, alpine grassland, bushland and insular agriculture, the mean annual runoff contribution is increasing to about 900 mm (28.54 l s^{-1} km^{-2}). High runoff generation is concentrating on the windward side of the Himalaya in the North Eastern Region (NER) of India where heavy monsoon rainfall produces runoff contribution of up to 5200 mm (164.89 l s−1 km−2).

Figure 6.16 indicates that most of the runoff contribution that causes floods in Assam is generated within the NER of India with changing climate and LULC management due to deforestation and draining of wetlands.

6.4.2.5 Recommendations for ILWRM

The results of the DPSIR components delivered by the ISA, the modelling exercises and the WRRU analysis were input to the DIKS developed by Flügel and Busch (2011). They provided important knowledge about the UBRB for designing the ultimate adaptive ILWRM strategy options, which Giannini et al. (2015) priority ranked by means of stakeholder processes.

Flügel (2011c) analysed them with respect to (i) impacts from ongoing climate change; (ii) melting dynamics of glaciers, snowfields and permafrost; (iii) transformation of wetlands into agriculture and settlements; (iv) results obtained from the hydrological rainfall-runoff modelling; and (v) vulnerability analysis and governance assessment as follows:

(i) Station observations and the RCM confirm climate warming and increasing evapotranspiration in the UBRB. Semiarid parts of the UBRB will get dryer and will produce less runoff, which will affect water allocation management especially in Tibet. Higher precipitation in the monsoon-driven NER of India will increase the risk of flooding of valuable farmland as well as riverbank erosion and both need protecting ILWRM solutions.

(ii) Progressing meltwater discharge from glacier and snowfield storages will increase the risk of GLOFs. After the ice and snowfield storage balances have been adapted to the new climate conditions, their runoff contribution will likely be smaller, and water management has to account for these adaptations. With the disappearance of the cementing permafrost, slope stability will weaken, and landslides, mudslides and rock falls will add further sediment to the headwater tributaries in the Himalaya.

(iii) Impacts of wetland drainage for settlements or agriculture are hard to quantify in detail with respect to ILWRM and require detailed eco-hydrological analysis. For small wetlands, the impact on biodiversity might by far outweigh the effects on the hydrological dynamics. In the floodplains of Assam, the drainage of permanent flooded wetlands (Beels) will affect essential ecosystem functions (ESF), i.e. flood retention or biodiversity regeneration as well as ecosystem services (ESS), i.e. food supply from fishery and purification of floodwater or water supply for the local population.

(iv) The modelled future decline of the Brahmaputra River discharge in the NER of India is a serious ILWRM concern and needs further model validation. A comprehensive ISA is required to provide the knowledge for the development of an integrated rainfall-runoff and water allocation model (WAM) for the Brahmaputra River and its tributaries. The latter must comprise the ESF and ESS of Beels and has to integrate aspects of flood and bank erosion protection, hydropower generation, irrigation management, environmental flow requirements and biodiversity. Satellite-based channel braiding analysis and properly designed river training measures are appropriate means to address and analyse riverbank erosion and develop measures for river channelisation in Assam.

(v) IWRM strategy development must support the growth of GDP, which contributes to reduce the flood vulnerability, i.e. by a better maintenance and management of protection measures and water supply management operations. To support its implementation, adaptive ILWRM options must be functional within the governance system in place and have to account for the latter potential to support the 'what-if' scenario storylines.

6.5 Conclusions and Future Research

Changing LULC management and climate have impacts active on different scales, and assessing and analysing these effects on the hydrological dynamics of water balance components in river basins must account for this fact.

The ILWRM approach presented and discussed herein is offering a holistic concept for the assessment and analysis of impacts from changing LULC management and climate in river basins. The methodological ILWRM components are *firstly* the DIKS-enhanced

DPSIR approach that establishes a workflow chain for impact assessment, analysis and response option development and *secondly* the ILMS toolset, which implements the workflow chain by means of geoinformatics tools.

Multi-scale application projects in river basins from South Africa and SE Asia proved the usefulness of the ILWRM approach and its implementing ILMS toolset and due to their NCA design provided details of their methodological potential. Insight into the changing hydrological river basin dynamics was improved and transferred into 'what-if?' scenario storylines by means of integrated indicators providing input for modelling exercises. Modelled time series analysis in turn allowed for the development of ILWRM recommendations for water managers and land use planners.

Future research, e.g. in the Brahmaputra River basin, is required to fine-tune the ILWRM approach and the NCA potential. Funding programmes should put a focus on multi-scale holistic river basin systems projects as they can provide the required synergy from interdisciplinary cooperation to access and analyse impacts of changing LULC management and climate in a global perspective.

Acknowledgement The author acknowledges the financial funding and logistic support received from the European Commission (EC), the German Research Association (DFG), the German Federal Ministry of Education and Research (BMBF), the Volkswagen Association (VW-Stiftung), Mondi Forest Ltd. South Africa and the Deutsche Stifterverband for the research projects presented from South Africa and SE Asia. The support from the research team at the Department of Geoinformatics, Hydrology and Modelling (DGHM) at the Friedrich-Schiller University of Jena in Germany and my cooperating colleagues in the respective research projects is acknowledged as well.

References

Ahrens B (2003) Rainfall downscaling in an Alpine watershed applying a multiresolution approach. J Geophys Res 108:D88388. doi:10.1029/2001JD001485

Alcamo J, Henrichs T (2002) Critical regions, a model-based estimation of world water resources sensitive to global changes. Aquat Sci 64:352–362

Bende-Michl U, Kemnitz D, Helmschrot J, Krause P, Cresswell H, Kralisch S, Fink M, Flügel W-A (2007) Supporting natural resources management in Tasmania through spatially distributed solute modeling with JAMS/J2000-S. In: Kulasiri D, Oxley L (eds) MODSIM 2007 International congress on modelling and simulation. Modelling and Simulation Society of Australia and New Zealand, December 2007: 2354–2360, ISBN: 978-0-9758400-4-7

Bossa AY, Diekkrüger B, Agbossou EK (2015) Scenario-based impacts of land use and climate change on land and water degradation from the meso to regional scale. Water 6:3152–3181. doi:10.3390/w6103152

Calder IR (2005) Integrated land and water resources management. In: Anderson MG (ed) Encyclopedia of hydrological sciences, vol 16. Wiley, Chichester, pp 1–15

Dahlke H, Helmschrot J, Behrens T (2005) A GIS-based terrain analysis approach for inventory of wetland in the semi-arid headwaters of the Umzimvubu basin, South Africa. Göttinger Geogr Abh 113:78–86

DeFries R, Eshleman KN (2004) Land-use change and hydrologic processes: a major focus for the future. Hydrol Process 18:2183–2186

Dobler A, Yaoming M, Sharma N, Kienberger S, Ahrens B (2011) Regional climate projections in two alpine river basins: upper Danube and Upper Brahmaputra. Adv Sci Res 7:11–20

EC, European Commission (2000) Directive 2000/60/EC of the European Parliament and the Council, Official Journal of the European Communities, Brussels, L327/1-L327/72

EEA, European Environmental Agency (1999) Environmental indicators: typology and overview, European Environmental Agency, EEA, Copenhagen, Technical Report No. 25, p 19

EEA, European Environmental Agency (2005) Agriculture and environment in EU-15 – the IRENA indicator report. European Environmental Agency, EEA, Copenhagen, Report No. 4: p 128

EEA, European Environmental Agency (2008a) Modelling environmental change in Europe: towards a model inventory (SEIS/Forward). European Environmental Agency, EEA, Copenhagen, Technical Report, No. 11: p 69

EEA, European Environmental Agency (2008b) Impacts of Europe's changing climate – 2008 indicator-based assessment. European Environmental Agency, EEA, Copenhagen, Report No. 4: p 246

Exler N, Wagner I, Janauer GA (2015) Wetlands and their dynamics. In: Sharma N, Flügel W-A (eds) Applied geoinformatics for sustainable integrated land and water resources management (ILWRM) in the Brahmaputra River basin. Springer, New Delhi, pp 31–35. doi:10.1007/978-81-322-1967-5_6

Fink M, Beisecker R, Kralisch S, Mauden R (2007a) Strategien zur Reduktion des diffusen Stickstoffaustrages aus landwirtschaftlich genutzten Flächen. Forum Hydrol Wasserbewirtsch 1(20.07):63–69

Fink M, Krause P, Kralisch S, Bende-Michl U, Flügel W-A (2007b) Development and application of the modelling system J2000-S for the EU-water framework directive. Adv Geosci 11:123–130

Fink M, Fischer C, Führer N, Firoz AMB, Viet TQ, Laux P, Flügel W-A (2013) Distributive hydrological modelling of a monsoon dominated river system in central Vietnam. In: Piantadosi J, Anderssen, RS, Boland J (eds) MODSIM2013, 20th International congress on modelling and simulation. Modelling and Simulation Society of Australia and New Zealand, December 2013: 1826–1832. ISBN: 978-0-9872143-3-1

Fischer G, Shah M, van Velthuizen H (2002) Climate change and vulnerability. International Institute for Applied Systems Analyis, IIASA, Laxenburg, Austria, p 152

Flügel W-A (1993) River salination due to non-point contribution of irrigation return flow in the Breede River, Western Cape Province, South Africa. Water Sci Technol 28:193–197

Flügel W-A (1995a) Delineating hydrological response units by geographic information system analyses for regional hydrological modelling using PRMS/MMS in the drainage basin of the river Bröl, Germany. Hydrol Process 9:423–436

Flügel W-A (1995b) River salination due to dryland agriculture in the Western Cape Province, Republic of South Africa. Environ Int 21(5):679–686

Flügel W-A (1996) Hydrological Response Units (HRU) as modelling entities for hydrological river basin simulation and their methodological potential for modelling complex environmental process systems Results from the Sieg catchment. DIE ERDE 127:42–62

Flügel W-A (2000) Systembezogene Entwicklung regionaler hydrologischer. Modellsysteme Wasser Boden 52(3):14–17

Flügel W-A (2010) Climate impact analysis for IWRM in Man-made landscapes: applications for Geoinformatics in Africa and Europe. Initiativen Umweltschutz 79:101–134

Flügel W-A (2011a) Development of adaptive IWRM options for climate change mitigation and adaptation. Advances in Science & Research 7:91–100. doi:10.5194/asr-7-1-2011, http://www.adv-sci-res.net/7/index.html

Flügel W-A (2011b) Twinning European and South Asian river basins to enhance capacity and implement adaptive integrated water resources management approaches – results from the EC-project BRAHMATWINN. Advances in Science & Research 7:1–9. doi:10.5194/asr-7-1-2011, http://www.adv-sci-res.net/7/index.html

Flügel W-A (2011c) Geoinformatics for comprehensive impact assessment and analysis of climate change for integrated water resources management. In: Joshi PK, Singh TP (eds) Geoinformatics for climate change studies. TERI Press, p 492, chapter 6, ISBN 9788179934098

Flügel W-A, Bartosch A (2011) Analysis of present IWRM in the Upper Brahmaputra and the Upper Danube River Basins. Adv Sci Res 7:21–36

Flügel W-A, Busch C (2011) Development and implementation of an integrated water resources management system (IWRMS). Adv Sci Res 7:83–90

Flügel W-A, Märker M (2003) The response units concept and its application for the assessment of hydrological related erosion processes in catchments of Southern Africa. ASTM STP 1420:163–177

Flügel W-A, Busch C, Sharma N (2015a) Integrated land and water resources management system (ILWRMS). In: Sharma N, Flügel W-A (eds) Applied geoinformatics for sustainable integrated land and water resources management (ILWRM) in the Brahmaputra River basin. Springer, New Delhi, pp 67–70. doi:10.1007/978-81-322-1967-5_6

Flügel W-A, Pechstädt J, Flemming A (2015b) Applying the Response Units (RU) Concept for ILWRM. In: Sharma N, Flügel W-A (eds) Applied geoinformatics for sustainable integrated land and water resources management (ILWRM) in the Brahmaputra River Basin – results from the EC-project BRAHMATWINN. Springer, India, p 45–52. ISBN 978-81-322-1966-8

Giannini V, Giupponi C (2011) Integration by identification of indicators. Adv Sci Res 7:55–60

Giannini V, Ceccato L, Hutton CW, Allan AA, Kienberger S, Flügel W-A, Giupponi C (2011) Development of responses based on IPCC and "what-if?" IWRM scenarios. Adv Sci Res 7:71–81

Giannini V, Allan A, Hutton CW, Giupponi C (2015) Adaptive IWRM responses to cope with "What-If?" scenarios. In: Sharma N, Flügel W-A (eds) Applied geoinformatics for sustainable integrated land and water resources management (ILWRM) in the Brahmaputra River basin. Springer, New Delhi, pp 61–66. doi:10.1007/978-81-322-1967-5_6

Gordon TJ, Pease A (2006) RT Delphi: an efficient, "Round-less", almost real time Delphi method. J Technol Forecast Soc Chang 73(4):321–333

GWP-TAC (2000) Integrated water resources management. TAC Background Paper, No. 4: p 67

GWP-TEC, Global Water Partnership – Technical Committee (2004) "...Integrated Water Resources Management (IWRM) and Water Efficiency Plans by 2005" why, what and how? TEC Background Paper, No. 10: p 45

Heathcote IW (1998) Integrated watershed management. Wiley, New York, p 414

Helmschrot J (2006a) An integrated, landscape-based approach to model the formation and hydrological functioning of wetlands in semiarid headwater

catchments of the Umzimvubu River, South Africa. Sierke Verlag, Göttingen, p 314. ISBN 3-933893-75-5
Helmschrot J (2006b) Assessment of temporal and spatial effects of landuse changes on wetland hydrology: a case study from South Africa. In: Kotowski W, Maltby E, Miroslaw–Swiatek D, Okruszko T, Szatylowicz J (eds) Modelling, monitoring, management, in wetlands. Taylor & Francis/A.A. Balkema Publisher, London/Dordrecht, pp 197–204
Helmschrot J, Flügel W-A (2002) Land use characterization and change detection analysis for hydrological model parameterisation of large scale afforested areas using remote sensing. Phys Chem Earth 27:711–718
Hoffmann T, Thorndycraft VR, Brown AG, Coulthard TJ, Damnati B, Kale VS, Middelkoop H, Notebaert B, Walling DE (2011) Human impact on fluvial regimes and sediment flux during the Holocene: review and future research agenda. Glob Planet Chang 72:87–98
Hutton CW, Kienberger S, Johnson FA, Allan A, Giannini V, Allen R (2011) Vulnerability to climate change: people, place and exposure to hazard. AVR 7:37–45
IPCC, Intergovernmental Panel on Climate Change (2000) Emissions scenarios. A special Report of IPCC Working Group III, p 27
IPCC, Intergovernmental Panel on Climate Change (2007a) Climate Change (2007) The physical science basis. Contribution of Working Group I to the Fourth Assessment Report of the IPCC. http://www.ipcc.ch/ipccreports/ar4-wg1.htm
IPCC, Intergovernmental Panel on Climate Change (2007b) Climate Change 2007, Impacts, adaption and vulnerability. Contribution of Working Group II to the Fourth Assessment Report of the IPCC. http://www.ipcc.ch/ipccreports/ar4-wg2.htm
IPCC, Intergovernmental Panel on Climate Change (2007c) Climate Change 2007, Mitigation of climate change. Contribution of Working Group III to the Fourth Assessment Report of the IPCC. http://www.ipcc.ch/ipccreports/ar4-wg3.htm
IUCN (2003) Change. Adaptation of water management to climate change. IUCN, Gland, p 53
Jain SK, Singh VP (2003) Water resources system planning and management. Elsevier, Amsterdam, p 858
Kääb A, Frauenfelder R, Sossna I (2015) Glacier changes and permafrost distribution. In: Sharma N, Flügel W-A (eds) Applied geoinformatics for sustainable integrated land and water resources management (ILWRM) in the Brahmaputra River basin. Springer, New Delhi, pp 25–30. doi:10.1007/978-81-322-1967-5_6
Kienberger S, Hutton CW, Johnson FA (2015) Vulnerability assessment and scenarios. In: Sharma N, Flügel W-A (eds) Applied geoinformatics for sustainable integrated land and water resources management (ILWRM) in the Brahmaputra River basin. Springer, New Delhi, pp 53–59. doi:10.1007/978-81-322-1967-5_6
Klein J, Frei C, Gurtz J, Lüthi D, Vidale PL, Schär C (2005) Hydrologic simulations in the Rhine basin driven by a regional climate model. J Geophys Res 110:D04102. doi:10.1029/2004JD005143
Kralisch S, Fischer C (2012) Model representation, parameter calibration and parallel computing – the JAMS approach. In: Seppelt R, Voinov AA, Lange S, D. Bankamp D (eds) Proceedings of the international congress on environmental modelling and software, Sixth Biennial Meeting. Leipzig. http://www.iemss.org/iemss2012/proceedings/D3_1_0802_Kralisch_Fischer.pdf
Kralisch S, Krause P (2007) JAMS – a framework for natural resource model development and application. In: Voinov A, Jakeman A, Rizzoli AE (eds) Proceedings of the iEMSs third biannual meeting "Summit on Environmental Modelling and Software", Burlington, USA, July 2006. Int. Env. Modelling and Software Society
Kralisch S, Fink M, Flügel W-A, Beckstein C (2003) A neural network approach for the optimization of watershed management. Environ Model Softw 18(8–9):15–23
Kralisch S, Fink M, Beckstein C (2005) Neural network based sensitivity analysis of natural resource models. In: Zerger A, Argent RM (eds) Proceedings MODSIM 2005, December 2005: 2498–2504
Kralisch S, Krause P, Fink M, Fischer C, Flügel W-A (2007) Component based environmental modelling using the JAMS framework. In: Kulasiri D, Oxley L (eds) Proceedings of the MODSIM 2007 international congress on modelling and simulation. Modelling and Simulation Society of Australia and New Zealand, December 2007
Kralisch S, Böhm B, Böhm C, Busch C, Fink M, Fischer C, Schwartze C, Selsam P, Zander F, Flügel W-A (2012) ILMS – a software platform for integrated water resources management. In: Seppelt R, Voinov AA, Lange S, D. Bankamp D (eds) Proceedings of the international congress on environmental modelling and software, Sixth Biennial Meeting. Leipzig. http://www.iemss.org/iemss2012/proceedings/I2_2_0734_Kralisch_et_al.pdf
Kralisch S, Zander F, Flügel W-A (2013) OBIS – a data and information management system for the Okavango Basin. In: Oldeland J, Erb C, Finckh M, Jürgens N (eds) Biodiversity and ecology, 5: 213–220. doi:10.7809/b-e.00276
Krause P (2002) Quantifying the impact of land use changes on the water balance of large catchments using the J2000 model. Phys Chem Earth 27:663–673
Krause P, Flügel W-A (2005) Model integration and development of modular modelling systems. Adv Geosci 4:1–2
Krause P, Hanisch S (2009) Simulation and analysis of the impact of projected climate change on the spatially

distributed water balance in Thuringia, Germany. Adv Geosci 7:1–16

Krause P, Bende-Michl U, Bäse F, Fink M, Flügel W-A, Pfennig B (2006) Investigations in a Mesoscale Catchment. – Hydrological modelling in the Gera Catchment. Adv Geosci 9:53–61

Krause S, Bronstert A, Zehe E (2007) Groundwater–surface water interactions in a North German lowland floodplain–implications for the river discharge dynamics and riparian water balance. J Hydrol 347(3):404–417

Kumar KR, Sahai AK, Kumar KK, Patwardhan SK, Mishra PK, Revadekar JV, Kamaka K, Pant CB (2006) High-resolution climate change scenarios for India for the 21st century. Curr Sci 90(3):334–345

Künne A, Fink M, Kipka H, Krause P, Flügel W-A (2012) Regionalization of meso-scale physically based nitrogen modelling outputs to the macro-scale by the use of regression trees. Adv Geosci 31:15–21

Lang S, Kääb A, Pechstädt J, Flügel W-A, Zeil P, Lanz E, Kahuda D, Frauenfelder R, Casey K, Füreder P, Sossna I, Wagner I, Janauer G, Exler N, Boukalova Z, Tapa R, Lui J, Sharma N (2011) Assessing components of the natural environment of the Upper Danube and Upper Brahmaputra river basins. Adv Sci Res 7:21–36

Marke T (2008) Development and application of a model interface to couple regional climate models with land surface models for climate change risk assessment in the Upper Danube Watershed. Dissertation LMU München: Fakultät für Geowissenschaften, Digitale Hochschulschriften der LMU München: p 188, http://edoc.ub.uni-muenchen.de/9162

Mustafa YM, Amin MSM, Lee TS, Shariff ARM (2005) Evaluation of land development impact on a tropical watershed hydrology using remote sensing and GIS. J Spat Hydrol 5(2):16–30, Fall

Nejadhashemi AP, Wardynski BJ, Munoz JD (2011) Evaluating the impacts of land use changes on hydrologic responses in the agricultural regions of Michigan and Wisconsin. Hydrol Earth Syst Sci Discuss 8:3421–3468

Nepal S, Krause P, Flügel W-A, Fink M, Fischer C (2014) Understanding the hydrological system dynamics of a glaciated alpine catchment in the Himalayan region using the J2000 hydrological model. Hydrol Process 28(3):1329–1344. doi:10.1002/hyp.9627

Oleson KW, Lawrence DM, Bonan GB, Flanner MG, Kluzek E, Lawrence PJ, Levis S, Swenson SC, Thornton PE, Dai A, Decker M, Dickinson R, Feddema J, Heald CL, Hoffman F, Lamarque J-F, Mahowald N, Niu G-Y, Qian T, Randerson J, Running S, Sakaguchi K, Slater A, Stockli R, Wang A, Yang Z-L, Zeng X, Zeng X (2010) Technical description of version 4.0 of the Community Land Model (CLM). NCAR technical note NCAR/TN-478+STR. National Center for Atmospheric Research, Boulder, p 257

Pfennig B, Kipka H, Wolf M, Fink M, Krausem P, Flügel W-A (2009) Development of an extended spatially distributed routing scheme and its impact on process oriented hydrological modelling results. IAHS Publ 333:37–43

Prasch M (2010) Distributed process oriented modelling of the future impact of glacier melt water on runoff in the Lhasa River Basin in Tibet. Dissertation, LMU München: Fakultät für Geowissenschaften, Digitale Hochschulschriften der LMU München. p 206, http://edoc.ub.uni-muenchen.de/13031/

Prasch M, Marke T, Strasser U, Mauser W (2011) Large scale integrated hydrological modelling of the impact of climate change on the water balance with DANUBIA. Adv Sci Res 7:61–70

Prasch M, Marke T, Strasser U, Mauser W (2015) Large scale distributed hydrological modelling. In: Sharma N, Flügel W-A (eds) Applied geoinformatics for sustainable integrated land and water resources management (ILWRM) in the Brahmaputra River basin. Springer, New Delhi, pp 37–43. doi:10.1007/978-81-322-1967-5_8

Querner EP (2002) Analysis of basin response resulting from climate change and mitigation measures. IAHS Publ 274:77–84

Rahmann MM, Varis O, Kajander T (2004) EU water framework directive vs. integrated water resources management: the seven mismatches. Water Resour Dev 20(4):565–575

Schmidli J, Frei C (2005) Downscaling from GCM precipitation: a benchmark for dynamical and statistical downscaling methods. Int J Climatol 26(3):679–689

Seidel K, Martinec J (2002) NOAA/AVHRR monitoring of snow cover for modelling climate-affected runoff in Ganges and Brahmaputra Rivers. Proc. of EARSeL-LISSIG-Workshop Observing our Cryosphere from Space, Bern, March 11–13:188–200

Sharma N, Flügel W-A (eds) (2015) Applied geoinformatics for sustainable integrated land and water resources management (ILWRM) in the Brahmaputra River basin. Springer, New Delhi, p 70. doi:10.1007/978-81-322-1967-5_3

Steudel T, Bugan R, Kipka H, Pfennig B, Fink M, de Clercq W, Flügel W-A, Helmschrot J (2013) Implementing contour bank farming practices into the J2000 model to improve hydrological and erosion modelling in semi-arid Western Cape Province of South Africa. Hydrol Res 46(2):192–211. doi:10.2166/nh.2013.164

Subba B (2001) Himalayan waters. The Panos Institute, South Asia, 286 p

Thanapakpawin P, Richey J, Thomas D, Rodda S, Campbell B, Logsdon M (2006) Effects of land use change on the hydrologic regime of the Mae Chaem river basin, NW Thailand. J Hydrol 334:215–230

Thapa R, Lang S, Schöpfer E, Kienberger S, Füreder P, Zeil P (2015) Land use/land cover classification of the natural environment. In: Sharma N, Flügel W_A (eds) Applied geoinformatics for sustainable integrated land

and water resources management (ILWRM) in the Brahmaputra River basin. Springer, New Delhi, pp 17–23. doi:10.1007/978-81-322-1967-5_6

Uppala SM, Kållberg PW, Simmons AJ, Andrae U, da Costa B, Fiorini M, Gibson JK, Haseler J, Hernandez A, Kelly GA, Li X, Onogi K, Saarinen S, Sokka N, Allan RP, Andersson E, Arpe K, Balmaseda MA, Beljaars ACM, Van de Berg L, Bidlot J, Bormann N, Caires S, Chevallier F, Dethof A, Dragosavac M, Fisher M, Fuentes M, Hagemann S, Hólm E, Hoskins BJ, Isaksen L, Jansen PAEM, Jenne R, McNally AP, Mahfouf J-F, Morgrette J-J, Rayner NA, Saunders RW, Simon P, Sterl A, Trenberth KE, Untch A, Vasiljevici D, Viterbo P, Woollen J (2005) The ERA-40 re-analysis. Q J R Meteorol Soc 131(612):2961–3012. doi:10.1256/qi.04.176

van Roosmalen L, Sonnenborg TO, Jensen KH (2009) Impact of climate and land use change on the hydrology of a large-scale agricultural catchment. Water Resour Res 45(W00A15):18. doi:10.1029/2007WR006760

Wallace JS, Gregory JP (2002) Water resources and their use in food production systems. Aquat Sci 64:363–375

Wijesekaraa GN, Guptab A, Valeoc C, Hasbanid JG, Marceaue DJ (2010) Impact of land-use changes on the hydrological processes in the Elbow river watershed in southern Alberta. In: Yang DA, Voinov W, Rizzoli AA, Filatova T (eds) International environmental modelling and software society (iEMSs), 2010 International congress on Environmental Modelling and Software Modelling for Environment's Sake, Fifth Biennial Meeting, Ottawa, Canada, Swayne. http://www.iemss.org/iemss2010/index.php?n=Main.Proceedings

Wilby RL, Charles SP, Zorita E, Timbal B, Whetton P, Mearns LO (2004) Guidelines for use of climate scenarios developed from statistical downscaling methods. Supporting material of the intergovernmental panel on climate change, p 27. http://www.ipcc-data.org/guidelines/dgm_no2_v1_09_2004.pdf

Zander F, Kralisch S, Flügel W-A (2013) Data and information management for integrated research – requirements, experiences and solutions. 20th International Congress on Modelling and Simulation, Adelaide, Australia, 1–6 December 2013. Adelaide: 2201–2206. http://www.mssanz.org.au/modsim2013/K5/zander.pdf

Wolfgang-Albert Flügel
Department of Geoinformatics, Hydrology and Modelling, Friedrich-Schiller-University Jena, Jena, Germany

Analysis of Climate Variability in a Part of Brahmaputra River Basin in India

7

Pratibha Warwade, Nayan Sharma, Ashish Pandey, and Bodo Ahrens

Abstract

Regional specific study of the most important climatic variables is essential to reduce the adverse effects of climate change in developing countries. In the present study, an attempt has been made to detect the trends in rainfall and temperature (past and future) time series and the possibility of any rational relationship between the trends and elevation over study region. The study was mainly focused on a part of Brahmaputra river basin in Northeast India, i.e., the Dikhow catchment (area = 3100 km^2). The Dikhow catchment is a set of unique composite catchment topography comprising fairly higher altitude hills as well as alluvial plains at much lower elevation. Historically (from the year 1901 to 2002), annual precipitation and monsoon precipitation are decreasing significantly (5 % level of significance), whereas significant rise was observed on annual and seasonal basis for both maximum and minimum temperature time series. Further, during the entire time series (1901–2002), significant negative correlation was obtained between precipitation (annual and monsoon) trend magnitude and elevation of the study region. Significant rise for future precipitation was identified over the region for both A2 and B2 SRES scenarios, and winter precipitation is likely to increase more (59.18–69.44 % under A2 and 47.71–54.90 % under B2 over the catchment) than summer and monsoon precipitation. The maximum temperature has higher rate of rise than the minimum temperature for annual, summer, and monsoon temperature time series, while winter season shows higher rate of rise rather than the maximum temperature (1.02 °C over 102 years). For diurnal temperature range (DTR), monsoon season shows the significant rising which varies from 0.0016 to 0.0021 °C/year.

P. Warwade (✉) • N. Sharma • A. Pandey
Department of Water Resources Development and Management, Indian Institute of Technology Roorkee, Roorkee, Uttarakhand, India
e-mail: pratibhawarwade@gmail.com

B. Ahrens
Mesoscale Meteorology and Climate Institute of Atmosphere and Environment Sciences/Geozentrum Riedenberg, Goethe-University, Frankfurt am Main, Germany

N. Sharma (ed.), *River System Analysis and Management*, DOI 10.1007/978-981-10-1472-7_7

Extreme temperature indices (TXx, TXn, TNx, and TNn) also show the significant warming picture over the catchment. Trend magnitude of extreme temperature indices and elevation shows significant relationship, while trend magnitude of the average maximum and minimum temperature and DTR does not show the noticeable connection between warming rate and elevation of the study area. Future Tmax and Tmin are increasing sharply for both A2 and B2 scenarios. A2 scenario shows higher rate than the B2 scenario.

7.1 Introduction

India is expected to feel severe impacts of climate change. The country's northeastern region, for instance, is highly vulnerable. Anticipated impacts include melting of glaciers, more floods, and extended droughts (www.dandc.eu). At the same time, the region's natural resources are under enormous pressure due to population growth and rising prosperity. Emphasis of hilly region is on improving the prevailing agricultural practice (Jhum cultivation). In a shifting cultivation system called "Jhum," according to which, plots are used for a while but then abandoned again so they can regain their products. Elevation dependency has been widely cited (Chen et al. 2003; Frauenfeld et al. 2005) and used to explain the rapid warming derived from ice cores (Kang et al. 2007; Tian et al. 2007). Scientific research and field observations confirm that the region is suffering from the impacts of climate change already. Due to its unique location and topography, it has distinct precipitation and drainage patterns. Because of climate change, rainfall is becoming more unpredictable and erratic. The impact on people, fields, and livestock is devastating and set to get worse.

Variations in temperature and rainfall pattern due to anthropogenic activity have raised the interest of researchers as well as apprehension regarding the present and possible future changes at regional scale. Extreme precipitation events and increase in temperature tend to have adverse impacts on agricultural production, ecology, and biodiversity. Temperature and precipitation are the most important climatic parameters due to their influence in the hydrological cycle, quantity of evapotranspiration (Barnett et al. 2005; Song et al. 2009), and both quantity and quality of runoff. Temperature is important for runoff estimation which is also necessary for developing watershed management plans involving soil and water conservation measures (Pandey et al. 2008). Rainfall and temperature are pertinent parameters that are affected by climate change which in turn influences the hydrological cycle. Temperature is a key factor in shaping up climate or weather of a region, and it drives crop growth and crop duration. Local temperature is one of the major climatic elements to record environmental changes brought about by industrialization and urbanization (Gadgil and Dhorde 2005). Rainfall is another important parameter which is quite variable, particularly in the monsoon season (Singh et al. 2010) which regulates the circulation of water in the atmosphere and determines sowing time of rain-fed crops.

Several studies were undertaken to analyze the trends in long-term precipitation and temperature and its inter-annual, seasonal, and decadal variability at different scales such as local, regional, national, and continental spatial scales (Chen et al. 1992; Chaudhari 1994; Kadioglu 1997; Izrael et al. 1997; Mirza and Dixit 1997; Rankova 1998; Ren et al. 2000; Salinger and Griffiths 2001; Wibig and Glowicki 2002; Lu et al. 2004; Domroes and El-Tantawi 2005; Gadgil and Dhorde 2005; Tomozeiu et al. 2006; ElNesr et al. 2010). Literature review reveals that it is essential to know and predict the exact trend of rainfall and temperature pattern for optimal utilization of existing water resources, short-term development goals, and the mitigation of existing social, economic, and environmental problems. Hence, this study investigated the trends in rainfall and temperature (both past and future) and the possibility of any rational relationship between the trends and elevation over study region.

7.2 Materials and Methods

7.2.1 Study Area Description

This study is mainly focused on the Dikhow catchment as a part of larger Brahmaputra river basin; Dikhow River which is originated from the hills of the state Nagaland is adopted as representative of the largest Brahmaputra river basin. It is a south bank tributary of Brahmaputra river contributing 0.7 % runoff. Lower Brahmaputra river basin is a region where the hydrological impact of climate change is expected to be particularly strong and population pressure is high (Gain and Giupponi 2015). Brahmaputra river is the biggest trans-Himalayan river basin (Sharma and Flügel 2015). Figure 7.1 presents the location map and demonstrates with the digital elevation map (DEM) of the study area situated between 94° 28′49″E to 95° 09′ 52″ E longitude and 26° 52′ 20″N to 26° 03′ 50″ N latitude. The SRTM (Satellite Radar Topographical Mission) data of the Dikhow catchment was downloaded from Global Land Cover Facility (GLCF) Internet site and used in the preparation of the DEM. Digital elevation map of the study area depicted the elevation which ranges from 28 to 2651 m from m.s.l. in the selected catchment. The catchment is a set of unique composite catchment topography comprising fairly higher altitude hills as well as alluvial plains at much lower elevation. Geographical area of Dikhow catchment is about

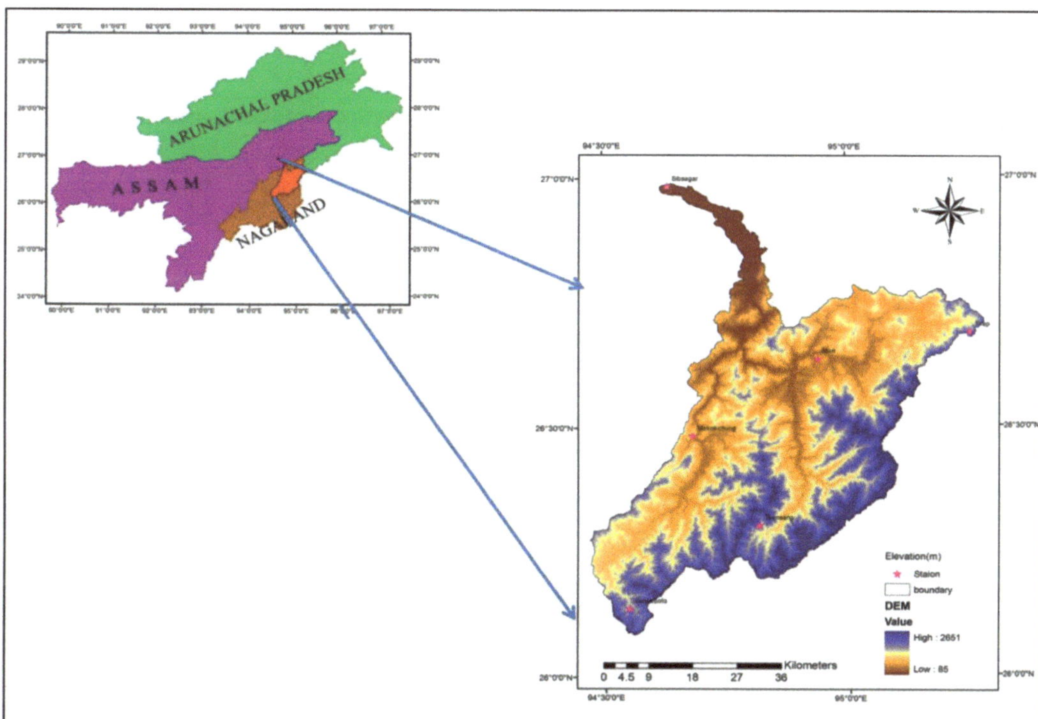

Fig. 7.1 Location map of Dikhow catchment

Table 7.1 Selected stations of the study area categorized under different zones of elevation

S.N.	State	Station	Latitude	Longitude	Altitude (m)	Zone
1	Assam	Sibsagar	26° 5′ 42″	94° 37′ 42″	98	LEZ
2	Nagaland	Mokokchung	26° 19′ 20″	94° 30′ 53″	1323	HEZ
3	Nagaland	Mon	26° 43′ 21″	94°01′ 52″	720	MEZ
4	Arunachal Pradesh	Tirap	26° 59′ 43″	95° 32′ 27″	776	MEZ
5	Nagaland	Tuensang	26° 14′ 9″	94° 48′ 46″	1570	HEZ
7	Nagaland	Zunheboto	26° 00′ 35″	94° 31′42″	1818	HEZ

3100 km^2 which encompasses around 85 % of Nagaland, 10 % of Assam, and 5 % of Arunachal Pradesh.

7.2.2 Climate

The average annual rainfall of the study area varies from 1960 to 2156 mm and 1448–1536 mm during the monsoon season. The mean monthly maximum temperature (Tmax) and minimum temperature (Tmin) of the study area range from 25.4–32 °C to 10–23 °C, respectively. August is the hottest and January is the coldest month in the study area. In the study area, three seasons, viz., winter (November–February), summer (March–May), and monsoon (June–October), are experienced. The prevailing winds at 8:30 and 17:30 h generally blow from N–NE sector toward S–SW sector throughout the year. The annual average wind speed is 3.8 km/h with April and July having the highest mean wind speed of 5.4 km/h and December having the lowest mean wind speed of 1.7 km/h.

7.2.3 Elevation Zones in Study Area

The study area is remote and largely inaccessible, highly diversified in terms of elevation (elevation ranges from 28 to 2651 m from m.s.l.) as dotted with hills and small mountains, and land cover changes by practicing *shifting* cultivation (slash and burn) predominantly. And hence a composite catchment (study area) is divided into three physiographic zones, namely, low elevated zone (LEZ), moderate elevated zone (MEZ), and high elevated zone (HEZ), because it is expected that each zone will respond differently. Stations below 100 m from m.s.l. fall under LEZ also known as alluvial plain, stations between 101 and 1000 mm elevation from m.s.l. cover the MEZ, and stations between 1001 and 2000 m elevations from m.s.l. were classified as HEZ. Table 7.1 presented the stations considered for the present study with their geographical location, classified topographic zone, and other details. Their zones were specifically defined to identify the relationship between trend magnitude (temperature and precipitation) and elevation.

7.2.4 Details of Data

7.2.4.1 Meteorological Data

Monthly precipitation and temperature data of six stations covering 102 years (1901–2002) were downloaded from the India Meteorological Department (IMD) site India Water Portal (http://www.indiawaterportal.org/metdata). Location of meteorological stations is presented in Fig. 7.1.

Outliers play a key role in parametric test and in assessing the magnitude of possible changes by mean and the linear regression. In the present research, data series are plotted to detect the outliers. After visual detection of outliers, the suspected values are calculated using the normal ratio method. November, December, January, and February were considered as winter, while computing the values of winter season, and December of the previous year was included. March, April, and May were considered as the summer season and June to October considered monsoon season. For precipitation, monthly values were summed to obtain annual and seasonal values, and for temperature, the monthly data were averaged to provide annual and seasonal values of each year.

7.2.4.2 Large-Scale Atmospheric Variables

A monthly reanalysis dataset of the National Centers for Environmental Prediction (scale of 2.5° latitude × 2.5° longitude) re-gridded on scale of 2.5° latitude × 3.75° longitude and simulated data of HadCM3 on a scale 2.5° latitude × 3.75° longitude were downloaded from the website http://www.cics.uvic.ca/scenarios/sdsm/select.cgi. The data were extracted for two different socioeconomic scenarios both characterized by regionally focused development but with priority to economic issues in one (A2 scenario) and to environmental issues in the other (B2 scenario).

7.2.5 Methodology

Steps of the procedure adopted are discussed below:

1. The preliminary analysis of annual precipitation and temperature series were computed for each station over the period of 1901–2002.
2. All times series were checked for autocorrelation (Anderson 1941).
3. Mann-Kendall (MK) test and Sen's slope estimator (Duhan et al. 2013; Warwade et al. 2015) were applied to the whole time series (as per need) to detect the direction and magnitude of a trend.
4. CUSUM, cumulative deviations (CD) test, and linear regression were applied to detect the abrupt changes in series.
5. MK test was applied again in the subseries identified by CUSUM test.
6. Magnitude of changes was quantified by Sen's slope estimator.
7. Sen's slope estimator was used for computing the percentage change.
8. Precipitation magnitude of change in percentage was computed by multiplying the slope and length of years, and it was divided by the mean value of precipitation for each station for all series. Spatial distribution of trends on annual and seasonal basis was carried out by Kriging interpolation technique using Arc GIS 9.3.

Multiple linear regression (MLR) is used as downscaling technique of GCM data (Murphy 1999; Schoof and Pryor 2001; Hay and Clark 2003) to predict future precipitation and temperature time series. Multiple linear regression (MLR) is a statistical method that is used to model a linear relationship between a dependent variable (predictand) and one or more independent variables (predictors). MLR is a least-square-based method, and predictand is a continuous variable. MLR assumes that the relationship between variables is linear.

7.3 Results and Discussion

7.3.1 Precipitation

7.3.1.1 Preliminary Analysis

Nonsignificant autocorrelation was identified in all time series; Fig. 7.2 shows the sample of annual and monsoon autocorrelation test for all stations, and it is indicated that the nonsignificant autocorrelation (significance if test statistics – z-values more than ± 1.96) for all stations in the study area. Hence, MK test was applied directly in all the time series.

Monsoon precipitation contributed highest to annual precipitation among all seasons in the study area. Contribution of average monsoon precipitation to average annual precipitation over 102 years for six stations is represented in Fig. 7.3. Figure 7.3 shows that the average precipitation contribution of the monsoon season is 73 % of the total average annual precipitation with individual stations experiencing not less than 71 %. There is a strong influence of monsoon precipitation to the annual average precipitation as observed in past 102 years over the catchment. Contribution of monsoon precipitation to annual precipitation is highest at HEZ (Tuensang) and lowest at LEZ (Sibsagar).

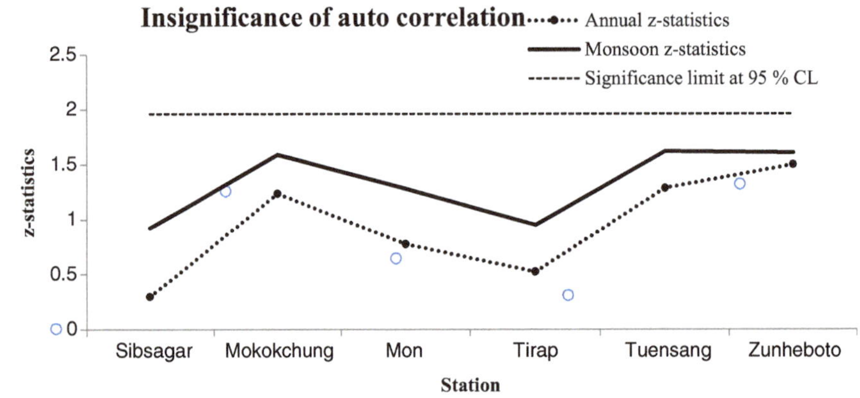

Fig. 7.2 Sample of nonsignificant autocorrelation obtained in annual and monsoon precipitation series (1901–2002) for all stations

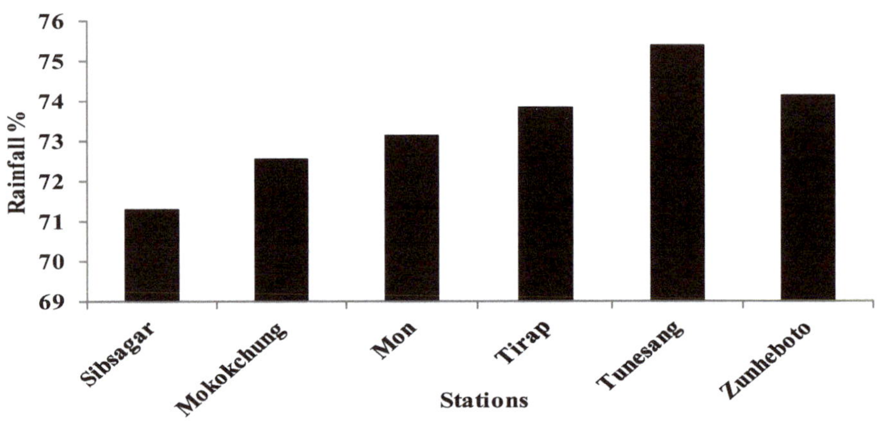

Fig. 7.3 The average monsoon precipitation contribution to average annual precipitation

7.3.1.2 Trends in Historical Precipitation Over 1902–2002

7.3.1.2.1 Direction and Magnitude of Trend

Table 7.2 illustrates the result obtained from the Mann-Kendall and Sen's slope estimator test to find out the direction and magnitude of trends for annual and seasonal precipitation (Chung and Yoon 2000) in the study area. Bold values of z-statistics in the table show the significant trend at 5 % significance level.

Annual Precipitation

Annual trend analysis of historical precipitation of the entire period (1901–2002) indicated that there are three stations, namely, Sibsagar, Mon, and Tirap, having the significant decreasing trend. Significantly falling rate of annual precipitation trend was highest at Sibsagar which is 1.846 mm/year followed by 1.817 mm/year at Tirap and 1.764 mm/year at Mon. Sibsagar is the representative of the low elevated zone of the study area, having the highest rate of decreasing trend (1.817 mm/year), and then moderate elevated zone (Tirap and Mon) shows the next higher rate of decreasing trend. Results of analysis also show that the stations situated over the higher elevation zone (HEZ) have not indicated any significant trend, although it shows the nonsignificant decreasing trend in annual

Table 7.2 Results of Mann-Kendall test and Sen's slope estimator test (1901–2002)

	Annual		Monsoon		Summer		Winter	
Station/season	*z*-value	*β*	*z*-value	*B*	*z*-value	*β*	*z*-value	*β*
Sibsagar	**−2.036**	−1.846	**−2.278**	−1.787	0.266	0.023	−0.596	−0.043
Mokokchung	−1.787	−1.596	**−2.088**	−1.593	0.058	0.065	−0.572	−0.048
Mon	**−2.024**	−1.764	**−2.359**	−1.721	0.179	0.006	−0.636	−0.043
Tirap	**−2.128**	−1.817	**−2.348**	−1.761	0.376	0.078	−0.775	−0.056
Tuensang	−1.619	−1.295	−1.949	−1.332	−0.133	−0.088	−0.619	−0.041
Zunheboto	−1.509	−1.294	−1.839	−1.371	0.052	0.065	−0.515	−0.033

precipitation. Trend magnitude of annual precipitation shows that the elevation plays a key role in changing rate of historical rainfall.

Seasonal Precipitation

From Table 7.2, the seasonal precipitation especially during monsoon season, the trends of precipitation are significantly negative (Subash and Sikka 2014) for four stations, i.e., Sibsagar, Mon, Tirap, and Mokokchung. The rate of decreasing trend is highest at Sibsagar (1.787 mm/year) followed by Tirap (1.761 mm/year), Mon (1.721 mm/year), and then Mokokchung (1.593 mm/year). Monsoon precipitation also showed the same behavior like annual precipitation, as the low elevated station (Sibsagar) has the highest magnitude of decreasing trend than MEZ. In addition, the next elevated station Mokokchung (HEZ) also shows the significantly decreasing trend, but the rate of trend is smaller than MEZ and LEZ.

Furthermore, no significant trend in winter and summer precipitation was observed for all the stations. However, nonsignificant negative and positive trends are obtained for winter and summer precipitation, respectively, for almost all the stations (all zones) in the study area.

7.3.1.2.2 Spatial Analysis of Annual and Seasonal Precipitation

Spatial analysis of annual and seasonal precipitation was carried out using GIS tool (Chowdary et al. 2008; Chowdhury et al. 2009; Warwade et al. 2014) which is an excellent alternative to conventional mapping techniques in monitoring and mapping. As per the above discussion, the monsoon precipitation is playing a major role in contribution of the average annual precipitation (discussed in Sect. 3.1.1 and Fig. 7.3). Figure 7.4a–d denotes the spatial distribution of annual, monsoon, winter, and summer precipitation in terms of percentage change, and Fig. 7.4a, b shows spatial distribution of annual and monsoon precipitation over the catchment and displayed similar spatial pattern over the region; magnitude was higher over the northeast part and less over the southwest part of the catchment. It indicates whether any change in monsoon precipitation will directly affect the annual precipitation over the region. Spatial distribution of winter and summer precipitation is differing with spatial behavior of annual and monsoon precipitation (Fig. 7.4c, d).

Table 7.3 shows the percentage change in precipitation over 102 years, bold value shows the changes greater than 10 %. The maximum percentage change in precipitation was found to be −12.41 at Tirap station in the monsoon season over 102 years which has overall impact of decreasing the annual precipitation by −9.45 %. Further, decreasing trends of −11.78 %, −11.75 %, and −10.78 % in the monsoon precipitation are obtained for the Sibsagar, Mon, and Mokokchung.

The zone of low elevation experienced more reduction of precipitation followed by moderate elevated zone, while high elevated zone is not affected by such a greater amount of reduction in precipitation over the catchment.

7.3.1.2.3 Probable Change Year During 1901–2002

Figure 7.5a–d shows the deviation of annual and seasonal precipitation (monsoon, summer, and

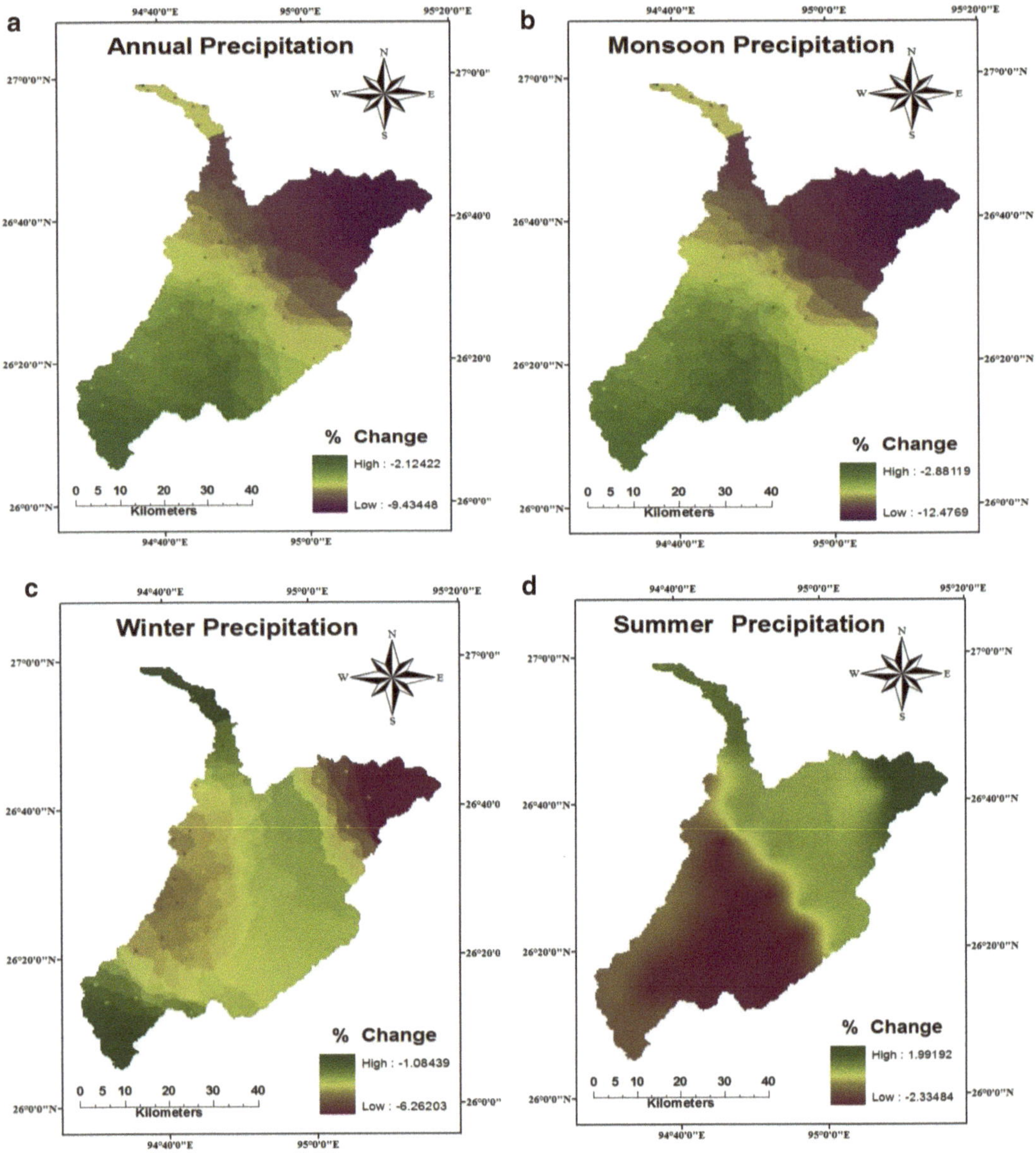

Fig. 7.4 Spatial distribution of annual (**a**) and seasonal precipitation (monsoon (**b**), winter (**c**), and summer (**d**)) in terms of % change over 102 years

Table 7.3 Percentage change by mean over 102 years of precipitation

S.N.	Station	Annual	Monsoon	Summer	Winter	Remarks (zone)
1	Sibsagar	−8.74	**−11.86**	0.46	−4.28	LEZ
2	Mokokchung	−7.83	**−10.78**	−1.39	−5.42	HEZ
3	Mon	−8.80	**−11.75**	−0.13	−4.70	MEZ
4	Tirap	−9.45	**−12.41**	1.88	−6.29	MEZ
5	Tuensang	−6.74	−9.15	−2.25	−4.86	HEZ
6	Zunheboto	−6.49	−9.28	−1.52	−3.88	HEZ

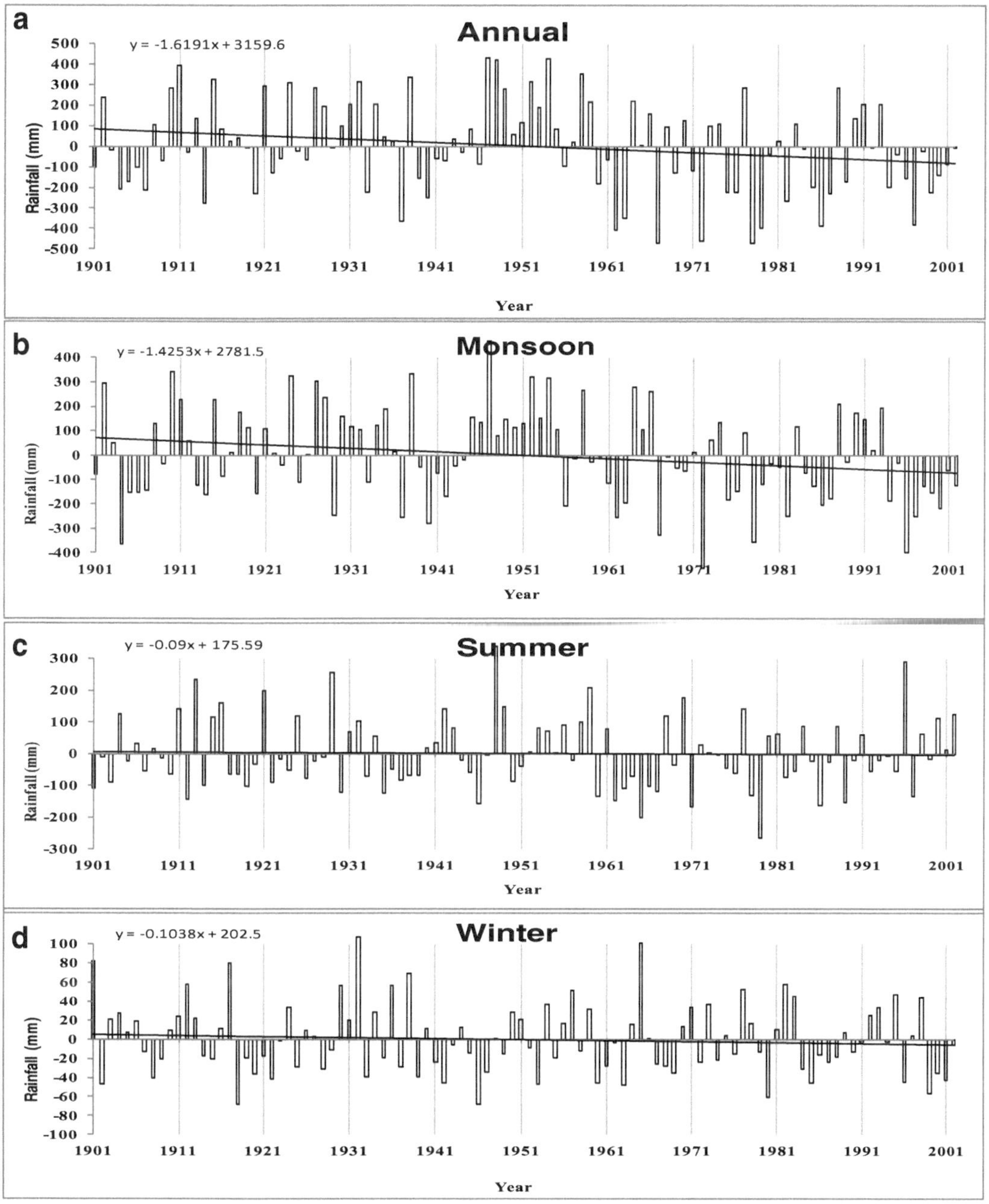

Fig. 7.5 Trend of annual (**a**) and seasonal precipitation (**b–d**) anomalies during 1901–2002 by the linear regression method

winter) from the mean value using linear regression method. Figure 7.5a, b depicts the trend of annual and monsoon precipitation anomalies, clearly showing prominent decreasing trend after 1959. Winter season (Fig. 7.5d) also shows decrease in precipitation, while for summer season (Fig. 7.5c) trend is not so clear. Before 1959 the anomalies were positive and after that anomalies are negative in annual and seasonal (monsoon and winter) precipitation.

Table 7.4 shows the results of probable change point obtained from CUSUM and cumulative

Table 7.4 Change point in precipitation series from 1901 to 2002

	CUSUM			Cumulative deviation		
Station	Year	Maximum deviation	Significance at $\alpha = 0.05$ LS	Year	Q/Sqrt (n)	Significance at $\alpha = 0.05$ LS
Sibsagar	**1959**	15	Sign	**1959**	1.7	Sign
Mokokchung	1959	11	Non sign	**1959**	1.7	Sign
Mon	1959	13	Non sign	**1959**	1.83	Sign
Tirap	**1959**	17	Sign	**1959**	1.774	Sign
Tuensang	1959	9	Non sign	**1959**	1.7	Sign
Zunheboto	1974	10	Non sign	**1959**	1.676	Sign

Table 7.5 Results of trend test in divided time series of precipitation

	Annual		Monsoon		Summer		Winter	
Stations	z-value	β	z-value	β	z-value	β	z-value	β
1901–1959								
Sibsagar	1.36	2.013	0.562	0.654	1.569	1.186	−0.026	−0.102
Mokokchung	**2.433**	3.295	1.857	2.456	1.556	1.184	−0.275	−0.011
Mon	1.74	2.611	1.09	1.644	1.543	0.963	−0.052	−0.064
Tirap	0.772	1.605	0.63	0.202	**2.533**	2.866	0.021	−0.18
Tuensang	**2.537**	3.614	1.792	2.796	1.308	0.804	−0.314	−0.015
Zunheboto	**2.747**	4.087	**2.093**	3.168	1.465	1.001	−0.314	−0.021
1960–2002								
Sibsagar	−0.712	−1.72	−0.167	−0.154	**2.302**	3.409	−0.251	0.039
Mokokchung	−0.188	−0.083	−0.963	−2.286	1.884	2.523	0.356	0.008
Mon	−0.44	−0.083	−0.523	−2.286	**2.177**	2.523	0.272	0.008
Tirap	−0.772	−1.605	0.63	0.202	**2.533**	2.866	0.021	0.18
Tuensang	−0.021	−0.162	−1.067	−1.241	1.905	2.892	0.502	0.08
Zunheboto	0	−0.029	−1.277	−2.617	1.926	2.176	0.523	0.148

deviation test for precipitation; values in bold show significant at 5 % level of significance. According to both the tests, year 1959 is identified as the most probable change point in the annual rainfall during the entire time series (1901–2002); results are similar to the study of Goyal (2014) who found that the most probable year of change was 1959 in annual precipitation which was carried out in Assam, Northeast India, while linear regression method also showed the agreement on the 1959 as a probable change point in the entire time series. The changing pattern of rainfall is dissimilar before and after the change point, these results having agreement with Basistha et al. (2009) who also found an increasing trend up to change point followed by a decreasing trend after change point over Indian Himalayas. Hence, to quantify the trend of annual and seasonal precipitation with respect to most probable change point, all the precipitation time series (1901–2002) were divided into two segments, i.e., the years 1901–1959 as first and years 1960–2002 as second.

Table 7.5 shows the results of trend test obtained for both the segments (before and after the change point). Significance of trend indicated by bold values is marked in Table 7.5. During the first segment (years 1901–1959), for annual analysis of precipitation, positive trend was observed, significant at three stations (i.e., Mokokchung, Tuensang, and Zunheboto) or high elevation zone (HEZ). The maximum increase in precipitation (4.087 mm/year) was found at Zunheboto (station of HEZ) station. In seasonal analysis, only Zunheboto (station of HEZ) observed significant positive trend in monsoon (2.093 mm/year) and Tirap (2.553 mm/year) in summer season. Overall positive trends were observed in first times series (years 1901–1959) except winter season.

Table 7.6 Precipitation trend over different zones

Zones	Before change point (1901–1959)	After change point (1960–2002)
Annual precipitation		
LEZ	Increases nonsignificantly	Decreases nonsignificantly
MEZ	Increases nonsignificantly	Decreases nonsignificantly
HEZ	**Increases significantly for all stations**	Mixed trend but nonsignificant
Monsoon precipitation		
LEZ	Increases nonsignificantly	Decreases nonsignificantly
MEZ	Increases nonsignificantly	Decreases nonsignificantly
HEZ	**Increases significantly for one station only**	Decreases nonsignificantly
Summer precipitation		
LEZ	Increases nonsignificantly	**Increases significantly**
MEZ	**Increases significantly at one station**	**Increases significantly**
HEZ	Increases nonsignificantly	Increases nonsignificantly
Winter precipitation		
LEZ	Decreases nonsignificantly	Increases nonsignificantly
MEZ	Decreases nonsignificantly	Increases nonsignificantly
HEZ	Decreases nonsignificantly	Increases nonsignificantly

In the second segment (years 1960–2002), three stations of LEZ and MEZ (Sibsagar, Mon, Tirap) showed significant increasing trend in summer season with a maximum increase of 2.533 mm per year, whereas annual and remaining seasonal precipitation series were not indicating any significance of trend over the catchment. Although in annual and monsoon, nonsignificant decreasing trends were observed in all stations. Variation of precipitation trend over different zones is summarized in Table 7.6.

Table 7.6 (bold values show significance) shows the significant rise of annual and monsoon precipitation over HEZ and one station of MEZ before change point, whereas after change point significant rise of summer precipitation was observed over LEZ and MEZ. Variation of precipitation trend in relation with elevation and time was clearly identified over the region, and it is the indication of changing pattern of rainfall with respect to topography. Specially, annual precipitation increases earlier over HEZ, and later on summer precipitation increases over LEZ and MEZ.

Moreover, 10 % rise or fall of annual precipitation is also a matter of concern; thus, extreme events (in terms of more or less than 10 % of precipitation from mean) were identified before and after the change point and found that during the years 1901–1959, 23.3 % positive extreme events (greater than 2317 mm) and 15 % negative extreme events (less than 1897 mm) were observed, while during 1960–2002, positive extreme events were 23.3 %, but the negative events increased up to 18.6 %; results confirm that of annual precipitation decreased have been more in recent years from the former years.

Table 7.7 Correlation between rates of change of precipitation trend with altitude

Parameter	Annual	Monsoon	Summer	Winter
1901–2002				
r	**−0.921**	**−0.911**	−0.119	−0.449
P-value	**0.009**	**0.012**	0.058	0.372
1901–1959				
r	**0.878**	**0.872**	−0.299	−0.014
P-value	**0.021**	**0.023**	0.564	0.798
1960–2002				
r	−0.079	**−0.636**	−0.751	0.456
P-value	0.059	0.175	0.085	0.363

7.3.1.3 Relationship Between Rainfall Trend and Elevation

Possibility of any rational relationship between the rainfall trend magnitude and elevation is discussed in this section. Table 7.7 shows the results obtained from the correlation analysis between rate of change of precipitation and elevation in Dikhow catchment; bold value shows the significant correlation for three time slices 1901–2002, 1901–1959, and 1960–2002. During the entire time series (1901–2002), significant negative correlation was obtained between precipitation (annual and monsoon) trend magnitude and elevation; further, no significant correlation was observed between summer and winter precipitation trend with elevation during 1901–2002.

Correlation analysis of segment one (1901–1959) shows the significantly positive correlation of elevation with annual and monsoon precipitation trend. And for the second segment (1960–2002), monsoon precipitation indicates the significant negative correlation with elevation; further, no significant correlation was observed between elevation and remaining annual and seasonal precipitation time series trend magnitude.

Scatter plot between precipitation trend magnitude and elevation for both segments (1901–1959 and 1960–2002) and for the entire time series (1901–1959) is presented in Figs. 7.6, 7.7, and 7.8, respectively.

For the first segment (1901–1959), Fig. 7.6 shows the magnitude of annual and monsoon precipitation trend is positive and it increases from LEZ to HEZ, i.e., trend magnitude is increasing with increase in elevation. In case of summer and winter precipitation, this relationship was just opposite in which it was decreasing from LEZ to HEZ, although the correlation between elevation and precipitation trend was significant for annual and monsoon season only.

For the second segment (1960–2002), Fig. 7.7 shows the trend magnitude of monsoon precipitation significantly increases with increase in elevation (from LEZ to HEZ). Magnitude of annual and summer precipitation decreases with increase in elevation, while trend magnitude of winter precipitation increases with increases in elevation, although for annual, summer, and winter, this relationship was nonsignificant.

Figure 7.8 represents linear relationship between the rate of change of precipitation trend magnitude and elevation for the entire time series 1901–2002. Trend magnitude of annual and monsoon precipitation decreases

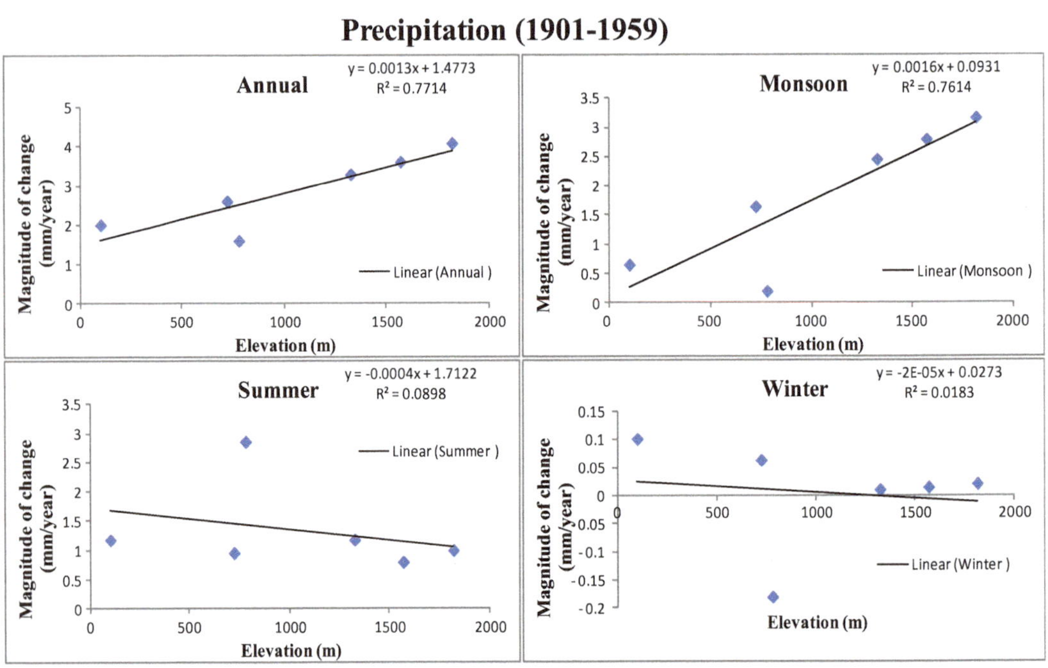

Fig. 7.6 Relationship between rainfall trends and elevation during 1901–1959

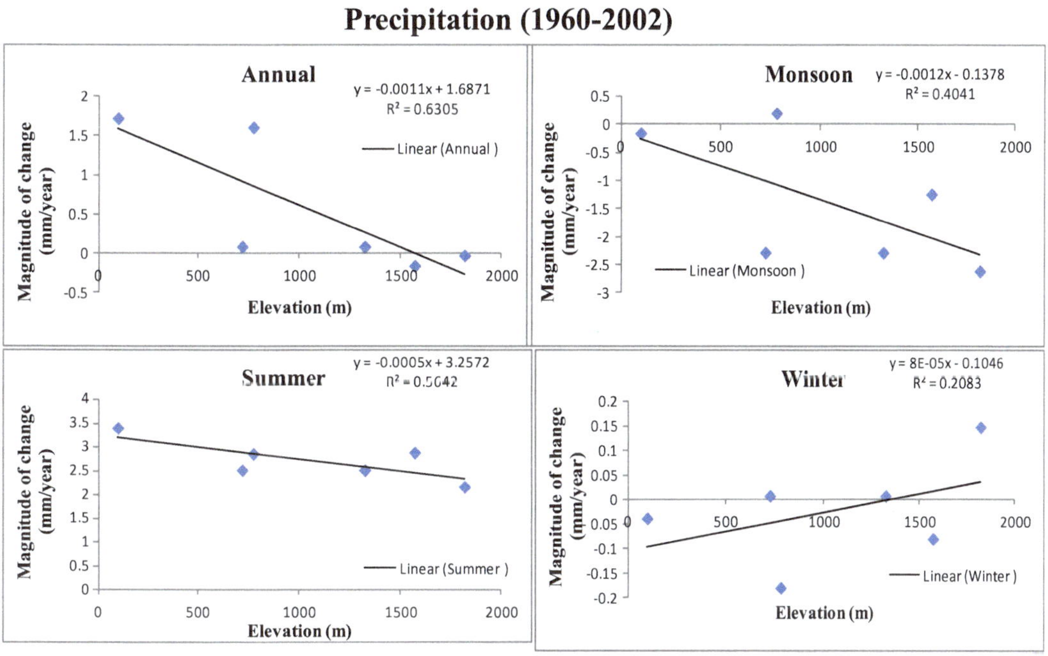

Fig. 7.7 Relationship between rainfall trends and elevation during 1960–2002

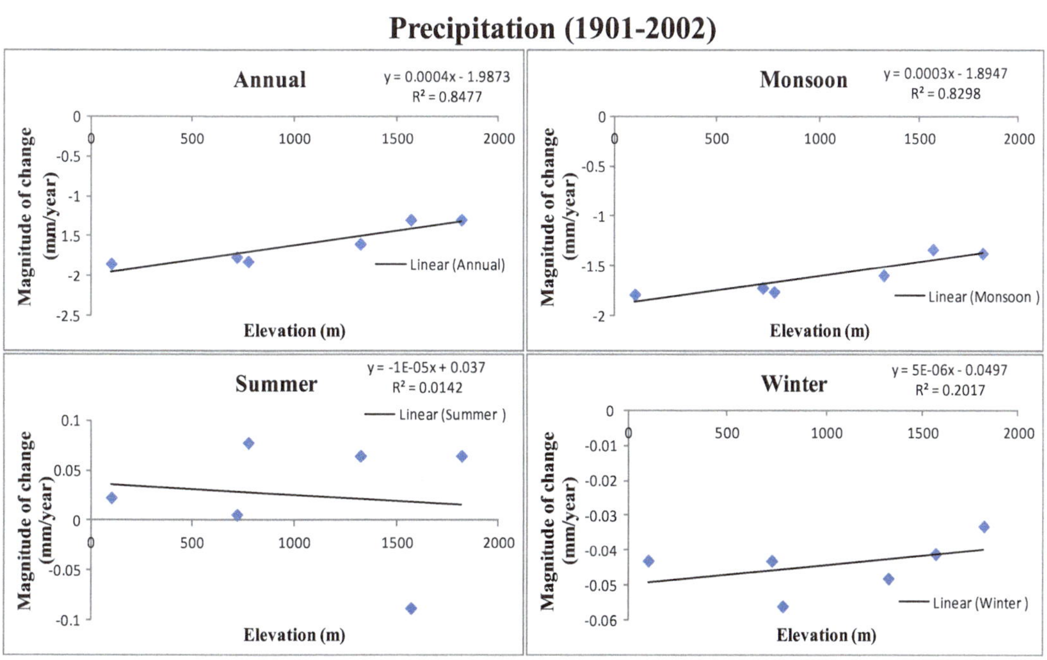

Fig. 7.8 Relationship between rainfall trends and elevation during 1901–2002

Table 7.8 Results of trend test for annual and monsoon precipitation for the years 2003–2099

	Annual A2			Annual B2		
Station	Mean (mm)	Z-value	% change	Mean (mm)	Z-value	% change
Sibsagar	2480.8	**10.692**	30.98	2480.8	**9.933**	28.01
Mokokchung	2449.6	**10.685**	30.68	2449.6	**9.855**	26.87
Mon	2409.1	**10.748**	31.95	2409.1	**9.799**	28.01
Tirap	2432.7	**10.564**	31.67	2432.7	**9.763**	27.34
Tuensang	2345.9	**10.692**	34.10	2345.9	**9.699**	28.50
Zunheboto	2403.6	**10.678**	31.69	2403.6	**9.813**	27.47
	Monsoon A2			Monsoon B2		
Station	Mean (mm)	Z-value	% change	Mean (mm)	Z-value	% change
Sibsagar	1875.9	**10.240**	28.12	1731.8	**9.703**	27.76
Mokokchung	1849.4	**10.219**	27.75	1730.7	**9.787**	27.34
Mon	1826.9	**10.302**	28.73	1707.4	**9.675**	27.96
Tirap	1888.7	**10.254**	28.15	1736.8	**9.842**	27.96
Tuensang	1843.5	**10.268**	29.71	1680.3	**9.675**	28.59
Zunheboto	1860.9	**9.650**	28.12	1708.2	**9.787**	27.84

Table 7.9 Results of trend test for summer and winter precipitation from 2003 to 2099

	Summer A2			Summer B2		
Station	Mean (mm)	Z-value	% change	Mean (mm)	Z-value	% change
Sibsagar	647.7	**5.158**	24.76	596.9	**5.158**	20.68
Mokokchung	619.2	**6.029**	24.53	574.7	**4.810**	20.32
Mon	602.1	**6.280**	26.32	555.5	**4.949**	20.99
Tirap	594.5	**5.820**	26.29	553.5	**4.642**	21.16
Tuensang	574.6	**6.190**	28.59	523.6	**4.642**	22.44
Zunheboto	599.7	**6.190**	25.58	554.2	**4.719**	21.16
	Winter A2			Winter B2		
Station	Mean (mm)	Z-value	% change	Mean (mm)	Z-value	% change
Sibsagar	151.8	**6.385**	59.188	151.5	**5.068**	54.90
Mokokchung	132.2	**6.503**	69.443	143.6	**4.580**	51.81
Mon	140.6	**6.371**	66.832	145.5	**4.893**	54.71
Tirap	134.7	**6.260**	68.731	141.8	**4.147**	47.71
Tuensang	138.1	**6.329**	68.961	141.4	**4.614**	51.32
Zunheboto	141.3	**6.404**	66.640	140.5	**4.454**	51.83

with increase in elevation significantly. Further, summer and winter trend magnitude also decreases with increase in elevation nonsignificantly. Decreasing rate was highest over LEZ and lowest over HEZ for annual and summer precipitation.

7.3.1.4 Trends in Future Precipitation (2003–2099)

Trend analysis was also carried out for the future annual and seasonal precipitation time series of the catchment for all stations considering A2 and B2 scenarios, and results are presented in Tables 7.8 and 7.9. Results of Mann-Kendall test show that trend for both A2 and B2 scenarios showed significantly positive trend for annual and seasonal precipitation. Trends have been more pronounced to increase in A2 scenario rather than B2 scenario.

Table 7.8 also shows the percentage change of future precipitation from the mean value of precipitation. 30.98–34.1 % enhancement for A2 and 26.87–28.50 % for B2 scenario have been projected for future annual precipitation. Annual

Table 7.10 Comparison between past and future trend of precipitation

Percentage change of future average precipitation from past average precipitation					
Season	Average past (1901–2002)	Average (future A2, 2003–2099)	Average (future B2, 2003–2099)	% change from avg. past A2	% change from avg. past B2
Annual	2038.67	2520.28	2420.28	**19.11**	**15.77**
Monsoon	1496.33	1857.55	1715.87	**19.45**	**12.79**
Winter	91.83	139.78	144.05	**36.30**	**36.25**
Summer	451.67	606.30	559.73	**25.50**	**19.31**

precipitation is likely to enhance more under A2 scenario than B2 scenario in the study area.

In monsoon precipitation, the rise of 27.75–29.71 % for A2 and 27.34–28.59 % for B2 scenario from the mean value of precipitation has been projected in the future. Monsoon precipitation may rise almost equal for both A2 and B2 scenarios.

In summer precipitation, the rise of 24.54–28.59 % for A2 and 20.32–22.44 % for B2 scenario from the mean value of precipitation has been projected in the future. The rise in summer precipitation is larger in A2 rather than B2 scenario.

Greater enhancement has been predicted in winter precipitation, and it may likely to increase by 59.18–69.44 % under A2 and 47.71–54.90 % under B2 over the catchment. Noticeable difference was observed in the increase of winter precipitation for A2 and B2 scenarios.

Table 7.10 illustrates the comparison between past and future precipitation pattern, and it is found that future trend of precipitation is totally different from the historical precipitation. Trends have been more pronounced to increase in A2 scenario rather than B2 scenario. The maximum percentage change in precipitation was observed in winter season (36.30 %), followed by summer (25.50 %), monsoon (19.45 %), and annual (19.11 %) with respect to the historical period. Depiction of historical precipitation reveals (Sect. 3.1.1) the monsoon rainfall has a major contribution in annual precipitation (monsoon season plays a significant role), but very small amount of winter and summer season precipitation. In contrast, Table. 7.10 shows that in the future, great increment of winter and summer precipitation was projected than the monsoon precipitation. It may change the constituents of the average annual rainfall, by which the small amount of winter and summer (in past average annual precipitation) is likely to come out as the prominent contributor to the average annual rainfall in the future.

According to the present analysis, precipitation is most likely to increase more in the future (until 2099), and this is reversed from the trends of historical precipitation analyzed for 102 years.

Agriculture is vulnerable to climatic changes (Haskett et al. 2000); analysis reveals that future precipitation will be more in winter season and it would be beneficial for the Rabi crops; the present study agreed with Giupponi et al. (1998) who stated that the change in the distribution of rainfall seems to favor autumn-winter crops (wheat) at the expense of summer crops (soybean and maize). This increase of precipitation in winter and summer can be utilized for the enhancement of agriculture production by changing the current cropping pattern and crop variety, with improved technology.

Present research revealed the spatial and temporal variability of annual and seasonal historical precipitation for a catchment diversified with hills and several lofty elevations. Hence, the present study has been carried out considering the three physiographic zones, LEZ, MEZ, and HEZ, based on the elevation of the stations; the results of the analysis are distinct for each three topographic zones.

The decreasing trend of annual rainfall is found significant in LEZ and MEZ, and there is no significance of annual precipitation trend observed over HEZ. The rate of change of significantly decreasing trend is higher in LEZ (1.846 mm/year) than MEZ. Trends of seasonal precipitation confirm the trend of precipitation being influenced by the elevation. In case of monsoon precipitation, LEZ shows the highest rate of deceasing trend, the declining rate of precipitation increases with the decrease in elevation, i.e., declining rate is 1.787 mm/year at LEZ having 98 m elevation (Sibsagar) followed by 1.761 mm/year to 1.721 mm/year at MEZ (below 1000 m) and 1.593 mm/year at 1323 m (Mokokchung) of HEZ, and there is no significant trend observed onward over HEZ.

In addition, future precipitation (2003–2099) is also the main focus of the present research, and temporal trend analysis of future precipitation shows the clear-cut rising trend, in the catchment for both annual and seasonal bases. The rising trend of precipitation may decrease the water scarcity or may increase the flood risk in the future. Timely rainfall may surely reduce the water shortage and can improve the economy of the region. The present study reveals that the winter precipitation is rising for all the stations, leading to reduction of the use of irrigation water; this may increase the groundwater level which has been reduced by the overexploitation of groundwater sources for irrigation and other purposes of water use. Rising of precipitation in the future needs proper planning and development of hydraulic structures to get benefited.

7.3.2 Temperature

Understanding of trend in climatic parameter at mountainous region is very important for climate change research; in the present study, six stations of Dikhow catchment are heterogeneous in terms of elevation which have been considered for the analysis of temperature trend from 1901 to 2002. In this section, this mixed topography is examined for the trend analysis of the maximum and minimum temperature, and observations based on the trends are made thereafter.

7.3.2.1 Preliminary Analysis

The statistical parameters, i.e., mean, standard deviation, skewness, kurtosis, and coefficient of variation of annual maximum and minimum temperature, have been computed for each station and are presented in Table 7.11. The annual maximum temperature varied from 22.26 to 27.03 °C and minimum temperature from 12.72 °C to 17.54 °C at Tuensang and Sibsagar in the study area. The station of HEZ shows lowest and LEZ shows highest temperature over the catchment.

Standard deviation varied from 0.41 to 0.52 for annual maximum temperature and 0.37–0.41 for minimum temperature over 102 years, respectively. The kurtosis varied from 0.08–0.57 to 0.67–2.46 for annual maximum and minimum temperature, respectively. Skewness ranged from 0.41–0.57 to 0.60–1.20 for annual maximum and minimum temperature. The skewness is positive for all the stations and higher for the annual minimum temperature than annual maximum temperature. Positive skewness for both the temperatures indicates asymmetry during the time period and it lies to the right of the mean.

The coefficient of variation (CV) varied from 1.71–2.34 % to 2.36–3.00 % for maximum and minimum temperature, respectively; the average CV of the study area is 2.71 % for the minimum

Table 7.11 Description of statistical parameters of annual maximum and minimum temperature for the study period (1901–2002)

	Annual Tmax					Annual Tmin				
Station	SD	Mean	KURT	SKEW	CV (%)	SD	Mean	KURT	SKEW	CV (%)
Sibsagar	0.46	27.03	0.48	0.55	1.71	0.41	17.54	2.44	1.11	2.36
Mokokchung	0.43	25.22	0.08	0.44	1.71	0.39	15.79	2.62	1.20	2.49
Mon	0.45	24.43	0.39	0.52	1.84	0.40	14.85	2.38	1.11	2.72
Tirap	0.52	22.46	0.57	0.57	2.34	0.37	12.83	0.67	0.60	2.86
Tuensang	0.42	22.26	0.19	0.41	1.89	0.39	12.72	2.27	1.13	3.00
Zunheboto	0.41	23.18	0.17	0.43	1.79	0.38	13.79	2.46	1.19	2.78

temperature which is highest. CV is maximum at high elevation zone of stations. From the statistical test, it can be concluded that the high elevation zone is more vulnerable to the climate change than the low elevation zone.

7.3.2.2 Analysis of Temperature Anomalies

The temperature anomalies have been computed for better understanding of the observed trends in the annual and seasonal (winter, summer, and monsoon) time series of minimum and maximum temperature over 102 years (1901–2002) of the study area, depicted in Figs. 7.9 and 7.10, respectively. Figure 7.9 shows the annual and seasonal maximum temperature anomalies during 102 years; it indicates the clear-cut warming trend over the catchment, approximate half of the period experienced negative anomalies, while remaining half of the period experienced positive anomalies for almost all the series of the maximum temperature. Figure 7.10 illustrates the rising trend of annual and seasonal minimum temperature over the Dikhow catchment; it also indicates the positive values of anomalies after the mid-century at both annual and seasonal time scales. Quantification and significance of trend are discussed in the next section based on the results obtained from Mann-Kendall and Sen's slope estimator test.

7.3.2.3 Historical Trend Analysis of Temperature

7.3.2.3.1 Annual Temperature Trend

Results obtained from trend analysis are presented in Table 7.12 (bold values show significant trend at 5 % significance level) for annual maximum and minimum temperature for each station of the study area. Significantly, the rising trend of maximum and minimum temperature (Subash and Sikka 2014; Jeganathan and Andimuthu 2013) was identified for each station. The rate of annual maximum temperature varied from 0.007 to 0.008 °C/year for all the stations, and there is no large variation in the rate of change of trend in annual maximum temperature over the study area. In the same way, the rate of annual minimum temperature varied from 0.005 to 0.006 °C/year over 102 years in the Dikhow catchment. A significantly average increasing rate of 0.76 °C and 0.56 °C over 102 years in the average annual maximum temperature and minimum temperature was observed. The rate of increasing trend of annual maximum temperature was higher than the rate of the annual minimum temperature in the study area. Results of the present analysis confirmed the investigation of Bhutiyani et al. (2007) and Deka et al. (2009) in which they observed higher rate of increase in Tmax than Tmin over some parts of India.

7.3.2.3.2 Seasonal Temperature Trend

Table 7.12 also shows results of Mann-Kendall test and Sen's slope estimator test for seasonal temperature (maximum and minimum temperature). Seasonal maximum and minimum temperature shows the significantly rising trend at all stations of the study area. In the summer season, the magnitude of rising trend of maximum temperature varied from 0.004 °C/year (Tirap) to 0.009 °C/year (Sibsagar), and the trend magnitude of minimum temperature ranges from 0.006 °C/year (Tirap) to 0.008 °C/year (Sibsagar).

The monsoon season is the most important period for regional agricultural productivity, and the temperature pattern during this period is critical for the growth of crops. In the study area, the magnitude of trend varies from 0.003 °C/year (Mokokchung) to 0.008 °C/year (Sibsagar, Mon) in maximum temperature, while the trend magnitude varies from 0.004 °C/year (Mokokchung, Tirap, Tuensang, and Zunheboto) to 0.005 °C/year (Sibsagar and Mon) for minimum temperature. In the monsoon season, the trend of maximum temperature is also greater than the minimum temperature.

During the winter season, trend was significantly positive for both the temperature patterns.

The trend magnitude of winter maximum temperature varies from 0.004 °C/year (Tirap) to 0.009 °C/year (Sibsagar, Mokokchung, and Mon), while the trend magnitude of winter minimum temperature varies from 0.009 °C/year (Mokokchung and Zunheboto) to 0.012 °C/year

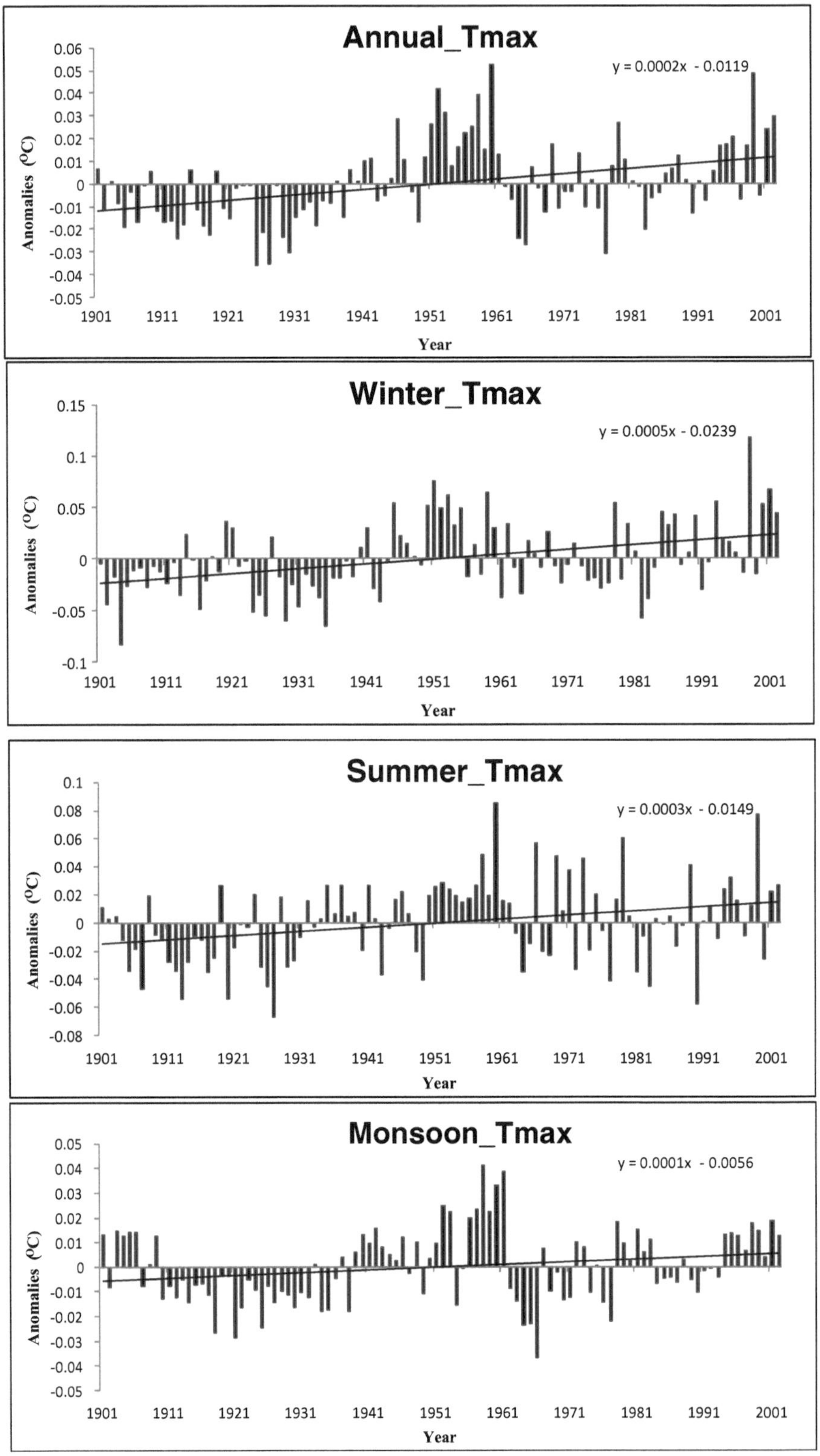

Fig. 7.9 Anomalies in annual and seasonal maximum temperature at Dikhow catchment

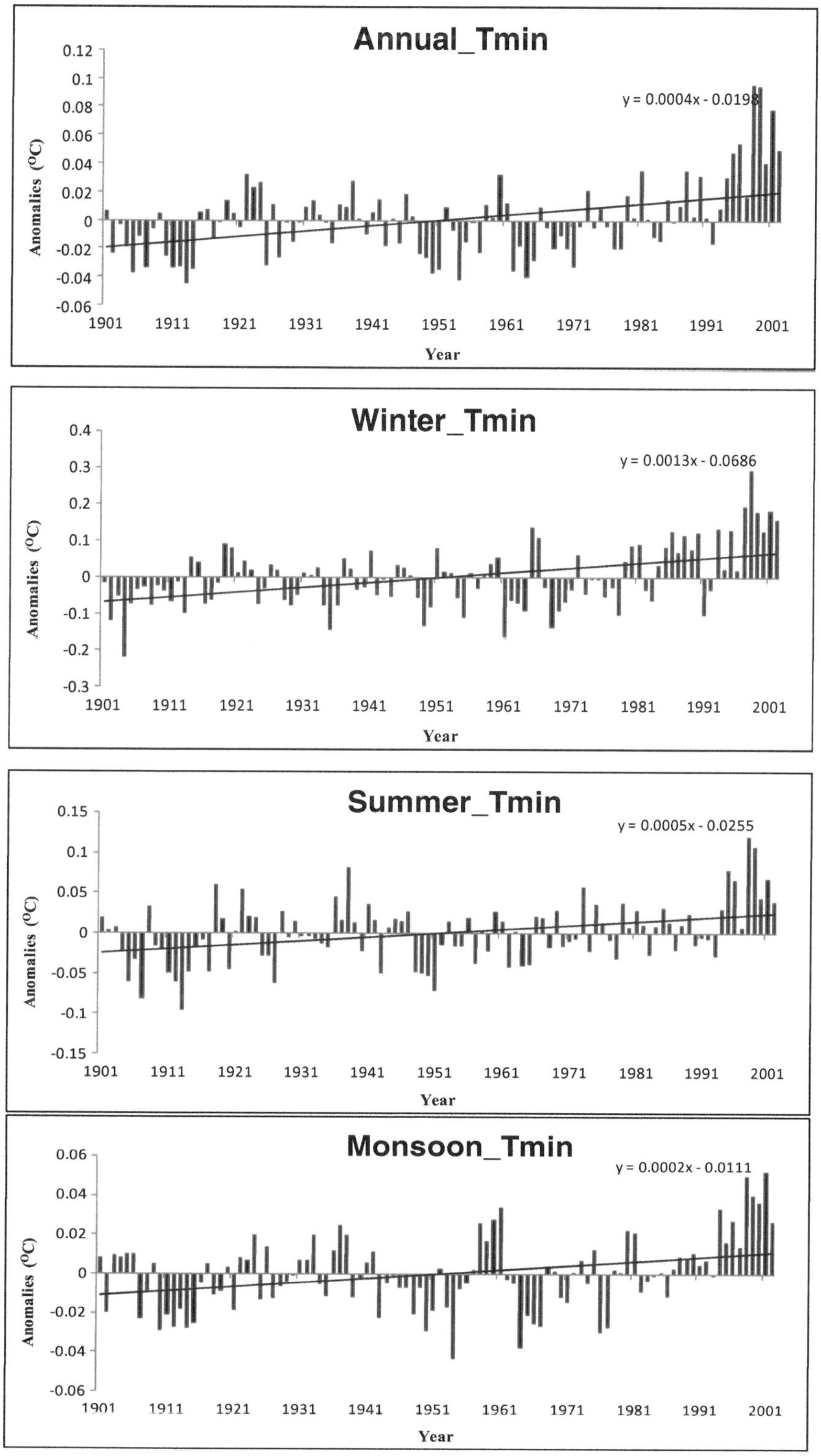

Fig. 7.10 Anomalies in annual and seasonal minimum temperature at Dikhow catchment

Table 7.12 Trend results for maximum and minimum temperature

Seasons	Annual		Winter		Summer		Monsoon	
Station	z-value	β (°C/year)	z-value	β (°C/year)	z-value	β (°C/year)	z-value	β (°C/year)
Maximum temperature (1901–2002)								
Sibsagar	**5.436**	0.008	**5.056**	0.009	**3.683**	0.009	**5.037**	0.008
Mokokchung	**5.314**	0.008	**3.645**	0.009	**4.898**	0.005	**3.152**	0.003
Mon	**5.401**	0.008	**3.745**	0.009	**3.735**	0.009	**5.170**	0.008
Tirap	**5.141**	0.007	**2.517**	0.004	**3.363**	0.004	**4.629**	0.007
Tuensang	**5.306**	0.007	**3.668**	0.008	**3.622**	0.008	**5.155**	0.007
Zunheboto	**5.158**	0.007	**3.565**	0.008	**3.596**	0.008	**5.121**	0.007
Minimum temperature trend (1901–2002)								
Sibsagar	**4.453**	0.005	**4.155**	0.01	**3.776**	0.008	**3.397**	0.005
Mokokchung	**4.239**	0.006	**3.935**	0.009	**3.496**	0.007	**3.154**	0.004
Mon	**4.453**	0.006	**4.111**	0.01	**3.715**	0.007	**3.166**	0.005
Tirap	**4.629**	0.006	**4.983**	0.012	**2.923**	0.006	**2.985**	0.004
Tuensang	**4.175**	0.005	**4.003**	0.01	**3.553**	0.007	**2.987**	0.004
Zunheboto	**4.356**	0.005	**3.918**	0.009	**3.553**	0.007	**2.981**	0.004

(Tirap). Winter temperature is showing a quite different behavior among other seasons in which the magnitude of the rising trend is higher than the magnitude of maximum temperature at most of the stations of the study area. The average rate of the rising trend of minimum temperature is 1.02 °C over 102 years observed in the Dikhow catchment which was highest among all.

Annual and seasonal scale analysis is very helpful to identify the more vulnerable epoch to climate change. The present study clearly indicates that the winter season is more prominent to warming. Further, winter minimum temperature having the highest rate of rising, supported by the increase in cloudiness, as per the proposition of Easterling et al. (1997), is possibly one of the main contributing factors for the observed increases in nighttime temperature. Roy and Balling (2005) observed that the rising Tmin in NE India was a result of increasing cloud cover in winter, and the view was also supported by diurnal temperature range (DTR) which decreased at Tocklai in February and Chuapara in winter.

In the entire catchment, the annual maximum and minimum temperature has been increased by 0.75 °C and 0.55 °C, respectively, over 102 years. The present study has reflected that the winter season is more prominent to the warming; it shows the greatest changes among all the seasons and even on annual time scale. Winter minimum and maximum temperatures have been increased by 1.02 °C and 0.83 °C over 102 years in the Dikhow catchment.

7.3.2.3.3 Diurnal Temperature Range (DTR) Trend

Results obtained from MK test and Sen's slope estimator test to identify trend in DTR are presented in Table 7.13. Bold values in Table 7.13 show the significant trend at 5 % level of significance. The trend analysis of diurnal temperature range (DTR) was also analyzed on annual and seasonal basis. Insignificant trends were observed for annual and seasonal DTR except the monsoon season. The monsoon season shows the significant trend for four stations of the study area, and the magnitude of significant rising DTR in the monsoon season varies from 0.0016 to 0.0021 °C/year. This outcome is confirming the results of Jhajharia and Singh (2011) and Kumar et al. (1994); they reported increase in DTR over the Indian subcontinent, and Fowler and Archer (2006) also documented the large increase in DTR over Hindu Kush mountains during the years 1961–2000.

Table 7.13 Trend results for diurnal temperature range (DTR)

Station	Diurnal temperature							
	Annual		Winter		Summer		Monsoon	
	z-value	Slope	*z*-value	Slope	*z*-value	Slope	*z*-value	Slope
Sibsagar	0.497	0.00012	−0.512	−0.00001	−0.233	0	1.801	0.0016
Mokokchung	0.709	0.00012	−0.674	−0.00060	−0.209	0	**2.113**	0.0017
Mon	0.456	0.00000	−0.630	−0.00026	−0.135	0	**2.125**	0.0021
Tirap	0.374	0.00000	−0.483	−0.00007	−0.377	0	**2.037**	0.0020
Tuensang	0.732	0.00018	−0.821	−0.00059	−0.112	0	**2.107**	0.0016
Zunheboto	0.644	0.00014	−0.803	−0.00075	−0.091	0	1.872	0.0017

Table 7.14 Trend results for extreme temperature

Station	Annual extreme values of maximum temperature					
	TXx			TXn		
	Mean	*z*-value	Slope	Mean	*z*-value	Slope
Sibsagar	30.69	**5.217**	0.011	21.64	**2.048**	0.005
Mokokchung	28.42	**5.211**	0.010	20.18	1.558	0.003
Mon	27.85	**5.243**	0.010	19.36	1.799	0.004
Tirap	26.43	**5.273**	0.011	17.67	**1.978**	0.004
Tuensang	25.31	**5.109**	0.009	17.46	1.576	0.003
Zunheboto	26.16	**5.167**	0.009	18.37	1.499	0.003
	Annual extreme values minimum temperature					
	TXx			TXn		
Station	Mean	*z*-value	Slope	Mean	*z*-value	Slope
Sibsagar	23.56	**3.034**	0.005	9.22	**2.347**	0.006
Mokokchung	21.52	**2.409**	0.003	7.79	1.837	0.005
Mon	20.80	**2.846**	0.004	6.78	**2.013**	0.006
Tirap	19.14	**2.826**	0.004	4.81	**2.195**	0.006
Tuensang	18.40	**2.016**	0.002	4.85	1.711	0.004
Zunheboto	19.36	1.822	0.002	5.97	1.631	0.004

7.3.2.3.4 Extreme Values of Minimum Temperature and Maximum Temperature

The above indices were chosen to reflect different aspects of climate extremes which are important from agricultural point of view used in the recent study of Guan et al. (2015). Numbers of studies were carried out to detect the trend in the average maximum and minimum temperature, and the study of trend detection in extreme values of maximum and minimum temperature has been lacking. In Table 7.14 maximum and minimum values of annual maximum temperature are denoted by TXx and TXn, respectively. Similarly, maximum and minimum values of annual minimum temperature are denoted by TNx and TNn, respectively. Extreme event indices (TXx, TXn, TNx, and TNn) are very important for crop growth and water availability and for overall food production. These indices are reflecting the significant warming picture over the composite catchment topography, the bold values indicate the significant trend presented in Table 7.14.

Each station of the catchment indicates that the warming trend in TXx ranges from 0.009 to 0.011 °C/year in the study area. TXx shows the distinct trend for different zones considered for the present research with the highest elevated zone showing the least rate of warming, while LEZ shows the highest rate of increasing trend.

For TXn, two stations Sibsagar and Tirap show the significant positive trend having the magnitude of 0.005 °C/year and 0.004 °C/year, respectively. LEZ also shows the highest rate of rise in TXn.

Significant warming trend for five stations was observed in TNx, the magnitude varies from 0.002 °C/year to 0.005 °C/year, and significant rising is maximum at LEZ followed by MEZ and HEZ.

In the case of TNn, only three stations Sibsagar (LEZ) and Mon and Tirap (MEZ) showed the significant rising trend. The magnitude ranges from 0.007 to 0.006 °C/year. Rising is maximum at LEZ followed by MEZ, and no significant warming was observed over HEZ.

The trend analysis of absolute indices including TXx, TXn, TNx, and TNn reflected connection between the rate of change of trend and topographic zone considered for the present study. TXx shows the greatest trend over LEZ (1.1 °C during 102 years), followed by MEZ (1 °C during 102 years) and HEZ (0.91 °C during 102 years). In the same manner, the rate of trend of TXn is highest over LEZ (0.5 °C during 102 years) followed by MEZ (0.4 °C during 102 years) and no significant trend identified over HEZ.

TNx shows the greatest magnitude over LEZ (0.5 °C during 102 years) followed by MEZ (0.4 °C during 102 years) and least over HEZ (0.34 °C during 102 years).

TNn also shows strong significant trend over LEZ (0.7 °C during 102 years) followed by MEZ (0.6 °C during 102 years) and no significant trend over HEZ.

Time series of TXx are the maximum value of Tmax from the whole year (including all seasons). It was found that the monsoon season has the highest value of Tmax in the study region. Hence, it can be stated that significant increasing rate of TXx means an increase in daytime temperature during Kharif season, and it may dry the crop plant before maturity during the crop season.

Similarly, time series of TNn are minimum values of Tmin from the whole year (including all seasons), and it was extracted from the winter season. Hence, the rising trend of TNn shows that warmer nights during growth stages of Rabi crops may affect the water supply and photosynthates.

Temporal variability of annual and seasonal temperature (minimum and maximum temperature) indicated that the minimum temperature of winter season has the highest rate of 0.012 °C per year of the rising trend. Mostly Rabi crops like gram, sunflower, wheat, etc. are cultivated during this season which are highly sensitive to the temperature. A rise in winter minimum temperature may lead to a rise in metabolic rate in crop plants leading to net loss of photosynthesis. It may also lead to shortened crop duration and/or increase of crop water demand.

Increasing rate of daytime temperature during the monsoon season (0.009–0.011 °C per year) may increase the water demand for agricultural activity by increasing the evaporation losses.

Overall increasing trend of temperature may increase the water demand for both agriculture and domestic purposes. Due to the rise in warmer trend of temperature, agricultural production may reduce which ultimately affects the economy of the study area.

7.3.2.4 Relationship Between Temperature Trend Magnitude and Elevation

The possibility of any rational relationship between the temperature trend magnitude and elevation (You et al. 2010) is discussed in this section. Results of correlation analysis are presented in Table 7.15; bold values of correlation coefficient show the significant correlation at 5 % level of significance between temperature trends and elevation. No significant correlation could be identified between Tmax, Tmin, and DTR trend with the elevation of the study area. The trend of the average maximum and minimum temperature does not show the noticeable connection between warming rate and elevation, and this result conforms to the general findings of Vuille and Bradley (2000), Pepin and Seidel (2005), Pepin and Lundquist (2008), and You et al. (2010). They also fail to find any relationship between

Table 7.15 Correlation between temperature trend and elevation

Tmax				
Parameter	Annual	Monsoon	Summer	Winter
r	−0.5804	−0.4004	−0.1472	−0.4587
P-value	0.22	0.43147	0.78082	0.36025
Tmin				
Parameter	Annual	Monsoon	Summer	Winter
r	−0.1914	−0.7812	−0.3369	−0.4559
P-value	0.71647	0.06657	0.51213	0.36357
DTR				
Parameter	Annual	Monsoon	Summer	Winter
r	0.19475	0.02218		−0.6322
P-value	0.71157	0.96673		0.178
Temperature extreme				
	TXx	TXn	TNx	TNn
r	**−0.8833**	**−0.9666**	**−0.9855**	**−0.9093**
P-value	0.01963	0.00165	0.00031	0.01196

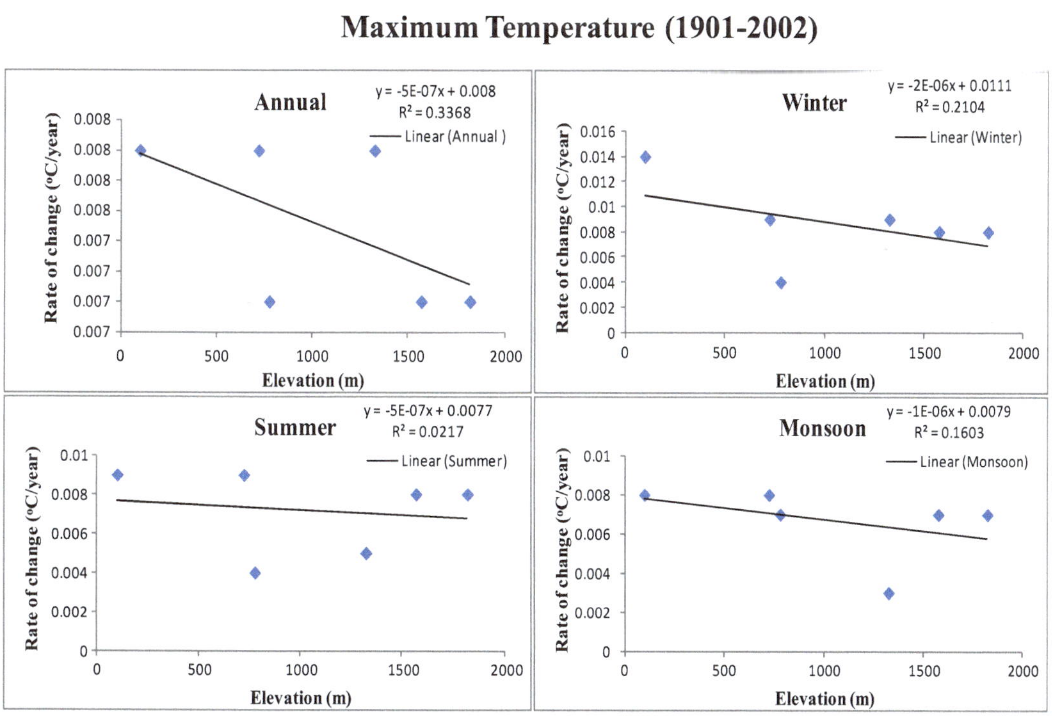

Fig. 7.11 Relationship between rates of change of maximum temperature and elevation

warming amplitude and elevation in various mountainous regions over the globe. Although the rate of maximum and minimum temperature of annual and seasonal basis is decreasing nonsignificantly with increase in altitude as presented in Figs. 7.11 and 7.12, increasing rate of annual DTR increases with elevation (Fig. 7.13), nonsignificant decreasing rate of winter DTR increases with increase in elevation, while the monsoon DTR does not show noticeable relationship between rate of change and elevation.

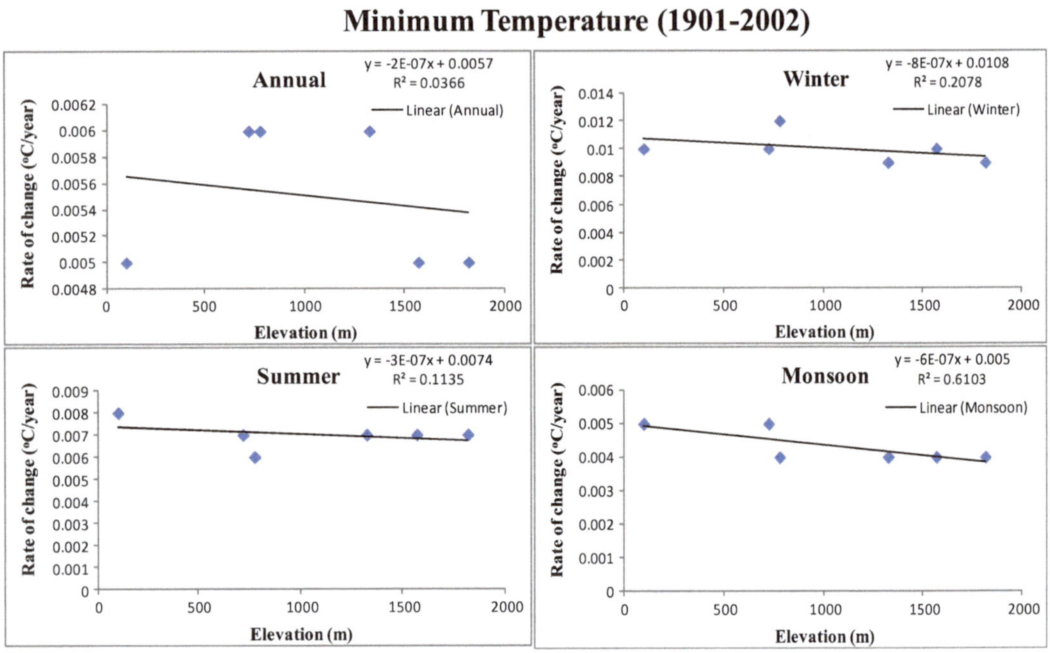

Fig. 7.12 Relationship between rates of change of minimum temperature and elevation

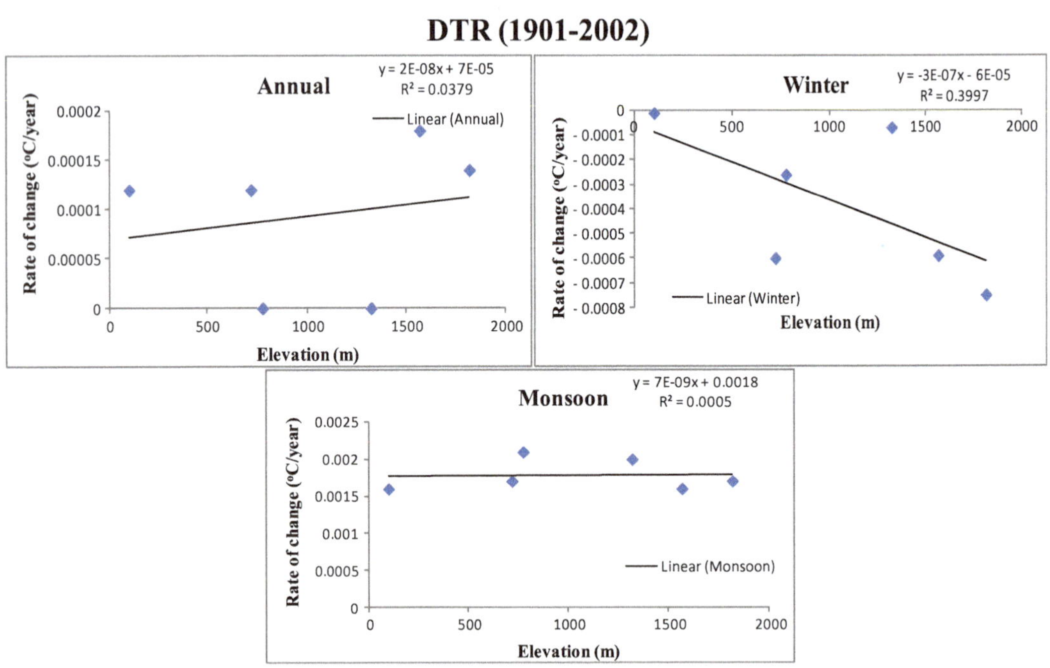

Fig. 7.13 Relationship between rates of change of DTR and elevation

On the other hand, the correlation analysis between the trend magnitude of extreme temperature indices and elevation shows significant relationship presented in Table 7.15; bold values show the significance at 95 % confidence level. Figure 7.14 shows that for all four extreme

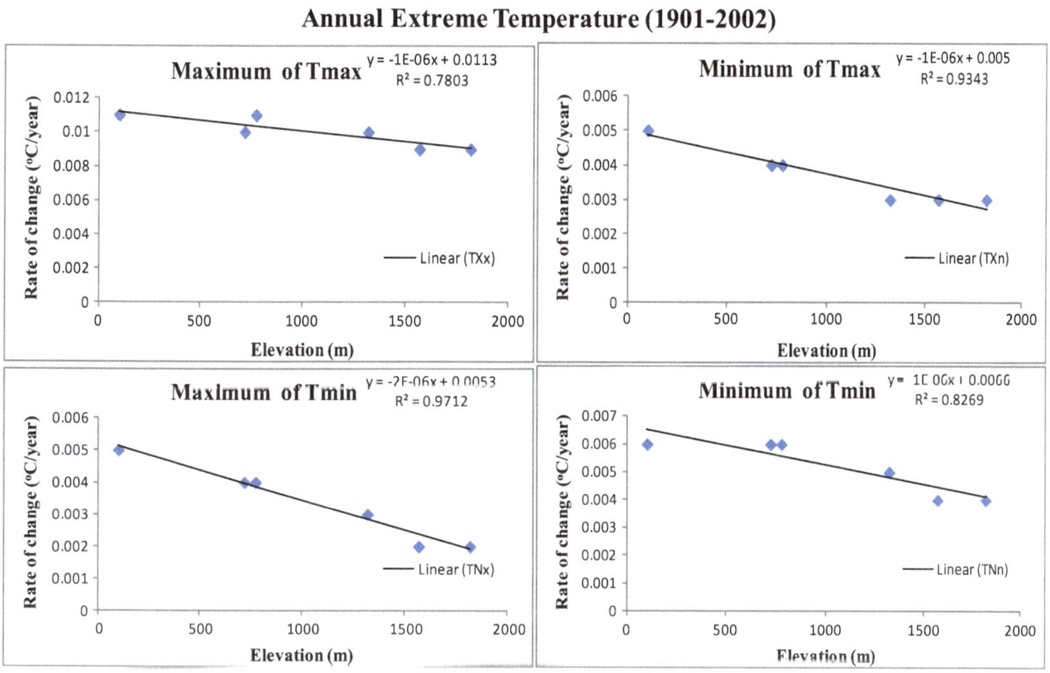

Fig. 7.14 Relationship between rate of change of extreme values of Tmax and Tmin and elevation

temperature indices (TXx, TXn, TNx, and TNn), the rate of increasing trend decreases with increase in elevation (i.e., LEZ experiences greater rate of increasing extreme temperature indices trend than MEZ and HEZ). Among all four indices, TXn and TNx show good relationship with the elevation of the study area.

7.3.2.5 Trends in Future Temperature (2003–2099)

7.3.2.5.1 Maximum Temperature Trend

Table 7.16 illustrates the results obtained from the MK test and Sen's slope estimator test for the direction and magnitude of the trend. Trend test was applied for the future maximum and minimum temperature times series at annual and seasonal basis for A2 and B2 SRES scenarios. Table 7.13 shows the significant positive trend for both A2 and B2 scenarios at annual and seasonal scales. A2 scenario shows higher rate than the B2 scenario, and Tmax is likely to increase by the average rate of 0.038 °C/year, 0.036 °C/year, 0.035 °C/year, and 0.01 °C/year for winter, summer, annual, and monsoon season, respectively, under A2 scenario.

7.3.2.5.2 Minimum Temperature Trend

Table 7.17 indicates the significant rising trend of minimum temperature at annual and seasonal scales for both A2 and B2 scenarios. Trends for A2 scenario were higher than B2 scenario for annual, summer, and monsoon season. The winter Tmin trend was almost the same for both A2 and B2 scenarios and highest among all the series including Tmax. The climate forecasts indicate that these trends will be exacerbated in the future. Average minimum temperatures are projected to increase about 0.053 °C per year, 0.051 °C per year, and 0.049 °C per year for winter, summer, and monsoon, respectively, for A2 scenario.

The analysis reveals that the future trend of temperature continues the effect of the past trend as well and also specifies that the minimum temperature is increasing by higher rate in all seasons than maximum temperature. Due to this

Table 7.16 Results obtained from the trend test for A2 and B2 scenario of Tmax (2003–2099)

Seasons	Annual		Winter		Summer		Monsoon	
Station	Z	β (°C/year)	Z	β (°C/year)	Z	β (°C/year)	Z	β (°C/year)
Tmax A2								
Mokokchung	12.553	0.027	11.59	0.038	11.209	0.037	11.473	0.019
Mon	12.707	0.08	11.823	0.040	11.142	0.036	11.798	0.022
Sibsagar	12.682	0.029	11.78	0.041	11.191	0.038	11.749	0.022
Tirap	12.474	0.024	11.43	0.036	11.191	0.052	11.749	0.016
Tuensang	12.59	0.026	11.676	0.038	11.099	0.035	11.479	0.019
Zunheboto	12.523	0.025	11.577	0.037	11.135	0.035	11.307	0.017
Tmax B2								
Mokokchung	11.559	0.017	9.734	0.022	9.687	0.021	10.939	0.015
Mon	10.006	0.016	10.197	0.022	10.018	0.021	11.215	0.017
Sibsagar	11.749	0.012	10.183	0.023	10.006	0.022	11.234	0.017
Tirap	11.479	0.096	9.926	0.020	9.558	0.019	10.534	0.013
Tuensang	11.135	0.05	10.005	0.021	9.828	0.02	11.010	0.015
Zunheboto	11.099	0.035	9.981	0.021	9.705	0.02	10.798	0.014

Table 7.17 Results obtained from the trend test for future minimum temperature (2003–2099)

Seasons	Annual		Winter		Summer		Monsoon	
Station	Z	β (°C/year)	Z	β (°C/year)	Z	β (°C/year)	Z	β (°C/year)
Tmin A2								
Mokokchung	12.916	0.053	11.185	0.053	11.774	0.05	12.253	0.05
Mon	12.909	0.053	11.215	0.053	11.755	0.052	12.24	0.053
Sibsagar	12.867	0.053	11.215	0.053	11.762	0.052	12.21	0.052
Tirap	12.952	0.023	11.209	0.052	11.805	0.049	12.234	0.048
Tuensang	12.934	0.022	11.191	0.052	11.737	0.051	12.247	0.051
Zunheboto	12.94	0.051	11.191	0.052	11.78	0.05	12.234	0.049
Tmin B2								
Mokokchung	10.11	0.032	11.089	0.053	10.11	0.05	11.602	0.037
Mon	10.126	0.033	11.153	0.053	10.141	0.053	11.596	0.038
Sibsagar	11.798	0.022	11.103	0.053	10.123	0.052	11.577	0.038
Tirap	11.665	0.037	11.183	0.052	10.178	0.048	11.608	0.035
Tuensang	11.749	0.022	11.102	0.052	10.116	0.051	11.633	0.037
Zunheboto	11.633	0.037	11.183	0.052	10.116	0.049	11.627	0.036

the agricultural cultivation of alluvial plains in the study area will be affected more than the other regions, because the Rabi crops are mainly irrigation based, which are generally practiced over alluvial plains under restricted temperature.

Table 7.18 compares the past and future trend of Tmax and Tmin. A huge difference has been observed between the past and future rate of increasing trend of both temperatures on annual and seasonal basis. The trend magnitude was higher for minimum temperature than maximum temperature. More warming is projected in A2 scenario than B2 scenario.

The analysis of past and projected future temperature trends has indicated that the study area needs urgent attention to improve the crop variety and crop pattern to sustain under changing climate regime. It is required to develop such an irrigation system to enhance the optimum utilization of water.

The present study provides some scenarios of pattern of maximum temperature, minimum temperature, diurnal temperature range (DTR), and rainfall, which may be used for sensitivity analysis of water demand-supply of the hydrological cycle of the region.

Table 7.18 Comparison between past and future temperature trend

Time scale	Past (1901–2002)	Future A2 (2003–2099)	Future B2 (2003–2099)
Rate of change Tmax (°C/year)			
Annual	0.0075	0.035	0.038
Monsoon	0.0067	0.019	0.015
Winter	0.0087	0.038	0.015
Summer	0.0088	0.036	0.021
Rate of change Tmin (°C/year)			
Annual	0.0055	0.0425	0.0305
Monsoon	0.0043	0.0505	0.0368
Winter	0.01	0.0525	0.0525
Summer	0.007	0.0507	0.0505

7.4 Conclusions

Conclusions drawn from the trend test study applied for various precipitation and temperature time series are as follows:

1. Overall significant decreasing trends were identified in the annual (significant Sibsagar, Mon, and Tirap) and monsoon (Sibsagar, Mon, Tirap, and Mokokchung) precipitation from 1901 to 2002. No significant trend was identified in the winter and summer precipitation time series.
2. Year 1959 was identified as the most probable change year during the entire time series (1901–2002). Before change point precipitation anomalies were positive, and after change anomalies were negative for annual and monsoon precipitation.
3. The correlation analysis between the monsoon precipitation trend magnitude and elevation shows significant relationship for all three time series (1901–1959, 1960–2002, and 1901–2002) in the study region.
4. Future precipitation increases significantly all over the catchment for both A2 and B2 scenarios. The precipitation trend is more pronounced to increase in A2 scenario rather than in B2 scenario. Winter precipitation is likely to increase by 36.25 % with respect to historical precipitation, which is highest among all. In the future great increment of winter and summer precipitation was projected than the monsoon precipitation. It may change the constituents of the average annual rainfall, by which the small amount of winter and summer (in past average annual precipitation) is likely to come out as the prominent contributor to the average annual rainfall in the future.
5. Trends were also significant for seasonal and annual time scale for both Tmax and Tmin. Tmax increases higher rate (average of 0.0075 °C/year) than Tmin for annual, summer, and monsoon series, while Tmin having higher rate than Tmax is the highest (0.009–0.012 °C/year) among all in winter.
6. DTR shows significant positive trend for the monsoon season at five stations. Extreme temperature indices also show almost positive trend, and TXx shows the greatest rate (maximum of 0.011 °C/year) of increasing trend among TXn, TNx, and TNn over the catchment.
7. The Tmax, Tmin, and DTR trend magnitude does not show any significant relationship with elevation of the study area, while extreme temperature indices' (TXx, TXn, TNx, and TNn) trend magnitude shows the significant relationship with elevation. LEZ experiences greater rate of increasing extreme temperature trend than MEZ and HEZ.
8. Future Tmax and Tmin show the significant rising trend for both A2 and B2 scenarios. A2 scenario has the higher rate of rise than B2

scenario. Tmin shows prominent increasing rate, out of which winter Tmin shows the highest rising trend by the average rate of 0.053 °C/year in the study area.

References

Barnett TP, Adam JC, Lettenmaier DP (2005) Potential impacts of a warming climate on water availability in snow-dominated regions. Nature 438(7066):303–309

Basistha A, Arya DS, Goel NK (2009) Analysis of historical changes in rainfall in the Indian Himalayas. Int J Climatol 29(4):555–572

Bhutiyani MR, Kale VS, Pawar NJ (2007) Long-term trends in maximum, minimum and mean annual air temperatures across the Northwestern Himalaya during the twentieth century. Clim Chang 85(1–2):159–177

Chaudhari QZ (1994) Pakistan's summer monsoon rainfall associated with global and regional circulation features and its seasonal prediction. In: Proceedings of the international conference on monsoon variability and prediction, Trieste

Chen LX, Dong M, Shao YN (1992) The characteristic of international variations on the east Asian Monsoon. J Meteorol Soc Jpn 70:397–421

Chen B, Chao WC, Liu X (2003) Enhanced climatic warming in the Tibetan Plateau due to doubling CO_2: a model study. Clim Dyn 20(4):401–413

Chowdary VM, Chandran RV, Neeti N, Bothale RV, Srivastava YK, Ingle P, Singh R (2008) Assessment of surface and sub-surface waterlogged areas in irrigation command areas of Bihar state using remote sensing and GIS. Agric Water Manag 95(7):754–766

Chowdhury A, Jha MK, Chowdary VM, Mal BC (2009) Integrated remote sensing and GIS-based approach for assessing groundwater potential in West Medinipur district, West Bengal, India. Int J Remote Sens 30(1):231–250

Chung YS, Yoon MB (2000) Interpretation of recent temperature and precipitation trends observed in Korea. Theor Appl Climatol 67(3–4):171–180

Deka RL, Mahanta C, Nath KK (2009) Trends and fluctuations of temperature regime of North East India. Impact Clim Change Agric 376–380

Domroes M, El-Tantawi A (2005) Recent temporal and spatial temperature changes in Egypt. Int J Climatol 25:51–63

Duhan D, Pandey A, Gahalaut KPS, Pandey RP (2013) Spatial and temporal variability in maximum, minimum and mean air temperatures at Madhya Pradesh in central India. Compt Rendus Geosci 345(1):3–21

Easterling DR, Horton B, Jones PD, Peterson TC, Karl TR, Parker DE, Folland CK (1997) Maximum and minimum temperature trends for the globe. Science 277(5324):364–367

ElNesr MM, Abu-Zreig MM, Alazba AA (2010) Temperature trends and distribution in the Arabian Peninsula. Am J Environ Sci 6(2):191–203

Fowler HJ, Archer DR (2006) Conflicting signals of climatic change in the Upper Indus Basin. J Clim 19(17):4276–4293

Frauenfeld OW, Zhang T, Serreze MC (2005) Climate change and variability using European Centre for Medium-Range Weather Forecasts reanalysis (ERA-40) temperatures on the Tibetan Plateau. J Geophys Res Atmos 110(D2):1–9

Gadgil A, Dhorde A (2005) Temperature trends in twentieth century at Pune, India. Atmos Environ 39(35):6550–6556

Gain AK, Giupponi C (2015) A dynamic assessment of water scarcity risk in the lower Brahmaputra River Basin: an integrated approach. Ecol Indic 48:120–131

Giupponi C, Rosato P, Rounsevell M (1998) Integrated model to predict European land use: climate change and land use in the Venice Lagoon Watershed. In University of Minnesota, Center for International Food and Agricultural Policy conference papers, no. 14485, August

Goyal MK (2014) Statistical analysis of long term trends of rainfall during 1901–2002 at Assam, India. Water Resour Manag 28(6):1501–1515

Guan Y, Zhang X, Zheng F, Wang B (2015) Trends and variability of daily temperature extremes during 1960–2012 in the Yangtze River Basin, China. Glob Planet Chang 124:79–94

Haskett JD, Pachepsky YA, Acock B (2000) Effect of climate and atmospheric change on soybean water stress: a study of Iowa. Ecol Model 135(2):265–277

Hay LE, Clark MP (2003) Use of statistically and dynamically downscaled atmospheric model output for hydrologic simulations in three mountainous basins in the western United States. J Hydrol 282(1):56–75

Izrael Y, Anokin Y, Eliseev AD (1997) Vulnerability and adaptation assessments. Final report of the Russian country study on climate problem. Russian Federal service for hydrometeorology and environmental monitoring, vol 3, Task 3, Roshydromet, Moscow, Russia, p 105

Jeganathan A, Andimuthu R (2013) Temperature trends of Chennai City, India. Theor Appl Climatol 111(3–4):417–425

Jhajharia D, Singh VP (2011) Trends in temperature, diurnal temperature range and sunshine duration in Northeast India. Int J Climatol 31(9):1353–1367

Kadioglu M (1997) Trends in surface air temperature data over Turkey. Int J Climatol 17:511–520

Kang S, Zhang Y, Qin D, Ren J, Zhang Q, Grigholm B, Mayewski PA (2007) Recent temperature increase recorded in an ice core in the source region of Yangtze River. Chin Sci Bull 52(6):825–831

Kumar KR, Kumar KK, Pant GB (1994) Diurnal asymmetry of surface temperature trends over India. Geophys Res Lett 21(8):677–680
Lu A, He Y, Zhang Z, Pang H, Gu J (2004) Regional structure of global warming across China during twentieth century. Clim Res 27:189–195
Mirza MQ, Dixit A (1997) Climate change and water management in the GBM Basins. Water Nepal 5:71–100
Murphy J (1999) An evaluation of statistical and dynamical techniques for downscaling local climate. J Clim 12(8):2256–2284
Pandey A, Chowdary VM, Mal BC, Billib M (2008) Runoff and sediment yield modeling from a small agricultural watershed in India using the WEPP model. J Hydrol 348(3):305–319
Pepin NC, Lundquist JD (2008) Temperature trends at high elevations: patterns across the globe. Geophys Res Lett 35(14):1–6
Pepin NC, Seidel DJ (2005) A global comparison of surface and free-air temperatures at high elevations. J Geophys Res: Atmos 110(D3)
Rankova E (1998) Climate change during the 20th century for the Russian Federation. In: Abstract book of the 7th international meeting on statistical climatology, Whistler, British Columbia, Canada, May 25–29, Abstract 102–198
Ren G, Wu H, Chen Z (2000) Spatial pattern of precipitation change trend of the last 46 years over China. J Appl Meteorol 11(3):322–330
Roy SS, Balling RC (2005) Analysis of trends in maximum and minimum temperature, diurnal temperature range, and cloud cover over India. Geophys Res Lett 32(12):1–4
Salinger MJ, Griffiths GM (2001) Trends in New Zealand daily temperature and rainfall extremes. Int J Climatol 21:1437–1452
Schoof JT, Pryor SC (2001) Downscaling temperature and precipitation: a comparison of regression-based methods and artificial neural networks. Int J Climatol 21(7):773–790
Sharma N, Flügel WA (eds) (2015) Applied geoinformatics for sustainable Integrated Land and Water Resources Management (ILWRM) in the Brahmaputra River Basin: results from the EC-project BRAHMATWINN. Springer, New Delhi, Heidelberg, New York, Dordrecht, London
Singh A, Krause P, Panda SN, Flugel WA (2010) Rising water table: a threat to sustainable agriculture in an irrigated semi-arid region of Haryana, India. Agric Water Manag 97(10):1443–1451
Song ZW, Zhang HL, Snyder RL, Anderson FE, Chen F (2009) Distribution and trends in reference evapotranspiration in the North China Plain. J Irrig Drain Eng 136(4):240–247
Subash N, Sikka AK (2014) Trend analysis of rainfall and temperature and its relationship over India. Theor Appl Climatol 117(3–4):449–462
Tian L, Yao T, MacClune K, White JWC. Schilla A, Vaughn B, Ichiyanagi K (2007) Stable isotopic variations in west China: a consideration of moisture sources. J Geophys Res Atmos 112(D10):1–12
Tomozeiu R, Pavan V, Cacciamani C, Amici M (2006) Observed temperature changes in Emilia-Romagna: mean values and extremes. Clim Res 31:217–225
Vuille M, Bradley RS (2000) Mean annual temperature trends and their vertical structure in the tropical Andes. Geophys Res Lett 27(23):3885–3888
Warwade P, Hardaha MK, Kumar D, Chandniha SK (2014) Estimation of soil erosion and crop suitability for a watershed through remote sensing and GIS approach. Indian J Agric Sci 84(1):18–23
Warwade P, Kumari D, Sharma N (2015) Trends in reference evapotranspiration of Dikhow catchment, North East India. Int J Trop Agric 33(2 (Part IV)):1571–1578
Wibig J, Glowicki B (2002) Trends of minimum and maximum temperature in Poland. Clim Res 20:123–133
You QL, Kang S, Flügel WA, Pepin N, Yan Y, Huang J (2010) Decreasing wind speed and weakening latitudinal surface pressure gradients in the Tibetan Plateau. Clim Res 42(1):57–64

Pratibha Warwade Department of Water Resources Development and Management, Indian Institute of Technology Roorkee, Roorkee, Uttarakhand, India

Nayan Sharma Department of Water Resources Development and Management, Indian Institute of Technology Roorkee, Roorkee, Uttarakhand, India

Ashish Pandey Department of Water Resources Development and Management, Indian Institute of Technology Roorkee, Roorkee, Uttrakhand, India

Bodo Ahrens Mesoscale Meteorology and Climate Institute of Atmosphere and Environment Sciences/ Geozentrum Riedenberg, Goethe-University, Frankfurt am Main, Germany

Part III

River Hydraulics

Local Scour

8

Walter H. Graf and Mustafa S. Altinakar

Abstract
Flow in a channel with a mobile bed is usually accompanied by a transport of sediments; erosion and deposition might be the consequence. Additional erosion of sediments will be caused, where there is a local change in the geometry of the channel or in the flow.

Different kinds of local scour are encountered in fluvial hydraulics. Positioned in the flow, a pier or an abutment will locally alter the flow and cause erosion (deposition) in the vicinity of the obstacle. Constriction scour is encountered, if the width of a channel is reduced. Flow over and/or under a hydraulic structure has a considerable potential to cause scour at the downstream.

8.1 General Remarks

1. Flow in an open channel with a mobile bed is usually accompanied by a transport of sediments. The latter is a result of the interplay between erosion and deposition of the transported sediments.
2. Any local change in the geometry of the channel or in the flow will produce additional erosion (deposition) of the sediments; this is called *local scour* (German: *Kolk*). One encounters different types of local scour:
   (i) Pier scour
   (ii) Abutment scour
   A pier or an abutment, positioned in the flow, will *locally* alter the flow in the

This keynote paper constitutes a considerable abridged version from our book *Fluvial Hydraulics*, Chapter 9, John Wiley & Sons, 1998. This is done with the permission of the copyright holder, Presses Polytechniques et Universitaires Romandes (PPUR), and the publisher, John Wiley & Sons.

W.H. Graf (✉)
Laboratoire d'Hydraulique Environnementale (LHE), EPF Lausanne, ENAC IIC LHE, Bâtiment GC, Station 18, CH-1015 Lausanne, Switzerland

University of Lausanne, Lausanne, Switzerland
e-mail: walter.graf@epfl.ch

M.S. Altinakar
National Center for Computational Hydroscience and Engineering (NCCHE), The University of Mississippi, Brevard Hall, Room 327, P.O. Box 1848, University, MS 38677-1848, USA

N. Sharma (ed.), *River System Analysis and Management*, DOI 10.1007/978-981-10-1472-7_8

channel, causing erosion (scour) and possibly deposition in the vicinity of the pier or the abutment.

(iii) Constriction scour

A change in the geometry of the channel, like a constriction, will alter the flow in the channel, thus its capacity of transporting sediments; scour might be the consequence.

(iv) Hydraulic structure scour

Since most hydraulic structures perturb at least locally the flow, erosion (scour) or possibly deposition in the vicinity of the structure might be encountered.

Local scour occurs when the capacity of the flow to erode and to transport the sediments is larger than the capacity to supply (replace) the sediments.

3. Knowledge about the topography of the erosion-deposition of the channel bed is of paramount importance for the hydraulic engineer. Continued scour could undermine the foundation of the structure. Notably, excessive discharges (floods) are liable to produce also severe local scour. Protection (structural) against erosion-deposition might become necessary.
4. In the presence of an obstacle (pier, abutment, constriction, etc.), the unidirectional flow in the channel gets to be a three-dimensional one. Hydraulic investigations become rather complex, and numerical modeling may be necessary. For such a reason, it became customary to communicate the results of studies in laboratories and in the field dealing with scour, with *dimensionless parameters*.

   To decide which of the dimensionless parameters are of importance in scour studies, a dimensional analysis must be performed.
5. Two types of scour should be distinguished:
   - *Clear-water scour*: when the sediments are removed from the scour hole and are not replaced
   - *Sediment transport (live-bed) scour*: when the scour hole is continuously supplied with sediments from the sediment transport in the channel
6. Despite the volume of investigations available—most of them come from laboratory studies and only a few from field tests—the hydraulics of local scour is as yet not well established. Consequently, the formulae to be presented herein will only offer a guideline for the engineer.

8.2 Pier Scour

1. Local scour around a pier is a complex phenomenon resulting from the strong interaction of the three-dimensional turbulent flow field around the pier and the erodible sediment bed.
2. The presence of a (isolated) pier changes the flow pattern, which in turn modifies the surface of the mobile bed (see Fig. 8.1). In the vicinity of the pier, scouring is the consequence. The intensity of scour, thus the scour depth, d_s, will depend on the channel flow, the sediments of the bed, and the geometry and alignment of the pier.

8.2.1 Scour Process

1. The evolution of the scour depth, d_s, during an increase of the flow velocity, U—evidenced by Chabert and Engeldinger in 1956 (see Shen 1971, p. 23.3)—is shown in Fig. 8.2.

   Upon reaching a certain flow velocity in the channel, the sediment particles close to the pier begin to move; scour is initiated. The eroded particles will follow the flow pattern and are carried from the front of the pier toward the downstream. Upon an increase in the flow velocity, more and more particles will get dislodged, forming a scour hole increasing in size and depth. Eventually a maximum scour depth, $(d_s)_{\max}$, is attained, which corresponds to a flow velocity being close to the critical velocity, $U + U_{cr}$, for initiation of sediment transport in the channel.

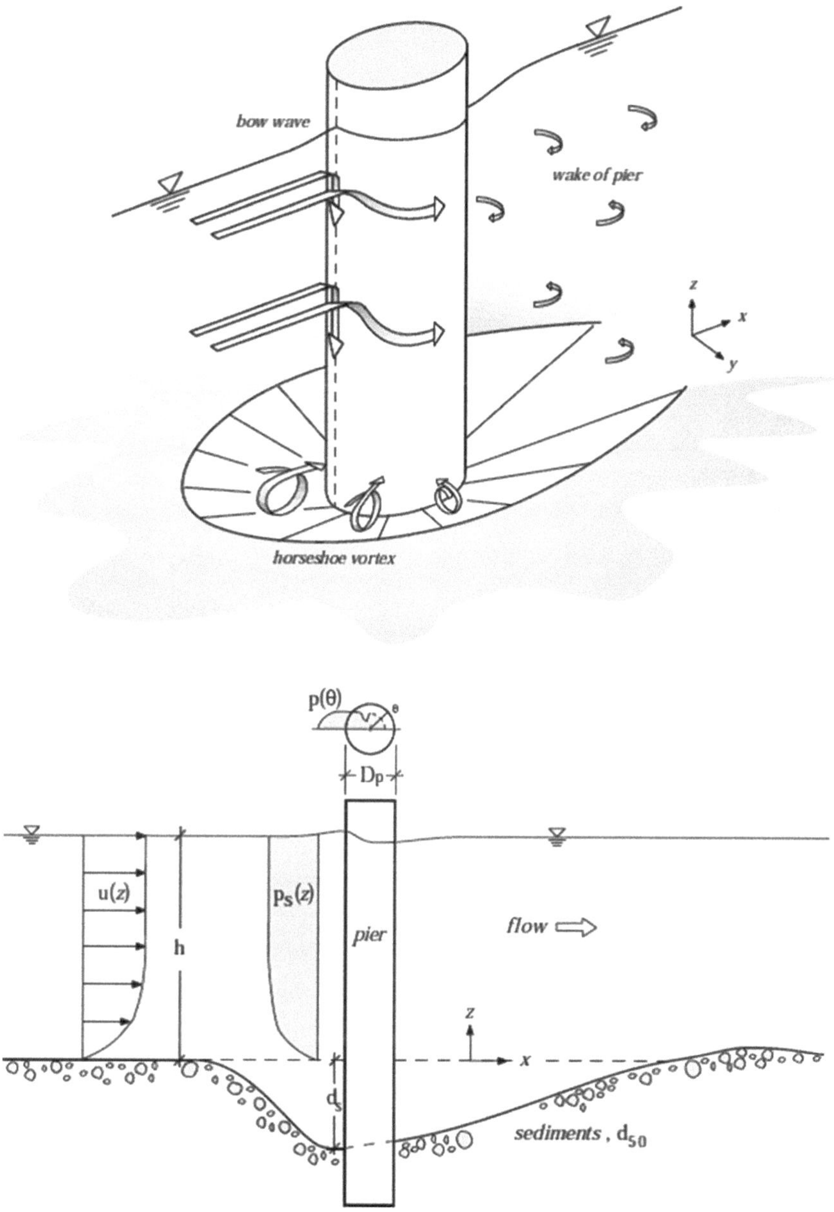

Fig. 8.1 Schema of flow pattern and local scour around a cylindrical pier

For nonuniform sediments, the larger sizes are less likely to be eroded, and an armoring layer forms itself in the scour hole.

A subsequent further increase in the flow velocity, $U > U_{cr}$, is responsible for a transport of sediments *in* and *out* of the scour hole, but the scour depth remains essentially constant. Thus, an average *equilibrium scour depth*, d_s, establishes itself, being slightly smaller than the maximum scour depth, $(d_s)_{\max}$.

2. The equilibrium scour depth, d_s, will usually be reached during high water flow of a long duration.

In steady flow conditions, the time necessary to attain this equilibrium depth depends

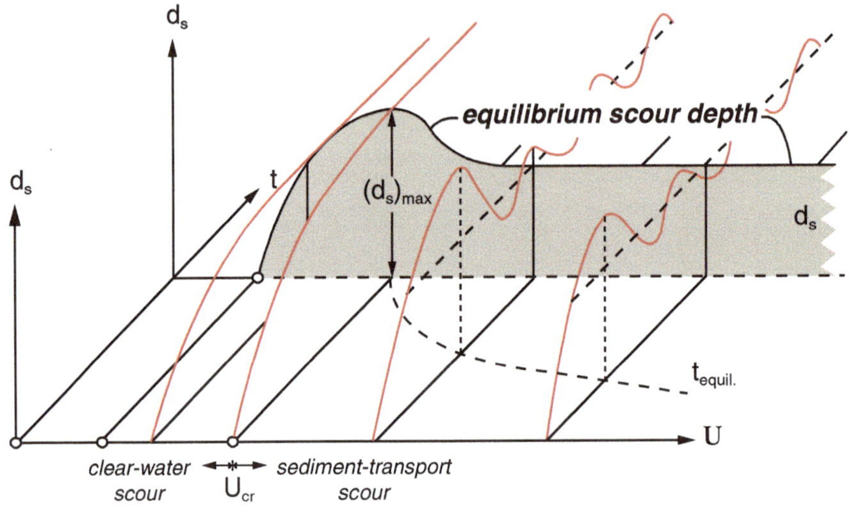

Fig. 8.2 Scheme of evolution of the scour depth, d_s, with respect to flow velocity, U, and time, t

on whether the flow is able to transport sediments, $(U > U_{cr}) > 1$, or not, $(U > U_{cr}) < 1$; thus one distinguishes (see Fig. 8.2):

- *Clear-water scour*: the scour depth increases gradually and approaches an asymptotic value, when the capacity of transport *out* of the scour hole is zero (laboratory experiments have to run continuously for several days before equilibrium conditions are reached).
- *Sediment transport scour*: the scour depth increases rapidly and attains an equilibrium value, when the capacity of sediment transport *out* of the scour hole is equal to the one *into* the scour hole.

3. Scour is initiated at, or close to, the nose of the pier (actually, two small lateral scour holes begin to form on the sides of the pier. They rapidly work their way around the pier to meet at the nose of the pier). The scour hole grows in depth and in volume by forming a groove. The upstream portion of the scour hole has the approximate shape of an inverted cone-like surface, stretching around the pier with side slopes about equal to the angle of repose of the sediments. Eroded material is transported—often in the form of bursts—toward the rear of the pier, where it may or not be deposited.

The maximum scour depth for a *cylindrical* pier is in general located in front of the pier (see Fig. 8.1); for a *rectangular* pier, it is found to occur at the nose of the pier, while for a *streamlined* pier at the sides of the pier.

8.2.2 Flow Pattern

1. Unidirectional flow in an erodible channel which encounters a protruding obstacle, like a pier, becomes three dimensional. The resulting flow pattern around the pier becomes complex and difficult to assess hydrodynamically.

   Results of various studies (see Shen 1971, p. 23/24; Raudkivi 1991, p. 63; Graf and Yulistiyanto 1998) have shown that different components of the flow pattern might play a role in local scouring (see Fig. 8.1).

2. Flow in a channel, being a boundary-layer flow, $u(z)$, approaches the pier, and a stagnation pressure, $p_s(z)$, decreasing with depth, establishes itself. This will produce a (weak) pressure gradient along the front of the pier and induce a *downward flow*, namely, from high to low velocities. Since there is also a (strong) pressure gradient around the pier, p

(θ), the downstream flow will be laterally diverted. However, it is generally agreed upon that the vertical component of the flow is responsible for the initiation of the scour. Due to the stagnation pressure in front of the pier (cylinder), the water surface increases, forming a *bow wave* (roller).
If this pressure increase becomes sufficiently strong, the three-dimensional boundary layer undergoes a separation. A *horseshoe vortex system* forms itself at the base of the pier. This vortex stretches into the downstream direction, diminishing its strength, and is very active in the local scour process.
A trailing *wake vortex system* is formed in the rear of the pier over the entire flow depth. There the turbulence intensity is increased, and consequently erosion and transport of sediments are enhanced.

8.2.3 Functional Relations

1. Considering an isolated single pier in a wide and rectangular open channel, where the flow is unidirectional, uniform, and steady and whose mobile bed is made up of cohesionless sediments, the equilibrium *scour depth*, d_s, depends on:
   - The *fluid*: the density, ρ, and viscosity, ν
   - The *sediments*: the density, ρ_s, and a characteristic diameter, d
   - The channel *flow*: the flow depth, h; the channel slope, S_f; and the gravity, g; thus the average velocity, $U = \sqrt{h\ S_f}$, or the friction velocity, $u_* = \sqrt{gh\ S_f}$
   - The geometry of *pier*: a characteristic dimension, D_c; for a cylindrical pier, the diameter of the pier is used, $D_c \equiv D_p$.

   This allows to write:

$$d_s = f_1(\rho,\ \nu;\ \rho_s,\ d;\ h\ ,\ U \text{ or } u_*,\ g;\ D_p) \tag{8.1}$$

2. Using dimensional reasoning the following dimensionless parameters (see Breusers et al. 1977, p. 219) are obtained:

$$\frac{d_s}{D_p} = f_2\left(\frac{u_* d}{\nu},\ \frac{\rho u_*^2}{g(\rho_s - \rho)},\ \frac{\rho_s}{\rho},\ \frac{h}{D_p},\ \frac{d}{D_p}\right) \tag{8.2}$$

   The particle Reynolds number, Re_*, and the dimensionless shear stress, τ_*, are related according to a relation developed by *Shields* (see sect. 3.4.2 in Graf and Altinakar 2002) such as:

$$\frac{\rho u_*^2}{g(\rho_s - \rho)\ d} = \tau_* = f_3(\mathrm{Re}_*) = f_3\left(\frac{u_* d}{\nu}\right) \tag{8.3}$$

   This relation can also be used in expressing the critical condition, when initiation of sediment transport takes place, or:

$$\tau_{*_{cr}} = f_4(\mathrm{Re}_*) \tag{8.3a}$$

   and consequently

$$\frac{\tau_*}{\tau_{*_{cr}}} = f_5(\mathrm{Re}_*) \tag{8.3b}$$

   Considering now the proportionalities (see Eq. 8.3) of:

$$\tau_* \quad \propto \quad u_* \quad \propto \quad U$$

   and realizing that the relative density, ρ_s/ρ, is already used in the definition of τ_* (see Eq. 8.3), the above relation, Eq. (8.2), becomes:

$$\frac{d_s}{D_p} = f_6\left(\frac{\tau_*}{\tau_{*_{cr}}} \text{or} \frac{u_*}{u_{*_{cr}}} \text{or} \frac{U}{U_{cr}},\ \frac{h}{D_p},\ \frac{d}{D_p}\right) \tag{8.4}$$

   where the subscript, $_{cr}$, indicates the critical condition, implying the beginning of the sediment transport in the channel (see sect. 3.4.2 in Graf and Altinakar 2002).
3. This relation, Eq. (8.4), can still be generalized to include some dimensionless correction factors such as:
   - A coefficient of sediment grading: ξ_g
   - A coefficient for pier shape: ξ_s

– A coefficient of angle of approach: ξ_α

A more general functional form for the scour depth reads now:

$$\frac{d_s}{D_p} = f_7\left(\frac{U}{U_{cr}}, \frac{h}{D_p}, \frac{d}{D_p}, \xi_g, \xi_s, \xi_\alpha\right) \quad (8.5)$$

where D_p is the width (diameter) of the pier.

4. Local scour, parameterized by the scour depth, d_s , is seen to be a complex phenomenon, where the various parameters interact. The influence of each of the above parameters (see Fig. 8.3) will be examined next by using experimental data from various investigations.
5. Influence of flow velocity
   (i) The influence of the relative flow velocity, U/U_{cr}, on the dimensionless scour depth, d_s/D_p, is shown in Fig. 8.3a (see Raudkivi 1991, p. 76) for a limited series of experiments. The overall tendency for other sediments, d, and different flow depths, h, would be the same.
   (ii) In Fig. 8.3a, the following zones are to be distinguished:
   - $(U/U_{cr}) < 0.5$: there is no local erosion (scour) and no sediment transport in the channel.
   - $0.5 < (U/U_{cr}) < 1.0$: there is active local erosion (scour), but no sediment transport in the channel; one talks of *clear-water scour* (see Fig. 8.2).
   - $(U/U_{cr}) > 1.0$: there is almost no net local erosion (scour), but sediment transport in the channel takes place; one talks of *sediment transport scour* (see Fig. 8.2); local erosion and deposition are in equilibrium due to the sediment transport in the channel; the scour depth might fluctuate in the presence of moving bed forms.

   (iii) For clear-water scour ($(U/U_{cr}) < 1.0$) where the flow velocity is of importance, an approximate relation proposed by Shen (1971, p. 23/9)—relating the strength of the horseshoe vortex to the pier Reynolds number—reads:

$$d_s = 0.00022\left(\frac{U\ D_p}{\nu}\right)^{0.619} \quad (8.6)$$

For sediment transport scour—taking $(U/U_{cr}) > 2.0$—where the flow velocity is no more of great importance, it is generally admitted that:

$$\frac{d_s}{D_p} = 2.0 \quad \text{to} \quad 2.3 \quad (8.7)$$

remains constant. This has been called the average equilibrium scour depth.

The above relations have been tested for subcritical flow, mainly in laboratory channels. For large Froude numbers, Fr >0.8, the value of Eq. (8.7) should be taken as 3.0 (see Johnson 1995). For small Froude numbers, Fr <0.2, this value could be taken as being less than 2.0.

6. Influence of flow depth
   (i) The influence of the relative flow depth, h/D_p, on the dimensionless scour depth, d_s/D_p, is shown in Fig. 8.3b (see Breusers et al. 1977, p. 242).
   (ii) As can be seen in Fig. 8.3b, the various investigations present often conflicting conclusions. Here is a selection of some relations (see Breusers et al. 1977, p. 241) presently in use:
   - A relation proposed by da Cunha (1973):

$$\frac{d_s}{D_p} = 1.35\left(\frac{h}{D_p}\right)^{0.3} \quad (8.8a)$$

   - A relation proposed by Breusers (1965):

$$\frac{d_s}{D_p} = 1.4 \quad (8.8b)$$

   valid for $h/D_p > 1.3$.
   - A relation proposed by Breusers et al. (1977):

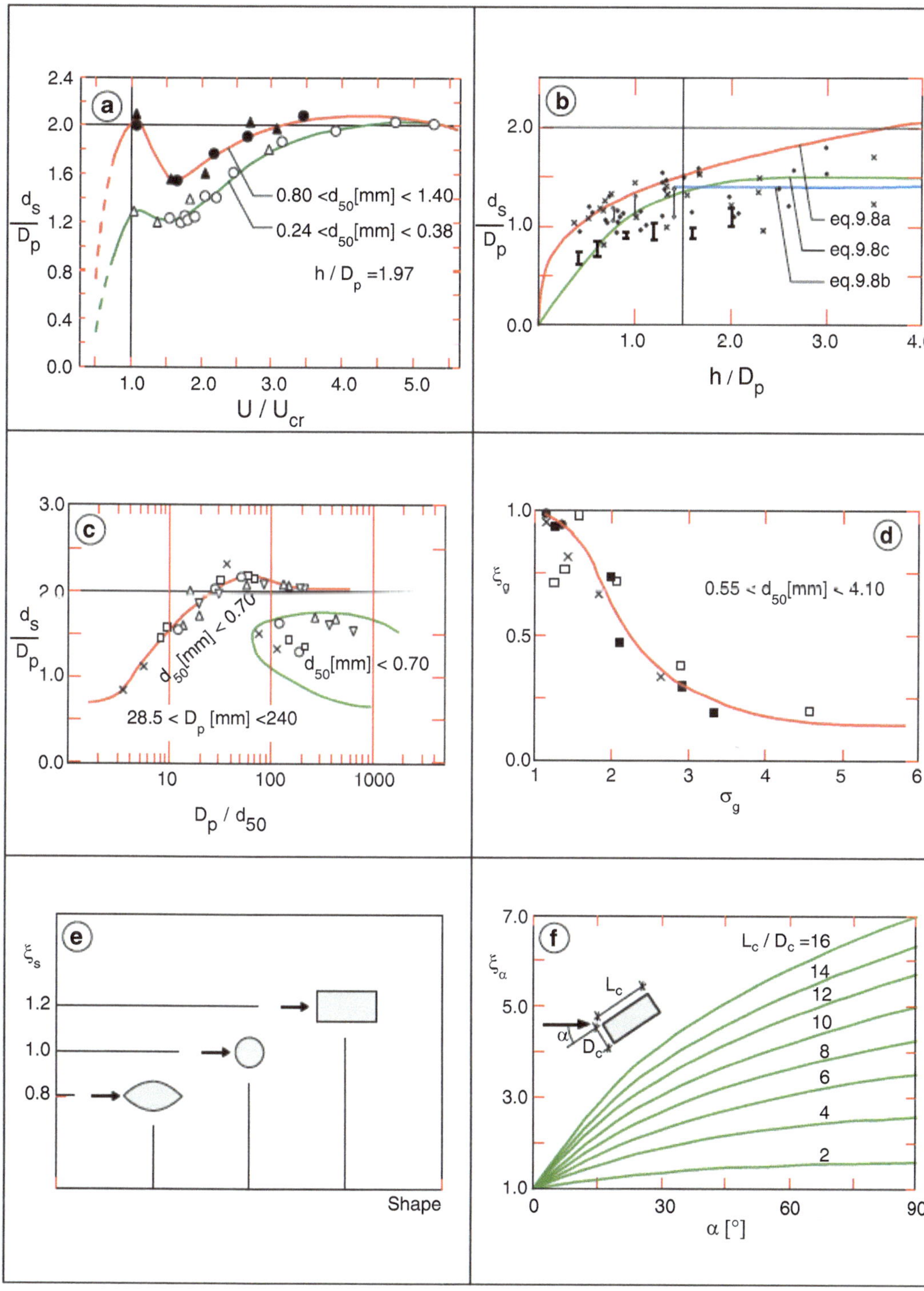

Fig. 8.3 Influence of the different parameters (see Eq. 8.5), on the dimensionless scour depth, d_s/D_p, of a pier

$$\frac{d_s}{D_p} = 1.5 \tanh\left(\frac{h}{D_p}\right) \qquad (8.8c)$$

whose constant can be taken as 2.0 instead of 1.5, in order to be on the safe side.

(iii) For shallow flow, when $(h/D_p) < 1.5$, the relative scour depth increases with the flow depth. It is however evident that for $(h/D_p) > 1.5$, the influence of the flow depth has little effect on the relative scour depth, d_s/D_p.

The experimental evidence, compared with the above relations, Eq. (8.8), is indeed not conclusive, as seen in Fig. 8.3b.

7. Influence of sediments
   (i) The influence of a sediment of uniform diameter, expressed by the relative diameter, D_p/d, on the dimensionless scour depth, d_s/D_p, is shown in Fig. 8.3c (see Raudkivi 1991, p. 69).
   (ii) As can be seen in Fig. 8.3c, the limited data arrange themselves rather well, notably if $d > 0.7$ [mm]. No relationship was proposed. However, it is evident that for $D_p/d > 25$—when the particle diameter, d, is small compared with the pier diameter, D_p—the influence of the relative diameter, D_p/d, has little effect on the relative scour depth, d_s/D_p.
   (iii) The influence of a sediment with a granulometric distribution, expressed with a coefficient of sediment grading, ξ_g, which in turn depends on the geometric standard deviation, $\sigma_g = (d_{84}/d_{16})^{0.5}$, is shown with Fig. 8.3d (see Raudkivi 1991, p. 67).

The coefficient of sediment grading, which is $\xi_g = 1$ for uniform sand, may be considerable reduced, $\xi_g < 1$, for graded sand. This implies that the relative scour depth, d_s/D_p, will also be reduced. This reduction in the scour depth can be partially attributed to the armoring effect in the scour hole.

8. Influence of pier shape
   (i) The influence of the geometry of the pier, expressed with a coefficient of pier shape, ξ_s, is given in Fig. 8.3e (see Raudkivi 1991, p. 73).
   (ii) Depending on the geometry of the nose of the pier, this coefficient may vary, being for:
   - A cylindrical pier: $\xi_s = 1.0$
   - A rectangular pier: $\xi_s \cong 1.2$
   - A lenticular pier: $\xi_s \cong 0.8$

By streamlining the pier—itself well aligned with the flow—the disturbance of the flow pattern around the pier is reduced.

9. Influence of pier alignment
   (i) The influence of the pier alignment, expressed with the coefficient of the angle of approach of flow, ξ_α, is given in Fig. 8.3f (see Raudkivi 1991, p. 73), using the experimental data of Laursen (1962).
   (ii) Depending on the pier alignment, α, but also upon the geometry of the pier, this coefficient may vary substantially, being for:
   - A cylindrical pier: $\xi_\alpha = 1$
   - A rectangular pier: $\xi_\alpha = f(\alpha, L_c/D_c)$ where D_c is the width and L_c is the length of the cross section of the pier. It should be noted that not the actual width, D_c, counts here, but the projected width, D_{cn}, increases with the angle of approach, α.
10. Influence among piers
   (i) Until now an isolated, single pier was considered. A group of piers will exhibit a mutual influence on the hydraulic behavior and on scouring. However, research on this topic is extremely scarce.
   (ii) Research results on scour at two piers (see Raudkivi 1991, p. 85) are shown in Fig. 8.4. The scour depth, d_s, of the reference pier is obviously influenced by the spacing of the piers; one observes:
   - When the two piers touch each other, $a/D_p = 1$, the scour depth of the front pier is not influenced.
   - When the two piers are separated, the scour depth of the front pier begins to increase but eventually falls off.

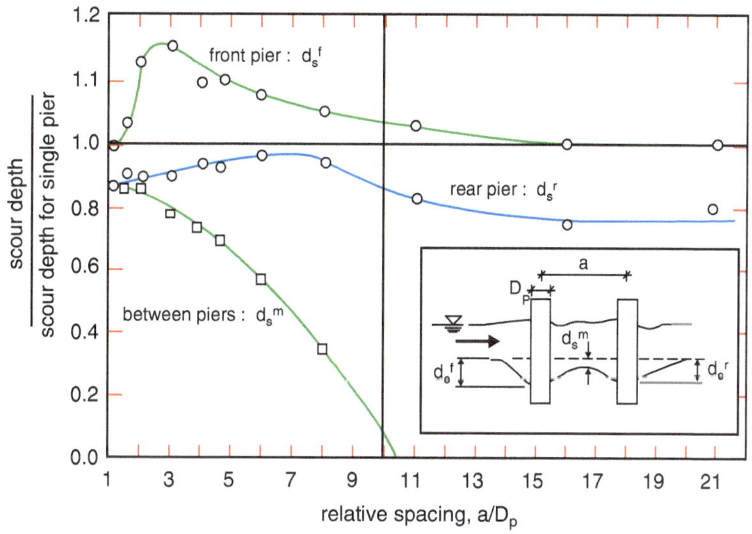

Fig. 8.4 Influence of two piers on the scour depth

- When the two piers are very separated, $a/D_p \approx 11$, the front pier is not influenced.
- The scour depth, d_s, of the rear pier is always smaller than the one of the front pier.
- The scour depth, d_s, *between* the piers diminishes rapidly and is negligible for $a/D_p > 10$.

11. Influence of unsteady flow
Unsteady flow in channels, such as encountered during the passage of floods, of translation waves or of wind waves, complicates the scour process considerably. Such waves may cause considerable pressure fluctuations on the bed, which in turn may help to displace or loosen particle in the scour hole.
Research on the effect of unsteady flow is scarce (see Raudkivi 1991, p. 83) and inconclusive.

8.2.4 Formulae for Design

1. Although it is possible to present a good deal of experimental data, it is rather difficult to propose a scour formula for design. Field data are often lacking. Theoretical developments, which would help to organize better the data of the experiments, are limited.
2. From the material presented above (see Sect. 8.2.3), the scour depth for a single pier, positioned in a wide channel, can be estimated using the following relations:
   (i) For *clear-water scour* ($U/U_{cr} < 1$):
   - The relation of Shen (1971):

$$d_s = 0.00022\left(\frac{U\ D_p}{\nu}\right)^{0.619} \tag{8.6}$$

   (ii) For *sediment transport* and *clear-water scour*:
   - The relation of Breusers et al. (1977, p. 248):

$$\frac{d_s}{D_p} = g\left(\frac{U}{U_{cr}}\right)\left(2.0\ \tanh\ \frac{h}{D_p}\right)\ \xi_s\ \xi_\alpha \tag{8.9}$$

$$g\left(\frac{U}{U_{cr}}\right)\begin{cases} = 0 & \text{for} \quad \dfrac{U}{U_{cr}} < 0.5 \\ = \left(2\ \dfrac{U}{U_{cr}} - 1\right) & \text{for} \quad 0.5 < \dfrac{U}{U_{cr}} < 1.0\,. \\ = 1 & \text{for} \quad \dfrac{U}{U_{cr}} > 0.5 \end{cases}$$

where ξ_s is the coefficient of pier shape and ξ_α is the coefficient of the angle of approach.

(iii) For *sediment transport scour* ($U/U_{cr} > 1$):
- A relation proposed by Raudkivi (1991 p. 88):

$$\frac{d_s}{D_p} = 2.3\ \xi_\alpha \quad (8.10)$$

- A relation of the type of Eq. (8.5):

$$\frac{d_s}{D_p} = 2.0\ \xi_g\ \xi_s\ \xi_\alpha \quad (8.11)$$

valid for $(h/D_p) > 1.5$ and $(D_p/d_{50}) > 25$.

In the above relations, the scour depth is the average limiting (equilibrium) one, which occurs usually after an extended flooding.

3. Evaluation of field measurements (see Nordin et al. 1989) of 104 cylindrical piers, covering a range of pier diameter, $0.2 < D_p$ [m] < 13.0; of scour depth, $0.17 < d_s$ [m] < 8.5; of flow depth, $0.3 < h$ [m] < 17.1; of approach velocity, $0.35 < U$ [m/s] < 2.72; of Froude number, $0.06 < \text{Fr} < 0.86$; and of sediment size, $0.3 < d$ [mm] < 10, showed that about 10 % of the observed scour depth, d_s, is larger than the design scour depth, using Eq. (8.8a). This represents an encouraging result.

   A large set of 515 field data—including clear-water and sediment transport scour—has been recently (see Johnson 1995, p. 628) evaluated. It was found that the above relation, Eq. (8.9), appears to work best.

4. All the above relations—in the light of their limitations and complications—must be considered as (conservative) guideline estimate for hydraulic design.

8.2.5 Scour Prevention

1. An important task for the hydraulic engineer is the concern of the protection against local scour. Prevention of scour can be achieved by diminishing the erosion effect of the downward flow and of the horseshoe vortex.

   Different methods have been suggested.

2. Riprap apron (see Fig. 8.5a)
   A most effective method for remedying local scour is the dumping of stones into the (potential) scour hole.

   For the determination of the size of the stones, d [m], a simple working relationship was proposed (see Breusers et al. 1977, p. 249):

$$U_d = 2.4\ \sqrt{d}$$

   where U_d [m/s] is taken to be the cross-sectional average velocity of the approach flow at the design discharge.

   The horizontal dimension of the loose stone riprap apron should cover two to three times the width of the obstacle. Its thickness should be three times the stone diameter.

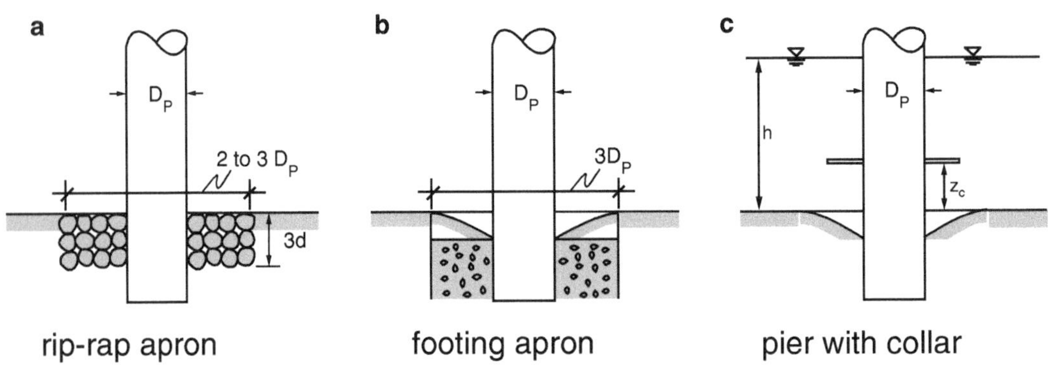

Fig. 8.5 Methods of scour prevention

3. Footing apron (see Fig. 8.5b)
   Footings or foundation blocks installed around the base of the obstacle and set below the bed level are effective to counteract the downward current.

   Pilings and sheet piles driven below the footing afford a degree of further protection against failure by scour.
4. Pier with collar (see Fig. 8.5c)
   The placement of a horizontal ring-formed shield placed on the pier proved to diminish the scour depth. The positioning at $z_c = 0.2$ h, where h is the flow depth, reduces the scour depth by 50 %. The presence of the collar disturbs the downward current, thus the erosion around the pier.
5. Pile arrangement
   A wedge-shaped arrangement of small piles, positioned upstream of the obstacle, can reduce considerably the local erosion. These piles disturb the flow and consequently weaken the horseshoe vortex.
6. Streamlining
   If the upstream nose of the pier is streamlined and/or properly aligned to the flow (see Fig. 8.3e, f), the scour depth can be reduced. In this way the horseshoe vortex and its effects are diminished.

8.3 Abutment Scour

1. The presence of an obstruction, protruding into the flow, such as an abutment, an embankment, a spur dike, or a groin—just like the one of a pier (see Sect. 8.2)—changes the flow pattern and subsequently modifies the mobile channel bed (see Fig. 8.6).

   In the vicinity of the abutment, scour will occur. The intensity of scour will depend upon the channel flow, the sediments of the bed, and the geometry of the obstruction.
2. The geometry of the abutment is given by its length, L_A, measured perpendicular to the

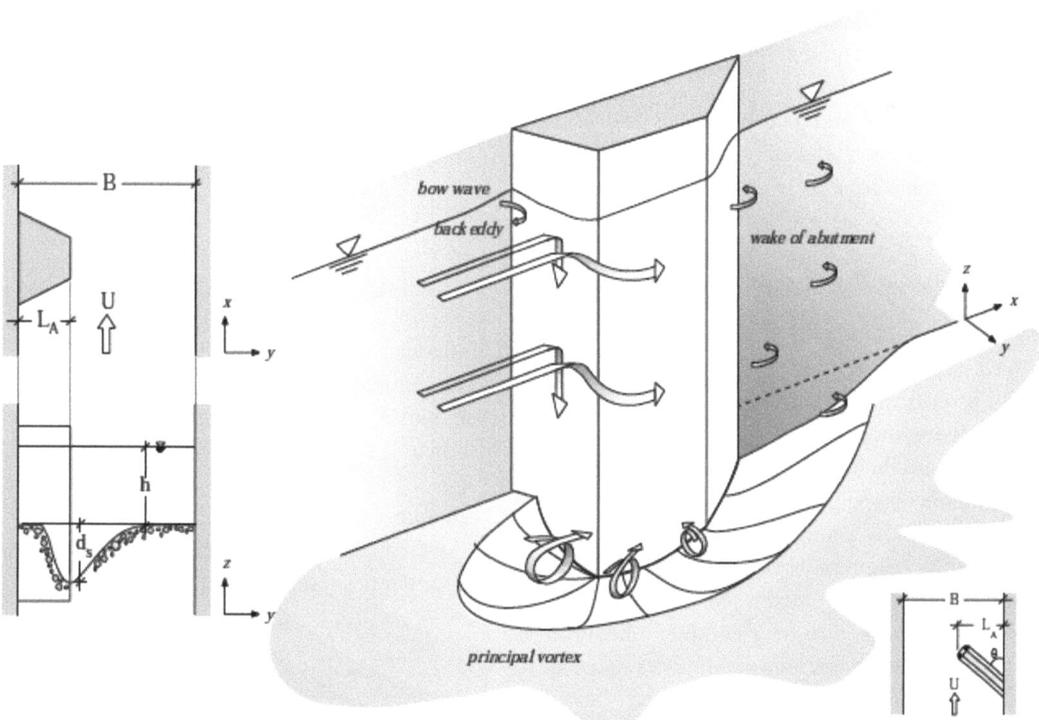

Fig. 8.6 Scheme of flow pattern and local scour around an abutment

flow, its shape, and its alignment with the flow.

8.3.1 Flow Pattern and Scouring

1. The flow pattern and scouring at abutments are rather similar to the ones around piers (see Fig. 8.6 as well as Fig. 8.1). A *downward flow* produces a *principal* (and a secondary) *vortex*, which is very active in the scour process.
2. When the length of the abutment, L_A, is considerable, a back eddy in the dead-water zone establishes itself on the upstream side of the abutment; this may influence the strength of the principal vortex. Furthermore a possible channel constriction, $(B - L_A)/B$, might have to be taken into account (see Sect. 8.4).

8.3.2 Functional Relations

1. To obtain functional relations, the same arguments as applied for a single pier (see Sect. 8.2.3) can be used.
2. Using dimensional reasoning the following dimensionless parameters are obtained (see Eqs. 8.4 and 8.5):

$$\frac{d_s}{L_A} = f_8\left(\frac{U}{U_{cr}}, \frac{h}{L_A}, \frac{d}{L_A}, \xi_g, \xi_s, \xi_\alpha\right) \quad (8.12)$$

   where h and U are the flow depth and average velocity of the approach flow, U_{cr} is the velocity at critical erosion condition, and d is the average sediment particle diameter. ξ_g, ξ_s, and ξ_α are coefficients of sediment grading, of abutment shape, and of angle of approach. Note that flow depth, h, and abutment length, L_A, could be interchanged in Eq. (8.12).
3. Local scour, parameterized by the scour depth, d_s, is seen to be a complex phenomenon, where the various parameters interact. The influence of each of the above parameters will be examined next, using experimental data.
4. Many of the available laboratory data have been systematically evaluated by Melville (1992), whose results will be presented herewith.

   It must be mentioned, however, that the existing information was obtained from laboratory tests at greatly reduced scales; extrapolation to full-scale conditions may be questionable.
5. Influence of flow velocity:
   The influence of the relative flow velocity, $U/U_{cr,}$ on the dimensionless scour depth, d_s/L_A, is shown in Fig. 8.7a (see Melville 1992, p. 624).

   In Fig. 8.7a, it is again (see also Fig. 8.3a) apparent that *clear-water scour* takes place at $U/U_{cr} < 1$ and *sediment transport scour* at $U/U_{cr} > 1$.

   The maximum scour depth, $d_{s_{max}}$, occurs at flow velocities being identical to the critical one; this was also observed for scour around piers (see Fig. 8.3a). Subsequently, upon an increase in the flow velocity, the average equilibrium scour depth, d_s, establishes itself, being indistinguishable the same as the maximum one.

   For sediment transport scour, where the flow velocity is of no further importance, it can be admitted that:

$$\frac{d_s}{L_A} \approx 2 \quad (8.13)$$

6. Influence of flow depth:
   The influence of the relative flow depth, h/L_A, on the dimensionless scour depth, d_s/L_A, is shown in Fig. 8.7b (see Melville 1992, p. 619), obtained from extensive laboratory data.

   As can be seen in Fig. 8.7b, the relative scour depth increases such as:

$$\frac{d_s}{L_A} = 2\sqrt{\frac{h}{L_A}} \quad 0.04 < \frac{h}{L_A} < 1 \quad (8.14)$$

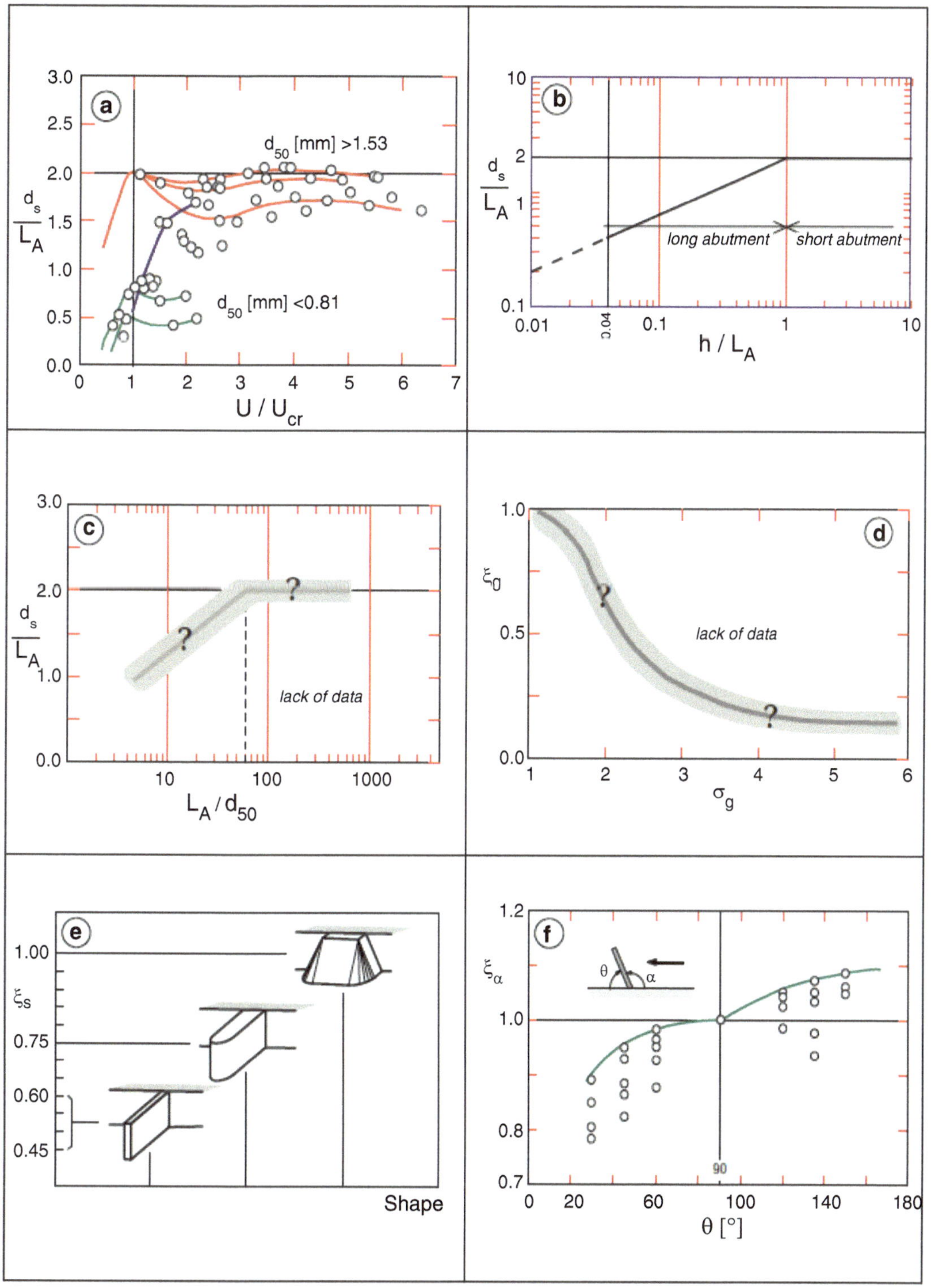

Fig. 8.7 Influence of the different parameters (see Eq. 8.12) on the dimensionless scour depth, d_s/L_A, for an abutment

and attains for *short* abutments a limiting value of:

$$\frac{d_s}{L_A} \cong 2 \quad \text{for } \frac{h}{L_A} > 1 \tag{8.14a}$$

For very *long* abutments, one takes:

$$\frac{d_s}{h} \approx 10 \quad \text{for } \frac{h}{L_A} < 0.04 \tag{8.14b}$$

Consequently, for abutments, which are neither very long nor very short, the scour depth, d_s, is influenced by the flow depth, h, and the abutment length, L_A; it is given by the above relations, Eq. (8.14), which are similar to a relation proposed by Laursen (1958).

7. Influence of sediments:
   No successful evaluation of the influence of sediments on the scouring is available.
   Using data obtained from pier scour studies (see Fig. 8.3c), it was deduced (see Melville 1992 p. 620) that for a relative sediment diameter, $L_A/d > 50$, there is little effect on the scour depth (see Fig. 8.7c). Consequently a value of:

$$\frac{d_s}{L_A} \approx 2$$

   represents a very conservative value.
   For sediments with a grain-size distribution, σ_g, it is reported (see Melville 1992, p. 620) that there is consistently produced a lower scour depth, as was also reported for pier scour (see Fig. 8.3d).
   However, since no conclusive experiments are available, the coefficient of sediment grading (see Fig. 8.7d) is to be taken as $\xi_g \approx 1$; once again this is a very conservative value.
8. Influence of abutment shape:
   The influence of the geometry of the abutment, expressed with a coefficient of shape, ξ_s, is given in Fig. 8.7e (see Melville 1992, p. 617).
   Depending on the geometry of the abutment, the coefficient may vary, being for:
   - Narrow vertical walls: $\xi_s \cong 1$
   - Vertical walls with rounded ends: $\xi_s \cong 0.75$
   - Spill-through dikes: $0.45 < \xi_s < 0.6$
9. Influence of abutment alignment:
   The influence of the alignment, expressed with the coefficient of the angle of approach of flow, ξ_α, is given in Fig. 8.7f (see Melville 1992, p. 623).
   Depending on the alignment angle of the abutment, α, the scour depth, d_s, increases with an increase in the angle; the perpendicular aligned abutment, $\alpha = 90°$, serves as a reference. An upper envelope curve delimits the data points.

8.3.3 Formulae for Design

1. A good deal of experimental data is available, but they come exclusively from laboratory measurements. There is a great shortage of field data.
2. Preliminary design recommendation—to be used with great care—is given, notably valid for sediment transport scour, as:

$$\frac{d_s}{L_A} = 2.0 \ \xi_s \ \xi_\alpha \quad \text{for } \frac{h}{L_A} > 1 \tag{8.15a}$$

$$\frac{d_s}{L_A} = 2\sqrt{\frac{h}{L_A}} \ \xi_s \ \xi_\alpha \quad \text{for } \frac{h}{L_A} < 1 \tag{8.15b}$$

The above relations—in the light of the above limitations and complications—must be considered as (conservative) guideline estimates for hydraulic design.

8.4 Constriction Scour

8.4.1 Hydraulic Considerations

1. A constriction (contraction) of a watercourse, due to a reduction in width, $B_2 < B_1$, will lead in *fluvial* regime to an increased flow velocity, $U_2 > U_1$, in the reduced cross section. If the channel has a mobile bed, erosion of the bed may be the consequence (see Fig. 8.8).

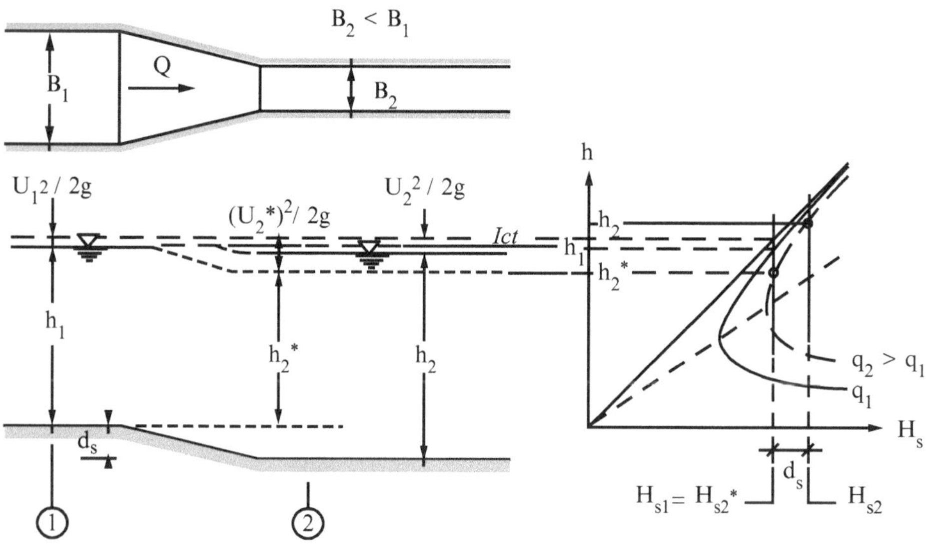

Fig. 8.8 Definition sketch for a long constriction

Considered will be one-dimensional flow in a rectangular channel. It is supposed that uniform flow exists in both the approach and the constricted cross section. If the reach of the constriction is a *long* and gradual one, the resulting head loss will be negligible.

2. This problem is one of a channel of variable width (see sect. 4.5.2 in Graf and Altinakar 2002) and with a variable bed floor (see sect. 4.5.1 in Graf and Altinakar 2002).
   (i) The equation of continuity for a constant discharge, Q, reads:

$$q_1 B_1 = Q = q_2 B_2 \qquad (8.16)$$

   For a channel of variable width, the unit discharge, q, will be variable.

   (ii) The specific energy, H_s, will be maintained throughout the transition, thus:

$$H_{s_1} = H_{s_2}^{*} = H_{s_2} - d_s$$

   A graphical solution is to be followed (see Fig. 8.8):
   - In fluvial regime, a *constriction*, $B_2 < B_1$, causes an increase of the unit discharge, $q_2 > q_1$, and consequently a decrease in the flow depth, $h_2^* < h_1$, and an increase in velocity, $U_2^* > U_1$.
   - This increase in flow velocity may cause erosion in the constricted section; consequently a *drop*, d_s, in the bed floor will appear.
   - Downstream of the floor change, the specific energy, H_{s_2}, is changed by the height of the floor change, d_s; it has now a value of $H_{s_2} = H_{s_2}^{*} + d_s$, thus a shift of d_s to the right. Consequently, an increase in the flow depth, $h_2 > h_2^*$, and a decrease in the flow velocity, $U_2 < U_2^*$, will install itself.
   - Comparing now the hydraulic situation in the approach channel, h_1 and U_1, and in the constricted channel, h_2 and U_2, the following are observed:
     If d_s is small: $h_2 < h_1$; $U_2 > U_1$
     If d_s is large: $h_2 > h_1$; $U_2 \approx U_1$ (see Fig. 8.8)

This should be kept in mind, when approximating the terms in the equation of energy, Eq. (8.20), but also in the interpretation of the values obtained from Fig. 8.9.

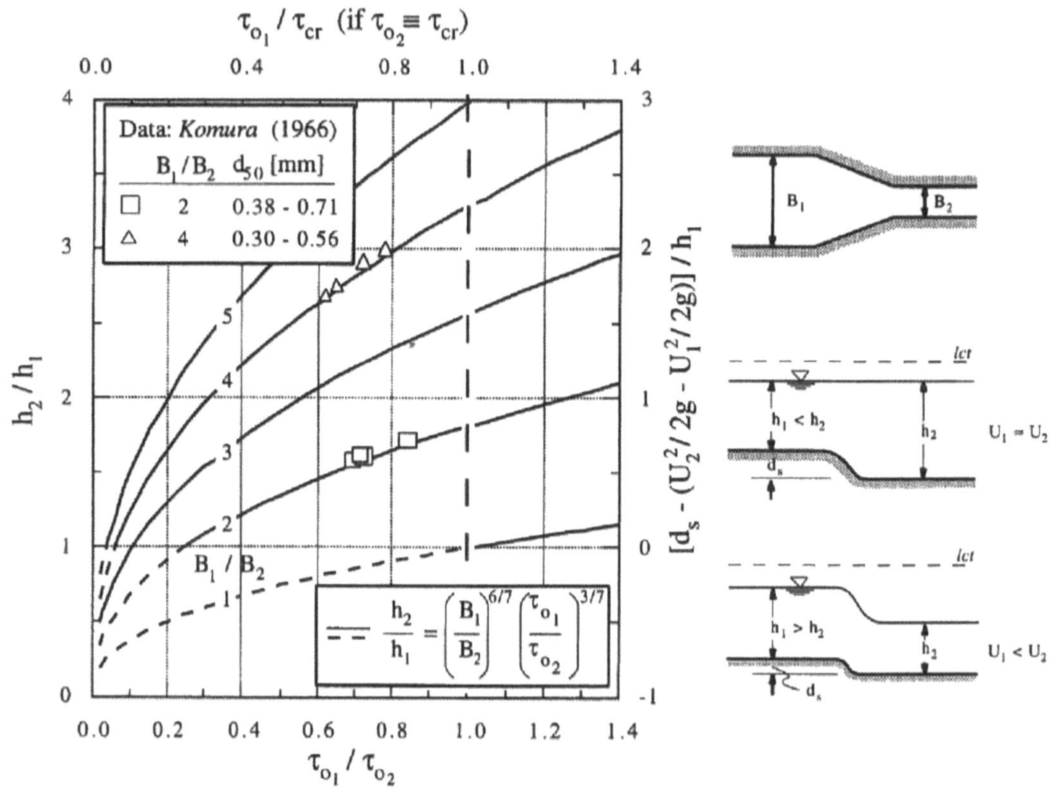

Fig. 8.9 Relative flow depth, h_2/h_1, or scour depth, d_s/h_1, as a function of the relative shear stress, τ_{o_1}/τ_{o_2}, or the relative shear stress, τ_{o_1}/τ_{cr}, (if $\tau_{o_2} \equiv \tau_{cr}$), both for different width reductions, B_1/B_2

3. The equation of continuity, for the water discharge, Q, between the two cross sections, reads:

$$B_1 U_1 h_1 = B_2 U_2 h_2 \tag{8.16a}$$

Using the Manning formula (see Eq. 3.16) for wide channels, this yields:

$$B_1 \frac{1}{n_1} h_1^{5/3} \; S_{e_1}{}^{1/2} = B_2 \frac{1}{n_2} h_2^{5/3} \; S_{2_2}{}^{1/2} \tag{8.17}$$

If the granulate is the same, $d_1 = d_2$, thus the roughness coefficient, $n_1 = n_2$, is also the same, one obtains:

$$B_1 \; h_1^{5/3} \; S_{e_1}{}^{1/2} = B_2 h_2^{5/3} \; S_{2_2}{}^{1/2} \tag{8.18}$$

$S_e \equiv S_f$ may be taken, if the flow is a uniform one.

4. The shear stress in a wide rectangular channel is given by:

$$\tau_o = \rho g h S_e \tag{8.19}$$

where S_e is the slope of the energy grade line.

5. The equation of energy, for the two cross sections under consideration (see Fig. 8.8), can be written as:

$$d_s + h_1 + \frac{U_1^2}{2g} = h_2 + \frac{U_2^2}{2g} + h_s + h_r \tag{8.20}$$

Assuming the head loss due to friction, h_r, and to the constriction, h_s, is negligible, an expression for the scour depth reads:

$$d_s = (h_2 - h_1) + \left(\frac{U_2^2}{2g} - \frac{U_1^2}{2g}\right) \tag{8.20a}$$

For fluvial flow, it seems often reasonable (see Laursen 1963, p. 97) to assume that the variation in the velocity head is small, which renders:

$$d_s \approx (h_2 - h_1) \text{ or } h_2 \approx h_1 + d_s \qquad (8.21)$$

8.4.2 Scour Depth Relations

1. The above relation, Eq. (8.18), shall be used to express the relative flow depth:

$$\frac{h_2}{h_1} = \left(\frac{B_1}{B_2}\right)^{3/5} \left(\frac{S_{e_1}}{S_{e_2}}\right)^{3/10} \qquad (8.22)$$

Furthermore the definition of the shear stress, Eq. (8.19), yields:

$$\frac{\tau_{o_1}}{\tau_{o_2}} = \frac{\rho g h_1 S_{e_1}}{\rho g h_2 S_{e_2}} \quad \Rightarrow \quad \frac{S_{e_1}}{S_{e_2}} = \frac{\tau_{o_1}}{\tau_{o_2}} \frac{h_2}{h_1} \qquad (8.23)$$

2. Combining the above relations, renders:

$$\frac{h_2}{h_1} = \left(\frac{B_1}{B_2}\right)^{3/5} \left(\frac{\tau_{o_1}}{\tau_{o_2}} \frac{h_2}{h_1}\right)^{3/10}$$

or rewritten:

$$\frac{h_2}{h_1} = \left(\frac{B_1}{B_2}\right)^{6/7} \left(\frac{\tau_{o_1}}{\tau_{o_2}}\right)^{3/7} \qquad (8.24)$$

This relation is plotted in Fig. 8.9, with the relative flow depth, h_2/h_1, as a function of the relative shear stress, τ_{o_1}/τ_{o_2}, for various values of width reduction, B_1/B_2. Here, the following is to be observed:

- For a given relative shear stress, τ_{o_1}/τ_{o_2}, an increase in the width reduction, B_1/B_2, is responsible for an increase in the relative depth, h_2/h_1.
- If the flow depth throughout the entire constriction does not change—the relative depths yield unity, $h_2/h_1 = 1$—it implies (see Eq. 8.24) that:

$$\frac{\tau_{o_1}}{\tau_{o_2}} = \left(\frac{B_2}{B_1}\right)^2 \text{ or } \frac{\rho g h_1 S_{e_1}}{\rho g h_2 S_{e_2}} = \frac{S_{e_1}}{S_{e_2}} = \left(\frac{B_2}{B_1}\right)^2 \qquad (8.25)$$

having a limiting value at:

$$\frac{\tau_{o_1}}{\tau_{o_2}} = 1 \quad \text{when} \quad \frac{B_2}{B_1} = 1$$

- For a relative flow depth below unity, $h_2/h_1 < 1$, scour is also possible; the complete equation of energy, Eq. (8.20a), must be considered.

3. An interesting relationship can be obtained, assuming that the same granulate, thus the same critical shear stress, $\tau_{cr_1} \equiv \tau_{cr_2}$, exists in the approach and the constricted channel. The above relation, Eq. (8.24), can now be rewritten:

$$\frac{h_2}{h_1} = \left(\frac{B_1}{B_2}\right)^{6/7} \left(\frac{\tau_{o_1}/\tau_{cr}}{\tau_{o_2}/\tau_{cr}}\right)^{3/7} \qquad (8.24a)$$

If the shear stress, τ_{o_2}, in the constricted channel is taken to be the critical one, $\tau_{o_2} = \tau_{cr}$, the above relation becomes:

$$\frac{h_2}{h_1} = \left(\frac{B_1}{B_2}\right)^{6/7} \left(\frac{\tau_{o_1}}{\tau_{o_2}}\right)^{3/7} \qquad (8.26)$$

already presented by Laursen (1963, p. 97). This relation is also given in Fig. 8.9, where the following is to be observed:

- For a given relative shear stress, τ_{o_1}/τ_{cr}, an increase in the width reduction, B_1/B_2, is responsible for an increase in the relative depth, h_2/h_1.
- For the special case, when the shear stress is the critical one throughout the entire channel, namely, $\tau_{o_1} = \tau_{cr} = \tau_{o_2}$, the above relations, Eqs. (8.24) or (8.26), reduce to :

$$\frac{h_2}{h_1} = \left(\frac{B_1}{B_2}\right)^{6/7} \qquad (8.27)$$

4. Allowing for a variation of the sediment sizes, $d_1 \neq d_2$, the above relation, Eq. (8.24), can be written (see Komura 1966, p. 23) as:

$$\frac{h_2}{h_1} = \left(\frac{B_1}{B_2}\right)^{6/7} \left(\frac{\tau_{o_1}}{\tau_{o_2}}\right)^{3/7} \left(\frac{d_2}{d_1}\right)^{1/7} \qquad (8.27a)$$

8.4.3 Formula for Design

1. From the material presented above (see sect. 4.2 in Graf and Altinakar 2002), the scour depth can be estimated using the following relation:

$$\frac{h_2}{h_1} = \left(\frac{B_1}{B_2}\right)^{6/7} \left(\frac{\tau_{o_1}}{\tau_{o_2}}\right)^{3/7} \quad (8.24)$$

which is given in Fig. 8.9.

The shear stress, τ_{o_1} and τ_{o_2}, defined with Eq. (8.19), can be evaluated by means of the friction coefficient, namely, according to the relations of Manning, Chézy, or Strickler (see section 3.2 in Graf and Altinakar 2002).

2. Due to a lack of laboratory data as well as a complete lack of field data, the above relation, Eq. (8.24), must be considered as a guideline estimate for hydraulic design (see Raudkivi 1990, p. 245).

8.5 Hydraulic Structure Scour

8.5.1 Notions

1. Weirs or low-head dams as well as underflow gates are installed in channels and waterways to control and/or measure the discharge (see sect. 4.4.1 and sect. 4.4.3 in Graf and Altinakar 2002). The flow passes over, q_o, and/or under, q_u, such a hydraulic structure toward the downstream (see Fig. 8.10).
2. Such a flow, which often appears in a form of jets, may have considerable hydraulic potential to scour on the downstream side of the structure; a scour hole is formed.
3. Since scour hole develops rather rapidly, the engineer takes interest in the equilibrium scour depth, d_s, and the length of the scour hole, L_s.
4. Scour will not only endanger the stability of the channel bed, but it might also have devastating effects on the hydraulic structure itself.
5. Considering a hydraulic structure positioned in an open channel, where the flow is unidirectional, uniform, and steady and where the bed is made of cohesionless sediments, the equilibrium *scour depth*, d_s, depends on:
   - The *fluid*: the density, ρ, and viscosity, ν
   - The *sediments*: the density, ρ_s, and a characteristic diameter, d
   - The channel *flow* (in the absence of the structure): the water depth, h;, the average velocity, U; and the gravity, g
   - The interference of the *structure* on the channel flow: resulting in a drop in the water depth, $\Delta h = h_1 - h_2$

This allows to write:

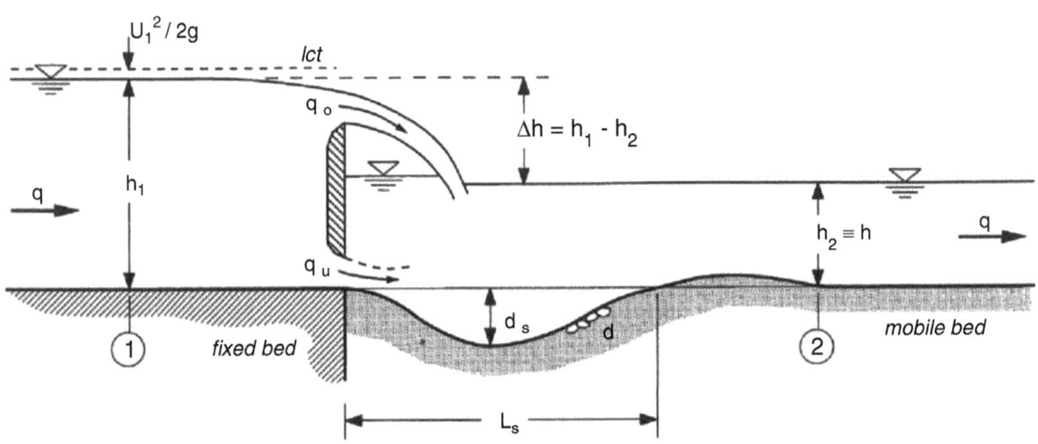

Fig. 8.10 Schema of hydraulic structure scour; a combined sluice gate allowing overflow and underflow

$$d_s = f_1(\rho,\ \nu;\ \rho_s,\ d;\ h,\ U,\ g;\ \Delta h) \quad (8.28)$$

6. Using dimensional reasoning the following parameters are obtained:

$$\frac{d_s}{d} = f_2\left(\frac{u_* d}{\nu},\ \frac{(Uh)^2}{g\ d^3},\ \frac{\rho_s}{\rho},\ \frac{h}{d},\ \frac{\Delta h}{d}\right) \quad (8.29)$$

   Usually it is assumed (see Kotoulas 1967 p. 38) that the particle Reynolds number, Re_*; the relative density, ρ_s/ρ; and the relative roughness, h/d, play a negligible role; thus:

$$\frac{d_s}{d} = f_3\left(\frac{q^2}{g\ d^3},\ \frac{\Delta h}{d}\right) \quad (8.30)$$

   where $q = Uh$ is the unit discharge in the channel.

   For flow *over* the structure, the discharge is $q = q_o$.

   For flow *under* the structure, the discharge is $q = q_u$.

   For flow *over* as well as *under* the structure, the discharge is $q = q_o + q_u$.

7. Stilling basins are sometimes used to *protect* against destructive scouring. The dimensioning will require knowledge of the possible length, L_s, of the scour hole.

   There might still be scouring downstream of the stilling basin (see Raudkivi 1990, p. 283) where rock blocks or riprap could be placed as a measure of protection (see Sect. 8.2.5).

8.5.2 Flow Over a Structure

1. Scour due to flow *over* a hydraulic (low-head) structure is schematically shown in Fig. 8.11. The scour depth, d_s, is the equilibrium maximum value, which develops rather fast. The unprotected channel bed below the free overfall is eroded. Downstream of the scour hole of a certain length, L_s, may appear a mound of deposition. The free overfall, plunging in the form of a jet into the downstream water, forms two vortices, whose erosive power often considerable is responsible for the scour hole.
2. For the prediction of the scour depth, d_s, various formulae have been developed. However, all these prediction formulae are based on small-scale laboratory studies; upscaling of these in order to predict scour on real structures has to be viewed with caution.

   Here is a selection of some relations presently in use.
3. Early attempts by Schoklitsch (1932) and Veronese (1937) have shown (see Eggenberger and Müller 1944) that the scour depth, d_s, cannot be expressed by a simple relation.
4. A limited set of laboratory data, using the similarity of Froude, allowed Eggenberger and Müller (1944, p. 31 and p. 45) to propose the following relation:

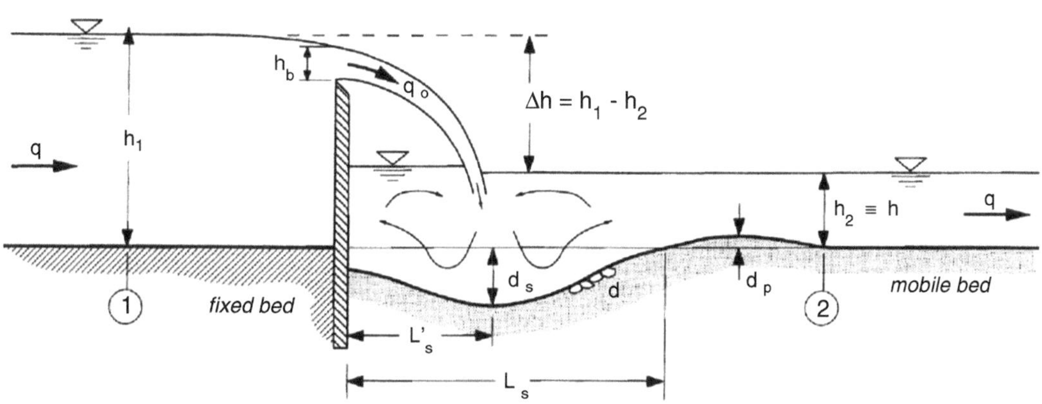

Fig. 8.11 Schema of flow over a structure

$$h + d_s = w \frac{\Delta h^{0.5}\ q^{0.6}}{{d_{90}}^{0.4}} \quad (8.31)$$

where the coefficient is:

$$w = 1.44\left[\mathrm{s}^{0.6}/\mathrm{m}^{0.3}\right], \text{ if } d_{90}[\mathrm{m}]$$

$$w = 22.88\left[\mathrm{s}^{0.6}/\mathrm{m}^{0.3}\right], \text{ if } d_{90}[\mathrm{mm}]$$

$$w'' = 3.6[-]$$

This relation was subsequently generalized to:

$$h + d_s = \frac{w''}{g^{0.3}}\left(\frac{\rho}{\rho_s - \rho}\right)^{4/9} \frac{\Delta h^{0.5}\ q^{0.6}}{{d_{90}}^{0.4}} \quad (8.32)$$

where d_{90} [m] is the diameter of the largest particles in the granulometry of the original bed; thus, an armoring (see Sect. 6.3.4) takes place.

It has been frequently remarked that the above formula, Eq. (8.31), overestimates the scour depth.

For the length of the scour hole, the following relations are advanced:

$$\frac{L_s}{h + d_s} \approx 1.8 \qquad \frac{L_s'}{h + d_s} \approx 0.5 \quad (8.33)$$

5. Performing laboratory test and evaluating these data according to the dimensional reasoning (see Eq. 8.30) Kotoulas (1967, p. 40) proposed a relation for the dimensionless equilibrium scour depth, such as:

$$\frac{h + d_s}{d_{95}} = 1.9\left(\frac{q^2}{g\ {d_{95}}^3}\ \frac{\Delta h}{d_{95}}\right)^{0.35}$$

or (8.34)

$$h + d_s = 1.9\ \frac{1}{g^{0.35}}\left(\frac{\Delta h^{0.35}\ q^{0.7}}{{d_{95}}^{0.4}}\right)$$

The temporal evolution of the scour depth, $(h + d_s)_t$, was given by:

$$(h + d_s)_t = \left[1.9 \frac{1}{g^{0.35}}\left(\frac{\Delta h^{0.35}\ q^{0.7}}{{d_{95}}^{0.4}}\right)\right]\left(1 - e^{-0.55t^{1/5}}\right) \quad (8.35)$$

Equilibrium scour is obtained in the laboratory after 24 [h].

A comparison of Eq. (8.34) with Eq. (8.30) was done and showed that the scour depth, d_s, obtained with Eq. (8.31) is usually 85 % larger.

6. Many of the available model and prototype data were analyzed by Mason and Arumugam (1985, p. 232). The following relation was proposed:

$$h + d_s = K_M \frac{1}{g^{0.35}}\left(\frac{\Delta h^y\ q^x}{{d_m}^{0.1}}\right) h^{0.15} \quad (8.36)$$

where d_m [m] is the mean particle size, rather than d_{90} [m]. The coefficient and the exponents are given, being for:

model data : $K_M = 3.27,\ y = 0.05$
$x = 0.6$
prototype and model data : $K_M = 6.42 - 3.10\ \Delta h^{0.1},\ y = 0.15 + \Delta h/200$
$x = 0.6 - \Delta h/300$

This formula is said to be applicable for bed material of $0.001 < d_m[\mathrm{m}] < 0.028$; it can also be applied for all types of rocks, by assuming $d_m = 0.25[\mathrm{m}]$.

Since the investigated data set included also data from high-head structures (spillway chute flip buckets and tunnel outlets), the above relation, Eq. (8.36), might be less valuable for low-head structure in fluvial hydraulics.

7. It should be remarked that the above presented formulae, Eqs. (8.31), (8.32), (8.34), and (8.36), maintain a rather simple combination of the parameters obtained through dimensional reasoning.

Some other formulae, having a complex form—these include the Russian relations (see Mason and Arumugam 1985, p. 224)—have not been considered. Also not included are formulae obtained from data of high-head structures (see Breusers 1991, p. 119).

8.5.3 Flow Under a Structure

1. Scour due to flow *under* a hydraulic structure is schematically shown in Fig. 8.12. The

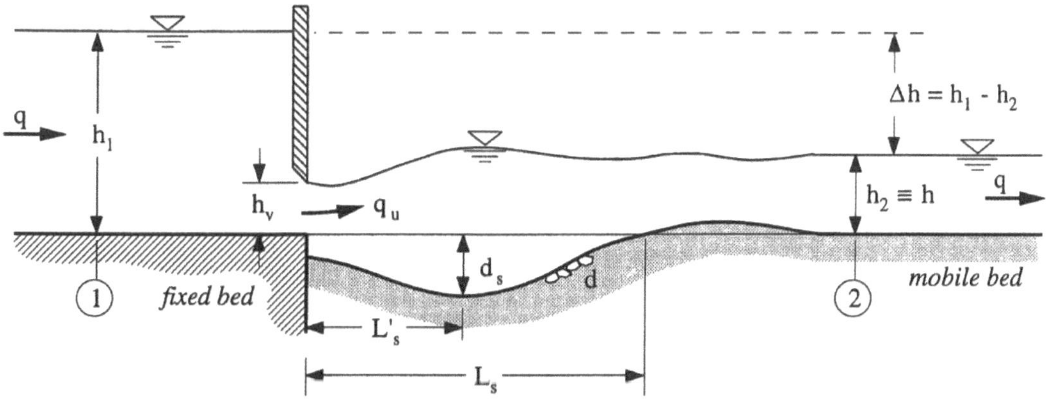

Fig. 8.12 Schema of flow under a structure

underflow forms a kind of horizontal jet moving toward the downstream, whose erosive power could become considerable.

2. For the prediction of the scour depth, d_s, a few formulae have been advanced; all are based on small-scale laboratory studies.

   Here is a selection of some relations presently in use.

3. The form of the relation, Eq. (8.31), established by Eggenberger and Müller (1944, p. 40) for overflow, $q = q_o$, was found valid for underflow, $q = q_u$, or:

$$h + d_s = w \frac{\Delta h^{0.5} \ q^{0.6}}{{d_{90}}^{0.4}} \tag{8.37}$$

   The coefficient for a submerged jet is $w = 10.35\left[\mathrm{s}^{0.6}/\mathrm{m}^{0.3}\right]$ and for a free jet it is $w = 15.40\left[\mathrm{s}^{0.6}/\mathrm{m}^{0.3}\right]$, if $d_{90}[\mathrm{mm}]$.

   For the same flow conditions, q and Δh, and the same sediment, d_{90}, it is seen that scour at underflow produces a smaller scour depth, $d_{s'}$.

   For the length of the scour hole, the following relations are advanced:

$$\frac{L_s}{h + d_s} \approx 6 \quad \frac{L_s'}{h + d_s} \approx 3 \tag{8.38}$$

   In comparison with the scour hole for overflow (see Eq. 8.33), the one for underflow is considerably larger.

4. Reviewing available laboratory data, the following relation was advanced by Breusers (1991, p. 107):

$$\frac{d_s}{h_v} = 0.008 \left(\frac{U_u}{u_{*_{cr}}}\right)^2 \tag{8.39}$$

   valid for $(U_u/u_{*_{cr}}) < 100$. Here $U_u = q_u/h_v$ is the underflow velocity and $u_{*_{cr}} = (\tau_{o_{cr}}/\rho)^{1/2}$ is the critical (scour) shear velocity of the bed material.

   For the length of the scour hole, an estimation is given as:

$$\frac{L_s}{d_s} \approx 5 \text{ to } 7 \tag{8.40}$$

   Above relations are developed for scour by submerged horizontal jets.

5. The scour resulting from a combination of flow *under* and *over* the hydraulic structure (see Fig. 8.10) was investigated by Eggenberger and Müller (1944, p. 43). The general form of the relation valid for overflow, Eq. (8.31), or for underflow, Eq. (8.37), was used:

$$h + d_s = w \frac{\Delta h^{0.5} \ q^{0.6}}{{d_{90}}^{0.4}} \tag{8.41}$$

   where $q = q_o + q_u$. The coefficient was determined as being:

$$w = 22.88 - \frac{1}{0.0049 \ {R_q}^3 - 0.0063 \ {R_q}^2 + 0.029 \ R_q - 0.064} \tag{8.42}$$

with $R_q = q_o/q_u$ as the ratio of the overflow to the underflow unit discharge. Note, for underflow only, when $R_q << 1$, $w = 10.35\left[s^{0.6}/m^{0.3}\right]$ and, for overflow only, when $R_q >> 1$, $w = 22.88\left[s^{0.6}/m^{0.3}\right]$.

References

Breusers HN (1991) Scour by jets. Ch. 6 in scouring. In: Breusers H, Raudkivi AJ (eds) IAHR, hydraulic structures design manual N° 2. A.A. Balkema, Rotterdam

Breusers HNC (1965) Scour around drilling platforms. IAHR Bull Hydraul Res 19:276

Breusers HN et al (1977) Local scour around cylindrical piers. J Hydrol Res 15(3):211–252

Da Cunha LV (1973) Local scour at obstacles protruding from lateral walls (in Portuguese). Mem. 428, Laboratório Nacional de Engenharia Civil, Lisbon

Eggenberger W, Müller R (1944) Experimentelle und theoretische Untersuchungen über das Kolkproblem, Mitteil. Versuchsanstalt f. Wasserbau, N° 5, Zürich

Graf WH, Altinakar MS (2002) Fluvial hydraulics: flow and transport in channels of simple geometry. Wiley, Chichester

Graf WH, Yulistiyanto B (1998) Experiments on flow around a cylinder; the velocity and vorticity fields. J Hydr Res 36(4):637–653

Johnson PA (1995) Comparison of pier-scour equations using field data. Proc Am Soc Civ Eng 121:JHE 8

Komura S (1966) Equilibrium depth of scour in long constrictions. J Hydraul Div 92(5):17–37

Kotoulas D (1967) Das Kolkproblem im Rahmen der Wildbachverbauung, Mitteil. Schweizer Anstalt f. forstliches Versuchswesen, vol 43/1, Birmensdorf

Laursen EM (1958) Scour at bridge crossings. Iowa Highway Research Board Bulletin 8, 53 p

Laursen EM (1962) Scour at bridge crossings. Trans Am Soc Civ Eng 127(3294 pt I):166–209

Laursen EM (1963) An analysis of relief bridge scour. Proc Am Soc Civ Eng 89:HY3

Mason PJ, Arumugam K (1985) Free-jet scour below dams and flip buckets. Proc Am Soc Civ Eng 111: JHE 2

Melville BW (1992) Local scour at bridge abutments. Proc Am Soc Civ Eng 118:JHE 4

Nordin C et al (1989) Testing design methods for local scour at bridge piers. Personal communication, USA

Raudkivi AJ (1976 and 1990) Loose boundary hydraulics. Pergamon Press, Oxford

Raudkivi AJ (1991) Scour at bridge piers. Ch. 5 in scouring. In: Breusers H, Raudkivi AJ (eds) IAHR, hydraulic structures design manual N° 2. A.A. Balkema, Rotterdam

Schoklitsch A (1932) Kolkbildung unter Oberfallstrahlen [Scour formation below overfall jets]. Wasserwirtschaft 25(24):341–343 (in German)

Shen HW (1971) River mechanics, vol II, Chap. 23, H.W. Shen Publications, Fort Collins

Veronese A (1937) Erosioni di fondo a valle di un scarico [Bottom erosions downstream of a dam]. Annali dei Lavori Pubblici 75(9):717–726 (in Italian)

Walter H. Graf Laboratoire d'Hydraulique Environnementale (LHE), EPF Lausanne, ENAC IIC LHE, Lausanne, Switzerland

University of Lausanne, Lausanne, Switzerland

Mustafa S. Altinakar National Center for Computational Hydroscience and Engineering (NCCHE), The University of Mississippi, University, MS, USA

Emerging Methodologies for Turbulence Characterization in River Dynamics Study 9

Harinarayan Tiwari, Amir Khan, and Nayan Sharma

Abstract

River engineering study consists of large variation in time and spatial scales. Timescale of river varies from years to second, and, similarly, the variation of spatial scales is from kilometre to millimetre. Spatial scales can be divided into river basin scale and hydraulic scale, and temporal scale can be divided into hydrological, hydraulic and turbulence. Each spatio-temporal scale has fixed contextual uses. In general, turbulence plays the most key role with respect to the influences that rivers have on their channels and beds. Turbulent flows are characterized by asymmetrical patterns, irregular behaviour and the existence of various spatio-temporal scales. To extract better turbulence events and flow structure using point velocity measurements (Eulerian approach) in river, we are proposing generalized three-dimensional octant events instead of conventional two-dimensional quadrant events. Beyond that, we characterize the transitional probability of octant event occurrence in the case of unsteady flow condition. In the field, there is the assumption of steadiness of the flow under high unsteady conditions. Basically, river discharges and all the associated processes are physically unsteady, and river channel flows are typically non-uniform. In this chapter we are mainly discussing the new emerging methodological aspects to characterize the river turbulence using state-of-the-art technology. In this chapter, some of the major issues and developments linked with river dynamics and turbulence study have also been discussed with two case studies. The case studies have been presented and discussed using experimental data and their interpretation in light of river dynamics. The study has significant importance because the turbulent motion is the natural state of river engineering problems.

H. Tiwari (✉) • A. Khan • N. Sharma
Department of Water Resources Development and Management, Indian Institute of Technology Roorkee, Roorkee, Uttarakhand, India
e-mail: haribit31@outlook.com

N. Sharma (ed.), *River System Analysis and Management*, DOI 10.1007/978-981-10-1472-7_9

9.1 Introduction

River dynamics is subject of infinite opportunity with multidimensional faces (Fig. 9.1). Considering river as one of the important and limited source of freshwater, it is our prime duty to capture state-of-the-art approaches for the river science research. Rivers are natural open channels, draining water from the land. Rivers are auto-shaped channels in which channel dimensions, outline, slope and bed material character are adjusted through scouring and deposition of the unbalanced sediments through which they flow. The sediment transport rate depends not only on water discharge but also on channel slope, morphology and bed characteristics (Church and Ferguson 2015). So, channel system state and process are connected by feedback loops. Despite increasing appreciation of the potential of turbulence on sediment and momentum transport, key knowledge gaps exist in relation to turbulent events for sediment dynamics at micro-scales.

9.2 Multiple Scales in Rivers

Studies of river engineering are characterized by many scales of time and space (Fig. 9.2). The river scales vary from years to second in time and kilometre to millimetre in length (Church 1995; Helder and Ruardij 1982; Whipple and Tucker 1999). A viable theory of river science research must accommodate a variety of processes that occur at diverse scales. Interlinking between different scales of the processes requires enormous amount of quality data derived from the field and laboratory studies. The data quality and their

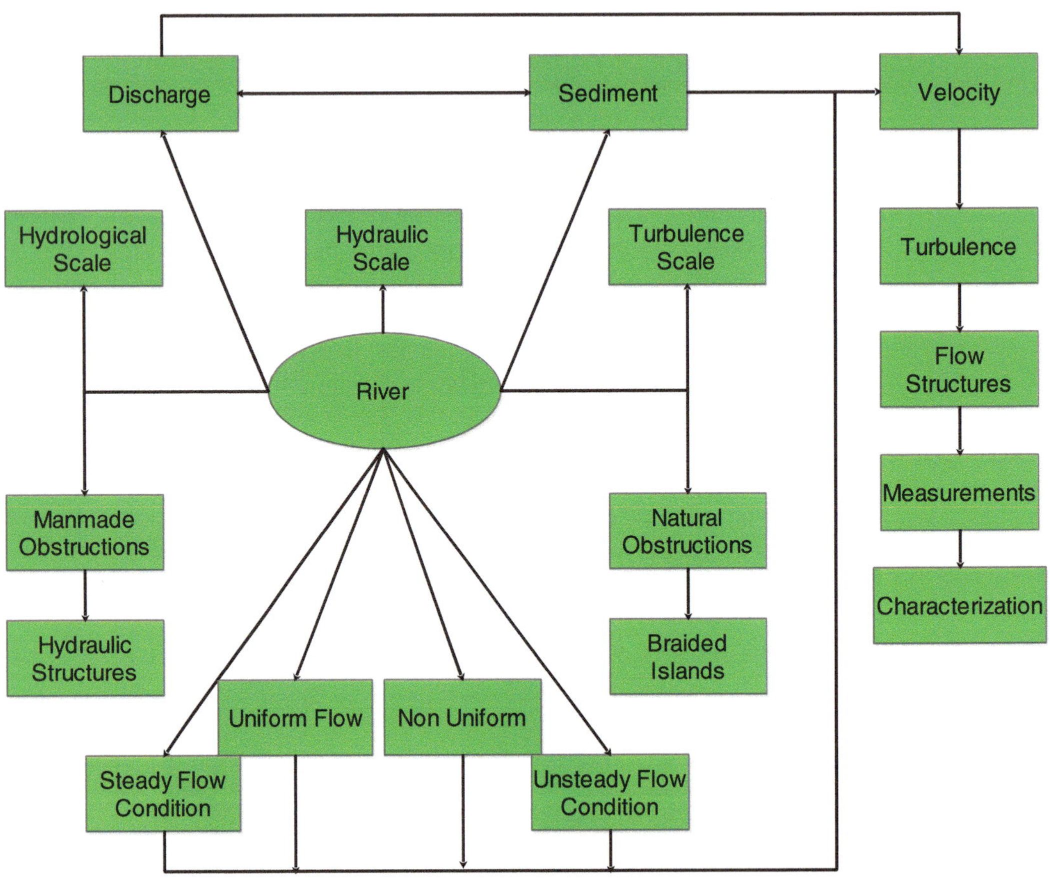

Fig. 9.1 River and some underlying components

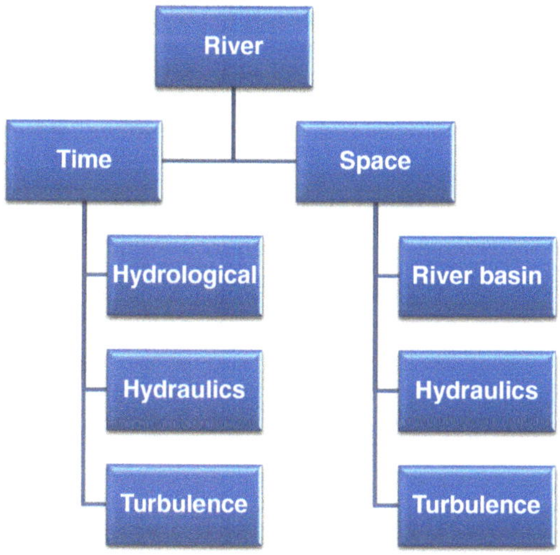

Fig. 9.2 Multiple scales for river science research

extraction are very much location specific, so the development of general river model based on multiple theoretical aspects is the major challenge for river engineers.

9.2.1 Hydrological Scale

The hydrological timescale of the river varies from years to hour which depends on the parameters of analysis. One concept of dealing with variability over many orders of magnitude is the notion of characteristic scales. The idea of a characteristic scale is that instead of dealing with a spectrum of lengths and times, one adopts a typical length and time that are representative of a particular process (Fig. 9.3). Often a

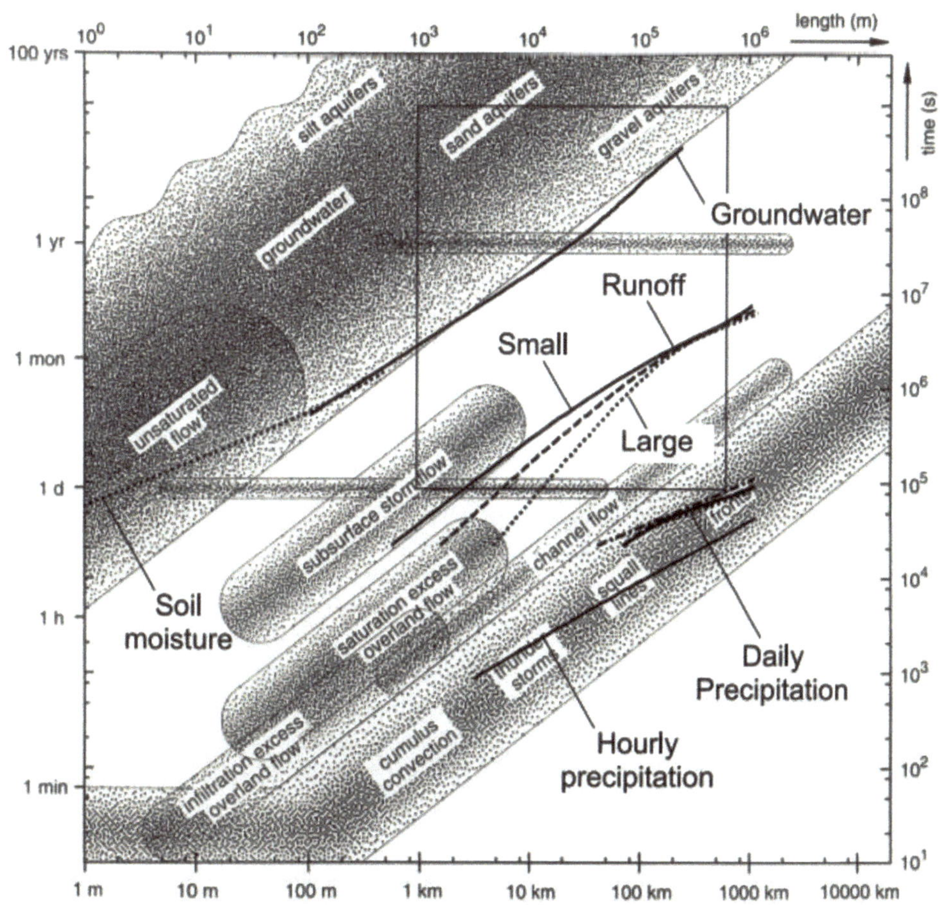

Fig. 9.3 Schematic relationship between spatial and temporal process scales in hydrology (Blöschl and Sivapalan 1995; Seyfried and Wilcox 1995)

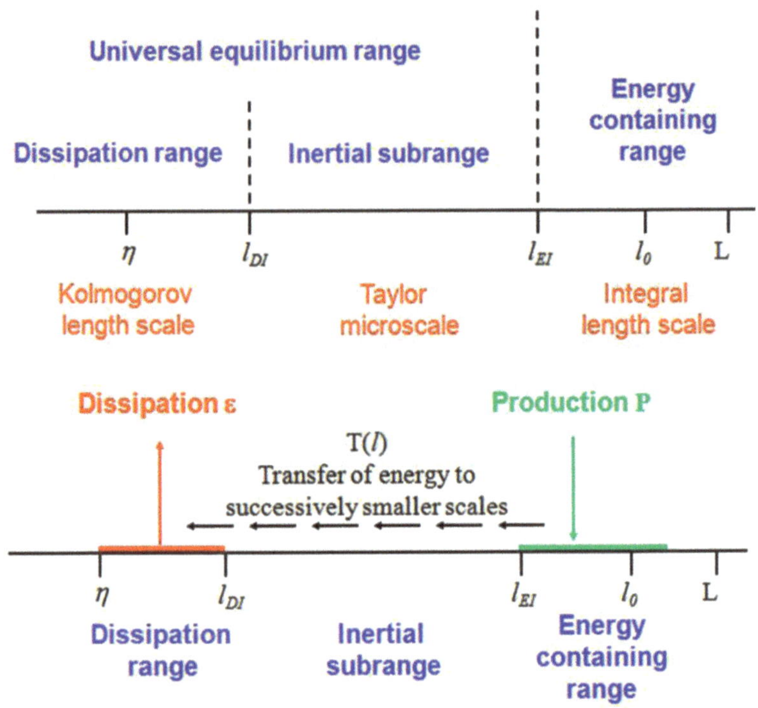

Fig. 9.4 Turbulence scale in hydraulics

characteristic scale is an order of magnitude figure, given as an integer power of ten, rather than a precise number.

9.2.2 Hydraulic Scale

Till date, there is no clear-cut discrimination between different timescales and space scales of river. Assumptions vary with the types of the problem and scope of the studies. Under basic assumptions, hydraulic timescale may be considered to be hours to second.

9.2.3 Turbulence Scale

According to Richardson, 'Big whirls have little whirls which feed on their velocity; and little whirls have lesser whirls, and so on to viscosity in the molecular sense' (Richardson 1921). The smallest known scale is the Kolmogorov scale under the dissipation range. At integral length scale, there is the production of energy by movement of fluid particle and it dissipates at finer scales (Fig. 9.4).

9.3 Turbulence

Turbulence appears in the channel when inertial forces dominate over viscosity forces in a flow, which means that the Reynolds number evaluating the relative strength of these two forces must exceed a certain threshold (Bradshaw 2013). The measurement and characterization of turbulence is certainly important due to its intrinsic association with sediment transportation, suspension and deposition (Gordon 1975). As any flow in river has three dimensions, so three-dimensional flow structure characterizations add an improvement to sediment entrainment and transport mechanism (Keshavarzi and Gheisi 2006; Mianaei and Keshavarzi 2008). Bursting events and their transitional probability occurrence help to extract better information on flow mechanism for three-dimensional numerical model developments. In the three-dimensional bursting events, there is consideration of lateral velocity in concurrence with longitudinal and vertical velocity simultaneously. In context of four basic bursting events, addition of lateral velocity makes it three dimensional (internal and external). Flow structures in environmental turbulent flows contain a significant part of the turbulent energy. Hence, they are important for momentum exchange as well as mixing and scalar and sediment transport (Keylock et al. 2005, 2014; Nepf and Vivoni 2000). The results of bursting events (quadrant and octant) add to understanding of the hydrodynamics of sediment transport mechanism and may be used for the development of better parameterizations of small-scale processes for application in large-scale studies under different hydraulic conditions. In addition to the complexity of the near-bed flow field, there has been widespread recognition of the need to go beyond Reynolds stress formulations in order to understand the interaction between turbulence and sediment entrainment (Keylock et al. 2014; Sharma and Tiwari 2013; Tiwari and Sharma 2015a, b). In the light of importance of turbulence bursting events, characterization of it under several natural conditions is needed. This chapter presents the study related to natural river condition under laboratory investigations. In the first part of the study, attempt has been made to derive the character of bursting events (quadrant and octant) under quasi-steady and unsteady flow conditions. In the second part of the study, the authors explored connection between the bed scouring/depositions features and bursting events under braided river laboratory model.

9.4 Case Study: Bursting Events for Quasi-steady and Unsteady Flow

Open channel flow in the nature is generally unsteady and assumed to be steady for estimation ease. The effect of unsteadiness in open channel has been continued as critical area of hydraulic research due to its wide field applications. Bursting events (quadrant and octant) have specific features depending on type of flow along with their hydraulic conditions. Transition probability calculates the occurrence of specific event after another event. The one-step transition probability is the probability of transitioning from one state to another in a single step and also termed as first-order Markov chain process. This study estimates the transition probability matrix of bursting events for the case of steady and unsteady flows. The transition probability matrix has been calculated using first-order Markov chain process. Significant change has been observed for the transition probabilities under steady and unsteady flow condition.

9.4.1 General

Flow is defined as steady when the flow parameters (i.e. discharge, depth) are constant over time span. Presumption of time span may vary according to the needs of the study. In nature, no open channel flow can be considered as purely steady. Quasi-steady is the state where unsteady flow over certain time interval can be assumed as steady. Nearly all geophysical flows, including open channel flows, are also turbulent. Bursting events are principal component of turbulent events. The coherent structures related with the bursting process are random in space and highly three dimensional, and their distinctive form and development have been hard to account and model in particular terms. Unsteady flow itself is very complex and an intensive area for a hydraulic researcher. Bursting events, especially quadrant events under quasi-steady flow, have relatively more literature than in unsteady flow. Quadrant events have been studied extensively along with Reynolds stress. The effects of the turbulent events on the sediment entrainment, transport and deposition have been comprehensively considered by the researchers. Literature review indicates that the main focus of bursting events is centred on two dimensional by considering two velocity components (u and w), where u is longitudinal velocity and w is the vertical velocity. Availability of high-quality velocity measuring instruments (i.e. acoustic Doppler velocimeter (ADV), laser Doppler velocimeter, etc.) has been deployed for observation of the three-dimensional velocity components. So, the combination of turbulent events with unsteadiness presents large complexities and needs experimental investigation.

9.4.2 Methodology

The bar chart (Fig. 9.5) has been presented to show the basic steps for first case study. The number of samples has been collected on very higher side (90,000). The experimental design was kept simple as shown in Fig. 9.6. For the first two conditions (E1 and E2 – Table 9.1), flow parameters (average discharge, average velocity, average depth, etc.) have been kept constant, while variation comes into picture for the third experimental condition (E3 – Table 9.1).

For each case, the total number of velocity samples has been taken equal. There are five general steps, which have been held during this study. It includes laboratory set-up, data collection, data modification, bursting events and then estimation of transition probability matrix.

9.4.3 Experimental Set-Up

The experimental study was conducted in the controlled open channel of the river engineering laboratory (Department of Water Resource Development and Management, Indian Institute of Technology, Roorkee, India). The channel is about 3 m long and 0.5 m wide. Figure 9.6 shows the L-section of the channel. The velocity measurements were taken at the mid of the channel from the channel bed using an acoustic Doppler velocimeter (ADV). ADV is a competent instrument for conducting turbulence study. SonTek 16 MHz ADV has been used to take

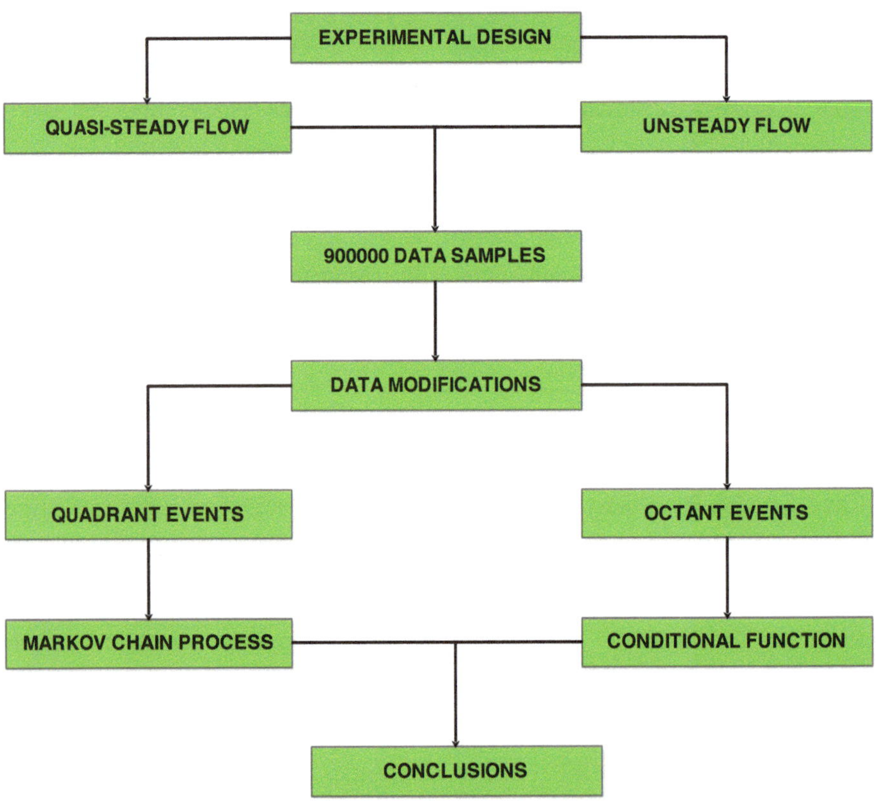

Fig. 9.5 Methodological charts for the case study

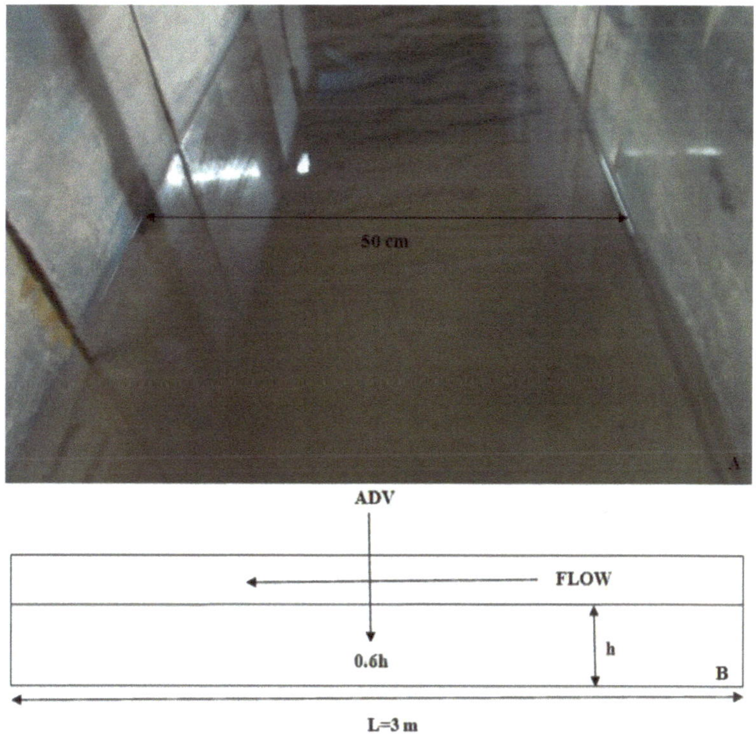

Fig. 9.6 (**a**) Cross section of experimental flume and (**b**) schematic diagram of L-section

Table 9.1 Experimental condition for case study 1

Experimental condition	Discharge (cc/s)	No. of velocity samples	Type of flow
E1	6055	90,000	Quasi-steady
E2	8400	90,000	Quasi-steady
E3	4670–9550	90,000	Unsteady

Table 9.2 Bursting events in *uw* plane

Sign of fluctuating velocities			
u'	w'	Quadrant events	Quadrant
+	+	Outward interaction (OI)	1
−	+	Ejection (EJ)	2
−	−	Inward interaction (II)	3
+	−	Sweeps (SW)	4

velocity measurements. Table 9.1 presents discharge variations for the experimental conditions, which have been considered in this part of study.

9.4.4 Theory of Bursting Events

Kline et al. (1967) divided two-dimensional bursting events into four-part (outward interaction, ejection, inward interaction and sweep) events based on longitudinal and vertical velocities (Table 9.2). Results of Kline et al. (1967) evaluate sweep and ejection events as important among the four bursting events on sediment motion (Kline et al. 1967). Reynolds shear stress is also affected by the occurrence of bursting events.

9.4.4.1 Bursting Events (Quadrant Events)

Bursting event research has continued using generally two-dimensional velocities (u and w). The study can be extended to analogous events (Table 9.3) in the transverse plane (v and w) and horizontal plane (u and v) to extract better information and their effect. Estimation of probability of occurrence (in %) has been carried out using Eq. (9.1) for every time series:

$$Q_k = \frac{\text{Number of event in } k\text{th quadrant } (k = 1 \text{ to } 4)}{\text{Total number of events } (N)} \times 100 \quad (9.1)$$

9.4.4.2 Theory of Transitional Probability

The probability based on the change of state with respect to time can be termed as transitional probability. For any time series having set of events, there are the chances of occurrence of any particular event after the other events. Q_{1-1} can be defined as the probability of occurrence of first quadrant events after first quadrant event. In this case where we have four sets of events (quadrant events), there is estimation of 4 × 4 matrix of transitional probability (T_{px}) of random (x) time series (Eq. 9.2):

$$T_{px} = \begin{bmatrix} Q_{1-1} & Q_{1-2} & Q_{1-3} & Q_{1-4} \\ Q_{2-1} & Q_{2-2} & Q_{2-3} & Q_{2-4} \\ Q_{3-1} & Q_{3-2} & Q_{3-3} & Q_{3-4} \\ Q_{4-1} & Q_{4-2} & Q_{4-3} & Q_{4-4} \end{bmatrix} \quad (9.2)$$

9.4.4.3 Results and Discussions

Transitional probability of bursting events using Eq. 9.1 has been estimated for three experimental conditions mentioned above (Table 9.1). It is estimated for all three planes (uw, uv and vw) under every experimental condition (Tables 9.4, 9.5 and 9.6).

Table 9.3 Analogous bursting events (quadrant events)

Sign of fluctuating velocities		Quadrant
v'	w'	
+	+	1
−	+	2
−	−	3
+	−	4
u'	v'	
+	+	1
−	+	2
−	−	3
+	−	4

Table 9.5 Transitional probabilities of quadrant events in horizontal plane (uv)

Expt. condition		Q1	Q2	Q3	Q4
E1	Q1	28.37	25.17	21.24	25.22
	Q2	23.26	30.02	24.66	22.06
	Q3	20.88	26.11	27.67	25.34
	Q4	24.50	22.22	23.50	29.78
E2	Q1	27.72	23.86	23.17	25.25
	Q2	24.91	25.93	25.81	23.35
	Q3	22.82	25.87	27.74	23.56
	Q4	24.87	23.46	25.46	26.22
E3	Q1	44.57	15.58	10.92	28.92
	Q2	15.15	37.95	32.42	14.49
	Q3	11.10	33.36	39.41	16.13
	Q4	28.35	14.40	15.70	41.55

Table 9.4 Transitional probabilities of quadrant events in vertical plane (uw)

Expt. condition		Q1	Q2	Q3	Q4
E1	Q1	26.08	21.93	23.75	28.24
	Q2	22.25	26.21	27.69	23.85
	Q3	23.47	27.91	26.58	22.04
	Q4	26.32	23.23	23.12	27.33
E2	Q1	27.00	25.22	24.33	23.45
	Q2	23.74	27.11	27.38	21.77
	Q3	23.88	23.70	27.29	25.13
	Q4	26.80	22.90	23.39	26.90
E3	Q1	34.60	23.75	15.13	26.51
	Q2	22.13	44.77	20.58	12.52
	Q3	13.66	25.95	41.15	19.24
	Q4	22.02	14.25	24.67	39.06

Table 9.6 Transitional probabilities of quadrant events in transverse plane (vw)

Expt. condition		Q1	Q2	Q3	Q4
E1	Q1	22.42	25.43	20.99	31.17
	Q2	19.50	29.03	23.85	27.63
	Q3	20.15	31.26	22.47	26.12
	Q4	22.88	26.86	20.11	30.15
E2	Q1	26.47	24.46	23.57	25.50
	Q2	25.56	26.59	23.89	23.96
	Q3	23.27	25.39	27.12	24.22
	Q4	24.80	23.76	25.87	25.56
E3	Q1	33.45	31.62	14.07	20.85
	Q2	26.48	34.83	19.01	19.68
	Q3	14.81	21.51	33.21	30.46
	Q4	19.25	20.25	27.29	33.20

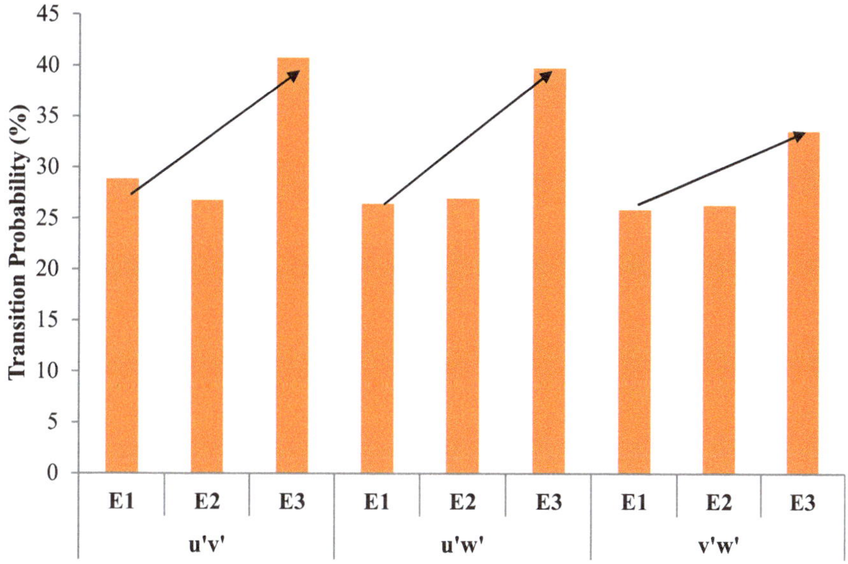

Fig. 9.7 Average diagonal transitional probabilities of quadrant events

The detailed analysis of Tables 9.4, 9.5, and 9.6 extracts the very notable point that at every plane, average probability of occurrence of diagonal event has been increasing under unsteady-state conditions (Fig. 9.7). The results also reveal that the effect of unsteadiness is maximum in horizontal plane (*uv*) and minimum in transverse plane (*vw*). Although this result may not be universal, but it can easily bring out some noticeable point that the contribution of unsteadiness is not similar in every plane, as in the case of steady flow.

9.4.5 Three-Dimensional Bursting Events (Octant Events)

Ali Reza Keshavarzi proposed a method of three-dimensional octant analysis, and it enables to take the effect of secondary flow (Keshavarzi and Gheisi 2006). As any flow in open channel has three dimensions, so three-dimensional bursting events may add an improvement to sediment entrainment and transport mechanism (Tiwari and Sharma 2014). In the three-dimensional bursting events, there is the consideration of lateral velocity in concurrence with longitudinal and vertical velocity simultaneously. In the context of four basic bursting events, addition of lateral velocity makes it three dimensional (internal and external – Table 9.7). Internal element of any of the event contributes into main flow domain and external element out to flow domain. Octant events and their transitional probabilities have been estimated using Eqs. 9.3 and 9.4:

$$O_k = \frac{\text{Number of events in } k\text{th octant } (k = 1 \text{ to } 8)}{\text{Total number of events } (N)} \times 100 \qquad (9.3)$$

$$T_{px} = \begin{bmatrix} O_{1-1} & O_{1-2} & O_{1-3} & O_{1-4} & O_{1-5} & O_{1-6} & O_{1-7} & O_{1-8} \\ O_{2-1} & O_{2-2} & O_{2-3} & O_{2-4} & O_{2-5} & O_{2-6} & O_{2-7} & O_{2-8} \\ O_{3-1} & O_{3-2} & O_{3-3} & O_{3-4} & O_{3-5} & O_{3-6} & O_{3-7} & O_{3-8} \\ O_{4-1} & O_{4-2} & O_{4-3} & O_{4-4} & O_{4-5} & O_{4-6} & O_{4-7} & O_{4-8} \\ O_{5-1} & O_{5-2} & O_{5-3} & O_{5-4} & O_{5-5} & O_{5-6} & O_{5-7} & O_{5-8} \\ O_{6-1} & O_{6-2} & O_{6-3} & O_{6-4} & O_{6-5} & O_{6-6} & O_{6-7} & O_{6-8} \\ O_{7-1} & O_{7-2} & O_{7-3} & O_{7-4} & O_{7-5} & O_{7-6} & O_{7-7} & O_{7-8} \\ O_{8-1} & O_{8-2} & O_{8-3} & O_{8-4} & O_{8-5} & O_{8-6} & O_{8-7} & O_{8-8} \end{bmatrix} \quad (9.4)$$

9.4.5.1 Results and Discussions

Results of octant events (Fig. 9.8) show some basic intrinsic relationship between octant events. It shows that there is positive intimacy between internal outward events ($O1$) and external inward events ($O6$). It also presents the similar positive intimacy among external outward interaction ($O2$), external ejection ($O4$)

Table 9.7 Three-dimensional bursting events (octant events)

SL no.	Sign of fluctuating velocities			Octant events
	u'	v'	w'	
1	+	+	+	Internal outward interaction (IOI)
2	+	−	+	External outward interaction (EOI)
3	−	+	+	Internal ejection (IEJ)
4	−	−	+	External ejection (EEJ)
5	−	+	−	Internal inward interaction (III)
6	−	−	−	External inward interaction (EII)
7	+	+	−	Internal sweeps (ISW)
8	+	−	−	External sweeps (ESW)

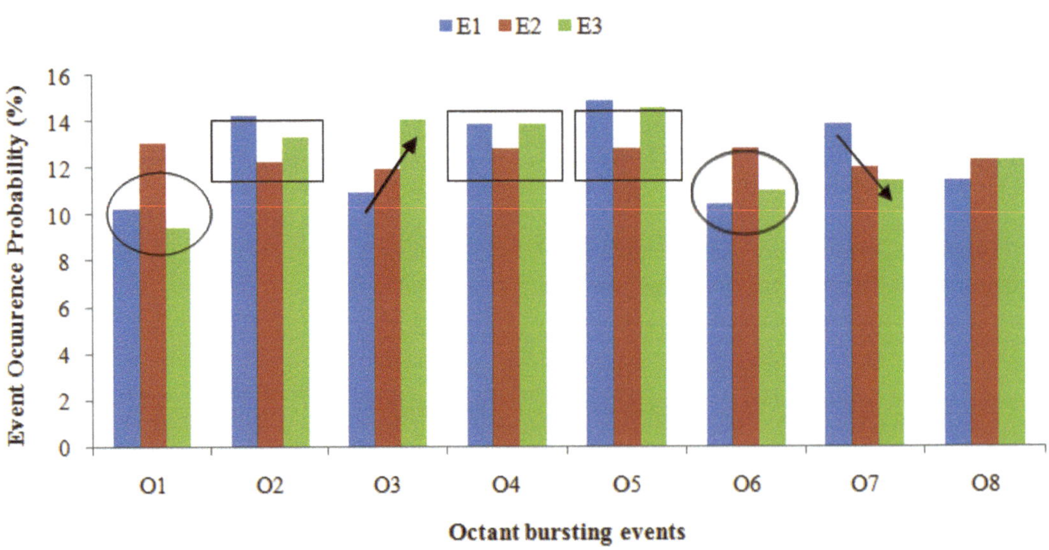

Fig. 9.8 Octant events under different experimental conditions

Table 9.8 Transitional probabilities of octant events for condition E1

11.66	14.06	10.37	11.72	14.73	9.56	16.41	11.49
10.83	15.25	9.49	12.43	13.89	9.86	15.23	13.01
10.50	14.18	10.59	12.70	15.00	10.13	14.76	12.13
10.05	14.10	10.71	13.37	15.09	10.65	14.06	11.97
10.08	13.80	11.22	13.72	15.44	10.58	13.77	11.40
9.99	13.93	11.21	14.17	15.31	10.71	13.37	11.31
10.27	13.92	11.16	13.85	15.15	10.46	13.89	11.30
10.28	14.25	10.96	13.88	14.87	10.41	13.91	11.45

Table 9.9 Transitional probabilities of octant events for condition E2

14.88	12.68	12.27	12.00	12.40	11.61	12.82	11.34
14.24	12.76	12.31	12.91	12.48	11.85	11.93	11.52
13.68	12.27	12.60	12.94	13.01	12.33	11.85	11.32
13.28	12.11	12.74	13.42	13.23	12.60	11.49	11.13
13.20	12.01	12.41	13.08	13.27	12.79	11.75	11.48
12.91	11.97	12.25	13.07	13.28	13.04	11.72	11.76
13.09	12.09	12.13	12.83	13.04	12.83	11.94	12.04
13.07	12.27	11.95	12.78	12.83	12.79	11.98	12.33

Table 9.10 Transitional probabilities of octant events for condition E3

16.48	18.73	14.04	10.75	9.37	5.29	13.88	11.46
14.55	20.05	12.09	11.66	9.04	6.09	12.97	13.55
13.12	16.97	16.88	15.25	10.40	6.42	10.55	10.40
11.74	15.98	17.97	17.36	10.85	7.29	9.37	9.43
10.51	14.21	17.06	16.30	13.66	9.46	9.41	9.40
9.65	13.36	16.21	15.98	14.79	11.06	9.26	9.69
9.72	13.31	15.15	14.78	14.73	10.91	10.55	10.86
9.47	13.31	14.06	13.87	14.55	11.02	11.43	12.29

and internal inward interaction ($O5$), while internal ejection ($O3$) and internal sweep ($O7$) have been found to be opposite in relationship. Increase in one event causes decrease in other event for all three considered experimental conditions.

Transitional probability matrix (Eq. 9.4) has been estimated using first-order Markov chain process for all three experimental conditions (Tables 9.8, 9.9 and 9.10).

To get the insight of the findings from the transitional probability matrix, average of column matrix (Fig. 9.9) and average of diagonal matrix (Fig. 9.10) have been plotted for each experimental condition.

9.4.6 Comparison Between Quadrant and Octant Analysis

Pattern of average diagonal probability has been found to be the same from quadrant and octant event analysis (Figs. 9.7 and 9.10). The average contribution of diagonal events from steady to unsteady is increasing for both the cases (quadrant and octant). Magnitude of increment has been found different in the different planes for two-dimensional quadrant analyses. For horizontal (uv) and vertical (uw) planes, the increment of average diagonal probability for unsteady flow has been estimated around 46–48 % in comparison with steady flow. The same was found to be

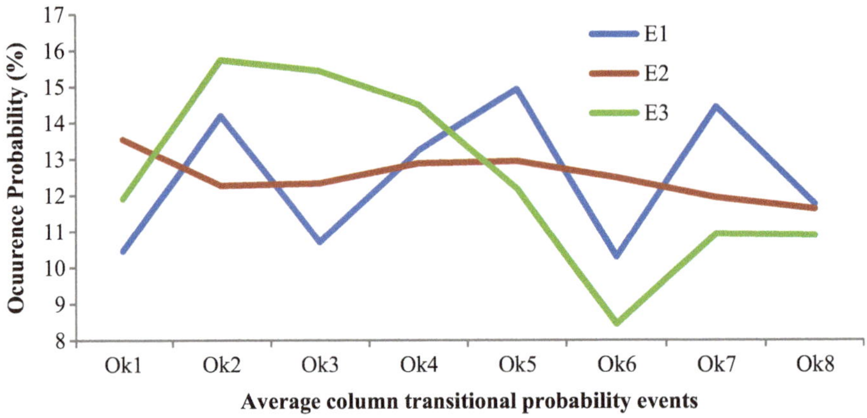

Fig. 9.9 Plot of average column transitional matrix for three experimental conditions (E1, E2 and E3)

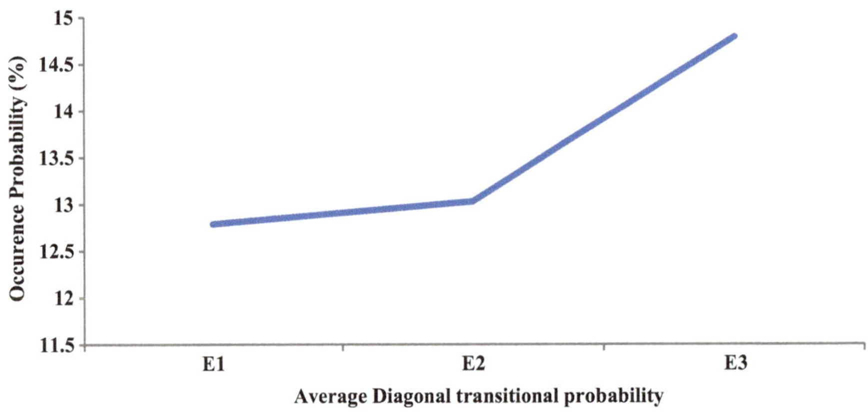

Fig. 9.10 Plot of average diagonal transitional matrix for three experimental conditions (E1, E2 and E3)

around 28 % in the transverse plane (*vw*) (Tables 9.4, 9.5 and 9.6). But in the case of three-dimensional event analysis (octant event), it was found that average diagonal contribution in unsteady flow is increased by 14 % compared to the steady flow conditions. It also shows that there is some factor which relate the events in isolation (three different quadrant events at different planes) and events occur simultaneously (octant events). Indeed it is of greater need to approximate the factor which may transfer the two-dimensional analysis to three-dimensional analysis.

9.4.7 Case Study 2: Bursting Events Under Braiding River Condition

A braided river is a river with a number of smaller channels, branched by islands. The islands may be small or large and temporary or permanent (Fig. 9.11).

9.4.7.1 General

River braiding usually occurs in the rivers having variable discharge with sufficient amount of sediment. Braided channel configuration usually forms on rivers with variable flow (wet and dry

Fig. 9.11 Braided Brahmaputra River in the upstream of Guwahati (India) (Ref: Google Earth)

season or snow melt season) and high quantities of sediment load. When a river flow is at maximum discharge, it is able to transport most of its loads. However, when the discharge falls along with the velocity (i.e. energy or power to carry sediment) of the river, deposition starts to take place.

Braided rivers comprise numerous, shifting channels that widen and become more numerous in response to increased river discharge. Factors which are directly involved in the braiding process of river are as follows:

1. Water discharge (in terms of stream power)
2. Sediment load (amount and variations)
3. Fluvial and geological characteristics (soil strata and land use and land cover status)

Braided channels occur in rivers with steeper bed slope when threshold level of sediment load or slope is reached; any slope over this threshold creates a braided stream. This occurs due to rapid evaporation and infiltration after period of heavy rainfall. It can lead to deposition of sediment load. Braiding is caused by fluctuations in discharge levels and low river stream power with sufficient sediment particles.

The generation of stream power depends on the velocity, so the relationship between velocity and different types of sediment particle movement can evolve very promptly for the braiding river. Hjulström proposed one of the very important curves named as Hjulström curve (Fig. 9.12) to define erosion, transportation and deposition zones based on the velocity and type of particles.

The majority of theories and studies have been attempting to explain the reasons for braiding in a river system which involve establishing an empirical relationship between braiding and hydrological characteristics such as watershed area, discharge, channel or valley slope, sinuosity, width to depth, sediment supply, sediment particle size and bank resistance. Slope and water discharge can be quantified more readily and accurately than sediment supply and usually form the bases for empirical meander/braided thresholds.

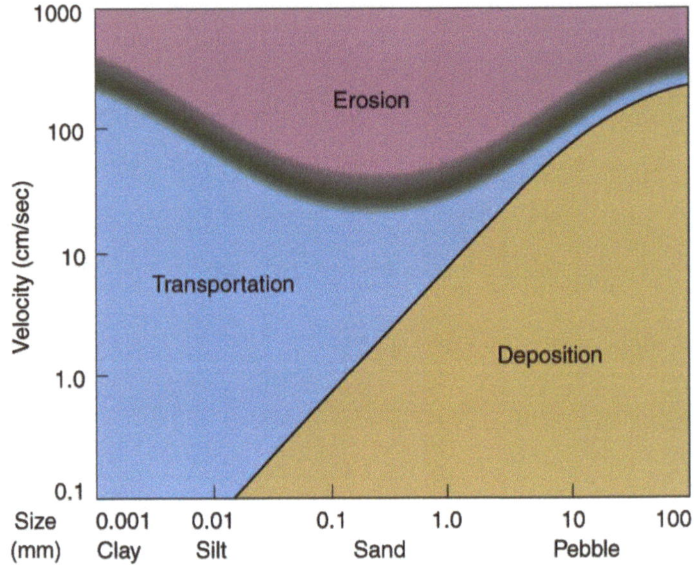

Fig. 9.12 Standard Hjulström curve (http://greenfieldgeography.wikispaces.com/Floodplain+management)

The understanding of the basic mechanisms operating the large-scale morphodynamics of river systems has been substantially improved in the last 50 years. The latter progress, though accomplished with the aid of a large number of field observations and laboratory investigations, is mainly the result of theoretical analyses based on a mechanical approach, coupled with reasonable assumptions on relative importance of the temporal and spatial scales involved. This part of the study involved with laboratory investigation of turbulence parameters under braided condition.

Alluvial river channels are self-formed as they flow over the alluvial lands having sandy-silty-clay materials. Their morphology results from the entrainment, transportation and deposition of the unconsolidated sedimentary materials of the valley fill and floodplain deposits across which they flow (Richards et al. 1993). In the natural condition, there are several river patterns such as a straight river, meandering river and braided river that exist. The structure of secondary current, bed shear stress in the meandering channel has been thoroughly studied by conducting the experiment and numerical simulation. Some research has been done to understand the turbulence characteristics of flow in the braided river; however, these works are not sufficient to fully understand the turbulence characteristics of flow around the braid bar (Murray and Paola 1994; Richardson and Thorne 2001; Roy and Bergeron 1990). A turbulent phenomenon in the braided river is much more complex than that in straight and meandering rivers, and turbulent flow characteristics around the braid bar are not well known till now (Ferguson 1993; Parker 1976).

Braided streams are associated with high stream power and unstable braid bars formed from sediments, which are often unvegetated. Braiding is the division of a single channel into two or more channel ways. Braiding river may be envisaged as a series of channel segments, which divide and rejoin around bars in a regular and repeatable pattern. Lane (1957) stated that a braided stream is characterized by 'having a number of alluvial channels with bars or islands between meeting and dividing again and presenting from the air the intertwining effect of a braid'. Schumm and Lichty (1963) expressed the braided channels as single bedload rivers which at low water have islands of sediment or relatively permanent vegetated islands in contrast to multiple channels in which each branch may have its own individual pattern.

9.4.7.2 Causes of Braiding

Cause and effect of braid bar development is difficult to determine but Leopold and Wolman (1957) in their flume experiment reported that a bar of coarse sand diverts flow to cause bank erosion and positive feedback then accentuates bar development and widening. Miller (1958) noted that the adjustment of width is the inverse of behaviour at tributary junctions. Generally, high-energy stream may create braided channel patterns as a result of the instability of bedload transport in wide, shallow channel (Parker 1976). Parker (1976) also has given an interpretation of the criterion for braiding and characteristics scale of meandering rivers. Process-based explanations of mid-channel bar initiation and growth, based on observation of flow patterns and sediment dynamic downstream of channel confluences, have been proposed by Ashmore (1991) and Ashworth Philip (1996) on the basis of measurements in the flumes and in the streams. While the models of Ashmore (1991) and Ashworth Philip (1996) may explain why bifurcation is likely to occur immediately downstream of an anabranch confluence, processes are observed at the confluence-influence unit. However, this does not fully explain the initiation of braiding in a single channel.

If the results of analyses based on fluid mechanics are accepted, it follows that the bifurcation of a single channel that results from inherent instability in the flow and deposition of bed material load at the centre is driven by flow instability. That is, sediment plays an essential, but responsive role in the bifurcation process, if braiding is division of a single flow stream into two or more threads. Morphological bifurcation of the channel can, indeed, occur subsequently by one of the five mechanisms described by Ferguson (1993).

9.4.7.3 Characteristics of Braided Channel

A braided stream occurs in high-energy environments of large and variable discharges, heavy sediment load and steeper gradient with erodible banks. Braided streams are characterized by wide and shallow cross sections with random bar formations creating flow divisions. Sediment transport takes place over the bar surface while incisions in the lateral channels lower the water surface to expose the bar, which then dissected. The complex of islands is stabilized by vegetation in natural streams and experience further high stage sedimentation (Richardson and Thorne 2001). The distributary channels formed by braiding are less hydraulically efficient than parent single channels. Braided reaches are characterized by steeper slope for maintenance of stream power necessary for sediment transport. As per Smith (1974), general characteristics of braided stream deposits include abundant ripple and dune, cross stratification, thin lenticular shales, many intraclasts, thin sedimentary units indicating variable flow regimes and numerous cut-and-fill structures.

9.4.7.4 Braiding and Turbulence

Both turbulence and stream braiding are represented by a hierarchy of scales. In turbulence the objects constituting these scales are referred as eddies; for braiding the bar, the term is used analogously for all types of sediment islands, including the whole zoology of bar types described by the previous researchers. Turbulence and braiding both exhibit fractal behaviour within the hierarchy of scales (Berge et al. 1984; Kozioł 2015). Also, both types of system interactions between structures give rise to short-lived events (ejection and sweep in turbulence and confluences in braiding) that contribute disproportionately to the overall net transport (momentum in turbulence; sediment in braiding).

9.4.7.5 Characterizing Turbulence

Turbulent eddies create fluctuations in velocity, and these fluctuations represent the chaotic motion. Turbulent flow can be characterized by the statistical concepts, and theoretically velocity is assumed continuous and mean velocity is calculated by integration. However, in practice velocity records consist of discrete samples u_i and w_i; hence, mean velocity (longitudinal and transverse) is calculated by summation of all discrete velocities divided by the number of discrete velocity sample as shown in Eqs. (9.5) and (9.6):

$$\overline{u} = \frac{1}{N}\sum_{i=1}^{N} u_i \tag{9.5}$$

$$\overline{w} = \frac{1}{N}\sum_{i=1}^{N} w_i \tag{9.6}$$

$$u_i^{'} = u_i - \overline{u} \tag{9.7}$$

$$w_i^{'} = w_i - \overline{w} \tag{9.8}$$

The quadrant events depend on sign of Reynolds stress components (longitudinal and vertical fluctuation components – Eqs. 9.7 and 9.8).

9.4.7.6 Experimental Programme

The experiments were conducted at the River Engineering Laboratory, Department of Water Resource Development and Management, Indian Institute of Technology, Roorkee, India. The experiments were carried out in a flume 2.6 m wide, 1 m deep and 10 m long. Velocity is measured at 12 points around the island (as shown in Fig. 9.13) in a braided river model with the help of ADV. The ADV uses the principle of Doppler shift to measure the velocity of water in three dimensions. The device sends out a beam of acoustic waves at a fixed frequency from a transmitter probe. These waves bounce off moving particulate matter in the water, and three receiving probes 'listen' for the change in frequency of the returned waves. The ADV then calculates the velocity of the water in the x, y and z directions. A braided river model with a central island is constructed in the River Engineering

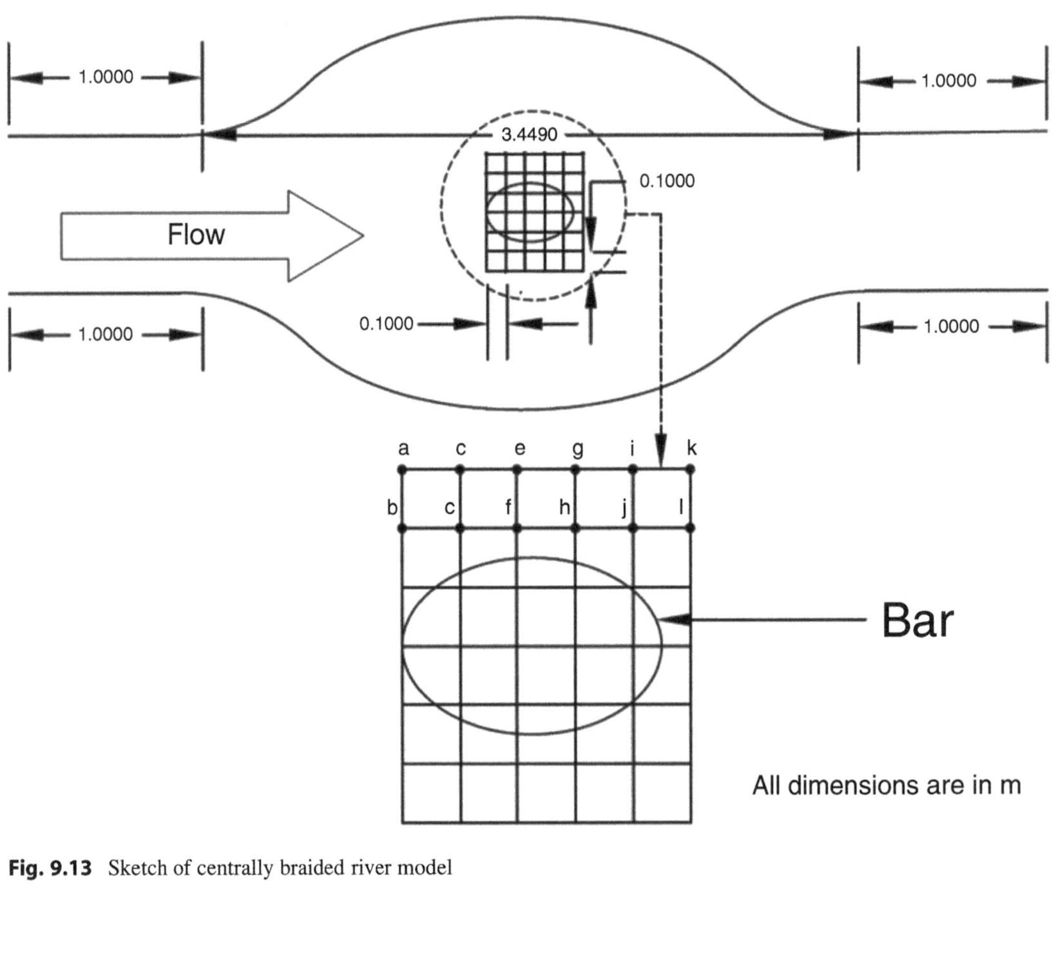

Fig. 9.13 Sketch of centrally braided river model

Lab, IIT, Roorkee. The flow characteristics are studied around central braided bar using quadrant analysis. The conventional quadrant method involves studying the relationship between temporal fluctuations of velocity components, u' and w', particularly their distribution between four quadrants numbered as shown in Table 9.2. u' and w' are fluctuating velocity components in longitudinal and vertical direction, respectively. These fluctuating velocity components are calculated using Eqs. (9.7) and (9.8). After studying the fluctuation behaviour of u′ and w′, it is found that turbulent phenomenon exhibits coherent structure. Experiments were carried out at two different water flow rate of 60,000 and 40,000 cm ^{3}/s. The velocity was measured at the distance of 0.6 cm from the bed at these points. The scouring/deposition data are collected around the island at these selected 12 points for two different discharges (Fig. 9.13). The objective of this experimental programme is to relate the bursting ratio (a new defined parameter by the authors) to the deposition and scouring patterns around the island in a braided river model.

9.4.7.7 Analysis of Experimental Data

9.4.7.7.1 Conditional Probability

Based on two-dimensional velocity fluctuations, the probability of the occurrence of each quadrant bursting event is determined by Eqs. (9.9) and (9.10):

$$P_k = n_k/N \tag{9.9}$$

$$N = \sum_{k=1}^{k=4} n_k \tag{9.10}$$

where P_k is the probability of occurrence of the events belonging to quadrant k, n_k is the number of events belonging to k quadrant, and N is the total number of events. Using the equation given above, the probability of occurrence of sweep and ejection quadrant events was calculated at given points of flow within the depth. The conditional probability of occurrence of sweep and ejection quadrant events is calculated at 12 different points at the distance of 0.6 cm from the bed. The contributions of coherent structure mainly ejection ($Q2$) and sweep ($Q4$) are intensively studied by the researcher. Nakagawa and Nezu (1977, 1978) and Grass (1971) have extensively studied quadrant events, and they found that sweep is the most important event that is related to transfer of momentum into the boundary layer. In addition, they found that frequency of sweep and ejection are more as compared to outward interaction ($Q1$) and inward interaction ($Q3$) events. A new turbulence parameter bursting ratio is defined in this study; it is defined as the ratio of probability occurrence of even events to the odd events. The bursting ratio is given by Eq. (9.11):

$$B_r = \frac{Q2 + Q4}{Q1 + Q3} \tag{9.11}$$

9.4.7.8 Results and Discussions

The bursting ratio has been found to be high at points of scouring, and their values are low at the points of deposition (Fig. 9.14), which also shows that the bursting ratio is related to the scouring/depositions around the bar. From Table 9.11, it has been encountered that the values of bursting ratios decreased when the discharge decreases from 60,000 to 40,000 cm ^{3}/s. For discharge of 60,000 cm^3/s, the maximum value of bursting ratio is 2.92, and for 40,000 cm^3/s it was decreased to 2.52. This shows that the even events $Q2$ and $Q4$ weaken down by lowering the discharge. Since these events are primarily responsible for scouring around the island, thus lower discharge will enhance depositions around the island in the braided river model.

Analysis of scouring/depositional pattern with bursting ratio under braided river model (Fig. 9.14) discloses a remarkable point. The results constitute threshold point of bursting ratio, and it indicates that if the bursting ratio is greater than 1, chance of scouring at that point will be high, and if bursting ratio is less than 1, chance of deposition will be more. This may prove a better parameter for scouring/deposition under different hydraulic conditions.

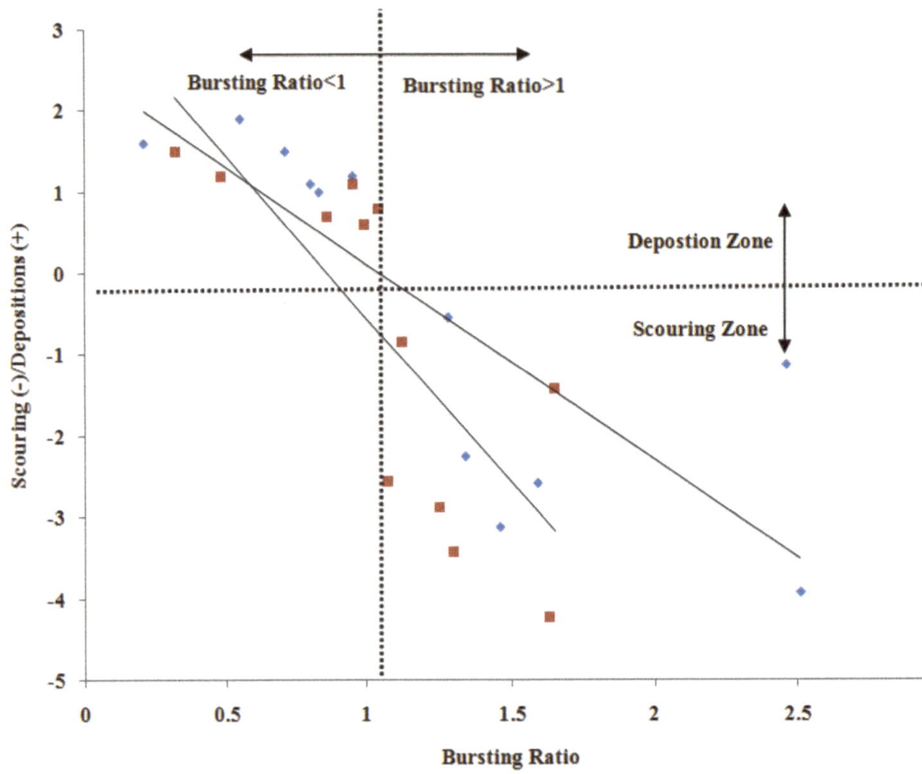

Fig. 9.14 Graph between scouring/depositions (cm) and bursting ratio

Table 9.11 Scouring and deposition patterns at different positions around bar

Point	Experiment A scouring/deposition	Experiment B scouring/deposition
a	−3.12	−3.42
b	−3.92	−4.22
c	−2.25	−2.55
d	−2.58	−2.88
e	−0.54	−0.84
f	−1.12	−1.42
g	+1	+0.6
h	+1.2	+0.8
i	+1.1	+0.7
j	+1.5	+1.1
k	+1.6	+1.2
l	+1.9	+1.5

9.4.8 Conclusions

The main aim of this chapter is to provide the idea of newer methodology to study the river dynamics. Under turbulence scale conditions, there is the enormous possibility to the river behaviour study under several hydraulic conditions experimentally. This study is supported by controlled laboratory investigations under the state-of-the-art methodological analysis.

In the first part of the study, turbulence behaviour of river has been investigated for steady and unsteady flow conditions. Bursting events (quadrant and octant) with their first-order transitional probability have been looked into. Transitional probability of diagonal bursting events has been found to be increased for unsteady flow condition under both the methodologies (quadrant and octant). Change of effect of unsteadiness in different planes has also been estimated, and it was found that transverse plane is the least affected with unsteadiness of flow. The larger sets of experiments under field conditions have been proposed to evaluate the better understanding of these events.

In the second part of the study, turbulence behaviour of river has been investigated for centrally braided river model. A new parameter named bursting ratio is proposed in this study. Bursting ratio has been compared with the scouring and deposition events around the braided bar. Authors have concluded that the parameter has potential to separate the scouring and deposition phenomenon based on threshold value of bursting ratio. Threshold value of bursting ratio was estimated as one in this case study. The further refinement is possible based on larger datasets and several field experimentations.

References

Ashmore PE (1991) How do gravel-bed rivers braid? Can J Earth Sci 28(3):326–341

Ashworth Philip J (1996) Mid-channel bar growth and its relationship to local flow strength and direction. Earth Surf Process Landf 21:123

Berge P, Pomeau Y, Vidal C (1984) Order within chaos, Wiley and Sons, NY

Blöschl G, Sivapalan M (1995) Scale issues in hydrological modelling: a review. Hydrol Process 9(3–4):251–290

Bradshaw P (2013) An introduction to turbulence and its measurement: thermodynamics and fluid mechanics series. Elsevier, Burlington

Church M (1995) Geomorphic response to river flow regulation: case studies and time-scales. Regul Rivers Res Manag 11(1):3–22

Church M, Ferguson R (2015) Morphodynamics: rivers beyond steady state. Water Resour Res 51(4):1883–1897

Ferguson R (1993) Understanding braiding processes in gravel-bed rivers: progress and unsolved problems. Geol Soc Lond Spec Publ 75(1):73–87

Grass AJ (1971) Structural features of turbulent flow over smooth and rough boundaries. J Fluid Mech 50(02):233–255

Gordon CM (1975) Sediment entrainment and suspension in a turbulent tidal flow. Mar Geol 18(1):M57–M64

Helder W, Ruardij P (1982) A one-dimensional mixing and flushing model of the Ems-Dollard estuary: calculation of time scales at different river discharges. Neth J Sea Res 15(3–4):293–312

Keshavarzi AR, Gheisi AR (2006) Stochastic nature of three dimensional bursting events and sediment entrainment in vortex chamber. Stoch Env Res Risk A 21(1):75–87

Keylock C, Hardy R, Parsons D, Ferguson R, Lane S, Richards K (2005) The theoretical foundations and potential for large-eddy simulation (LES) in fluvial geomorphic and sedimentological research. Earth Sci Rev 71(3):271–304

Keylock CJ, Lane SN, Richards KS (2014) Quadrant/octant sequencing and the role of coherent structures in bed load sediment entrainment. J Geophys Res Earth Surf 119(2):264–286

Kline S, Reynolds W, Schraub F, Runstadler P (1967) The structure of turbulent boundary layers. J Fluid Mech 30(4):741–773

Kozioł A (2015) Scales of turbulent eddies in a compound channel. Acta Geophys 63(2):514–532

Lane EW (1957) A study of the shape of channels formed by natural streams flowing in erodible material. US Army Engineer Division, Missouri River

Leopold L, Wolman M (1957) River channel patterns: braided, meandering, and straight: US Geological Survey professional paper 282-B., 1960, River meanders. Geol Soc Am Bull 71:769–794

Mianaei SJ, Keshavarzi AR (2008) Spatio-temporal variation of transition probability of bursting events over the ripples at the bed of open channel. Stoch Env Res Risk A 22(2):257–264

Miller JP (1958) High mountain streams: effects of geology on channel characteristics and bed material. State Bureau of Mines and Mineral Resources, New Mexico Institute of Mining and Technology, Albuquerque

Murray AB, Paola C (1994) A cellular model of braided rivers. Nature 371:54–57

Nakagawa H, Nezu I (1977) Prediction of the contributions to the Reynolds stress from bursting events in open-channel flows. J Fluid Mech 80(01):99–128

Nakagawa H, Nezu I (1978) Bursting phenomenon near the wall in open-channel flows and its simple mathematical model. Kyoto Univ Fac Eng Mem 40:213–240

Nepf H, Vivoni E (2000) Flow structure in depth-limited, vegetated flow. J Geophys Res Oceans 105 (C12):28547–28557

Parker G (1976) On the cause and characteristic scales of meandering and braiding in rivers. J Fluid Mech 76(03):457–480

Richards K, Chandra S, Friend P (1993) Avulsive channel systems: characteristics and examples. Geol Soc Lond, Spec Publ 75(1):195–203

Richardson LF (1921) Some measurements of atmospheric turbulence. Philos Trans R Soc Lond Ser A Cont Pap Math Phys Character 221:1–28

Richardson WR, Thorne CR (2001) Multiple thread flow and channel bifurcation in a braided river: Brahmaputra–Jamuna River, Bangladesh. Geomorphology 38(3):185–196

Roy A, Bergeron N (1990) Flow and particle paths at a natural river confluence with coarse bed material. Geomorphology 3(2):99–112

Schumm SA, Lichty RW (1963) Channel widening and flood-plain construction along Cimarron River in southwestern Kansas. US Geological Survey, Washington, DC

Seyfried M, Wilcox B (1995) Scale and the nature of spatial variability: field examples having implications for hydrologic modeling. Water Resour Res 31(1):173–184
Sharma N, Tiwari H (2013) Experimental study on vertical velocity and submergence depth near Piano Key Weir. Labyrinth Piano Key Weirs II-PKW 93–100
Smith ND (1974) Sedimentology and bar formation in the upper Kicking Horse River, a braided outwash stream. J Geol 82:205–223
Tiwari H, Sharma N (2014) Statistical study of turbulence near piano key weir: a review. J Exp Appl Mech 5(3):16–28
Tiwari H, Sharma N (2015a) Interaction between flow hydrodynamics and bed roughness in alluvial channel. ISH J Hydraul Eng 22(1):1–10
Tiwari H, Sharma N (2015b) Turbulence study in the vicinity of piano key weir: relevance, instrumentation, parameters and methods. Appl Water Sci 1–10
Whipple KX, Tucker GE (1999) Dynamics of the stream-power river incision model: implications for height limits of mountain ranges, landscape response timescales, and research needs. J Geophys Res: Solid Earth 104(B8):17661–17674

Harinarayan Tiwari Department of Water Resources Development and Management, Indian Institute of Technology Roorkee, Roorkee, Uttarakhand, India

Amir Khan Department of Water Resources Development and Management, Indian Institute of Technology Roorkee, Roorkee, Uttarakhand, India

Nayan Sharma Department of Water Resources Development and Management, Indian Institute of Technology Roorkee, Roorkee, Uttarakhand, India

Part IV

River Modelling

Prospects of Modeling and Morpho-dynamic Study for Brahmaputra River

10

Nayan Sharma and M.P. Akhtar

Abstract

The Brahmaputra River in the Northeastern State of Assam in India is characterized by its exceedingly high discharge with enormous sediment load, spatio-temporal variation in channel morphology, substantially high bed aggradations, and severe bank line erosion. The river maintains almost continuous channel braiding in most of its segments in the alluvial floodplains of Assam. One-dimensional (1-D) flow models are insufficient to tackle problems of braided streams due to lack of information with regard to transverse flow field. Hence, for better and more realistic flow field assessment, two-dimensional (2-D) or three-dimensional (3-D) numerical models are to be used. 3-D models are numerically too expensive for macro scale river reaches. Hence, 2-D enhanced model with secondary flow corrections in governing equations may be used. After successful implementation of 1-D mathematical model for a braided stretch of Brahmaputra River, 2-D enhanced numerical model with boundary-fitted coordinate system for the same stretch has been developed and verified. A new planform index is proposed namely *braid power*. It increases with decrease in incoming discharge into the reach in a particular instance of time. The measured braiding indicator, namely, *Plan Form Index* (*PFI*) using high-resolution satellite data application, is found to be quite useful for the monitoring and analysis of persistent and complex braiding behavior of a large river like the Brahmaputra. The erosion study based on satellite data based analysis conclusively identifies three to four major geological channel nodal points present along the Brahmaputra River. The variability of stream power with bank erosion and braiding process is investigated. A distinct behavioral pattern between

N. Sharma (✉)
Department of Water Resources Development and Management, Indian Institute of Technology Roorkee, Roorkee, Uttarakhand, India
e-mail: nayan.fwt@gmail.com

M.P. Akhtar
State Water Resources Department, Government of Bihar, Patna 800015, India

N. Sharma (ed.), *River System Analysis and Management*, DOI 10.1007/978-981-10-1472-7_10

these is observed. For example, with low stream power, braiding appears to intensify which in result may reflect a higher possibility of bank erosion.

10.1 Introduction

The Brahmaputra River System is one of the largest in South Asia draining a catchment area of 580,000 km^2. It flows across China, India, Bhutan, and Bangladesh and covers almost 2880 km of distance along its course. Brahmaputra River is considered as the socio-economic lifeline of north-eastern India since ages. It is well established that an alluvial river of such magnitude usually presents complex problems of sediment erosion-deposition associated with fluvial process. Many related problems such as flood, erosion, and drainage congestion in the Brahmaputra basin are of enormous in magnitude. The river has an intense braided channel configuration in most of its course in the alluvial plains of Assam. The lateral changes in channels cause relentless bank erosion leading to a considerable loss of highly fertile land every year. Bank recession also causes deviations in outfall points of its tributaries resulting in creation of newer areas of submergence. Thousands of hectares of prime inhabited and agricultural land are encountering severe unabated bank erosion in the Brahmaputra catchment which covers major parts of Indian states namely Assam, Arunachal Pradesh, Meghalaya, Nagaland and Manipur.

In order to tackle the problem of floods and erosion, various agencies including state Government, central government, and autonomous institutions are engaged in planning and execution of flood management program in the north-east region. To achieve effective flood management programs, a variety of structural and non-structural measures are adopted. However, due to the inherent widening characteristic of the Brahmaputra River, these measures do not sustain and adversely affect the benefits anticipated while implementing the flood control and anti-erosion works. High flood causes large-scale breaches in the existing embankments bringing vast areas under flood inundation.

River bank erosion and associated channel evolution are of much significance in relation to geo-morphological research with relevance to many scientific and engineering fields. River bank erosion can devastate infrastructure such as highway, bridges, residential dwellings, and industrial establishments. Hydraulic characteristics of the channel may vary leading to significant problems in adjusting water discharge rating curves and may represent up to 80–90 % of the sediment load in streams and rivers (Simon and Rinaldi 2000). It contributes to *total maximum daily loads* (*TMDLs*), which can be a significant source of non-point source sediment and nutrient pollution, and may have an adverse effects on water quality and natural river habitat. Nevertheless, river bank erosion is also beneficial and an integral part of many river ecosystem processes (Florsheim et al. 2008). For example, coarse sediment from bank erosion can provide substrate for fish spawning (Flosi et al. 1998) and sediment for crane roosting habitat (We 1994). Irregular banks provide habitat for invertebrates, fish, and birds (Florsheim et al. 2008), and areas affected by erosion provide substrate for the establishment of riparian vegetation (Miller and Friedman 2009). Furthermore, understanding bank erosion dynamics is significantly useful for river management. It is also useful for designing river restoration projects that accommodate the natural river migration processes that erode banks and build floodplains (Moody and Meade 2008) on various time scales (Couper 2004). Therefore, a better understanding on river channel bank variation is of great importance for water resources development and environmental management.

Engineers and researchers have been consistently endeavoring to evolve better and efficient techniques in order to broaden our understanding and enhance the quality of our lives ever since the beginning of human civilization (Wang and Wu 2004). Understanding flows through open channels is of crucial importance for addressing numerous hydraulic engineering problems. A pre-requisite for arriving at such optimal solutions is that the complex physics of open channel flows be understood properly. These flows, however, are typically turbulent, unsteady, and highly 3-D. They often take place in stratified environments and can involve multiple phases. For that reason, their understanding continues to present hydraulic engineers with rather formidable challenge. Traditional approaches for studying natural river flows and morpho-dynamics study are predominently based on field measurements and laboratory experiments. Owing to site-specific concerns, field studies of natural channel flows are very expensive, complex, and time consuming. Similar problems although to a lesser extent are encountered in laboratory physical model studies, which are additionally induced with scale effects owing to non-similarity of one or more dominant non-dimensional parameters (Sinha et al. 1998). In order to overcome above shortcomings, the development of approaches that generally do not exhibit aforementioned difficulties was stressed upon to provide practicing engineers with effective technique in the form of numerical models for better understanding of natural river flows.

10.2 Mathematical Model Development for Brahmaputra River

Developing a mathematical model for natural river reaches with irregular and deformed bed and banks with multiple braided channel configuration is a numerically expensive and tedious task. Such fluvial domain poses numerous difficulties to even most advanced numerical methods. Problem becomes more intricate and complex when dealing with varying topography, small-scale roughness, vegetation, irregular shaped obstructions, and boundaries whose shapes depend upon local flow conditions. It is required to be the essentially determined as the part of complete solution to the problem encountered.

10.2.1 1-D Mathematical Modeling

Sharma (1995) has attempted an application of 1-D modeling to simulate the process representation of braided Brahmaputra River in a certain stretch and successfully validated it. The detailed is being explained in subsequent sections.

10.2.1.1 Governing Differential Equations of the Mathematical Model

The governing differential equations of the model adopted for the water and sediment flow representation are

$$\frac{\delta Q}{\delta x}+\frac{\delta A}{\delta t}-q_{\prime 1}=0 \tag{10.1}$$

$$\frac{\delta \rho Q}{\delta t}+\frac{\delta \rho Q V}{\delta X}+gA\frac{\delta \rho Y}{\delta X}=\rho g A\left(S_0-S_f+D_1\right) \tag{10.2}$$

$$\frac{\delta Qs}{\delta x}+p\frac{\delta A_c}{\delta t}+\frac{\delta A_s}{\delta t}-q_s=0 \tag{10.3}$$

where x = longitudinal distance along the channel, t = time, Q = channel discharge, A = cross-sectional area of the channel, A_c = volume of sediment deposited per unit length of channel (the value of A_c will be negative when bed erosion occurs and vice versa), A_s = volume of suspended sediment over the cross section per unit length of channel, Qs = sediment discharge, q_l = lateral flow of water per unit length of channel, and q_s = lateral flow of sediment per unit length of channel ($+q_s$ indicates "inflow," whereas $-q_s$ indicates "outflow"); ρ = density of sediment laden water=$\rho_W+C_S(\rho_S-\rho_W)$, where C_S = sediment concentration = Q_S/Q, ρ_W = density of water, ρ_S = density of sediment, p = volume of sediment in a unit volume of bed layer, V = mean velocity of flow, Y = depth of flow, S_o = channel bed slope, and S_f = energy slope; D_l = dynamic contribution of lateral discharge = $\frac{Q1.V1}{g.A}$.

The following assumptions have been taken into consideration while using the governing equations (Eqs. 10.1, 10.2 and 10.3) for developing a mathematical modeling for 1-D river flow:

(i) The flow is 1-D (flow velocity is uniform over the cross-section). (ii) The streamline curvature is small and vertical accelerations are negligible (pressure is assumed to be hydrostatic only). (iii) Average bed slope is small. (iv) Water surface slope is small. (v) The unsteady flow resistance coefficient is assumed to be same as steady flow in alluvial channels. (vi) Water surface across the cross section is horizontal. Furthermore, the model presumes uniform sediment size for computational purpose that involves sediment routing and channel adjustments.

10.2.1.2 Supplementary Equations

In order to obtain complete solution of governing differential equations as described in Eqs. 10.1, 10.2 and 10.3, the model requires mathematical functions for the supplementary relations as described below.

10.2.1.2.1 *Energy Slope*

Energy slope may be expressed as

$$S_f = \frac{Q^2}{K^2} \tag{10.4}$$

where $K(\text{conveyance}) = \frac{AR^{2/3}}{n}$, n = Manning's roughness coefficient, and Q = discharge.

The model facilitates options for either using Garde-Ranga Raju velocity-depth predictor for computing varying roughness or adopting a constant value Manning's "*n*" based on model calibration. Garde-Ranga Raju predictor requires information with regard to hydraulic characteristics such as depth, width, and energy slope. The mean value of S_f between adjacent cross sections "*i*" and "*i+1*" has been used in the solution algorithm for better and accurate representation of the energy slope of the reach under consideration.

10.2.1.3 Upstream and Downstream Boundary Conditions

Discharge versus time or stage versus time relationships are employed as upstream boundary condition, while stage versus discharge rating curve is implemented as downstream boundary condition. In order to represent the downstream boundary, the following power function is fitted to the observed data using multiple regression analysis for rising and falling segments of the hydrograph.

$$h = c.Q_m^a.Q^b. \tag{10.5}$$

where h = gauge level, Q_m =threshold discharge where the trend has changed, and a, b, and c = regression coefficients.

10.2.1.4 Multiple Channel Geometry Representation

Modeling of complex braided alluvial rivers poses multiple difficulties arising out of the complication in flow geometry. These difficulties primarily associated with hydraulic representation of the complex braided channel within the domain of numerical solution algorithm for governing differential equation to represent the physical flow process. A practical solution to the problem can be to simplify the braided geometry using a hydraulically equivalent rectangular channel for the purpose of the hydraulic routing. Subsequently the sediment routing can be carried out considering the actual river cross section with all its complexity. The equivalent rectangular channel substitution for the braided river cross section can be presumed as a hydraulically similar rectangular channel with its depth and width determined based on the weighted average of sub-channel conveyances to the total for the river.

The channel geometry of an alluvial river is essentially non-prismatic and irregular in shape. It varies from cross-section to cross-section. The fact is that most of the sediment laden alluvial streams exhibit multiple channel characteristics with varying cross-sectional profile. The computer program developed by Sharma (1995) was meant for computing the geometric properties of a braided river cross section for known subsection wise bed levels and water levels. Using trapezoidal rule, it computes wetted perimeter, channel top width, cross-sectional area, as well as deepest bed elevation of the stream.

The channel is subdivided into segments as per the input data. The actual flow areas and other geometric properties are worked out taking into consideration the island (if any) within the cross sections. Channel width has significant effect on wave propagation speed. Similarly flow depth has also considerable influence on celerity, Manning's n, as well as flow dynamics. Hence representative width and depth for a braided alluvial stream are required to be assigned with utmost care. Noticeably, top width of a braided stream as an active width further leads to oversimplification of highly irregular channel cross section. Therefore, to account somehow accurately for the irregular cross-sectional shape of a braided channel, a weighted depth or equivalent depth is computed as per the following correlations:

$$\text{Equivalent depth, } EQD = \frac{\sum_{i=1}^{n} D_{av}.A_i.D_{av}^{2/3}}{\sum_{i=1}^{n} A_i.D_{av}^{2/3}} \quad (10.6)$$

$$\text{Equivalent width, } EQD = \frac{\sum_{j=1}^{n} A_i.D_{av}^{2/3}}{(EQD)^{5/3}}. \quad (10.7)$$

where A_i = area of subsection i and D_{av} = average water depth at subsection.

Equivalent width or effective width is computed from equivalent effective depth to maintain the proper $A\ (D^{2/3})$ for the cross section in a similar manner to *HEC-6*.

10.2.1.5 Numerical Solution of the Governing Differential Equations

The numerical solution of the governing differential equations is obtained in which the flow continuity equation (Eq. 10.1) and the flow momentum equation (Eq. 10.2) are solved first. The solution is then refined by solving the sediment continuity equation (Eq. 10.3). The solution is numerically uncoupled. The partial derivatives in the governing equations discretized using finite difference implicit scheme method. The Linear Fully Implicit Scheme of Chen (1973) is used in the model formulated by Sharma (1995) for obtaining numerical solution of flow continuity and momentum equations. For stability and accuracy of the numerical solution algorithm of the model, iteration procedure has been employed in order to compute quotients of discretized equation at downstream boundary condition till the changes in discharge of successive iterations are within tolerance limit. Similar to that, the sediment routing and its subsequent channel adjustment processes are subjected to an intensive iteration procedure till the changes in computed water levels of successive iterations satisfy the accuracy limit.

10.2.1.6 Model Application Calibration and Verification

The flow model formulated by Sharma (1995) was applied to a Brahmaputra River stretch between Pandu (Guwahati) and Jogighopa situated within the state of Assam in India. At both Pandu and Jogighopa, well-defined riverbanks exist, whereas intermediate segment of the reach exhibits intermittent fanning out and narrow braided river cross sections. At both ends of the reach under study, two important towns are situated namely Guwahati at the upstream and Goalpara at the downstream. At those locations, gauge and discharge measurements are recorded regularly by Flood Control Department of Assam. The inflow from tributaries in the study reach constitutes hardly 0.50–1 % of the main stem peak flow at Pandu (Guwahati). Hence lateral inflow may be considered to be negligible with respect to main channel flow. This river study reach has stable configuration with constricted width at certain points with rock outcrops confining the river. The minimum width of the river within the reach is around 3.5 km. Since the cross-sectional geometries were available only for the beginning of the simulation period, model calibration and verification were done with measured water stages at Jogighopa site for the years 1977 and 1978 respectively. From Fig. 10.1, it can be readily observed that the model computed stages are in fair agreement with the observed stage data. Among the hydrological data, water stage can be considered as the most reliable primary data with optimally minimized error in comparison to

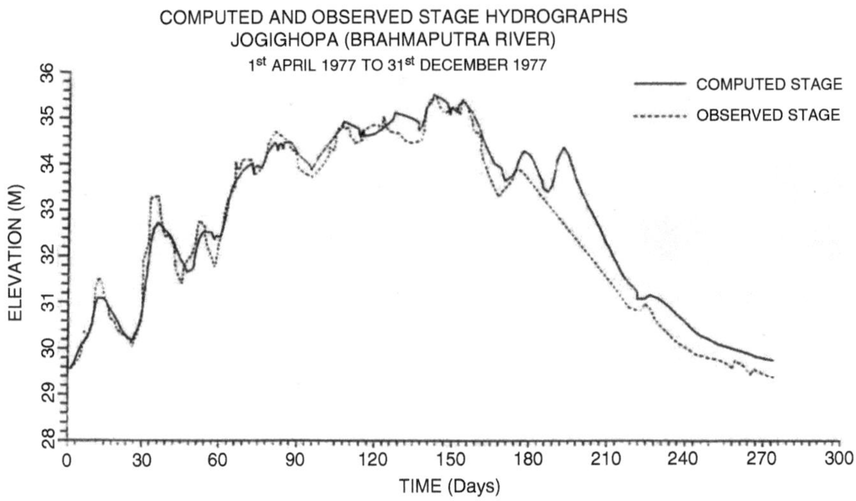

Fig. 10.1 Computed and observed stage hydrographs Jogighopa (Brahmaputra River) 1st April 1977 to 31st December 1977

other data. In the context above, reproduction of stages through the model is quite encouraging for the simplified model approach for such a highly braided alluvial stream, namely, Brahmaputra.

10.2.2 Need for 2-D or 3-D Mathematical Modeling for Braided Rivers

The hydro-graphic irregularities with the varying upstream and downstream flow field create essentially 3-D turbulent shear flows characterized by transverse flows, hydrodynamic pressure formation, flow reversal, and anisotropy effects. Natural river flow field is still needed to be represented accurately and completely. The fact behind this is that existing numerical models usually and primarily have focused rivers of simplified geometries. In 1-D mathematical modeling, a number of assumptions are made to obtain feasible solution; however, information with regard to secondary flow field remains absent. The secondary flow including transverse flow is one of the important causative factors of relentless bank erosion due to severe braiding process in the Brahmaputra. If the geometry is complex, at least 2-D or even a 3-D treatment is required to understand flow scenario within river reach. 1-D modeling is unable to generate bar pool riffle topography commonly encountered in alluvial rivers. Local variation in flow field as well as sediment transport simulation cannot adequately be modeled in 1-D mathematical models due to inherent limitations incurred due to over-simplifications and too many assumptions. The resulting flow field in alluvial streams has essentially topographic as well as bottom shear stress terms and is to adequately be incorporated in the proposed model formulation. It can be readily inferred that application of *CFD* with topographic 1-D modeling is practically inadequate where significant flow variability does exist in either the vertical or the transverse direction commonly associated with secondary circulation due to flow curvature or turbulence. It can be readily inferred that prediction of the outflow, 1-D models may be adequate (with reasonable consideration of cross-sectional spacing and model calibration), however, where the objective is to compute the flow field within a reach with cross-stream variability, one needs at least a 2-D, if not a 3-D treatment (Bates et al. 2005).

A number of 2-D numerical models have been formulated and developed to simulate braided rivers (Enggrob and Tjerry 1999; Lien et al. 1999 etc.). However, vast majority of existing numerical flow models have focused primarily on rivers of simplified geometries. More recently, numerous 2-D and 3-D numerical empirical models have been developed to simulate morphological changes in channels with mobile bed and variable bank lines, both in the laboratory and field. However, these models have certain limitations when employed to treat relatively shallow and wide braided rivers with highly irregular bed profile and complex bank lines resulting in dominant cross-stream flow field.

10.2.3 Need for Assessment of Secondary Flow for Braided Brahmaputra River Configuration

Secondary currents occur in the plane normal to the axis of the primary flow, which originate from interactions between the primary flow and gross channel features (Prandtl 1952). Secondary transverse flow results from the imbalance between transverse water surface gradient force and centrifugal forces over the depth due to vertical variation of the primary flow velocity (Lien et al. 1999). In braided rivers, most channel changes are associated with changes in bed morphology, which occur at high discharges when observation is very difficult (Smith 1974). Any mention of secondary currents in braided systems has been restricted to areas of channel confluence, and the effect of secondary currents in bifurcations and around braid bars has been largely neglected (Sankhua 2005). A number of attempts have been made to simulate the realistic flow field including transverse components in complex geometry like bends and curves (Lien et al. 1999; Odgaard 1989a; Duan 2004; Seo Won et al. 2008). However, assessment of flow field in braided river with "secondary flow correction" in complex geometry is hardly found in literature. Estimation of an improved flow field in braided river will lead to realistic assessment of bed changes and bank erosion in braided rivers. Akhtar (2011) developed a 2-D flow model for Brahmaputra River and successfully applied in the selected reach of Brahmaputra River. The salient features of the model formulated by Akhtar (2011) are presented in subsequent sections of this paper.

10.2.4 2-D Modeling Formulation for Process Representation

The full 3-D form of Navier-Stokes equations is modified. First Reynolds-averaged N-S equations (*RANS*) with variables decomposed into time-averaged and time-varying components. This is necessary as in spatial scales, direct numerical solution remains unfeasible. It gives additional terms in the momentum equations known as turbulent Reynolds stresses, which represent the transport of momentum due to turbulence. Second modification is that 3-D form is modified to 2-D depth-averaged form. The 2-D depth-averaged Navier-Stokes equations can be expressed in a horizontal Cartesian coordinate system as follows (Bates et al. 2005).

10.2.4.1 Continuity Equation

$$\frac{\partial h}{\partial t} + \frac{\partial}{\partial x^i}\left(h\overline{u}^i\right) = 0 \qquad (10.8)$$

10.2.4.2 Momentum Equations

$$\frac{\partial \overline{u}^i}{\partial t} + \frac{\partial\left(\overline{u}^i\overline{u}^j\right)}{\partial x^j} = -g\frac{\partial h}{\partial x^j} + \frac{1}{\rho}\frac{\partial \tau^{ij}}{\partial x^j} + \frac{\tau_s^i - \tau_b^i}{\rho h}$$
$$+\frac{1}{\rho h}\frac{\partial}{\partial x^j}\left(\int_0^h \rho\left(u^i - \overline{u}^i\right)\cdot\left(u^j - \overline{u}^j\right)dz\right) \qquad (10.9)$$

where $\overline{u}^i = \frac{1}{h}\int_0^h u^i dz$ are the depth-averaged velocity components of the fluid velocity and $\overline{u}^i$ is the x^i -direction of the horizontal Cartesian coordinate system ($i = 1, 2$). The turbulence shear stress $\tau_{i,j}$ can be determined using an appropriate turbulence model, the water surface, and bed stresses ($\tau_s{}^i$ and $\tau_b{}^i$), and the so-called dispersion terms, namely, the last term in above equation, and these terms usually require empirical formulae or models.

10.2.4.3 Transformed Governing Equations

The transformed depth-averaged governing equations in generalized curvilinear coordinate system (ξ, η, τ) for continuity and momentum equation are derived as follows:

$$\frac{\partial}{\partial \tau}(\rho hJ) + \frac{\partial}{\partial \xi}(\rho hJ\hat{u}_\xi) + \frac{\partial}{\partial \eta}(\rho hJ\hat{u}_\eta) = 0 \tag{10.10}$$

$$\begin{aligned}
&\frac{\partial}{\partial \tau}(\rho hJU_x) + \frac{\partial}{\partial \xi}[\rho hJ\hat{u}_\xi U_x] + \frac{\partial}{\partial \eta}[\rho hJ\hat{u}_\eta U_x] \\
&- \rho hJ\nu_t\left(\alpha_{11}\frac{\partial^2 U_x}{\partial \xi^2} + \alpha_{22}\frac{\partial^2 U_x}{\partial \eta^2}\right) \\
&= -\rho ghJ\left(\xi_x\frac{\partial H}{\partial \xi} + \eta_x\frac{\partial H}{\partial \eta}\right) \\
&- \rho JC_d(U_x)\sqrt{(U_x)^2 + (U_y)^2} \\
&+ \rho hJ\nu_t\alpha_{12}\frac{\partial^2 U_x}{\partial \xi \partial \eta} \\
&- \left(\xi_x\frac{\partial D_{xx}}{\partial \xi} + \eta_x\frac{\partial D_{xx}}{\partial \eta} + \xi_y\frac{\partial D_{xy}}{\partial \xi} + \eta_y\frac{\partial D_{xy}}{\partial \eta}\right)
\end{aligned} \tag{10.11}$$

$$\begin{aligned}
&\frac{\partial}{\partial \tau}(\rho hJU_y) + \frac{\partial}{\partial \xi}[\rho hJ\hat{u}_\xi U_y] + \frac{\partial}{\partial \eta}[\rho hJ\hat{u}_\eta U_y] \\
&- \rho hJ\nu_t\left(\alpha_{11}\frac{\partial^2 U_y}{\partial \xi^2} + \alpha_{22}\frac{\partial^2 U_y}{\partial \eta^2}\right) \\
&= -\rho ghJ\left(\xi_y\frac{\partial H}{\partial \xi} + \eta_y\frac{\partial H}{\partial \eta}\right) \\
&- C_d\rho J(U_y)\sqrt{(U_x)^2 + (U_y)^2} \\
&+ \rho hJ\nu_t\left(\alpha_{12}\frac{\partial^2 U_y}{\partial \xi \partial \eta}\right) \\
&- \left(\xi_x\frac{\partial D_{xy}}{\partial \xi} + \eta_x\frac{\partial D_{xy}}{\partial \eta} + \xi_y\frac{\partial D_{yy}}{\partial \xi} + \eta_y\frac{\partial D_{yy}}{\partial \eta}\right)
\end{aligned} \tag{10.12}$$

where $\xi_y = \partial\xi/\partial y, \tau, \xi_y = \partial\xi/\partial y, \eta_y = \partial\eta/\partial\eta$, $\alpha_{11} = \xi_x{}^2 + \xi_y^2$, $\alpha_{22} = \eta_x{}^2 + \eta_y^2$, $\alpha_{12} = 2(\xi_x\eta_y + \xi_y\eta_x), J = x_\xi y_\eta - x_\eta y_\xi$

D_{xx}, D_{yy}, and D_{xy} are dispersion terms

In Eqs. 10.10, 10.11 and 10.12, $\hat{u}_m(\mathrm{m} = \xi, \eta)$ are the velocity components in the curvilinear coordinate (ξ,η,τ) which relate to U_x,U_y as

$$\begin{pmatrix}\hat{u}_\xi \\ \hat{u}_\eta\end{pmatrix} = \begin{pmatrix}\xi_x & \xi_y \\ \eta_x & \eta_y\end{pmatrix}\begin{pmatrix}U_x \\ U_y\end{pmatrix} \tag{10.13}$$

10.2.4.4 Numerical Solution Procedure

The governing equations presented above are discretized using the finite volume method in curvilinear, non-staggered grid. In the curvilinear co-ordinate system, mass and momentum equations can be written in conservative tensor notation form as

10.2.4.4.1 Continuity Equation

$$\frac{\partial(\rho hJ)}{\partial \tau} + \frac{\partial}{\partial \xi^m}(\rho hJ\hat{u}_m) = 0 \tag{10.14}$$

10.2.4.4.2 Momentum Equation

$$\begin{aligned}
&\frac{\partial(\rho hJU_i)}{\partial \tau} + \frac{\partial}{\partial \xi^m}\left(\rho hJ\hat{u}_m U_i - \Gamma hJ\alpha_j^m\alpha_j^m\frac{\partial U_i}{\partial \xi_m}\right) \\
&= -\frac{\partial}{\partial \xi_m}\left(\rho ghJ\alpha_j^m H\right) + JS_{u_i}
\end{aligned} \tag{10.15}$$

where $\hat{u}_m(\mathrm{m} = 1,\ 2)$ are the velocity components in the curvilinear coordinate (ξ,η,τ) which relates to U_x, U_y, and others as follows, where $\alpha_i^m = \partial\xi_m/\partial x_i$:

$$\begin{aligned}
\begin{pmatrix}\hat{u}_\xi \\ \hat{u}_\eta\end{pmatrix} &= \begin{pmatrix}\alpha_1^1 & \alpha_2^1 \\ \alpha_1^2 & \alpha_2^2\end{pmatrix}\begin{pmatrix}U_x \\ U_y\end{pmatrix} \\
&= \begin{pmatrix}\xi_x & \xi_y \\ \eta_x & \eta_y\end{pmatrix}\begin{pmatrix}U_x \\ U_y\end{pmatrix}
\end{aligned} \tag{10.16}$$

In Eq. 10.15, $\Gamma = \rho v_t =$ *diffusivity*, U_i stands for depth-averaged velocities (i=x,y), S_{ui} is the corresponding source term in the equation for U_i, and J is the *Jacobian* of transformation between Cartesian coordinate system $x_i(x_1 = x,\ x_2 = y)$ and the computational curvilinear coordinate system $\xi_m(\xi_1 = \xi \text{and } \xi_2 = \eta)$. Source terms include cross derivative diffusive terms, dispersion stress terms, and external forces but exclude the second derivatives of co-ordinates (curvature terms) that are very sensitive to grid smoothness (Wu 2007).

The computational domain is discretized into finite number of control volumes by a computational body-fitted grid. The gridlines are identified as cell faces. The control volume centered at point P is embraced by four faces w, s, e, and n (see Fig. 10.2). It is connected with four adjacent control volumes centered at points W, E, S, and N. Here, W denotes west (the negative ξ direction) E east (positive ξ direction), S south (negative η direction), and N north (positive η direction). The convection terms in Eq. 10.2 are discretized by hybrid linear/parabolic approximation (*HLPA*) scheme (Zhu and Rodi 1991). The *HLPA* scheme is reported to be good at stability and accuracy (Wu 2007). The diffusion terms are discretized by central difference scheme. The time derivative term is discretized by first-order backward scheme.

To avoid the problem of spurious numerical oscillation, the momentum interpolation technique proposed by Rhie and Chow (1983) is used to evaluate the cell face variables from centered quantities, and *Semi-Implicit Method for Pressure Linked Equations-Consistent* (*SIMPLEC*) algorithm is then established as above to solve the governing equations (Majumdar et al. 1992; Wu 2007).

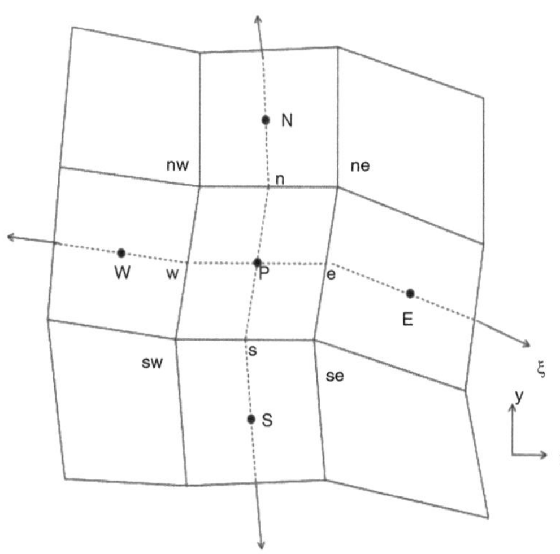

Fig. 10.2 2-D control volume

10.2.4.5 Boundary Implementation

Velocities at wall boundary are assumed to be nonslip and assigned zero. Near-wall boundary, wall function approach is employed. The resultant wall shear stress $\vec{\tau}_w$ is related to flow velocity $\vec{U}_P$ at the center P of the control volume close to wall as follows (Wu 2007):

$$\vec{\tau}_w = -\lambda_w \vec{U}p \tag{10.17}$$

For evaluating λ_w, Wu (2007) may be referred. The pressure (water level) at *south*(S) and *north* (N) boundary points can be extrapolated from values at adjacent internal points. Since the wall velocity is known, it is also unnecessary to perform a pressure correction here. Discharge is specified at the inlet boundary and water level (stage) is specified at the outlet downstream boundary. Flow velocity at the outlet can be extrapolated from the values at the adjacent internal points. Flux correction and pressure correction are kept zero respectively, at the outlet and inlet boundaries.

10.2.4.6 Computation of Bed Friction Coefficient

Using uncertainty approach, it is established that over a non-uniform river reach, U is likely to have far greater spatial variability than n, so will be the dominant control over spatial variation in shear stress (Bates et al. 2005). Physically based alternative to estimate n is used as mentioned in the following equation (Bates et al. 2005). The assumption is that law of wall holds throughout the full depth to estimate value of n at each node.

$$n = \frac{\kappa h^{\frac{1}{6}}}{\sqrt{g}\ln\left(\frac{h}{e \cdot z_0}\right)} \tag{10.18}$$

z_0(Roughness height) $= \frac{K_s}{30}$ or more recently $= 0.1D_{84}$, h=flow depth, $e = 2.71$.

10.2.4.7 Wetting and Drying Technique

In open channels with sloped banks, sand bars, and islands, the water edges change with time, and some part of the domain might be dry. A number of methods have been reported in the literature to handle this problem. The "fixed grid" that covers the largest wet domain (primary floodplain) and treats dry nodes as part of the solution domain is used (Wu 2007). It includes the "small imaginary depth" method. The "small imaginary depth" method uses a threshold flow depth (a low value, such as 0.02 m in natural rivers and 0.001 m in experimental flumes) to judge drying and wetting at each time step. The dry nodes are assigned zero velocity. The water edges between the dry and wet areas can be treated as internal boundaries, at which the wall function approach is applied.

10.2.4.8 Model Application and Validation

One of the main advantages of Brahmaputra River is that it has three to four constricted nodal points where the cross-sections remain unaltered and stable with time and space; moreover, around the vicinity within an extent, Brahmaputra follows uniform aligned channel configuration. This gives different segments of Brahmaputra separated with well-defined nodal points (with uniform channel width) which is adequately and favorably suited for applying 2-D hydrodynamic mathematical model with relative ease, so far as upstream and downstream boundary implementation is concerned. Still, process representation of fully developed braided stream is challenging on account of the presence of numerous 3-D flow structures within the flow domain.

The second advantage is that the Brahmaputra River is one of the rivers which are well under the observation of different stakeholders. The sediment discharges and flood discharges at certain locations have been continuingly recorded and the river cross sections have been periodically surveyed. However the actual data acquisition often remains offset by errors due to limitations in the human capacity, instrumentation, and the risk involved. The importance of the information that could be derived from the analysis of the data is very high in the design, management, and future risk and hazard strategies. Taking into account the situation as described above, the present study is a formative attempt to implement a 2-D flow simulation model based on the controlling equations and specified boundaries specially keeping in mind the flow behavior of Brahmaputra River on a chosen study reach. The algorithms established by the researchers/modelers in the various literature advocate success of flow simulation model application depending on the size of the data covering wide patterns of phenomena. Reliabilty of the result increases with increasing number of data sets. The technique is a data-driven model requiring gamut of data patterns representing the actual phenomena to accommodate all possibilities within the patterns of independent and dependent variables. In the near future, more work with more expertise on this line would be enhancing the dependability on the strength of the technique in more complex analysis. The study has been carried out for the study domain of the river channel (from Pandu to Jogighopa approx. 100 km) with 14 measured river cross-sections (1997) and hydrological data (Jogighopa-Pandu) for the same year.

(i) Pre-processing of hydrographic data
The Brahmaputra River basin in terms of its complexity calls for well-defined response models. In the study, some of the significant steps are followed. The first step is the abstraction of outliers and errors in the data sets. Conceptual or statistical tools such as regression and curve fitting were implemented on the variables pertaining to specific river/stream to identify the irrational data points; they were either discarded or rectified based on the earlier trends or pattern of the data.

(iii) Development of geometric data
Geometry of the physical system is represented by cross sections, specified by co-ordinate points and the distance between cross-sections. The 2-D surface is represented by grid with appropriately chosen nodal points with known x, y, and z co-ordinates. The river domain boundaries

(primary floodplain) for the study period are extracted in geographical co-ordinate system from the remote sensing imagery (Fig. 10.3) and transformed to Cartesian co-ordinate system to accurately represent the domain in Cartesian plane (Fig. 10.4). Furthermore, boundary grid points have been refined and evenly redistributed through algebraic method into 451 points along the positive x-axis (south and north boundaries) and 51 points along y-axis. The positive x-axis is aligned with the direction of the flow while applying the controlling equations for computation of flow field and water depth.

The contour plot for generated bed profile is shown in Fig. 10.5 for illustration. 2-D flow model was used for *20 Discharge Profiles* for receding flood of 1997 (12th July 1997 to 31st July 1997), and validation results are presented in Fig. 10.6. *Water surface level* (*WSL*) contour plots in

Fig. 10.3 Domain delineation through imagery (IRS *LISS-III*, year 1997)

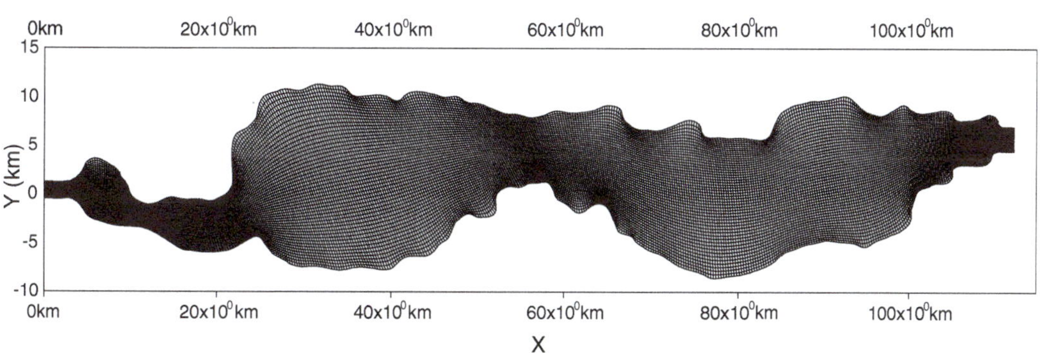

Fig. 10.4 Domain discretization in Cartesian coordinate system

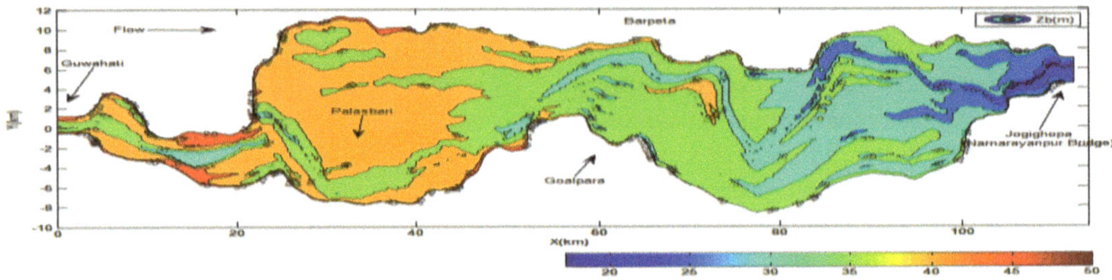

Fig. 10.5 Contour plot of bed elevation for Brahmaputra study reach

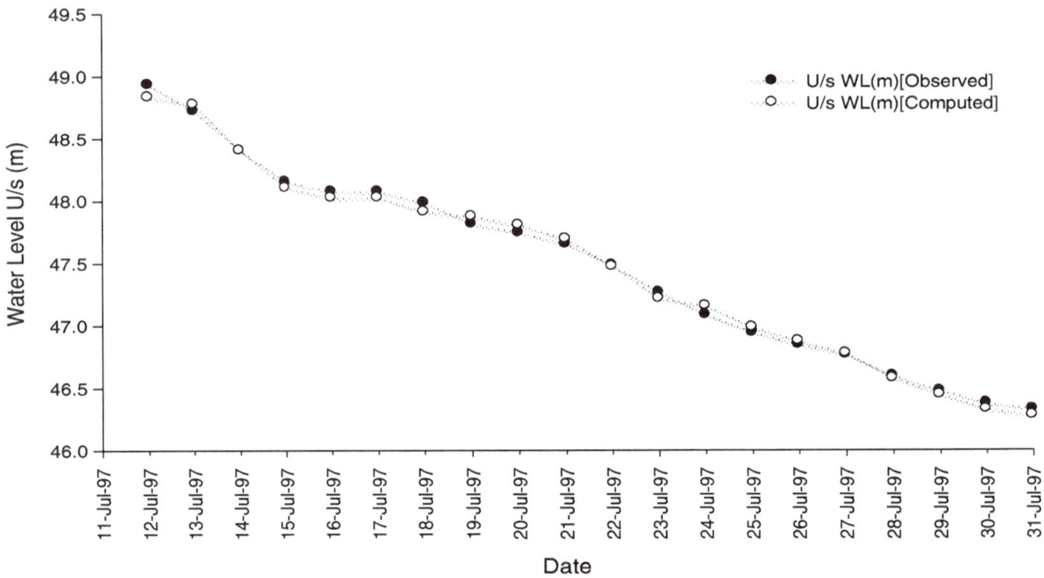

Fig. 10.6 *Observed* versus *computed water surface level* plot and discharge for *U/s* location for the study period

Fig. 10.7 show the development of no-flow zone (Braiding), while discharge decreases from *Discharge Profile-1* (49,389 m^3/s) and *Profile-3* (41,866 m^3/s) to *Discharge Profile-20* (21,562 m ^{3}/s). A terminology *braid power* is proposed (Akhtar 2011) to express the measure of braiding for a river reach as follows:

$$\text{Braid Power}(\text{N/m}^2 - \text{s}) = f_{nf} \cdot \frac{\gamma Q_{\text{inlet}} S}{\text{flow Area of Inlet of the Reach}} \quad (10.18)$$

where f_{nf} = ratio of no-flow zone area with respect to whole flow domain area (primary floodplain), γ = unit weight of water (N/m^3), and S = average longitudinal slope of the study reach. *Flow area* (m^2) is the cross-sectional flow area of the inlet boundary at the given discharge. Braid power increases with decrease in incoming discharge into the reach at a particular instance of time. It depends upon the geometry of the reach.

The simulated water surface elevation for the study period for receding flood for *Discharge Profiles 1*, *9*, and *20* is shown in Fig. 10.7. Plan form of the river domain is also simulated with application of wetting and drying technique as a supplementary boundary condition. As depicted in the prototype occurrences of Brahmaputra, braiding induces severe bank erosion due to dominant transverse flow field. So, improved

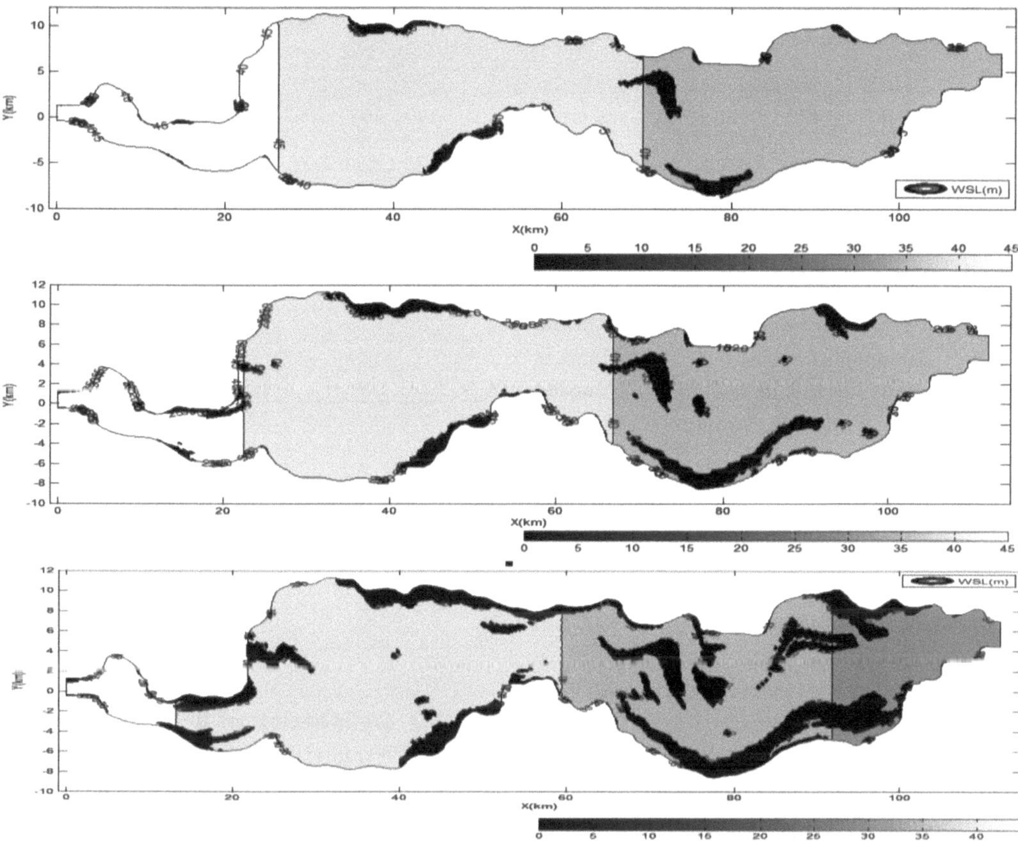

Fig. 10.7 Contour plots for *water surface level* for *Discharge Profiles-1*, *9*, and *20* in the developed 2-D flow model

and realistic flow field estimation will lead to realistic assessment of predictions of bank erosion and river bed evolution for braided alluvial rivers. Better erosion models can be developed with reasonable accuracy using estimated flow field as the prime input.

10.3 Morphological Change Impact of Brahmaputra River

10.3.1 Satellite-Based Assessment of Plan Form Index and Variability in Braiding Intensities (NDMA 2011; Akhtar et al. 2011)

Brahmaputra River and its tributaries controls the geomorphical processes of the entire region especially the Brahmaputra Valley in Assam floodplain. Significant areas of prime inhabited land are lost every year due to riverbank erosion, thereby impoverishing the local people with the intermittent loss of their dwellings and livelihood. Furthermore, unabated bank erosion has caused severe braiding resulting in creation of navigational bottleneck zones due to inadequate draft in lean period. With a view to manage unabated bank erosion problem spanning over hundreds of kilometers along Brahmaputra River, there is an imperative to evolve a braiding indicator that can quantify braiding rationally and assist to monitor complex fluvial landform alterations and also help to prioritize erosion-prone zones and facilitate maintenance of all-weather waterway.

In this study (Akhtar et al. 2011), it is endeavored to assess the channel morphological changes induced by process related to riverbank erosion. A relatively newer braiding indicator

Table 10.1 Characteristics of the remote sensing data used

Satellite/sensor	Acquisition period	Spatial resolution
IRS IA/LISS-I, IRS 1C, and D/LISS-III (standard product)	1990, 1997, 2007–2008	75 m, 23.5 m
IRS IA/LISS-I, IRS 1C, and D/LISS-III (standard product)	1990, 1997, 2007–2008	75 m, 23.5 m

namely Plan Form Indicator (*PFI*) for Brahmaputra River formulated by Sharma (2004), is employed to analyze the braiding behavior. Assessment of the temporal and spatial variation of braiding intensities along the whole stretch of Brahmaputra in Assam flood plains is carried out using satellite data analyses of chosen years within a time span of 18 years (1990–2008).

10.3.1.1 Study Area and Extraction of Channel Forms

Digital satellite images of *IRS LISS-III* sensor, consisting of scenes for the years 1990, 1997, and 2007–2008 are utilized for the present study. In order to bring all the images under one geometric coordinate system, these imageries are geo-referenced with Survey of India (1:50,000 scale) topo-sheets using second-order polynomial. *IRS-LISS* images for the years 1990, 1997, and 2008 are geometrically rectified with reference to the *LANDSAT* images of the same area. The *UTM* projection and *WGS 84* datum have been opted for geo-referencing. Rectification of the images was done with a residual *RMS* (root mean square error) of less than 1. The river stretch under study has been subdivided into 12 segments (reaches). Each segment (reach) is comprised of ten cross-sections. The bank line of the Brahmaputra River is delineated from each imagery's set and the channel plan forms are digitized using Arc*GIS* tool. The data used in the analysis are shown in Table 10.1. Then satellite images of the other chosen years were co-registered using image-to-image registration technique using image processing technique.

The study area of approximately 622.73 km of river stretch from Dhubri in Lower Assam to Kobo beyond Dibrugarh in Upper Assam along the Brahmaputra River is considered in view of data availability.

Table 10.2 Identification of reaches in respect of the location

Reach number	Locations
1	Dhubri
2	Goalpara
3	Palasbari
4	Guwahati
5	Morigaon (near Mangaldai)
6	Morigaon (near Dhing)
7	Tezpur
8	U/s of Tezpur (near Gohpur)
9	Majuli (near Bessamora)
10	U/s of Majuli (near Sibsagar)
11	Dibrugarh
12	U/s of Dibrugarh

10.3.1.2 Delineation of Riverbank Line

The main river has been subdivided into 120 strips for computation of braiding intensities using Planform Index (*PFI*) and reference cross sections were drawn at the boundary of each strip. Cross sections are grouped in ten adjacent cross-sections as a reach with numbers representing 1 to 12 from downstream to upstream of the river (of equal base length as presented in Table 10.2). Baseline of latitude 25.966°N and longitude 90° E has been considered as a permanent reference line. The data derived for each cross section from satellite images of years 1990, 1997, and 2008 have been analyzed. Bank lines along the study reach are also digitized for all the years considered. The length of arcs of both left and right banks is obtained using *GIS* tool. The years 1990, 1997, and 2008 have been used for erosion and deposition analyses along the riverbanks.

Intermediate channel widths and total widths of channel at each predefined cross-sections are measured using GIS software tools for computing Plan Form Indices for each cross-section for further analysis. Mosaic images of satellite data pertaining to study area for discrete years 1990 and 2007–2008 are presented as in Figs. 10.8 and 10.9 respectively.

Erosion in north and south banks of the adjacent area of the river during the study period (1990–2007–2008 and 1997–2007–2008) is estimated by *GIS* software tools through delineating riverbank lines. Areas under bank-line variations during the study period are computed using GIS tool.

Attempt has been made to assess the temporal as well as spatial variation of braiding intensities along the study stretch of Brahmaputra using remote sensing image analyses.

10.3.1.2.1 *Plan Form Index*

The formula for computing *PFI* can be given as follows with definition sketch shown in Fig. 10.10.

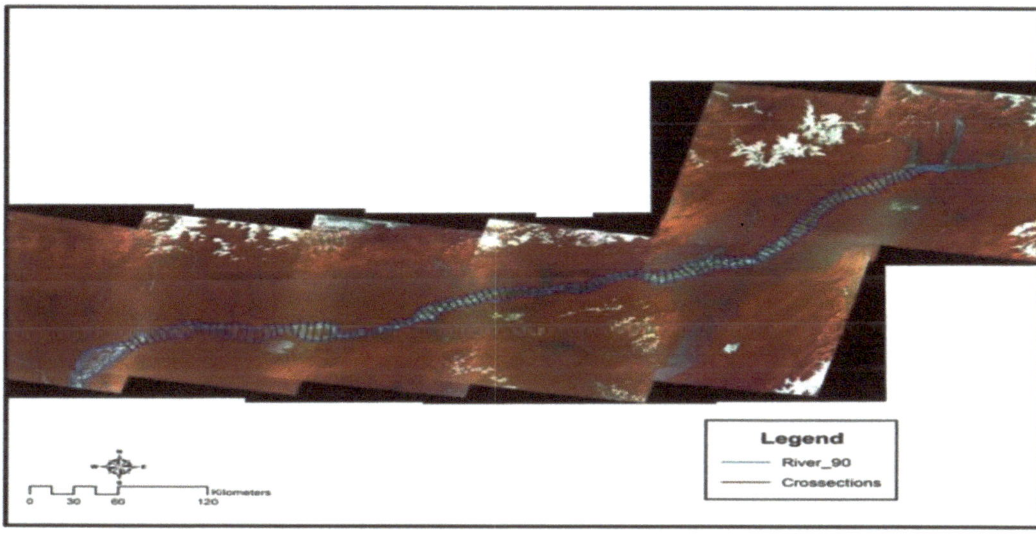

Fig. 10.8 Mosaic of images of Brahmaputra River in 1990

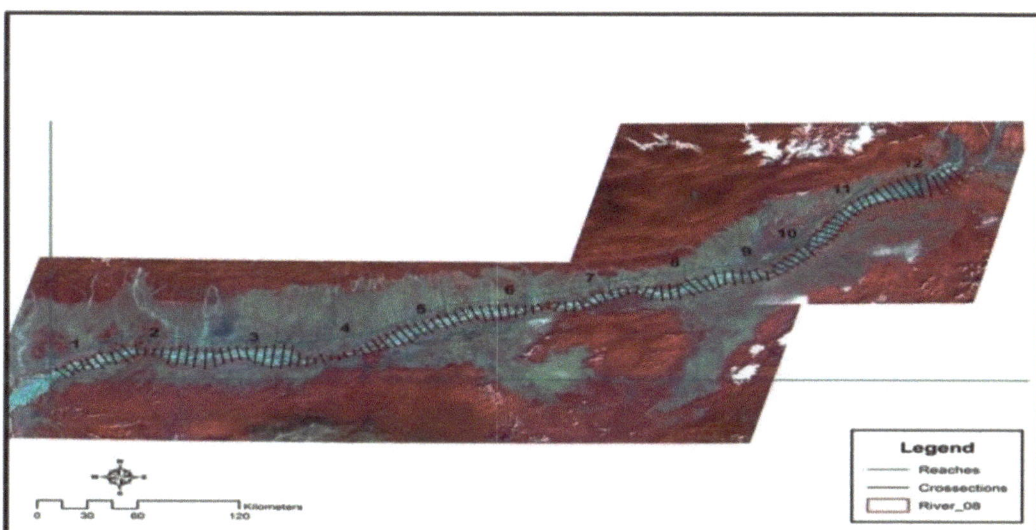

Fig. 10.9 Mosaic of images of Brahmaputra River in 2007–2008

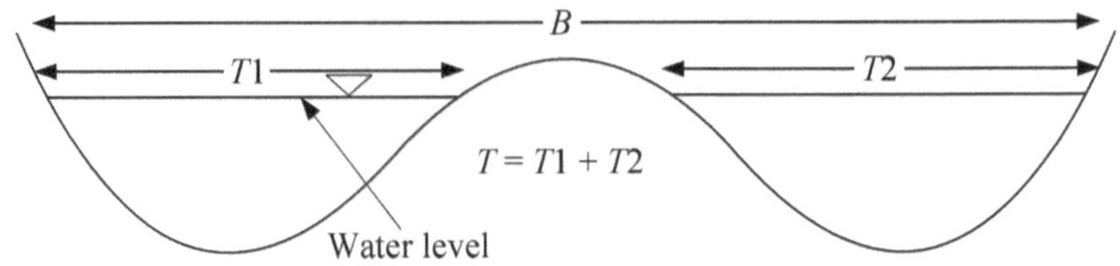

Fig. 10.10 Definition sketch of *PFI*

Table 10.3 Identification of reaches in respect of the location

S No.	Classification	*PFI* Value
1	Highly braided	$PFI < 4$
2	Moderately braided	$19 > PFI > 4$
3	Low braided	$PFI > 19$

$$\text{Plan Form Index} = \frac{\frac{T}{B} \times 100}{N} \quad (10.19)$$

Plan Form Index (*PFI*) in Eq. 10.19 (definition sketch as shown in Fig. 10.10) indicates fluvial landform disposition with respect to a known water level, and its lower value is indicative of higher degree of braiding.

A broad range of classification of the braiding intensity with threshold values for *PFI* is proposed by Sharma (1995) as follows (Table 10.3).

The computed *Plan Form Index* for each reference cross-section (120 in numbers across the study reaches) are plotted against reach cross section number and are shown in Figs. 10.11 and 10.12 (corresponding values presented in Table 10.4) for three discrete years. From the plot, it can be readily inferred that for the period 1990–2007–2008, the *PFI* values generally decrease significantly indicating a visible increase in braiding intensities in majority of cross-sections.

10.3.2 Variability of Unit Stream Power and Its Linkage with Braiding and Bank Erosion

The hydraulic parameters of an open channel flow such as hydraulic mean depth, wetted perimeter, flow area, bed gradient, channel discharge, and the associated physical properties γ $(= \rho g)$ of the fluid are correlated to express the unit stream power of the flow at certain point along the direction of flow (Yang 1976).

Unit stream power (N/m-s) denoted as Ω can be expressed as

$$\Omega = \frac{\gamma \times Q \times S}{w} \quad (10.20)$$

where γ is the specific weight of water, Q is the discharge in m^3/s, S is the slope, and w is the channel width (in meter).

To appreciate the linkage of stream power with braiding, certain considerations are important. Assessment of braiding index essentially depends upon the corresponding stages of the river at predefined locations. Care has been taken to measure this parameter at low flow periods to adequately represent the evolved channels of the river in a particular instance of time. However, it needs to correlate stage discharge at certain location of the river with corresponding measured braiding parameter to achieve meaningful analysis of the active fluvial process. Satellite data-based study enumerated in the previous section becomes much of relevance and likely to fetch significant and meaningful results when correlated to hydraulic and hydrological parameters such as stage discharge and sediment transport rates for the corresponding period of study.

For this, two most vulnerable sites along the river are selected based on the study carried out using satellite data as presented in the previous section. These selected segments are Guwahati-Palasbari (Reach 3–4) and Morigaon-Tezpur (Reach 6–7). These river segments have generally exhibited intensified braiding and considerable riverbank erosion in recent years. Each selected reach comprised of 20 cross-sections. It is observed that in the mid-reach at Guwahati-Palasbari, sediment flow rate increases with

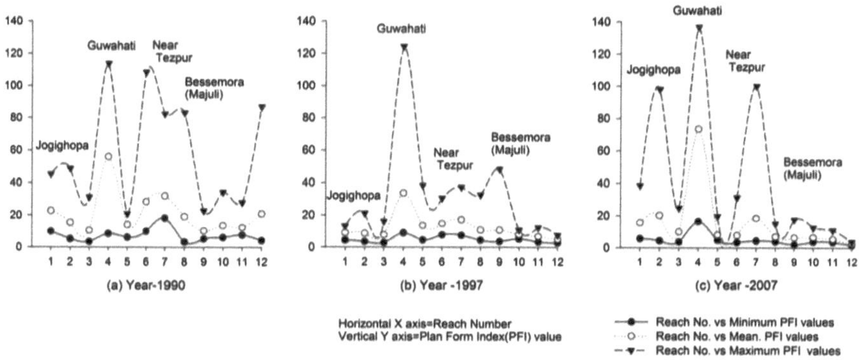

Fig. 10.11 Variation in extreme and mean *PFI* values along the river reaches in discrete years

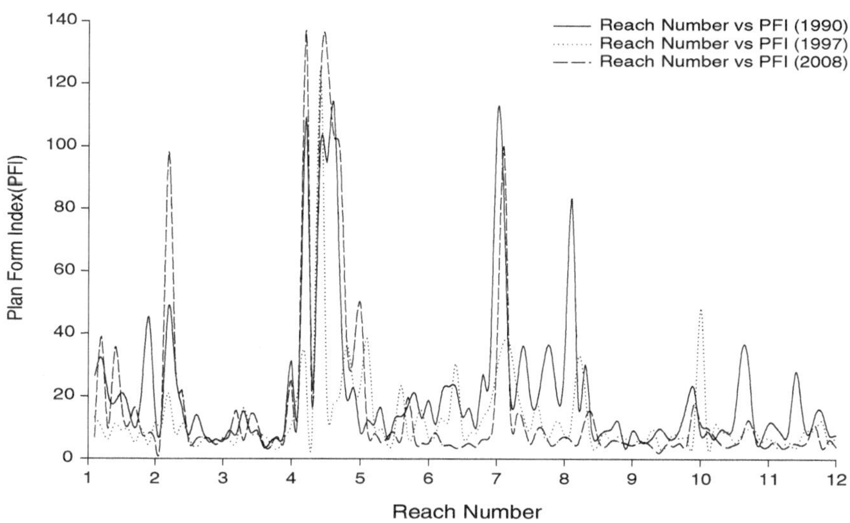

Fig. 10.12 Variation in mean *PFI* values along the river in discrete years

Table 10.4 Comparison of computed extreme and mean *Plan Form Index* (*PFI*) values for the year 1990, 1997, and 2007-08 for Brahmaputra River

	PFI (1990)			*PFI* (1997)			*PFI* (2007–2008)		
Reach number	Mean	Min.	Max.	Mean	Min.	Max.	Mean	Min.	Max.
1	22.69	9.90	45.16	8.94	13.22	4.37	15.66	5.74	38.65
2	15.39	5.23	48.67	8.60	3.50	21.09	20.30	4.42	98.05
3	10.55	3.50	31.15	7.71	2.83	16.22	9.99	3.75	24.79
4	55.97	8.46	113.89	33.62	8.94	124.48	73.64	16.47	136.85
5	14.19	6.18	20.92	13.55	4.52	38.39	8.08	4.88	19.67
6	28.34	9.98	108.61	14.74	7.73	30.46	7.78	3.40	31.30
7	31.94	18.23	82.47	17.21	7.48	37.27	18.50	4.37	100.00
8	19.12	3.28	83.19	10.77	4.39	32.43	6.89	3.84	14.95
9	10.14	5.25	22.71	10.69	3.64	48.37	6.34	2.00	17.55
10	13.61	6.09	34.31	7.87	5.25	10.65	6.54	3.77	12.57
11	12.38	7.65	27.93	6.81	3.36	12.23	5.41	3.52	10.91
12	20.95	4.27	87.05	4.89	2.76	7.57	2.61	1.63	3.70

considerable decrease in downstream reaches at Dhubri-Goalpara.

Variability of stream power at bankfull and low flow (lean) periods with estimated braiding index values for selected reaches namely Guwahati-Palasbari (see Fig. 10.13) and Tezpur-Morigaon (see Fig. 10.14), amply suggests that wherever *PFI* goes up, corresponding stream power goes down. It indicates a considerable decrease in potential energy expenditure per unit mass of water. This process triggers local river bed rise and formation of additional multiple channels. While the number of channels increases, concentration of flow becomes a usual phenomenon along the bank lines, which further triggers bank erosion which in result increases sediment load into the river, further causing the river to be prone to braiding. The process becomes complex as it hinders the clubbing of multiple channels to single stream at high stream power. Similarly bank erosion overloads the sediment to the river, thereby increasing sediment transport rate to the point where it becomes more than the sediment carrying capacity of the channel. The variability of flow behavior of Brahmaputra River is vividly noticeable in different reaches too. The geological control points as discussed earlier at Guwahati, Jogighopa, Bessamora, and Tezpur are due to the existence of rock outcrops. The tendency of the river on the rest of the stretch is highly dynamic and may be considered as one of the most challenging examples of fluvial hydrodynamics with complex morphology.

10.4 Conclusions

1-D flow models are insufficient to tackle problems of braided streams due to lack of information with regard to transverse flow field. In lieu of this, 2-D or 3-D numerical models are to be used. 3-D models are numerically too expensive for macro-scale river reaches. Hence, 2-D enhanced model with

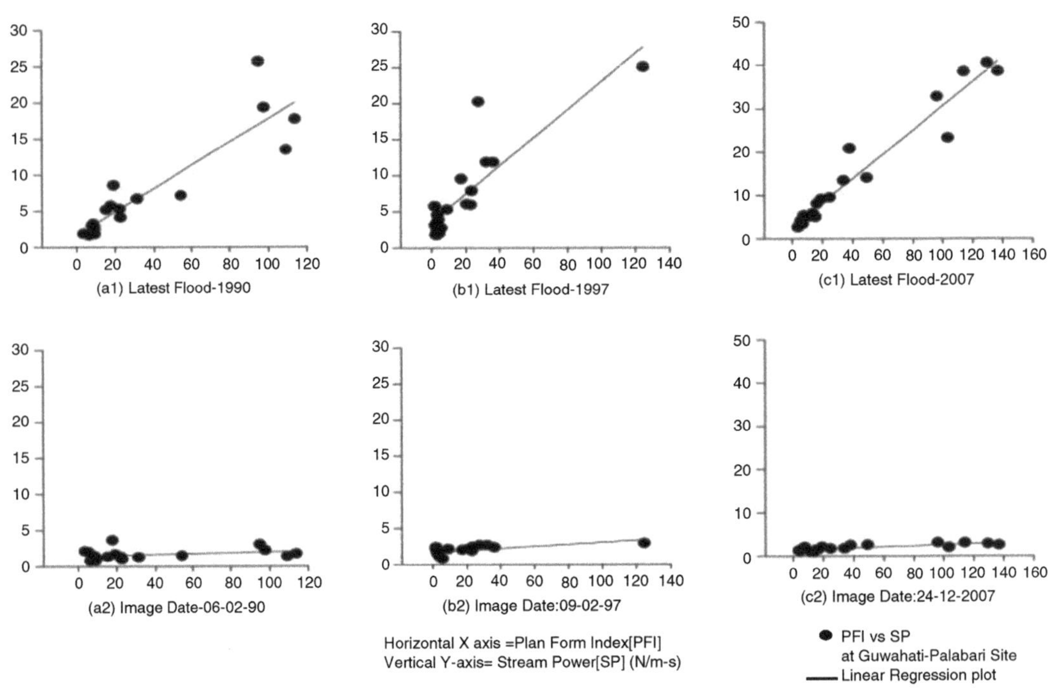

Fig. 10.13 Variability of stream power with *PFI* values during bankfull and low flow period for Guwahati-Palasbari site

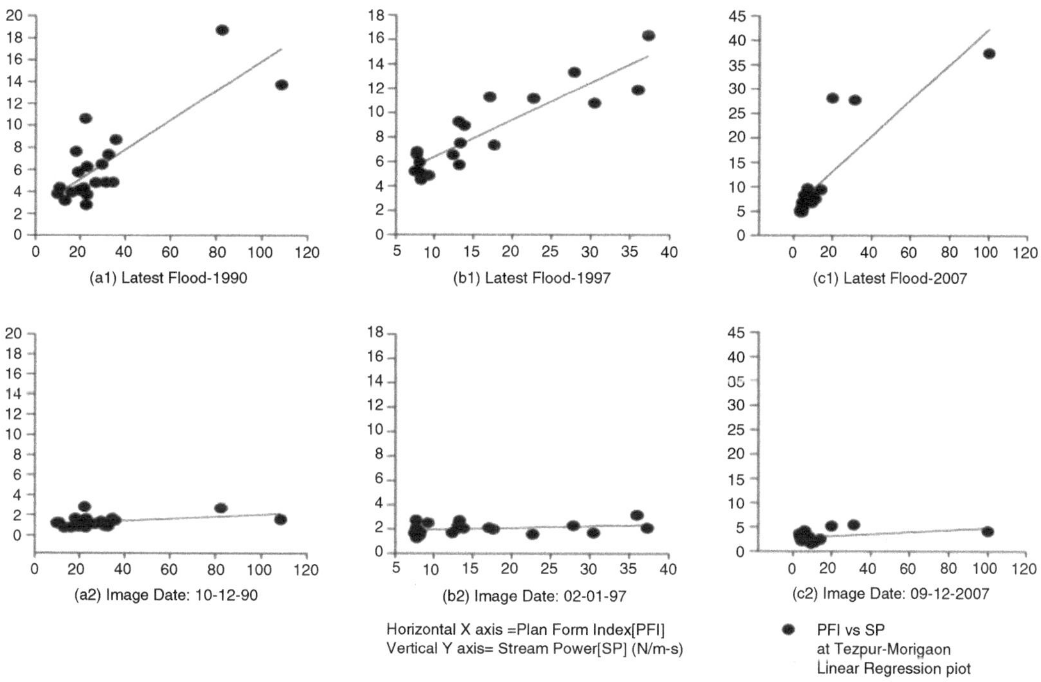

Fig. 10.14 Variability of stream power with *PFI* values during bankfull and low flow period for Tezpur-Morigaon site

secondary flow corrections in governing equations has been developed.

After successful 1-D mathematical modeling of a braided stretch of Brahmaputra River (Sharma 1995), 2-D enhanced numerical model with boundary-fitted coordinate system for the same stretch with highly braided configuration has been developed and verified (Akhtar 2011) for the first time.

Based on the obtained results and information from flow simulation for 20 discharge profiles at receding flood of 1997 for Brahmaputra River stretch under this study, an indicator namely *braid power* is proposed by Akhtar (2011) based on the model output to express the measure of braiding for a river reach as follows:

$$\text{braid power}\left(\mathrm{N/m^2-s}\right) = f_{nf}\cdot\frac{\gamma Q_{\text{inlet}}S}{\text{flow Area of Inlet of the Reach}}$$

where f_{nf} = ratio of no-flow zone area with respect to whole flow domain area, γ = unit weight of water (N/m^3), and S = average longitudinal slope of the study reach. Flow area (m^2) is the cross-sectional flow area of the inlet boundary at the given discharge.

It was observed that *braid power* increases with decrease in incoming discharge into the reach at a particular instance of time. The rate of decrease or increase of *braid power* depends upon geometric configuration of the reach at the particular instance of time along with other factors.

The braiding indicator namely *Plan Form Index* (*PFI*) proposed by Sharma (1995) with high-resolution remote sensing satellite data application, is found useful to analyze and monitor the complex braiding behavior of a large river like the Brahmaputra.

The erosion study identifies three to four major geological channel control points present along the Brahmaputra in Assam floodplains. These control points are in the vicinity of Jogighopa near Goalpara, Pandu near Guwahati, Tezpur, and Bessemora near Majuli. These channel control points usually have well-defined and stable hydrographical profiles. Intermittent fanning out behavior is displayed between these control points. They are being temporally severed in the geological time scale.

The variability of stream power with bank erosion and braiding process is investigated. A distinct behavioral pattern is observed. For example, with low stream power, braiding appears to intensify which in result may reflect a higher possibility of bank erosion.

10.5 Scope for Future Study

Better and improved erosion models incorporating sediment module can be developed using estimated flow field with reasonable accuracy, as the prime input.

Present research will be fruitful for the better prediction of land and environmental losses due to morphological changes of a well-developed braided river like Brahmaputra.

The improved 2-D flow model, upon scale up, could be integrated by relevant researchers with erosion mathematical module with inclusion of pronounced effect of transverse velocities.

Acknowledgment The present study is partly sponsored and funded by National Disaster Management Authority of India (*NDMA, Govt. of India*), New Delhi, which is gratefully acknowledged here.

References

Akhtar MP (2011) 2-D depth averaged modelling for a curvilinear braided stretch of Brahmaputra River. PhD Thesis, IIT Roorkee India

Akhtar MP, Sharma N, Ojha CSP (2011) Braiding process and bank erosion in Brahmaputra River in India. Int J Sediment Res (IJSR) Beijing, China 26(4):431–444

Bates PD, Lane SN, Ferguson RI (2005) Computational fluid dynamics, applications in environmental hydraulics. John Wiley, United Kingdom

Chen YH (1973) Mathematical modelling of water and sediment routing in natural channels. Dissertation Colorado State University

Couper PR (2004) Space and time in river bank erosion research: a review. Area 36(4):387–403

Duan Jennifer G (2004) Simulation of flow and mass dispersion in meandering channels. J Hydraul Eng 130(10):964–976

Enggrob HG, Tjerry S (1999) Simulation of morphological characteristics of a braided river. IAHR Symposium on River Coastal, and Estuarine Morphodynamics. IAHR Genova, Italy, pp 585–594

Florsheim JL, Mount JF, Chin A (2008) Bank erosion as a desirable attribute of rivers. BioScience 58(6):519–529

Flosi G, Downie S, Hopelain J, Bird M, Coey R, Collins B (1998) California salmonid stream habitat restoration manual. Department of Fish and Game, Sacramento California

Lien HC, Hsieh TY, Yang JC, Yeh KC (1999) Bend flow simulation using 2D depth-averaged model. J Hydraul Eng 125(10):1097–1108

Majumdar S, Rodi W, Zhu J (1992) Three-dimensional finite-volume method for incompressible flows with complex boundaries. J Fluids Eng 114:496–503

Miller JR, Friedman JM (2009) Influence of flow variability on floodplain formation and destruction Little Missouri River North Dakota. Geol Soc Am Bull 121:752–759

Moody JA, Meade RH (2008) Terrace aggradation during the 1978 flood on Powder River Montana USA. Geomorphology 99:387–403

NDMA (2011) Study of Brahmaputra river erosion and its control, Report of study conducted by I.I.T Roorkee for National Disaster Management Authority of India New Delhi www.ndma.gov.in/. . ./NDMA%20Final%20Report%20Brahmaputra%20River.pdf

Odgaard A (1989) River meander model I: development. J Hydraul Eng 115(11):1433–1450

Prandtl L (1952) Essentials of fluid dynamics. London, Blackie and Sons Press Trust of India Report (1995) The Sentinel 7th July 3

Rhie TM, Chow A (1983) Numerical study of the turbulent flow past an isolated air foil with trailing – edge separation. AIAA J 21:1525–1532

Sankhua RN (2005) ANN based spatio-temporal morphological model of the river Brahmaputra. PhD thesis I.I.T. Roorkee

Seo Won, Lee Myung Eun and Baek Kyong Oh (2008) 2D modeling of heterogeneous dispersion in meandering channels. J Hydraul Eng 134(2):196–204

Sharma Nayan (1995) Modelling of braided alluvial channels. PhD thesis, University of Roorkee, India

Sharma N (2004) Mathematical modelling and braid indicators. In: Singh VP (ed) The Brahmaputra basin water resources, vol 47. Kluwer Academic Publishers, Dordrecht, pp 229–260

Simon A, Rinaldi M (2000) Channel instability in the loess area of the midwestern United States. J Am Water Resour Assoc 36(1):133–150

Sinha SK, Sotiropoulos F, Odgaard J (1998) Three dimensional numerical model for flow through natural rivers. J Hydraul Eng 124(1):13–53

Smith ND (1974) Sedimentology and bar formation in the upper Kicking Horse River, a braided out wash stream. J Geol 82:205–223

Wang SSY, Wu W (2004) River sedimentation and morphology modeling- the state of the art and future development. In: Proceedings of the ninth international symposium on river sedimentation, October 18–21, 2004, Yichang, China. pp 71–94

We J (1994) Woodland expansion in the Platte River, Nebraska: patterns and causes. Ecol Monogr 64:45–84

Won S, Eun LM, Baek Kyong O (2008) 2D modeling of heterogeneous dispersion in meandering channels. J Hydraul Eng 134(2):196–204

Wu W (2007) Computational river dynamics. Taylor & Francis Group, London. ISBN 978-0-415-44961-8

Yang CT (1976) Minimum unit stream power and fluvial hydraulics. ASCE J Hydraul Div 102(7):919–933

Zhu J, Rodi W (1991) A low dispersion and bounded convection scheme. Comput Methods Appl Mech Eng 92:87–96

Nayan Sharma Department of Water Resources Development and Management, Indian Institute of Technology Roorkee, Roorkee, Uttarakhand, India

M.P. Akhtar State Water Resources Department, Government of Bihar, Patna, India

Development of a Fuzzy Flood Forecasting Model for Downstream of Hirakud Reservoir of Mahanadi Basin, India

11

Anil Kumar Kar, Anil Kumar Lohani, N.K. Goel, and G.P. Roy

Abstract

Floods occurring at delta of Mahanadi are mostly due to contribution of downstream catchment of Hirakud reservoir of Mahanadi basin. Controlling flood through other structural measures is inadequate and difficult. The downstream part is devoid of a sound flood forecasting method. In order to protect the life and property, a nonstructural measure like a workable flood forecasting model is needed to mitigate the destruction at delta by enhancing appropriate and timely relief measures. Establishment of a physical base model requires a lot of meteorological and physical information of the catchment. When such detailed information or data are not available, an alternative method based on soft computing technique may play an important role. Therefore, in this paper, a soft computing-based flood forecasting model using fuzzy inference system is attempted at Mundali station in Mahanadi river and the results compared with observed values. The input and output peaks are grouped into low, medium, high, medium high, and very high categories and operated by nine fuzzy rules.

The discharge at Mundali (forecasting station or FS) is forecasted using the discharge of Khairmal (base station or BS) and Barmul (intermediate station or IS) gauge and discharge sites. The base station is 115 km downstream of Hirakud reservoir and forecasting station is just before starting of delta zone. Forecasting stations are 200 km away from the base station. Peak discharges of 10 years (2001–2010) are considered as input

A.K. Kar
Water Resources Department, Government of Odisha, Bhubaneswar, India

A.K. Lohani (✉)
National Institute of Hydrology, Roorkee 247667, Uttarakhand, India
e-mail: lohani@nih.ernet.in

N.K. Goel
Department of Hydrology, Indian Institute of Technology Roorkee, Roorkee, Uttarakhand, India

G.P. Roy
FMIS Cell, Department of Water Resources, Government of Orissa, Bhubaneswar, India

N. Sharma (ed.), *River System Analysis and Management*, DOI 10.1007/978-981-10-1472-7_11

to the model. The output of the model is encouraging with good efficiency and lower RMSE values. The travel time has been finalized adopting clustering techniques, thereby differentiating as per five types of peaks. The model gives better forecasting for high peaks rather than medium and low peaks.

11.1 Introduction

The Mahanadi basin is known for its devastating floods. The major reservoir Hirakud is controlling most of the floods, but it is only 83,000 km^2 of upstream catchment draining into it. The delta part around 9500 km^2 has been the worst flood-hit part. The intercepted catchment of around 47,000 km^2 is contributing heavily toward the generation of floods in the delta. The inflow to Hirakud is being modeled by MIKE-11 models, but the downstream parts lack any sound flood forecasting model.

Flood is a problem not only to Orissa or India; in a global view, flood is a destructing calamity. Controlling the flood by structural measure is also not practicable in most of the time. Thus flood management by developing an accurate warning system plays an important role. Earlier many mathematical-, conceptual- and physical-based models (PBM) developed in this regard. Although the conceptual and PBMs have involved with the hydrological processes of the basin, these require a huge data and large time of computation. In this context, an application of soft computing techniques like fuzzy logic is useful.

In this study, a fuzzy logic-based model has been used. The classical fuzzy set theory was introduced by Zadeh in 1965. In the field of hydrology, fuzzy logic is being used invariably in the field of clustering, rainfall–runoff modeling, flood forecasting, and other related fields. In the applications of the fuzzy system in control and forecasting, there are mainly two approaches, the first one being the Mamdani approach and the other the Takagi–Sugeno approach (Kruse et al. 1994). The TS model has been applied successfully by Lohani et al. (2005a, b, 2006, 2007, 2010, 2014) and Kar et al. (2015). For the Mamdani approach (Mamdani and Assilian 1975), which has been used in some hydrological applications (Schulz and Huwe 1997; Schulz et al. 1999), Tareghian and Kashefipour (2007) have applied ANN and fuzzy logic models for prediction of daily reservoir inflow in Dez of southwest Iran. They had found the superiority of Mamdani model over Takagi–Sugeno fuzzy model.

11.2 Study Area

The study area Mahanadi basin is a major east-flowing river in the peninsular river system of India. It originates near Pharasiya village of Raipur district of Chhattisgarh state. The total drainage area of river basin is 141,589 km^2. The basin is encompassed within the geographical coordinate of 80° 30′ to 86° 50′ of east longitude and 19° 20′ to 23° 35′ of north latitude. In this study, the base station is at Khairmal, intermediate station is at Barmul and the forecasting station is at Mundali, just upstream of Cuttack city. The location of different stations with their corresponding travel time of flood peaks as currently being practiced by the Department of Water Resources, Government of Orissa, is shown in schematic diagrams (Figs. 11.1 and 11.2).

11.3 Data Availability

Discharge data for a period of 10 years from 2001 to 2010 in a 3-h basis at Khairmal, Barmul, and Mundali have been made available from the Department of Water Resources, Government of Orissa. A total 101 flood peaks of different magnitudes are being observed at three stations

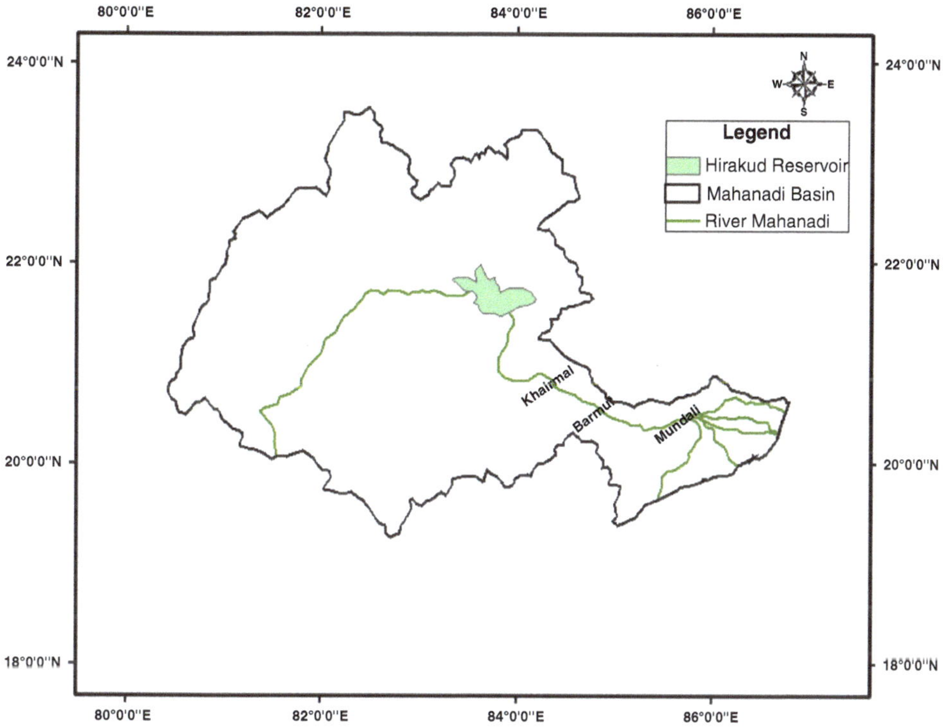

Fig. 11.1 Map of study area showing base, intermediate, and forecasting station

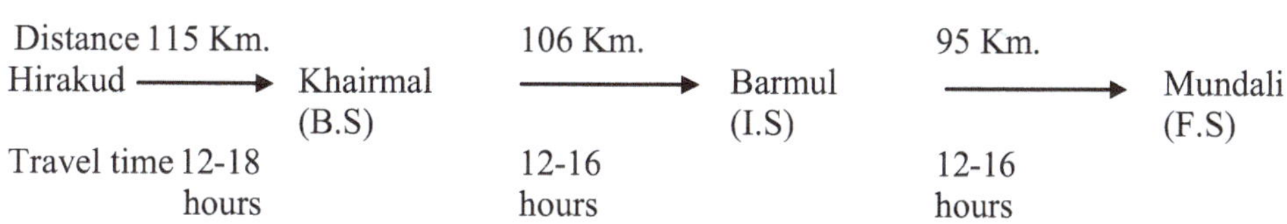

Fig. 11.2 Schematic diagram of BS, IS, and FS with distance and travel time (DOWR, Orissa)

along with their corresponding flood peaks. Initial 50 peaks are considered for calibration of the fuzzy Mamdani model and rest 51 for validation.

11.4 Methodology

The fuzzy Mamdani rule base takes a crisp input to deliver a crisp output. So it can be regarded as a crisp model system. It is basically used to model a complex non-linear system using fuzzy rules influenced by human common sense. Operation of Mamdani rule takes five steps such as fuzzification of input variables, application of fuzzy operators, mapping degree of matching to fuzzy outputs, aggregation of fuzzy outputs of all rules and defuzzification of the aggregated fuzzy output.

11.4.1 Fuzzification

The first step is to select inputs (antecedents) and determine the degree to which they belong to each of the appropriate fuzzy sets with the help of their membership function. The output of its fuzzification is the degree of membership corresponding to this numerical value defined by the linguistic qualifying set. The Mamdani model is based on crisp values but the rules provided in terms of fuzzy values.

11.4.1.1 Application of Fuzzy Operators

There may be a number of inputs in the left *if* part of the rule. The fuzzy logic operators that describe the relation between different inputs are then applied in order to combine them and obtain the result of the left if part of the rule.

11.4.1.2 Aggregation of Outputs

In aggregation, decision of all fuzzy rules is combined to get the final composite output fuzzy sets. This aggregation is done for each output variable.

11.4.1.3 Mapping of Degree to Fuzzy Outputs

The fuzzy operators are applied on the antecedent part to evaluate the consequence by mapping it on the consequent part of each rule. The methods like *clipping* and *scaling* have been used for this. In clipping the off part of the consequent is cut corresponding to the degree higher than that obtained by matching the *if* part of the rule. In scaling it scales down the membership function of the consequent in proportion to the matching degree obtained from the *if* part of the rule.

11.4.1.4 Defuzzification Process

The fuzzified inputs provide a fuzzified output. The output should be a crisp value for further analysis. The most popular defuzzification method is the centroid method, which returns the center of area under curve. Besides that bisector, middle of maximum, largest of maximum and smallest of maximum methods are also applied.

11.4.2 Clustering Approach

The inputs and output data set vary in a wide range, and it requires some distinction in particular when different peaks are available for the same travel time. In order to make proper justice for different peaks with respective travel time, it requires clustering. In this study K-mean method has been attempted in order to group the data sets into five possible divisions.

11.4.2.1 K-Mean

This method was developed by MacQueen (1967). It is best described as a partitioning method. It partitions the data into K mutually exclusive clusters and returns a vector of indices indicating to which of the K-clusters it has assigned each observation. The algorithm to clusters N objects based on attributes into K partitions, where K < N. The optimization function

$$V = \sum_{i=1}^{k} \sum_{x_j \in S_i} \left(x_j - \mu_i\right)^2 \tag{11.1}$$

It tries to achieve minimum intra-cluster variance or the squared error function. Where there are K-clusters, $S_i = 1, 2\ldots\ldots K$, and V_i is the centroid or mean point of all the points $X_j \in S_i$.

11.5 Results and Discussion

The three-hourly discharge data of base station (Khairmal), intermediate station (Barmul), and forecasting station at Mundali is collected for 10 years (2001–2010). The discharge above 2660 m^3/s is selected as a peak and maximum value of 44,740 m^3/s is found during 2008 flood. The peak discharge values at base station and its corresponding peaks at IS and FS are derived after plotting the time series of individual series. A fuzzy inference system has been developed with two inputs like peak discharge data at Khairmal and Barmul. The output is also the corresponding peak value at Mundali. The fuzzy inference system with inputs and output has been shown in Fig. 11.3.

Before categorizing the peak values to put the peaks in different categories, K-mean clustering has been applied over the data sets. It is found that the peaks can be placed in possibly five groups named as low (2660–8630 m^3/s), medium

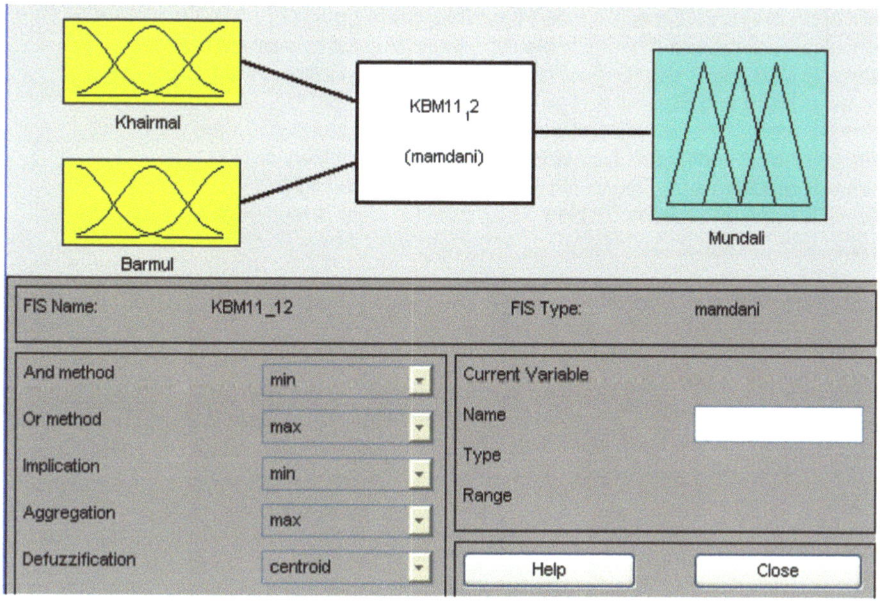

Fig. 11.3 Mamdani FIS showing inputs and output

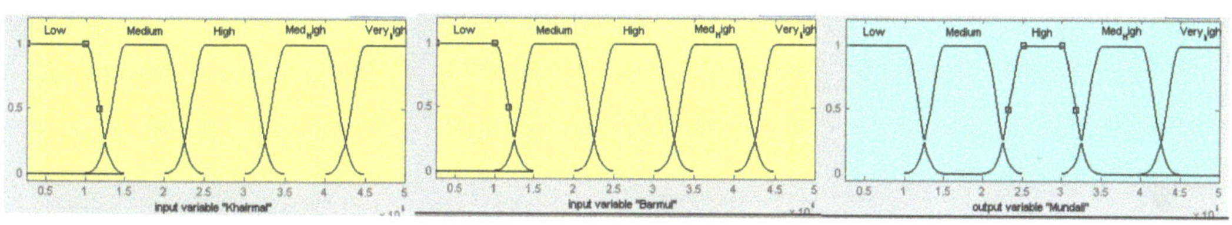

Fig. 11.4 Membership function plots for two inputs and one output

(6792–15,830 m^3/s), high (12,338–21,515 m^3/s), medium high (13,867–27,290 m^3/s), and very high (25,470–44,742 m^3/s). Different membership functions are trialed and finally Gauss2 membership function has been selected. The Gaussian2 function which is a combination of two Gaussian functions has two pairs of parameters as sig1, c1 and sig2, c2. The sig1 and sig2 are kept as 1510 throughout and c1 and c2 were varied. The membership functions for all variables are shown in Fig. 11.4.

To describe the relation between the magnitudes of the peaks, there are nine rules formed (Fig. 11.5). Every rule has also given a weightage of 1.

Basing on the inputs, membership functions and rules associated with a fuzzy output are computed. The input and fuzzy output are defuzzified to crisp output by using centroid method. The conceptual crisp outputs provide the flood forecast at forecasting site. The relation between the inputs and output is also represented by the three-dimensional plot named as surface view (Fig. 11.6.). It shows the variation of output with respect to inputs.

The comparison of crisp outputs is compared with the observed values and plotted in Fig. 11.7. It is revealed from the plot that the model is able to model properly the high peaks more effectively rather than medium or low peaks. The performance measure criteria like r^2, RMSE, and Nash efficiency have been derived for both calibration and validation periods (Table 11.1). The model is able to give a correlation of 0.857

1. If (Khairmal is Low) and (Barmul is Low) then (Mundali is Low) (1)
2. If (Khairmal is Medium) and (Barmul is Medium) then (Mundali is Medium) (1)
3. If (Khairmal is Med_High) and (Barmul is Med_High) then (Mundali is Med_High) (1)
4. If (Khairmal is High) and (Barmul is High) then (Mundali is High) (1)
5. If (Khairmal is Very_high) and (Barmul is Very_high) then (Mundali is Very_high) (1)
6. If (Khairmal is Low) and (Barmul is Medium) then (Mundali is Medium) (1)
7. If (Khairmal is Medium) and (Barmul is High) then (Mundali is High) (1)
8. If (Khairmal is High) and (Barmul is Med_High) then (Mundali is Med_High) (1)
9. If (Khairmal is Med_High) and (Barmul is Very_high) then (Mundali is Very_high) (1)

Fig. 11.5 Showing fuzzy rules adopted in the model

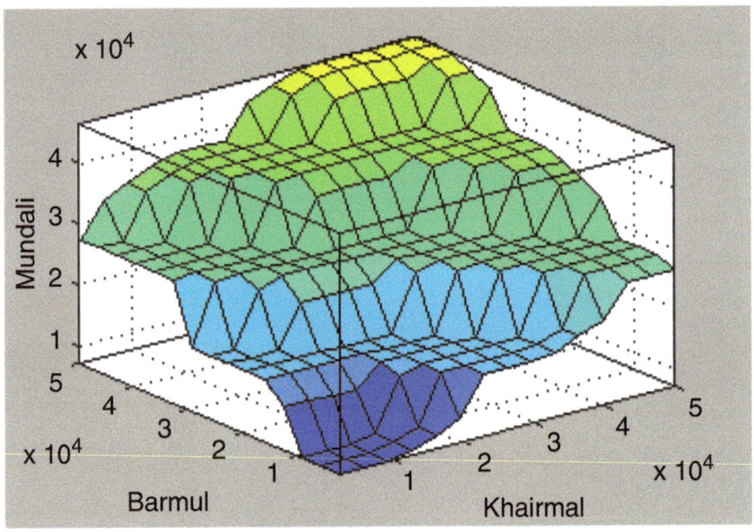

Fig. 11.6 Showing the surface view of inputs and output of fuzzy model

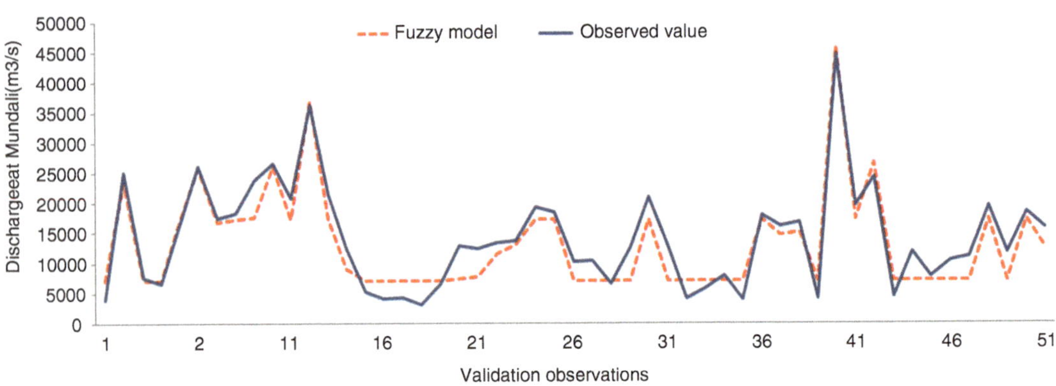

Fig. 11.7 Comparison of results at Mundali

and 0.901, RMSE of 2846.3 and 2872.16 m^3/s, and efficiency (%) of 95.6 and 89.6, respectively, for calibration and validation periods. Knowing the peak at Mundali and Barmul, the peak at Mundali can be calculated through the model. The efficiency in validation of r^2 and RMSE though acceptable is a bit low. The reason may be due to the ineffectiveness of the model toward

Table 11.1 Performance measures of fuzzy model

Performance measures	Calibration	Validation
R^2	0.857	0.901
RMSE (m3/s)	2846.3	2872.16
Efficiency (%)	95.6	89.6

Table 11.2 Travel times against the peaks

Peak category	Range of flood peak (m^3/s)	Tentative travel time (hours)
Low	2660–6835	34.85
Medium	7344–13357	30.50
High	12,338–18,734	27.50
Medium high	13,867–22,810	25.75
Very high	25,470–41,035	24.25

low and medium peaks. However, it gave fairly accurate result for very high peaks which is the foremost criteria of a flood forecasting model.

Again as per the clustered peaks of BS, the five categories of floods and their corresponding travel times are recorded. The range of flood peaks against their travel times is shown in Table 11.2. The intermediate and overlapped peaks may be interpolated to get the travel time.

11.6 Conclusion

The flood peaks are categorized into five categories depending upon the magnitude of peaks. The developed fuzzy Mamdani model is able to model the high peaks at FS more accurately rather than low and medium peaks. Earlier the Department of Water Resources, Government of Orissa, was adopting a travel time of 24–32 h for a peak traveling from Khairmal to Mundali. But in this study, travel time has been calculated using K-mean clustering. Once the magnitude of a peak reaching at base station is known, its corresponding peak and travel time to forecasting station can be calculated. Depending upon the magnitude of different peaks, the travel time varies from 24 to 35 h. As the base station is situated downstream of confluence of the Tel river, the contribution of one of the biggest tributaries is incorporated in the study. The flood data of further more years and less interval data might have produce better result.

References

Kar AK, Lohani AK, Goel NK, Roy GP (2015) Rain gauge network design for flood forecasting using multi-criteria decision analysis and clustering techniques in lower Mahanadi river basin, India. J Hydrol: Reg Stud 4(B):313–332

Kruse R, Gebhardt JE, Klowon F (1994) Foundations of fuzzy systems. Wiley

Lohani AK, Goel NK, Bhatia KKS (2005a) Real time flood forecasting using fuzzy logic. In: Perumal M (ed) Hydrological perspectives for sustainable development, vol 1. Allied Publishers Pvt. Ltd, New Delhi, pp 168–176

Lohani AK, Goel NK, Bhatia KKS (2005b) Development of fuzzy logic based real time flood forecasting system for river Narmada in Central India. In: International conference on innovation advances and implementation of flood forecasting technology, ACTIF/Floodman/Flood Relief, Oct 2005. Tromso, www.Actif.cc.net/conference 2005/proceedings

Lohani AK, Goel NK, Bhatia KKS (2006) Takagi-Sugeno fuzzy inference system for modeling stage-discharge relationship. J Hydrol 331:146–160

Lohani AK, Goel NK, Bhatia KKS (2007) Deriving stage—discharge—sediment concentration relationships using fuzzy logic. J Hydrol Sci 52(4):793–807

Lohani AK, Goel NK, Bhatia KKS (2010) Comparative study of neural network, fuzzy logic and linear transfer function techniques in daily rainfall-runoff modeling under different input domains. Hydrol Process. doi:10.1002/hyp.7831

Lohani AK, Goel NK, Bhatia KKS (2014) Improving real time flood forecasting using fuzzy inference system. J Hydrol 509:25–41

MacQueen JB (1967) Some methods for classification of multivariate observation. In: Proceedings of the 5th Berkeley symposium on probability and statistics. University of California Press, Berkely, pp 281–297

Mamdani EH, Assilian S (1975) An experiment in linguistic synthesis with a fuzzy logic controller. Int J Man–Mach Stud 7(1):1–13

Schulz K, Huwe B (1997) Water flow modeling in the unsaturated zone with imprecise parameters using a fuzzy approach. J Hydrol 201:211–229

Schulz K, Huwe B, Peiffer S (1999) Parameter uncertainty in chemical equilibrium calculations using fuzzy set theory. J Hydrol 217:119–134

Tareghian R, Kashefipour SM (2007) Application of fuzzy systems and artificial neural networks for flood forecasting. J Appl Sci 7(22):3451–3459

Anil Kumar Kar Water Resources Department, Government of Odisha, Bhubaneswar, India

N.K. Goel Department of Hydrology, Indian Institute of Technology Roorkee, Roorkee, Uttarakhand, India

Anil Kumar Lohani National Institute of Hydrology, Roorkee, Uttarakhand, India

G.P. Roy FMIS Cell, Department of Water Resources, Government of Orissa, Bhubaneswar, India

Distributed Hydrological Modelling Under Hypothetical Climate Change Scenario for a Sub-basin of the Brahmaputra River

12

Dheeraj Kumar, Ashish Pandey, Wolfgang-Albert Flügel, and Nayan Sharma

Abstract

A process-oriented hydrological model J2000 was employed on daily and monthly time steps for run-off simulation under Indian condition for the years 2003–2010. Further, a validated model was used for water balance assessment of the basin under both present scenario and hypothetical variations in precipitation and temperature. The regional sensitivity analysis (RSA) approach was employed for sensitivity analysis to determine the critical input parameter of the study area. It is inferred that recession coefficient for overland flow (soilConcRD1) is the most sensitive parameter, followed by recession coefficient for interflow (soilConcRD2) and linear reduction coefficient for AET (soilLinRed). Four evaluation criteria, i.e. coefficient of correlation (*R*), Nash-Sutcliffe efficiency (NSE), percent bias (PBIAS) and root-mean-square error (RMSE) observation standard deviation ratio (RSR), were adopted, for judging the model's performance. For daily time steps, the PBIAS, R, NSE and RSR were found to be 4.73, 0.85, 0.67 and 0.45, respectively, during calibration and −1.50, 0.92, 0.85 and 0.29, respectively, during validation of the model. For monthly time steps, the PBIAS, R, NSE and RSR were found to be 4.73, 0.88, 0.7 and 0.43, respectively, during calibration and −1.50, 0.94, 0.89 and 0.26, respectively, during validation of the model. Further, model-simulated run-off components are overland flow (RD1), model's interflow components (RD2 and RG1) and baseflow component (RG2), and the percent contribution was found to be 43.1 %, 35.8 % and 21.1 %, respectively. The water balance analysis carried out in the study demonstrated that the annual

D. Kumar (✉) • A. Pandey • N. Sharma
Department of Water Resources Development and Management, Indian Institute of Technology Roorkee, Roorkee, Uttarakhand 247 667, India
e-mail: kumar.3259@gmail.com

W.-A. Flügel
Department of Geoinformatics, Hydrology and Modelling, Friedrich-Schiller-University Jena, Jena, Germany

N. Sharma (ed.), *River System Analysis and Management*, DOI 10.1007/978-981-10-1472-7_12

average precipitation in the basin is about 1,388 mm, out of which about 21 % flows out as run-off, 22 % as groundwater and about 56 % as evapotranspiration. The higher value of evaluation criteria indicates that the model is efficient in imitating the hydrological phenomenon of the Kopili River basin fairly well and can be cogitated as a promising tool to understand the constantly changing hydrological processes of Northeast (NE) India. Moreover, the hypothetical temperature increases due to the lowest emission scenario (1.1–2.9 °C), and the highest emission scenario (2.4–6.4 °C) of the IPCC (Climate change 2007: the physical science basis. Contribution of working group I. In: Solomon S, Qin D, Manning M, Chen Z, Marquis M, Averyt KB, Tignor M, Miller HL (eds) Fourth assessment report of the intergovernmental panel on climate change. Cambridge University Press, Cambridge, UK, 2007) indicates that the higher evapotranspiration is mainly resulting in low surface run-off and groundwater contribution. The run-off fluctuations resulting from ±10 % variation in precipitation were also analysed and were found to be approx. ±18 % deviation from simulated run-off. This analysis will be useful to understand the verge of hydrological processes in anticipation of global climate change (CC) and finally for sustainable water resource management in the NE region of India.

12.1 Introduction

The per capita availability of water is decreasing swiftly all over the world, because of growing population, lack of conservation and increasing impact of CC. Understanding the potential climate change impacts which will certainly affect the hydrological balance and the future availability of water resources at continental and global scales for assessment and planning is one of the challenges hydrologists are facing nowadays. To overcome the above challenge, improved understanding with improved abilities is required to understand underlying processes and their impact on water availability. This entails employing a holistic approach which integrates hydrological phenomenon at the basin scale to determine an overall basin response to both user demands and changing climates (Singh and Woolhiser 2002). Hydrologic or watershed models are simplified, and theoretical representations of strong interactions between climate and surface hydrology are primary to water resource assessment, planning and management. These models are mainly used for understanding the physiognomies of a basin and its response to streamflow (Beven 2011; Krause 2002). Due to the merit of hydrological models over traditional methods, these models were used frequently for assessment of water resource. Studies carried out by previous researchers demonstrated the capability of hydrological models for simulation of run-off/discharge, sediment load, nutrient load, flash floods, etc. (Abbaspour et al. 2007; Murty et al. 2014; Srivastava et al. 2014; Kannan et al. 2007; Pandey et al. 2008; Bussi et al. 2014; Tolson and Shoemaker 2007; Chu et al. 2008; Blöschl et al. 2008; Braud et al. 2014).

In this chapter, the J2000 model which is a process-based distributed model (Bende-Michl et al. 2007; Fink et al. 2007; Gao et al. 2012; Krause 2002; Krause et al. 2009; Nepal et al. 2014; Rödiger et al. 2014) under the environment of Jena Adaptable Modelling System (JAMS) framework system (Fischer et al. 2009; Kralisch and Krause 2006; Kralisch et al. 2007, 2009) was employed to understand the hydrological processes of the Kopili River basin. The J2000 model has already been adapted and utilised for the Himalayan region (Gao et al. 2012; Nepal et al. 2014). The fundamental principle for selection of the J2000 model in this study involves its process-based distributed nature to imitate the geophysics in micro- to mesoscale river basins. As reported by IPCC (2007), the global average surface temperature has increased by 0.74 °C and ±0.18 °C during the last 100 years (1906–2005), and it is probable to increase more by 1.1–2.9 °C for their lowest emission scenario and 2.4–6.4 °C for their highest emission scenario by the twenty-first century. The change in atmospheric circulation and temperature due to CC certainly causes rapid movement of hydrological cycle and water resource redistribution on spatio-temporal scales and thus affecting the water balance of the basin.

The water balance equation is a fundamental hydrology equation that is valid for all temporal and spatial scales. Understanding the water balance of a watershed/basin is the most crucial aspect in sustainable water resource development, planning and management. The model application provides significant knowledge of watershed/basin attribute (such as run-off element and evapotranspiration) that is prerequisite information in simulation of water balance with certain degree of dependability. Therefore, the present study was accomplished with the specific objectives of calibration, validation, sensitivity analysis and evaluation of the J2000 model for analysing the water balance components of the Kopili River basin. Further, a way to quantify the impacts on hydro-meteorological variables based on the projected changes on different hydrological cycle components (e.g. total run-off volume) was assessed by applying hypothetical precipitation and temperature change scenarios.

12.2 Study Area

The Kopili River basin which is a part of Brahmaputra basin has been selected as study area (Fig. 12.1). The altitude of the basin ranges from 74 to 1967 m above m.s.l. (mean sea level). The Kopili River springs up from the north-western slope of the Shillong Peak, at an elevation of 1967 m, and enters the Assam Valley through its southern part.

The study area lies between 25.0979°N and 25.8627°N latitude and 92.1304°E and 93.4631° longitude. The basin area till Kherunighat gauging station is 7198 km^2 out of which 76 % is in Assam state and the other 24 % in Meghalaya. The outlet of the Kopili River basin which is at Kherunighat lies in Assam and has coordinates of 25° 51′ 2.6″ N latitude and 92° 53′10.9″ E longitude. The Kherunighat gauging station was considered as the outlet, due to the availability of observed discharge data required by the model. The basin was further divided into five sub-basins and is presented in Fig. 12.1.

The longest and second longest path of the river from its origin to Kherunighat discharge site is 190.4 km and 146.9 km, respectively. The longest path of the river originates within Assam state, and the origin of the second longest path of Kopili River lies in Meghalaya state. The climatic conditions of any river basin are often represented by rainfall and temperature variations and the level of humidity (Singh et al. 2004).

The climatic condition of the basin is moderate, the air being mostly dry except during the south-west monsoons. The annual precipitation of the basin varies from 980 to 1700 mm with an annual average precipitation of 1400 mm, and the maximum-minimum daily streamflow of the basin is about 858 and 16.1 m^3/s. The monsoon season generally starts from mid-May and continues till mid-October. The winter commences from mid-October and continues till February. Due to wide topography variation, Kopili River basin experiences different climate in different parts. During summer the atmosphere becomes sultry although the temperature ranges between 22 and 36 °C in summer and 9 and 25 °C in winter. Daily mean temperature of the study area ranges from 11.1 to 31.7 °C. The daily mean relative humidity varies from a minimum of 24 % (March and April) to a maximum of 95 % (July).

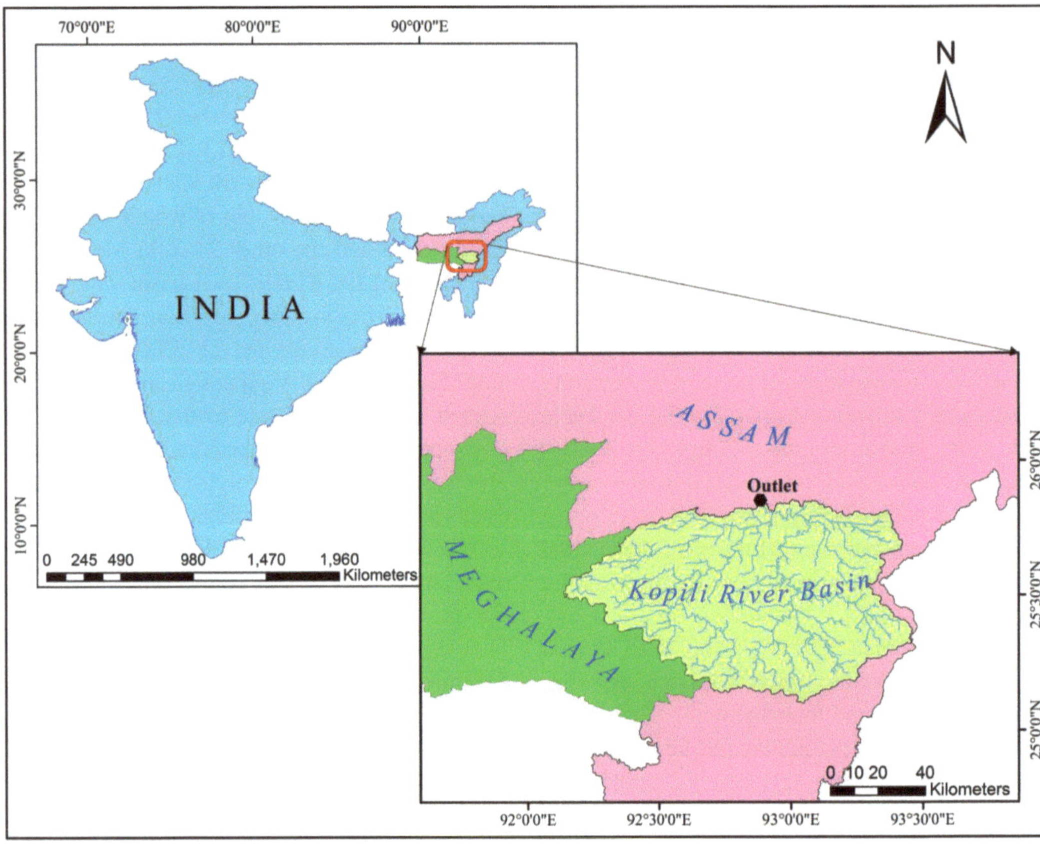

Fig. 12.1 Location map of Kopili River basin

12.3 Methodology

12.3.1 Hydro-meteorological Data

Historical daily precipitation data for 8 years (2003–2010) were collected and analysed for eight rain gauge stations (Kherunighat, Diphu, Golaghat, Kampur, Dharamtul, Laskein, Maibong, Umrangso) that were procured from India Metrological Department (IMD), Pune. Other meteorological data such as maximum and minimum air temperature, relative humidity, absolute humidity, sunshine hour and wind speed were collected from Umrangso meteorological observatory. Besides these, daily discharge data at the outlet of the Kherunighat gauging site (2003–2010) were also collected from the Central Water Commission (CWC) office.

12.3.2 Regionalisation of Climate and Precipitation Data

The input dataset regionalisation was conducted within the J2000 modelling system. The regionalisation process of the J2000 modelling system has been well reported in the model's technical manual (Krause 2010). A linear relationship between the elevation of the station and the daily station values is established, and the slope of the regression line (b_H) and the coefficient of determination (R^2) of this relation are calculated. It is assumed that the measurement value (MW) depends linearly on the terrain elevation (H) according to:

$$MW = a_H + b_H \cdot H \qquad (12.1)$$

According to the Gaussian method of least squares, the unknown values of a_H and b_H are defined:

$$b_H = \frac{\sum_{i=1}^{n} (H_i - \overline{H})(MW_i - \overline{MW})}{\sum_{i=1}^{n} (H_i - \overline{H})^2} \quad (12.2)$$

$$b_H = \overline{MW} - b_H \times \overline{H} \quad (12.3)$$

The coefficient of correlation (R) is calculated according to the following equation:

$$R = \frac{\sum_{i=1}^{n} (H_i - \overline{H})(MW_i - \overline{MW})}{\sqrt{\sum_{i=1}^{n} (H_i - \overline{H})^2 \times \sum_{i=1}^{n} (MW_i - \overline{MW})^2}} \quad (12.4)$$

A constant lapse rate for regionalisation of temperature was implemented, as only one temperature station is available in the study area.

In the present study, inverse distance-weighted (IDW) approach for localisation was applied, and the weightings of the n stations were defined with regard to their distances for every hydrological response unit (HRU) in a different module. By using the IDW method, the horizontal variability of the station data was taken into consideration according to its spatial position. The calculations were made by using the following equation:

$$W(i) = \frac{\left(\frac{\sum_{i=1}^{n} \omega Dist(i)}{\omega Dist(i)}\right)}{\left(\sum_{i=1}^{n} \frac{\sum_{i=1}^{n} \omega Dist(i)}{\omega Dist(i)}\right)} \quad (12.5)$$

12.3.3 Land Use and Land Cover of the Kopili River Basin

Land cover is a determining factor for the run-off. It influences evaporation as well as important characteristics of the soil, especially the rooting depth. In addition, land cover mainly controls the extent of the interception storage. Thus, the cloud-free geocoded data of LANDSAT was downloaded from Earth Explorer site. The imageries with spatial resolution of 30 m relevant to the basin were used in this study. The land use/land cover map of the study area was generated employing the satellite data of autumn season for the year 2010. The objective of categorisation is to assign a class or category to each cell in the study area. Visual survey of the study area was carried out to identify signatures of various objects and land use classes in the raw image. A hybrid classification approach, i.e. combining unsupervised ISODATA clustering and supervised maximum likelihood classification, was applied. The image classification module integrated in ArcGIS 10.1 was used for classifying the land uses. By default, all cells in the output raster are classified by MLC module, with each class having equal probability weights attached to their signatures. The classified land use/cover classes were deciduous forest (39.03 %), barren land/settlement (19.77 %), agricultural land (5.38 %), coniferous forest (34.83 %) and water bodies (less than 1 %). Area occupied by each land use classes in the study area is presented in Fig. 12.2.

12.3.4 Soils of the Kopili River Basin

According to Funke et al. (1999), soils predominantly influence the run-off process. The available soil storage of the upper soil zone is thus a function of porosity as well as of vegetation root depth. The soil storage is a determining factor for the infiltration, which is controlled by parameters like hydraulic conductivity and soil suction at the wetting front. The Kopili River basin falls under the Khasi group, Surma series and gneiss with old inliers Sela group and granites in Assam whereas Shella formation and Simsang formation in Meghalaya. Soil surveys of these regions have been carried out by the National Bureau of Soil Survey and Land use Planning (NBSS&LUP), Nagpur. Based on the data obtained from NBSS&LUP, soils of the basin were classified as coarse loamy, fine loamy, coarse silty, loamy skeletal and fine (Fig. 12.3). The chemical and physical characteristics of the soils are taken from soils of Assam (NBSS publication 101) and soils of Meghalaya (NBSS publication 121). Data related to various

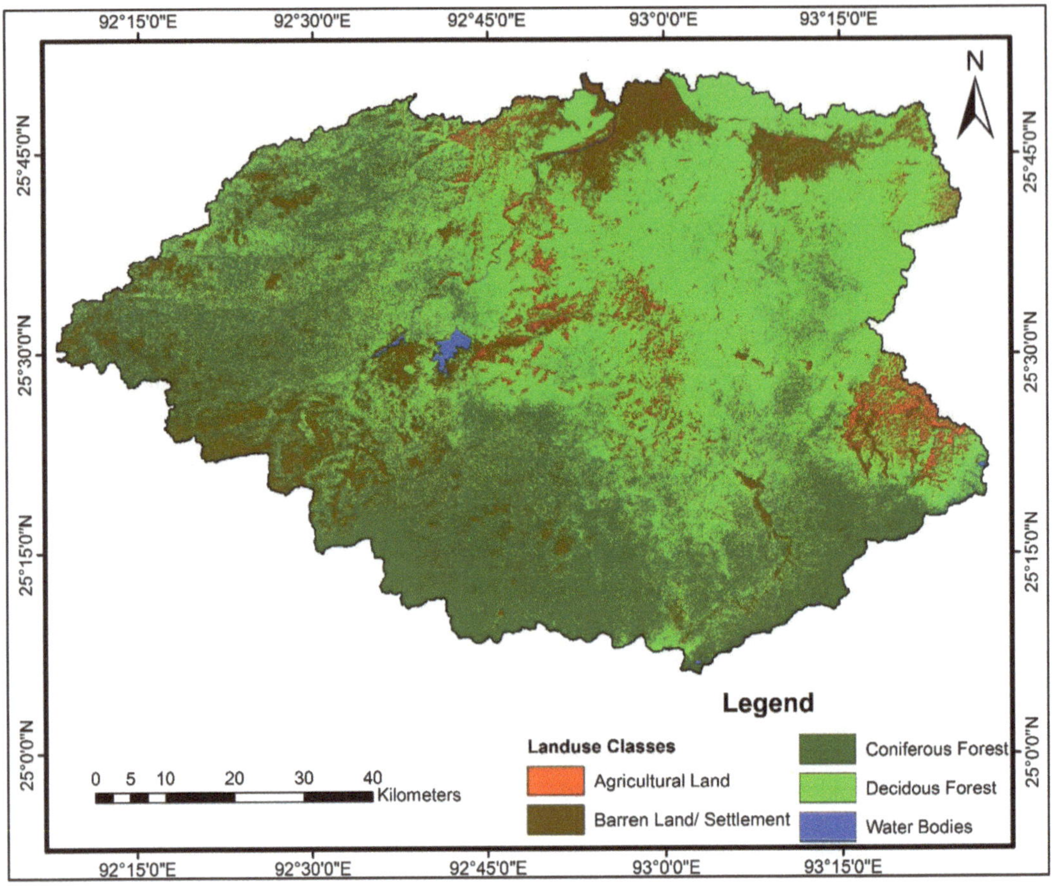

Fig. 12.2 Land use/land cover map of the study area

chemical and physical characteristics of different soils present in the basin area were collected for understanding the nature of different soils, and it was also used for the preparation of input file for the model. The Kopili River basin soil texture map is presented in Fig. 12.3. The soil map was prepared through GIS using available soil resource data of the respective area. An area occupied by different soil textures in the basin is presented in Table 12.1. Fine soil has occupied maximum area in the Kopili River basin followed by fine loamy soil (Table 12.1).

12.3.5 Topography

Topographical effect of the basin plays a substantial role in moisture regime dynamics (Abbaspour et al. 2007). Shuttle Radar Topography Mission (SRTM) (Farr et al. 2007) data was processed using ERDAS desktop 2010 (version 10.1.1) for generation of the digital elevation model (DEM) of the Kopili River basin (Fig. 12.4). Elevation data of SRTM is available on public domain (http://dds.cr.usgs.gov/srtm/) under joint operation of the National Geospatial-Intelligence Agency (NGA) and the National Aeronautics and Space Administration (NASA) and provides altitude with resolution of 3 arc seconds or 90 m for Indian subcontinent.

12.3.6 Hardware and Software Used

A work station equipped with Xeon processor, JAMS/JUICE version, ERDAS IMAGINE 2014

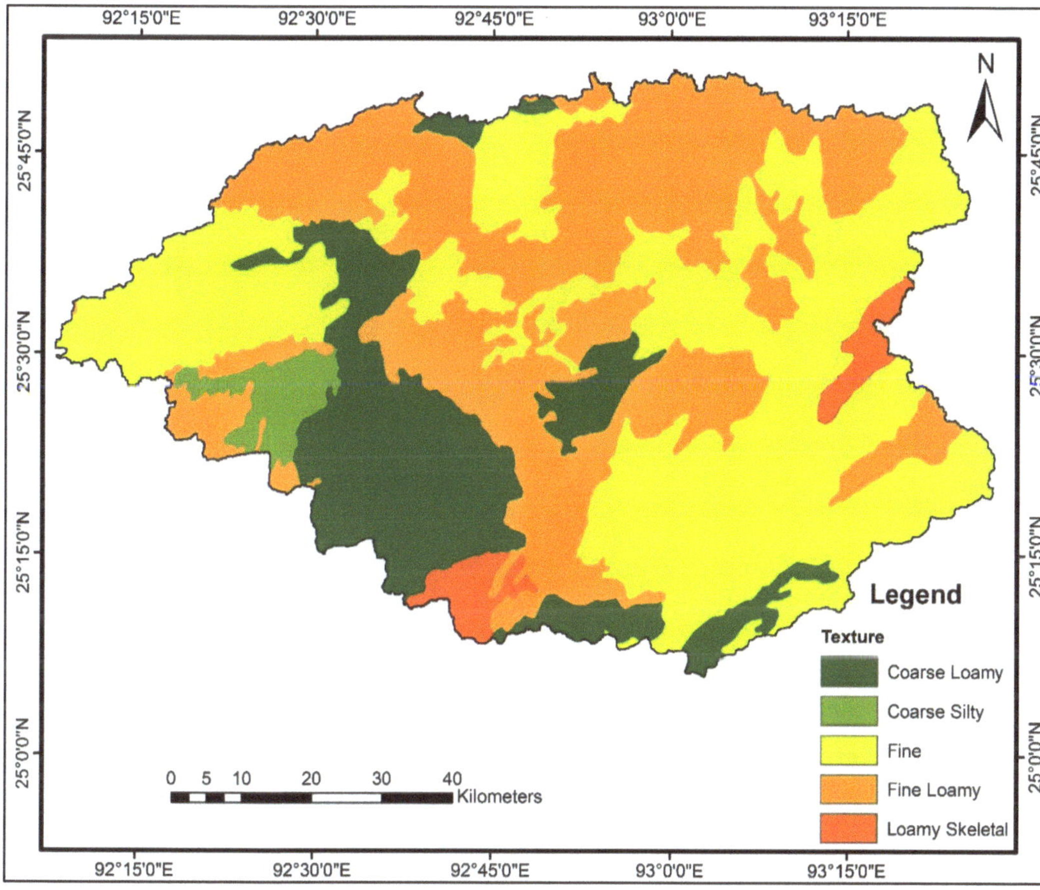

Fig. 12.3 Soil texture map of the study area

Table 12.1 Soil texture classification of the basin area

Soil texture	No. of pixels	Area (Km^2)	% image
Coarse loamy	1320389	1188.35	16.51
Fine loamy	2947019	2652.32	36.85
Coarse silty	185357	166.82	2.32
Loamy skeletal	209572	188.61	2.62
Fine	3335702	3002.13	41.71
Total	7998039	7198.2	100

image processing software, ArcGIS version 10.2, Contex full-scale scanner (FSS) 8000 and laser printer, available at the Institute Computer Centre and Department of Water Resource Development and Management Department, IIT Roorkee, India, was used in this analysis. These facilities were used for interpretation of DEM, preparation of land use/land cover and soil maps and for model simulation.

12.4 Description and Set-Up of the J2000 Model

12.4.1 The J2000 Modelling System

The J2000 model is a process-based hydrological model for simulation of meso- to macro-scale watershed/basin (Krause 2001). The J2000 model is an open source modelling system (Krause 2002) which is executed within the framework of Jena Adaptable Modelling System (JAMS) (Kralisch and Krause 2006; Kralisch et al. 2007). The JAMS is a fully modular structured, hydrological modelling framework

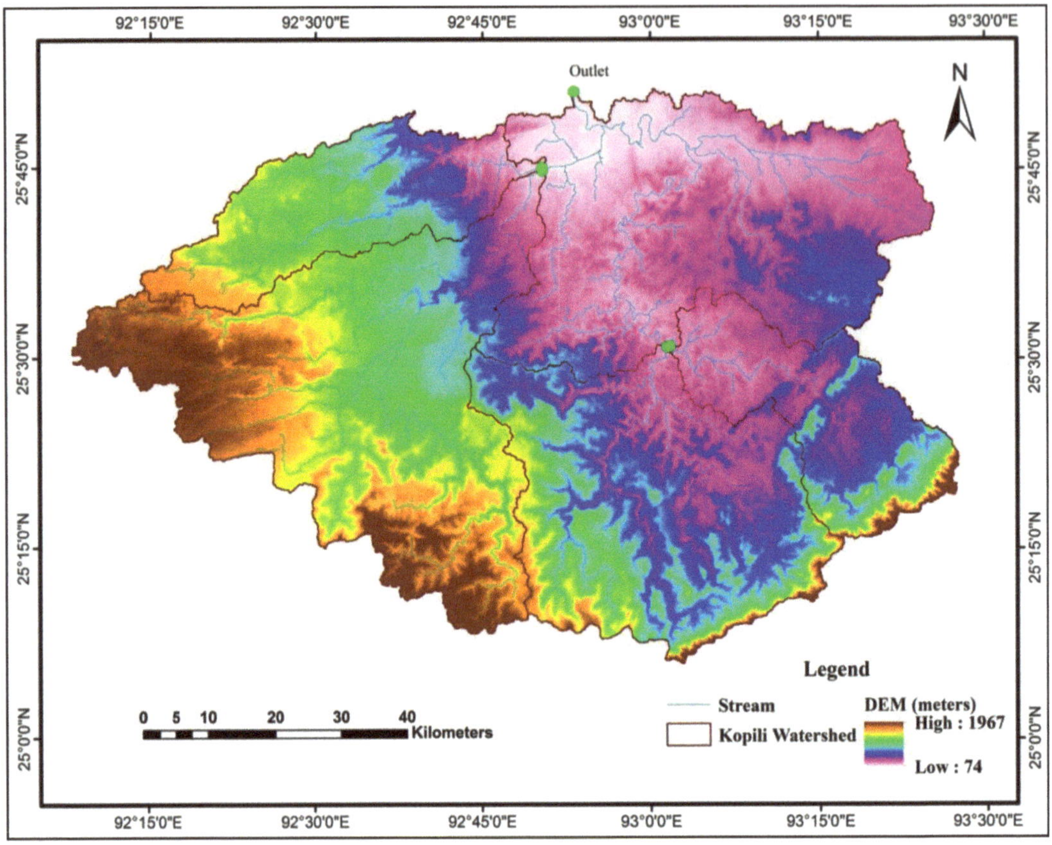

Fig. 12.4 Digital elevation model (DEM) of the study area

which was developed to meet the challenges in sustainable water resource management. JAMS is able to simulate environmental processes at distinct points in space and/or time. Simulations of different hydrological processes are executed independent of each other in an encapsulated process module in this model. Due to this flexibility, any single module can be changed, substituted or added without restructuring the model again from the start. The detailed description of the J2000 modelling system and its different modules can be found in Krause (2001) and Nepal (2012). JAMS and the J2000 hydrological model together with an example dataset can be downloaded from http://jams.uni-jena.de/index.php?id=5581&L=2.

The principal structure of the J2000 hydrological modelling system is presented in Fig. 12.5. The J2000 model generates four different run-off components. The component with the highest temporal dynamics is the fast direct run-off (RD1) which is known as overland flow. It consists of run-off from sealed areas, which is a saturated or infiltration access run-off which drains directly to a stream. The corresponding resultant to the lateral hypodermic run-off within soil zone is termed as slow direct run-off (RD2) component or interflow 1, which reacts slightly slowly than the RD1. Two further baseflow run-off components can be distinguished. The relatively fast baseflow run-off (RG1) (also known as interflow 2) simulates the run-off from the upper part of an aquifer, which is more permeable due to weathering, compared to the lower zone of the aquifer. The slow baseflow run-off component (RG2) can be seen as flow within fractures of solid rocks or matrix in homogeneous unconsolidated aquifers.

The hydrological processes in the J2000 model application are simulated in daily or

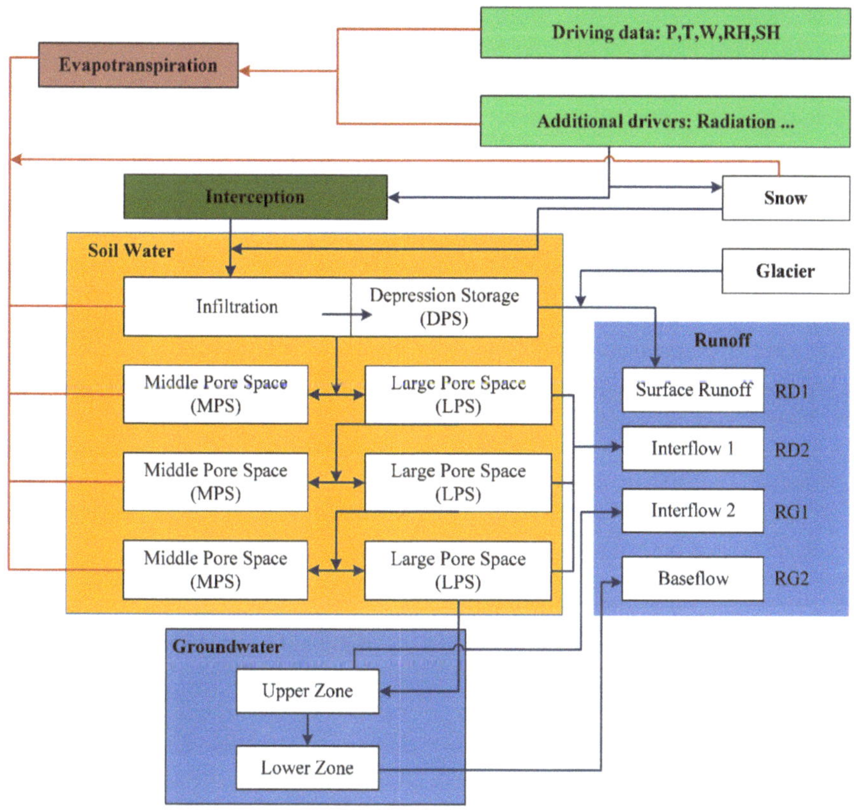

Fig. 12.5 Principal structure of the J2000 model

monthly time steps over each HRU using the following general water balance equation:

$$\Delta S = \sum_{i=1}^{t} P_{\text{day}} - Q_{\text{surf}} - E_a - Q_{gw} \qquad (12.6)$$

where

ΔS = change in water storage (mm)
P_{day} = precipitation depth (mm)
Q_{surf} = surface run-off depth (RD1 + RD2) (mm)
E_a = evapotranspiration depth (mm)
T = time (days)
Q_{gw} = return flow depth (RG1 + RG2) (mm)

12.4.2 Modules of the J2000 Modelling System

The J2000 is a fully modular process-based hydrological modelling system, which executes any single hydrological process as an encapsulated process module. Various modules can be employed depending on data availability for run-off simulation and on the purpose of the modelling application. In the present study, the following modules were employed:

- Distribution of precipitation
- Interception module
- Soil module
- Groundwater module
- Routing module

The most significant operations within the modules are described below.

12.4.2.1 Precipitation Distribution Module

Based on the air temperature, the precipitation is first distributed between rain and snow. The calibration parameters, i.e. *Trans* and *Trs*, are employed, where *Trs* is base temperature and *Trans* is a temperature range (upper and lower boundary) above and below the base temperature. To determine the amount of snow and rain, it is assumed that below a certain threshold, temperatures result in total snow precipitation exceeding a second threshold that results in total rainfall as precipitation. In between those threshold temperatures, mixed precipitation occurs. Between those thresholds, rain-snow mixtures with variable percentages for each component are calculated. These parameters are considered as non-flexible parameters and not necessarily placed in the JAMS framework as tuneable parameters. The actual amount of snow (P_s) of daily precipitation subjected to air temperature is calculated according to:

$$P_s = \frac{Trs + Trans - Temperature}{2 \cdot Trans} \quad [\text{mm}] \tag{12.7}$$

The daily amount of snow (P_s) or amount of rain (P_r) is calculated according to:

$$P_s = \text{Precipitation} \cdot P_s \quad [\text{mm}] \tag{12.8}$$

$$P_r = \text{Precipitation} \cdot (1 - P_s) \quad [\text{mm}] \tag{12.9}$$

12.4.2.2 Interception Module

Interception process is recognised as the most crucial component of hydrological cycle that can disturb the water balance components. In the interception module, the calculation of net rainfall from the gauged rainfall against the particular vegetation cover and their development in the annual variation is executed. The interception module calculates maximum interception storage capacity based on the leaf area index (*LAI*) of the particular type of land cover (Dickinson 1984). The four different *LAI* types are proposed in land use parameter file, for four different seasons and for each vegetation type (Nepal et al. 2014). The maximum interception capacity is calculated according to the following formula:

$$Int_{\max} = \alpha \times LAI \quad [\text{mm}] \tag{12.10}$$

where:

α = the storage capacity per m^2 leaf area against the precipitation
LAI = the leaf area index of the particular land use class

The emptying of the interception storage is carried out exclusively by evaporation.

12.4.2.3 Soil-Water Module

The soil-water module of unsaturated zone is the most complicated component of the J2000 model (Fig. 12.6), which controls the distribution and regulation of water movement (Nepal et al. 2014). The figure indicates the important process within the soil module. The soil module is structured in process units [evapotranspiration, infiltration] and store units [large pore storage (LPS), middle pore storage (MPS), depression storage] (Krause 2010). The abstraction of the soil-water is carried out by two parallel and interconnected storages, i.e. MPS and LPS. The infiltrated rain is distributed to both soil storage components (MPS and LPS). Any surplus water, if it exceeds the maximum infiltration capacity of the corresponding soil or saturation of the LPS, is stored as depression storage. The infiltration capacity is an important process which determines whether water can seep downward in the soil horizon entirely or whether it is stored for a short time at the surface and whether it generates depression storage at this location or results in surface run-off. An empirical calculation method is used in J2000 for the calculation of infiltration (*Inf*). A maximum infiltration rate (*maxINF* in mm/d) which is defined by the user, against the relative saturation deficit of the soil ($1 - soil_{\text{sat}}$), is taken into account:

$$Inf = (1 - soil_{\text{sat}}) \cdot (maxINF) \tag{12.11}$$

The equation governing the calculation of the relative saturation of the soil is:

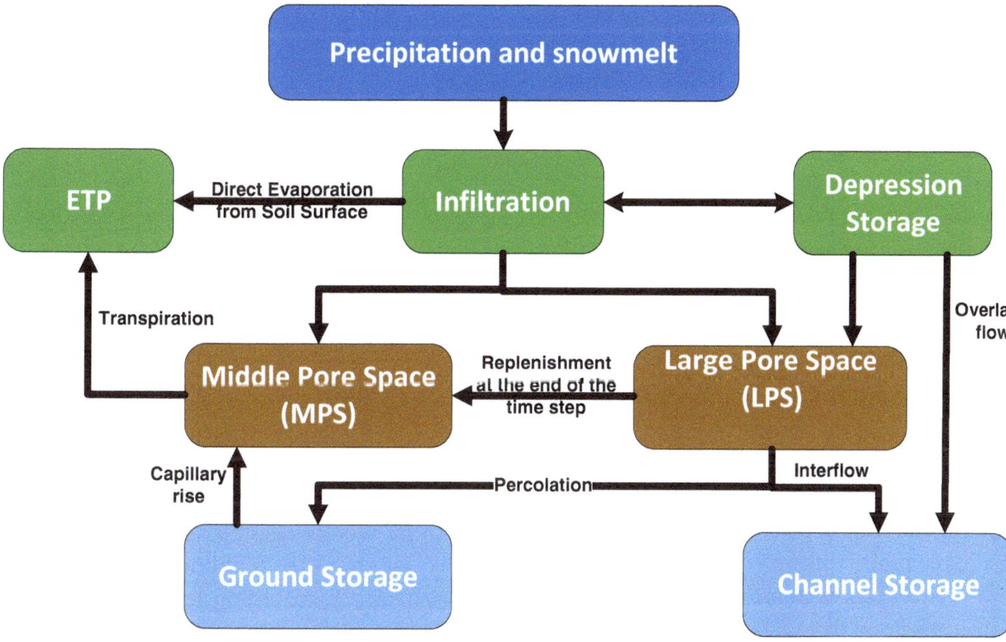

Fig. 12.6 Principal layout of the J2000 soil module (Source: Krause 2002)

$$soil_{sat} = \frac{MPS_{act} + LPS_{act}}{MPS_{max} + LPS_{max}} \quad [mm/day] \quad (12.12)$$

where:

MPS_{act} and MPS_{max} = actual and maximum filling of the middle pore storage

LPS_{act} and LPS_{max} = actual and maximum filling of the large pore storage

The maximum infiltration rate is defined by the user considering three infiltration scenarios, i.e. *Inf_winter* (normal case of the infiltration for the winter half year), *Inf_summer* (special infiltration conditions for the convective precipitation with short duration and high intensity) and *Inf_snow* (circumstance of decreased infiltration due to partly or complete frozen soil).

The actual infiltration (Inf_{act}) is disseminated between the LPS and MPS storages. The amount of water which is in every soil storage is subject to the saturation deficit of the MPS and is calculated using the calibration parameter *soilDistMPSLPS* as follows:

$$MPS_{in} = Inf_{act} \cdot \left(1 - e^{\left(\frac{-1 \cdot soilDistMPSLPS}{satMPS}\right)}\right) \quad [mm] \quad (12.13)$$

The infiltrated precipitation water which is not absorbed by the MPS enters into the large pore storage (LPS_{in}):

$$LPS_{in} = Inf - MPS_{in} \quad [mm] \quad (12.14)$$

Due to the water distribution in the model, the MPSs operate similar to a sponge and its potentiality for taking water increases with decreasing moisture condition. However, a certain amount always remains in the middle pore storage.

12.4.2.3.1 Depression Storage

As represented in the description of infiltration, the water amount which exceeds the maximum infiltration rate of the soil is transferred to the depression storage. This also applies if the soil is completely water saturated and no infiltration can take place. The water which is stored in the depression storage runs off partly as surface run-off. The maximum depression storage has to be defined during model simulation. According to Maniak (1997), the maximum depression storage decreases by 50 % in slope higher than 5°, and the volume of the depression storage is divided into two for areas exceeding this slope. The excess water is released as run-off when the maximum depression storage limit is exceeded on any particular area.

12.4.2.3.2 Middle Pore Storage (MPS)

Due to the adsorption powers, the water is held against gravity in the middle pore space of the soil, and for the extraction of water from the middle pore storage, an active soil-water potential is required. The potential for such a soil-water extraction is made available by evaporation. The two cases responsible for the extraction of soil-water from the MPS are the transpiration by the vegetation and the direct evaporation from the soil surface. The direct evaporation from the soil surface is proportionally lower since only a few millimetres of dry soil can cause an effective isolation of the underlying layers regarding the evaporation, whereas this isolation is shorted out by the vegetation cover, which makes a consistent exhaustion of water stored in the middle pores through transpiration. The calculation of reduction factor (*RF*) is carried out employing the threshold value and the actual water saturation of the middle pore storage (*satMPS*):

$$RF = 1 \text{ for } satMPS \geq soilLinRed \qquad (12.15)$$

$$RF = \frac{satMPS}{soilLinRed} \qquad (12.16)$$

12.4.2.3.3 Large Pore Storage (LPS)

The water which is available in the LPS component of the J2000 model is subject to gravitation and is therefore considered as the source of actual flow processes and run-off generation in the soil. The water amount which generates run-off from the LPS in the time interval is subject to the relative water saturation of the entire soil zone (LPS_{soil}) and is calculated according to:

$$Q_{LPS} = Sat_{soil}^{\alpha} \cdot LPS_{act} \qquad [\text{mm}] \quad (12.17)$$

where:

Q_{LPS} = LPS outflow
LPS_{act} = actual storage amount of the LPS (mm)
Sat_{soil} = relative water saturation of the soil at location (percent)
α = calibration coefficient

12.4.2.3.4 Diffusion

At the end of the time step, a deficit of the MPS resulting from evaporation can be balanced out by water from the LPS. The diffusion (*diff*) component is estimated in the model using the calibration coefficient *soilDiffMPSLPS* according to:

$$diff = actLPS \cdot \left(e^{-1 \cdot diff / satMPS}\right) \qquad [\text{mm}] \quad (12.18)$$

12.4.2.3.5 Percolation and Interflow Generation

The lateral (interflow) and vertical (percolation) water movement occurs in the LPS and is therefore dependent on the amount of the large pores. The following QLPS is distributed between vertical and lateral flow (interflow). The $slope_{weight}$ is computed using slope and a user-specific calibration factor (*soilLatVertDist*):

$$slope_{weight} = \left(1 - \tan\left(slope \cdot \frac{\pi}{180}\right)\right) \cdot soilLatVertDist \qquad (12.19)$$

where:

α = slope of the corresponding HRU (degree)
soilLatVertDist = calibration coefficient

The water depth which is available for percolation is computed in the model according to:

$$Percolation = \left(1 - slope_{\text{weight}}\right) \cdot LPS_{\text{out}} \quad (12.20)$$

This percolation rate is then set against the calibration coefficient *soilMaxPerc* which describes the maximum percolation rate per time step, and the interflow 1 (RD2) is calculated according to:

$$Interflow = slope_{\text{weight}} \cdot LPS_{\text{out}} \quad (12.21)$$

12.4.2.3.6 Run-off Detention

Direct run-off (RD1) and interflow (RD2) are delayed in time in order to take into account the areal expansion of the spatial model entity. The detention is computed according to:

$$RD1 = \frac{1}{soilConcRD1} \cdot RD1gen \quad [\text{mm}] \quad (12.22)$$

$$RD2 = \frac{1}{soilConcRD2} \cdot RD2gen \quad [\text{mm}] \quad (12.23)$$

Nepal (2012) suggested that due to non-linear behaviour of catchment, the delayed time may be different during high-flow periods in case of RD1. A new parameter (*concRD1Flood*) is used by the model when the $RD1_{\text{gen}}$ crosses a threshold value (*RD1FloodThreshold*) provided by a user. The value of (*concRD1Flood*) should be lower than concRD1 because it produces higher RD1 output flow:

$$RD1_{\text{flood}} = \frac{1}{soilConcRD1Flood} \cdot RD1gen \quad [\text{mm}] \quad (12.24)$$

12.4.2.4 Groundwater Module

In the individual geologic units, there is a distinction between the upper groundwater reservoir (RG1) in loose weathered material with high permeability and the lower groundwater reservoir (RG2) in fractures and clefts of the bedrock. Consequently, two basis run-off components are generated: a fast one from the upper groundwater reservoir and a slow one from the lower groundwater reservoir (Krause 2010). In the calculation of the groundwater run-off (outRG1 and outRG2), the storage retention coefficients have been taken into account as factors of the actual storage content, i.e. actRG1 and actRG2:

$$outRG1 = \frac{1}{gwRG1Fact \cdot recRG1} \cdot actRG1 \quad (12.25)$$

$$outRG2 = \frac{1}{gwRG2Fact \cdot recRG2} \cdot actRG2 \quad (12.26)$$

The *gwRG1Fact* and *gwRG2Fact* are the calibration parameters for the particular upper and lower groundwater reservoir.

12.4.2.5 Routing Module

The J2000 model has two routing components. The lateral routing in HRUs describes water transfer within a flow cascade from one HRU to another HRU from the upper catchment areas until the receiving stream (Krause 2010). The water transfer between the HRUs are seen as n:1 relation. Thus, an HRU can have several influxes but only one discharge. The reach routing describes flow processes in a stream channel by using the commonly applied kinematic wave approach and the calculation of velocity according to Manning and Strickler (Krause 2001). The only model parameter that has to be estimated by the user is a routing coefficient (*TA*). It represents the runtime of the run-off wave which moves in the channel until it reaches the catchment outlet after a precipitation event. Its value is required for the calculation of the restraint coefficient (*Rk*) together with the velocity of the river (v) and flow length (fl):

$$Rk = \frac{v}{fl} \cdot TA \cdot 3600 \quad (12.27)$$

Earlier to this, the velocity (v_{new}) has to be determined using the roughness factor by Manning (M), the slope of the riverbed (I) and the hydraulic radius (Rh) in metres. The hydraulic radius is calculated using the drained cross section (A) in m^2 of the reach resulting from the flow rate (q), velocity (v) and riverbed width (b). For this procedure, an initial starting velocity (v_{ini}) is assumed which is then iteratively adjusted with regard to the new calculated velocity (v_{new}) until both velocities differ by a value smaller than 0.001 m^{-1}.

$$Rh = \frac{A}{b + 2\frac{A}{B}} \quad [\text{m}] \quad (12.28)$$

$$A = \frac{q}{v_{\text{ini}}} \quad [\text{m}^2] \quad (12.29)$$

$$v_{\text{new}} = M \cdot Rh^{2/3} \cdot I^{1/3} \quad [\text{m}^3/\sec] \quad (12.30)$$

Finally, the amount of water of the corresponding reach (q_{act}) is calculated in the model which is allocated to run-off (q) using the run-off restraint coefficient (Rk) which has been generated:

$$q = q_{\text{act}} \cdot e^{-1/Rk} \quad [\text{m}^3/\sec] \quad (12.31)$$

The higher is the assumed value of TA, the faster does the run-off wave move within a certain time span and the less water remains in the channel.

12.4.3 Calculation of Evapotranspiration

The calculation of the evapotranspiration is carried out in J2000 according to the Penman-Monteith equation in several steps in regard to numerous parameters. The calculation for the day and night evapotranspiration is carried out according to the following equations, whereby the total value of the evaporation for the particular time step results as sum of these two values (Krause 2010):

$$ETP_d = \frac{1}{L_d} \cdot \frac{s_d \cdot (R_{N_d} - G_d) + \rho \cdot c_P \cdot \frac{e_{s_d} - e_d}{r_a}}{s_d + \gamma_d \cdot \left(1 + \frac{r_{s_d}}{r_a}\right)} \cdot \left(\frac{S_0}{24}\right) \quad (12.32)$$

$$ETP_n = \frac{1}{L_n} \cdot \frac{s_n \cdot (R_{N_n} - G_n) + \rho \cdot c_P \cdot \frac{e_{s_n} - e_n}{r_a}}{s_n + \gamma_n \cdot \left(1 + \frac{r_{s_n}}{r_a}\right)} \cdot \left(1 - \frac{S_0}{24}\right) \quad (12.33)$$

where:

$L_{d,n}$ = latent heat of evaporation $[\text{Wm}^{-2}]$ per $[\text{mmd}^{-1}]$
$s_{d,n}$ = slope of the vapour pressure curve $[\text{hPaK}^{-1}]$
$R_{Nd,n}$ = net radiation $[\text{Wm}^{-2}]$
$G_{d,n}$ = soil heat flux $[\text{Wm}^{-2}]$
P = density of the air $[\text{kgm}^{-3}]$
c_P = specific heat capacity of the air for constant pressure $[\text{Jkg}^{-1}\ \text{K}^{-1}]$
$e_{sd,n}$ = saturation vapour pressure [hPa]
$e_{d,n}$ = vapour pressure [hPa]
r_a = aerodynamic resistance of the land cover $[\text{sm}^{-1}]$
$\gamma_{d,n}$ = psychrometer constant $[\text{hPaK}^{-1}]$
$r_{sd,n}$ = surface resistance of the land cover $[\text{sm}^{-1}]$
S_0 = astronomic possible sunshine duration [h]

12.4.4 Watershed Delineation and Hydrological Response Unit (HRU) Definition

The study area comprises the entire Kopili River which is a tributary of river Brahmaputra. The delineation of the study area was done from the DEM of ASTER datasets. The area of the basin (Kopili River basin) up to Kherunighat gauging station is found to be 7198 km^2 from the DEM-based delineation. For this purpose, a complete process chain was implemented according to a service-oriented application in GRASS-GIS. A plug-in developed for QGIS creates an intuitive, wizard-driven and transparent environment for the execution of the process chain. The procedural step followed for watershed delineation and HRU creation are described below:

(i) The first step was to load the digital elevation model (DEM) in the GRASS-HRU interface. The DEM and land use, hydrogeology and soil classification datasets were resembled to 90 × 90 m spatial resolution and reprojected to a common projection system, i.e. Universal Transverse Mercator (UTM) with zone 46 north and datum as WGS-84.

(ii) A mask or a bounding box of the Kopili River basin was used to focus the basin area. The resulting bounding box should comprise the complete catchment area to be analysed.

(iii) The DEM was filled using the *default filling* methodology to delete local sinks. In this step the calculation of slope and slope orientation was carried out.

(iv) The DEM aspect and slope class generated were reclassified according to different indices.

(v) For the derivation of the stream network, flow accumulation, and flow direction and for partial catchment areas, a threshold value *minimum size of basins* is required. In this study, the default threshold value of 1500 pixel has been adopted to determine the derivable partial catchment area.

(vi) The locations of the different sub-basin outlets were added and selected along with the location of gauging site. This was to ensure that the model calibration was done at their exact locations. Once the outlets of the sub-basins were selected, the sub-basins as well as the entire basin were delineated.

(vii) The overlay of all data layers creates small areas which are not suitable as HRUs and should therefore be eliminated. For this purpose, a threshold value is defined (*size of smallest area to remove*) which determines the minimum size of an HRU (in cells). In this study a value of 300 pixels, i.e. 243 ha, was assumed, creating 587 HRUs spread over the five sub-basins of the study area (Fig. 12.7).

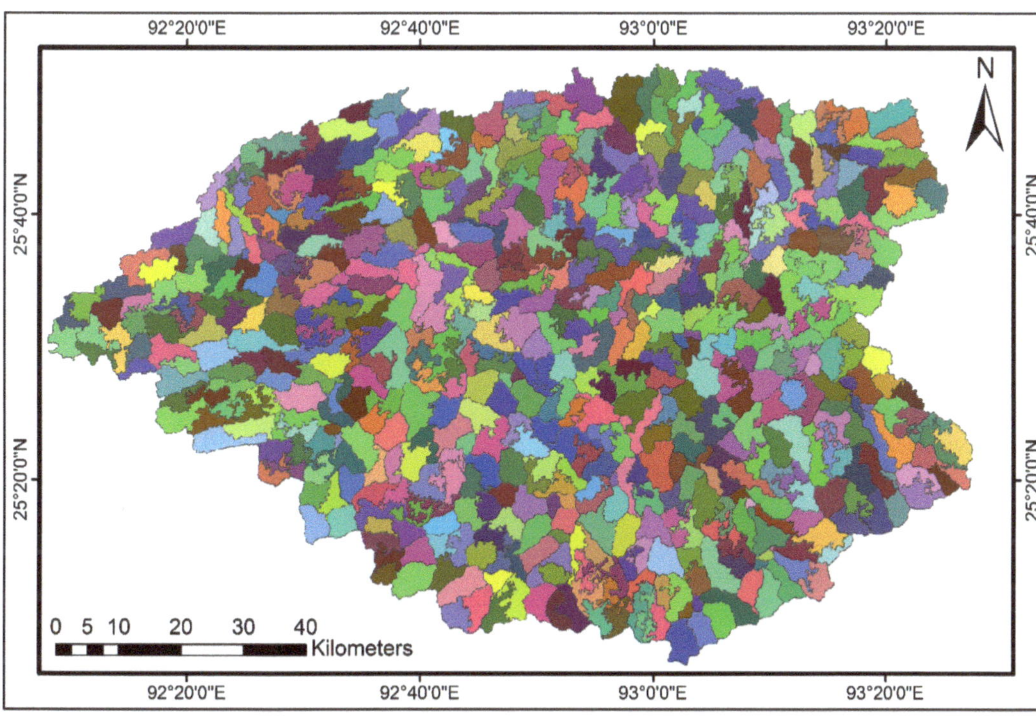

Fig. 12.7 HRU map of the Kopili River basin generated using GRASS-HRU

Hydrological response units (HRUs) are applied as model entities for the J2000 hydrological model. By this approach lateral flows between single HRUs can be modelled which is a preliminary for a better representation of the hydrological dynamics in general and for the integration of water quality aspects in particular (Krause 2002). The distribution concept of HRUs retains the heterogeneous nature of the catchment by keeping the high spatial resolution with many small polygons in higher hydrological dynamics (Nepal et al. 2014).

12.5 Criteria for Model Evaluation

Performance evaluation almost always involved a comparison of the model's output to some corresponding measured variable (ASCE Task Committee 1993). Haan et al. (1982) suggested that the graphical representation of the results could easily be interpreted, if the calibration is done for only one watershed at one stream gauge location. In this study, the evaluation criteria and its range suggested by Moriasi et al. (2007) were used to analyse the performance of the J2000 model, i.e. graphical technique-using hydrograph, Nash-Sutcliffe coefficient (NSE), percent bias (PBIAS) and RMSE-observations standard deviation ratio (RSR). Moreover, coefficient of correlation (R) has also been computed and compared, where Y_i^{obs} is the *i*th observed data, Y_{mean}^{obs} is mean of observed data, Y_i^{sim} is the *i*th simulated value, Y_{mean}^{sim} is the mean of model-simulated value and n is the total number of events.

12.5.1 The Coefficient of Correlation (R)

The correlation coefficient of two variables, sometimes simply called their correlation, is the covariance of the two variables divided by the product of their individual standard deviations. It is a normalised measurement of how the two variables are linearly related. It ranges from 0.0 to 1.0, with higher values indicating better agreement, and is given by:

$$R = \frac{n\left(\sum Y_i^{obs} * Y_i^{sim}\right) - \left(\sum Y_i^{obs}\right) * \left(\sum Y_i^{sim}\right)}{\sqrt{\left[n\sum \left(Y_i^{obs}\right)^2 - \left(\sum Y_i^{obs}\right)^2\right]\left[n\sum \left(Y_i^{sim}\right)^2 - \left(\sum Y_i^{sim}\right)^2\right]}} \quad (12.34)$$

12.5.2 Nash-Sutcliffe Coefficient (NSE)

The Nash-Sutcliffe efficiency (*NSE*) is a normalised statistic that determines the relative magnitude of the residual variance ("noise") compared to the measured data variance ("information") (Nash and Sutcliffe 1970). NSE indicates how well the plot of observed versus simulated data fits the 1:1 line. The NSE values can vary from $-\infty$ to 1, 1 indicating a perfect fit. NSE is computed from the following equation:

$$NSE = 1 - \left[\frac{\sum_{i=1}^{n} \left(Y_i^{obs} - Y_i^{sim}\right)^2}{\sum_{i=1}^{n} \left(Y_i^{obs} - Y_{mean}^{obs}\right)^2}\right] \quad (12.35)$$

12.5.3 Percent Bias (PBIAS)

Percent bias (PBIAS) measures the average tendency of the simulated data to be larger or

Table 12.2 General performance ratings for recommended statistics for a monthly time step

Performance rating	RSR	NSE	PBIAS (%)
Very good	0.00 < RSR < 0.50	0.75 < NSE < 1.00	PBIAS < ±10
Good	0.50 < RSR < 0.60	0.65 < NSE < 0.75	±10 < PBIAS < ±15
Satisfactory	0.60 < RSR < 0.70	0.50 < NSE < 0.65	±15 < PBIAS < ±25
Unsatisfactory	RSR > 0.70	NSE < 0.50	PBIAS > ±25

smaller than their observed counterparts (Gupta et al. 1999). The optimal value of PBIAS is 0.0, with low-magnitude value indicating accurate model simulation. Positive values indicate model underestimation bias, and negative value indicates model overestimation bias (Gupta et al. 1999). PBIAS is calculated by using following equation:

$$PBIAS = \left[\frac{\sum_{i=1}^{n} \left(Y_i^{\text{obs}} - Y_i^{\text{sim}}\right) \cdot (100)}{\sum_{i=1}^{n} \left(Y_i^{\text{obs}}\right)} \right] \quad (12.36)$$

12.5.4 RMSE-Observation Standard Deviation Ratio (RSR)

Based on the recommendation by Singh et al. (2005), a model evaluation statistic, named the RMSE-observation standard deviation ratio (RSR), was developed (Moriasi et al. 2007). RSR standardises RMSE using the observation standard deviation, and it combines both an error index and the additional information recommended by Legates and McCabe (1999). RSR is calculated as the ratio of the RMSE and standard deviation of measured data, as shown in the equation below:

$$RSR = \frac{RMSE}{STDEV_{\text{obs}}} = \frac{\left[\sqrt{\sum_{i=1}^{n} \left(Y_i^{\text{obs}} - Y_i^{\text{sim}}\right)^2}\right]}{\left[\sqrt{\sum_{i=1}^{n} \left(Y_i^{\text{obs}} - Y_{\text{mean}}^{\text{obs}}\right)^2}\right]} \quad (12.37)$$

In this study, criterion suggested by Moriasi et al. (2007) has been adopted to evaluate the performance of the J2000 model (Table 12.2).

12.6 Results and Discussions

In this study, the J2000 model was employed on monthly and daily basis for the years 2003–2010. The entire period of 8 years were divided into 2003–2007 for calibration and 2008–2010 for validation of the model. The first year, i.e. 2003, is considered as an initialisation period for the model. The range of parameters used in calibration process and specific values of the calibrated parameters (in bold) are presented in Table 12.3. The calibration was carried out using the combination of manual trial and error and of automatic or numerical parameter optimisation (Gupta et al. 2005).

12.6.1 Sensitivity Analysis of J2000 Model Parameters

Sensitivity analysis is the process of determining the rate of change in model output with respect to changes in model inputs variables. It is a necessary process to identify key parameters and parameter precision required for calibration (Ma et al. 2000). Therefore, for any model, it is necessary to find out which parameters are more sensitive to its outputs.

In this study, 16 parameters out of 36 were selected for the sensitivity analyses which were described by Nepal et al. (2014). All the selected parameters are presented in Table 12.3, and the most effective (12) parameters for this analysis are indicated in bold. The other parameters were

Table 12.3 Calibrated parameters of the J2000 hydrological model for Kopili River basin

Calibration parameters	Calibrated value	Range	Descriptions
a_rain	1.0	0–5	Interception storage for rain
soilLinRed	4.5	0–10	Linear reduction coefficient for AET
soilMaxInfSummer	25.0	0–200	Maximum infiltration in summer
soilMaxInfWinter	50.6	0–200	Maximum infiltration in summer
SoilImpLT80	0.27	0–1	Relative infiltration for impervious areas greater than 80 % sealing
soilOutLPS	9.0	0–10	Outflow coefficient for LPS
soilLatVertLPS	1.5	0–10	Lateral vertical distribution coefficient
soilMaxPerc	4.0	0–20	Maximum percolation rate to groundwater
soilConcRD1	8.0	0–10	Recession coefficient for overland flow
soilConcRD2	7.5	0–10	Recession coefficient for interflow
soilConcRD1Flood	1.3	0–10	Recession coefficient for flood event
gwRG1RG2dist	50	0–100	RG1-RG2 distribution coefficient
gwRG1Fact	4.0	0–10	Adaptation for RG1 flow
gwRG2Fact	2.5	0–10	Adaptation for RG2 flow
gwCapRise	0.4	0–10	Capillary rise coefficient
flowRouteTA	0.9	0–10	Flood routing coefficient

not included because their sensitivity was considered to be quite low during the trial-and-error process. A Monte Carlo analysis procedure combined with Latin hypercube sampling scheme (McKay et al. 1979) was conducted using 12 parameters with 1000 iterations to produce 12,000 simulations. The objective function NSE (Nash and Sutcliffe 1970) was employed for judging the results of the sensitivity analysis. The regional sensitivity analysis (RSA) (Hornberger and Spear 1981) has been used to analyse the sensitivity of the model parameters. The parameter ranking was carried out by normalising each parameter, on the basis of its sensitivity as suggested by Fischer et al. (2012). It was found (Fig. 12.8) that recession coefficient for overland flow (soilConcRD1) was the most sensitive parameter, whereas recession coefficient for flood event (soilConcRD1Flood) was the least sensitive parameters. The sensitivity analysis suggested the most significant parameters for run-off determination in the study area.

12.6.2 Evaluation of the J2000 Model for Kopili Basin

The J2000 hydrological model was calibrated and validated on daily and monthly time steps using gauged discharge data at Kherunighat gauging station. The twelve parameters which are used for model calibration and validation are presented in Table 12.3. The normal ranges of the adjusted parameters are also included in Table 12.3. The recommended evaluation criteria, viz. PBIAS, R, NSE and RSR, have been computed for daily and monthly values of observed and simulated run-off for both calibration and validation periods and are presented in Tables 12.4 and 12.5, respectively. The results of statistical analysis such as maximum run-off peak, average of observed and simulated run-off and standard deviation are also given for daily and monthly time steps in Tables 12.4 and 12.5, respectively.

12.6.2.1 Daily Time Steps

The daily observed and simulated run-off hydrograph (Figs. 12.9 and 12.10) shows that the simulated and observed low-flow conditions are reasonably comparable, though over-prediction and under-prediction can be found during the early years of calibration. Figures shows that the model over-predicts the recession period of the hydrograph for the years 2006 and 2007, whereas the peak flows are under-predicted for years 2004, 2005, 2006 and 2008. However, the simulated run-off is observed to be in close arrangement with the

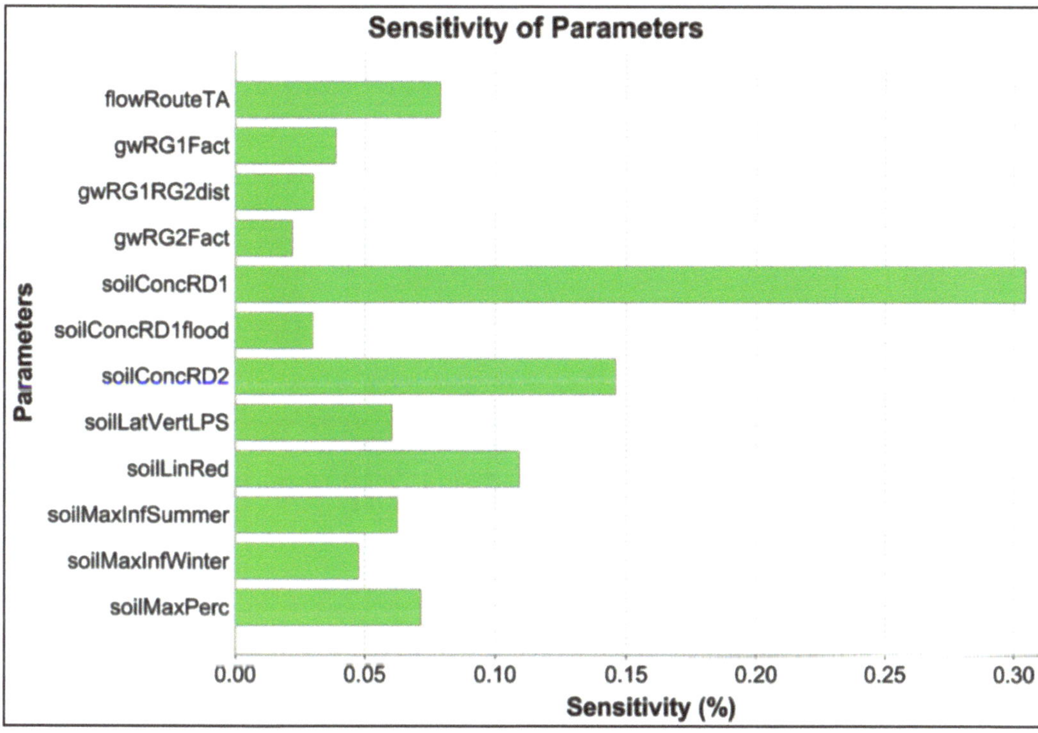

Fig. 12.8 Sensitivity of the selected calibration parameters with the Nash-Sutcliffe efficiency criterion

Table 12.4 Performance evaluation of J2000 model (daily)

		Calibration (2004–2007)			Validation (2008–2010)		
S. no.	Parameter	Observed	Simulated	Remark	Observed	Simulated	Remark
1.	Mean	142.18	148.90	–	116.42	114.67	–
2.	Standard deviation	127.39	137.31	–	88.00	82.76	–
3.	Maximum peak	858.79	788.78	–	415.11	364.33	–
4.	Count	1,461	1,461	–	1,096	1,096	–
5.	R	0.852		V. good	0.924		V. good
6.	NSE	0.672		Good	0.853		V. good
7.	PBIAS	4.731		V. good	−1.504		V. good
8.	RSR	0.455		V. good	0.295		V. good

observed flow for well-distributed rainfall events. Scarce density of rainfall stations in the hilly region could be the other possible reasons for lower simulation of run-off. The R value was found to be 0.85 and 0.92 during calibration and validation period, respectively, exhibiting a good correlation between the simulated and observed run-off. Similarly, the values of PBIAS, NSE and RSR were found to be 4.73, 0.67 and 0.46, respectively, during calibration and −1.50, 0.85 and 0.29, respectively, during validation period, exhibiting desirable model performance during calibration and validation on daily time step. Moreover, the satisfactory model's performance during calibration regarding NSE (0.67) may be due to the deficiency of rain gauges within the basin, i.e. only four, which are located near the outlet of the basin. Furthermore, a model that functioned very well during validation regarding NSE (0.85) is due to the availability of total eight rain gauges which are spatially distributed in the basin. From Fig. 12.9, it is evident that the observed run-off was significantly higher than the model-simulated run-off during the year

Table 12.5 Performance evaluation of J2000 model (monthly)

S. no.	Parameter	Calibration (2004–2007)			Validation (2008–2010)		
		Observed	Simulated	Remark	Observed	Simulated	Remark
1.	Mean	4327.54	4383.809	–	3544.33	3042.32	–
2.	Standard deviation	3629.23	3949.483	–	2587.94	2168.94	–
3.	Maximum peak	14,169.7	16,932.53	–	8798.04	8569.91	–
4.	Count	48	48	–	36	36	–
5.	R	0.879		V. good	0.944		V. good
6.	NSE	0.723		Good	0.891		V. good
7.	PBIAS	4.731		V. good	−1.504		V. good
8.	RSR	0.432		V. good	0.260		V. good

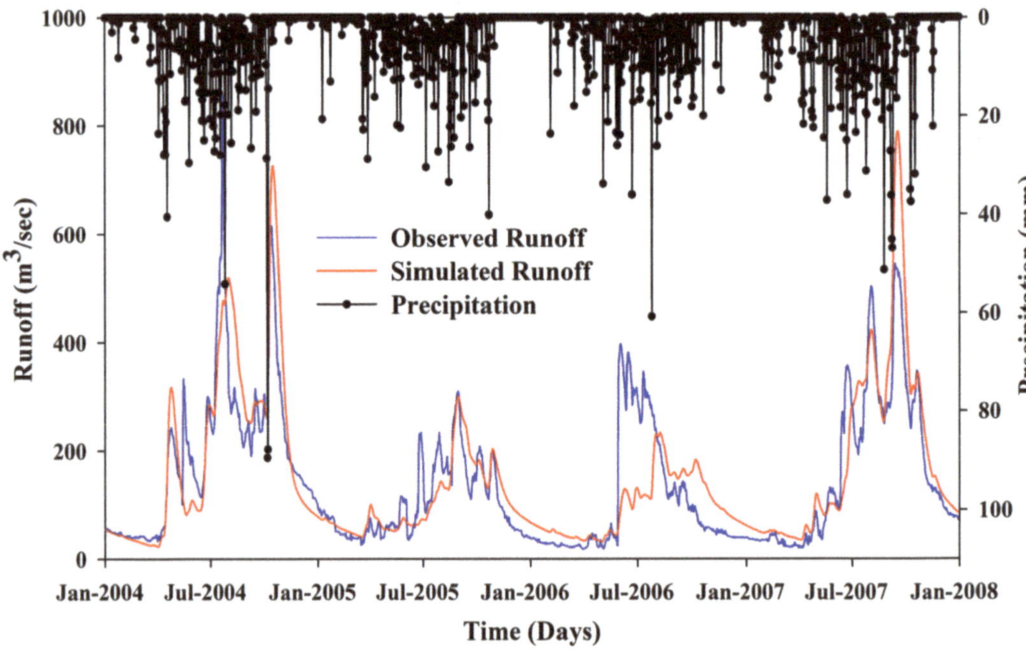

Fig. 12.9 Observed and simulated run-off during calibration (daily)

2006. This may be due to the plausible data uncertainty in the observed flow values for the year 2006; such conclusion is drawn because of absent rising limb and no corresponding high precipitation in the hydrograph.

12.6.2.2 Monthly Time Steps

The scatter graph of the observed and simulated run-off during the calibration period and validation periods is shown in Figs. 12.11 and 12.12, respectively. The values of R, NSE, PBIAS and RSR (Table 12.5) were found to be 0.88, 0.72, 4.73 and 0.43, respectively, during calibration, indicating good model prediction for monthly discharge. During validation, the values of R, NSE, PBIAS and RSR were found to be 0.94, 0.89, −1.50 and 0.26, respectively, which shows a very good simulation during validation and therefore accepted for further analysis.

The result of model efficiencies and statistical analysis suggests that the J2000 model is equally good for predicting the daily and monthly run-off of Kopili basin during calibration and validation period. It is concluded that the J2000 model can be successfully adopted for hydrological modelling of the basin in NE region of India with similar physiographical characteristics. Furthermore, it may also be concluded that the J2000

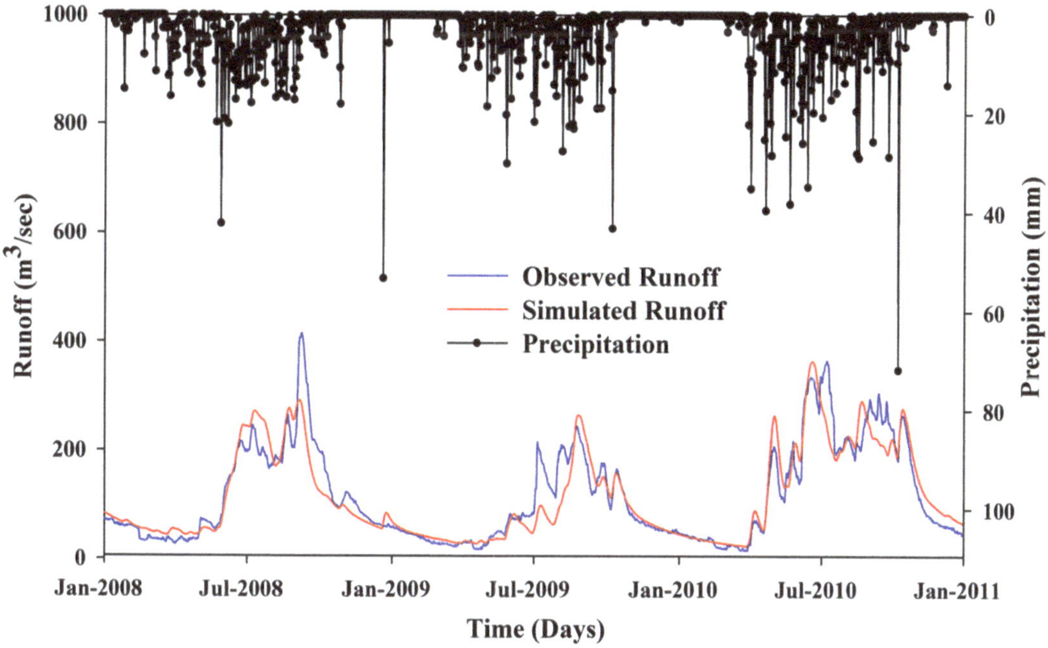

Fig. 12.10 Observed and simulated run-off during validation (daily)

Fig. 12.11 Scatter plot between observed and simulated run-off during calibration (monthly)

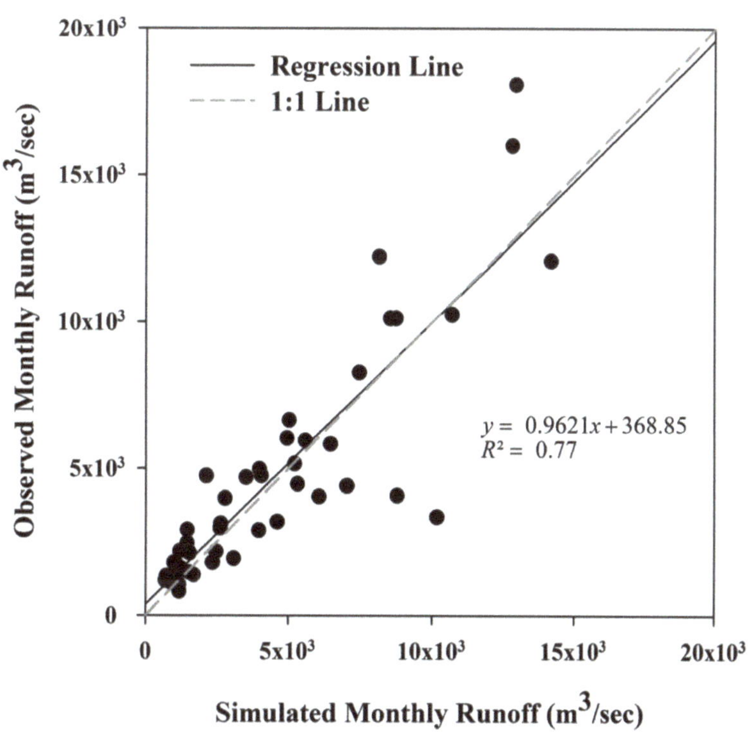

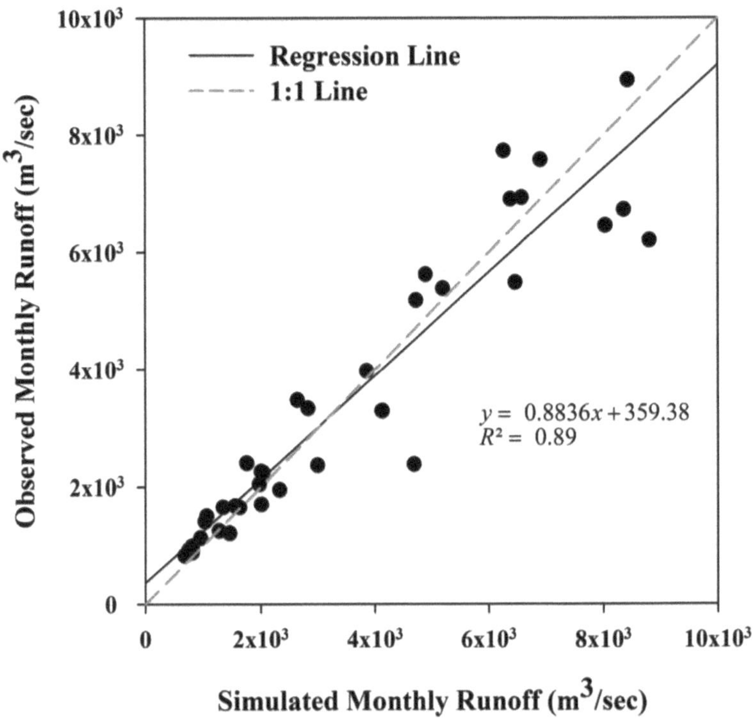

Fig. 12.12 Scatter plot between observed and simulated run-off during validation (monthly)

model can assess and represent the physical characteristics of the Kopili basin reasonably well and can be employed for further analysis of water balancing elements of the basin.

Four run-off components initiating from various sources were also generated from the J2000 model as presented in Fig. 12.5. The contributions from various run-off components are presented in Fig. 12.13. The figure reveals that the overland flow (RD1) is the most dominating component of the hydrograph and comprises about 43 % of the total run-off during the high-flow peak during the year, followed by interflow 1 (RD2) (21 % of the run-off). The high-intensity precipitation during the monsoon season results in saturated soil moisture condition, and most of the precipitation goes as surface run-off. The slow baseflow component (RG2) contributes continuously throughout the year (21 % of total run-off) and became the dominant component during low-flow conditions in the summer period. The fast groundwater storage (RG1) merely contributes 14 % of the total run-off generated, since it represents in the weathered loose materials on top of the bedrock with high permeability and short retention time.

12.6.3 Water Balance of Kopili Basin

Purview of foregone discussion, the J2000 model was calibrated and validated in daily time steps, and the sensitivity analysis was performed using RSA. The annual average water balance of the basin consisting of various components of the simulated hydrological processes in the study area is presented in Fig. 12.14. The aggregated sub-basin-wise water balance of the study area was also carried out and presented in Table 12.6. It can be observed (Fig. 12.14) that out of annual average rainfall of 1,388.5 mm, about 21 % flows out as surface run-off and 22 % as groundwater flow from the study area. Suitable management practices may be employed to reduce this amount of run-off from the basin which can be used for agricultural and drinking purposes. While performing sub-basin-wise analysis (Table 12.6), it was observed that the sub-basins 2, 3 and 5 outflow maximum run-off and groundwater of about 15–30 % and 25–30 % of the precipitation, respectively. It is inferred that there is scope for further utilisation of the water in most parts

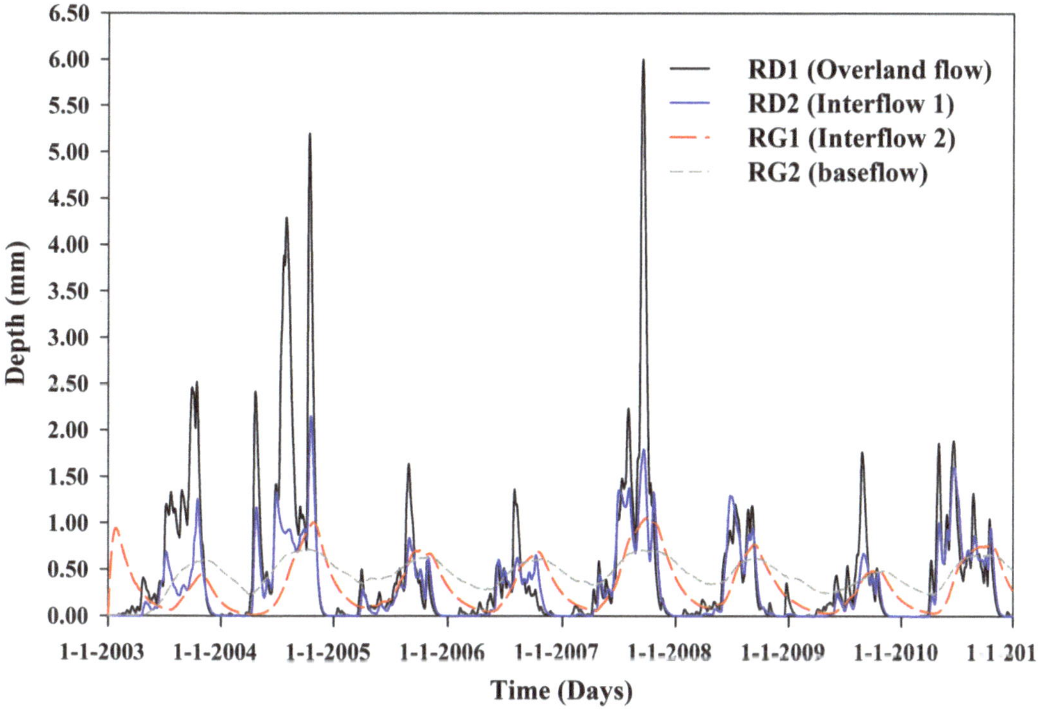

Fig. 12.13 Run-off components from simulated run-off (2003–2010)

Fig. 12.14 Average annual water balance in the study area

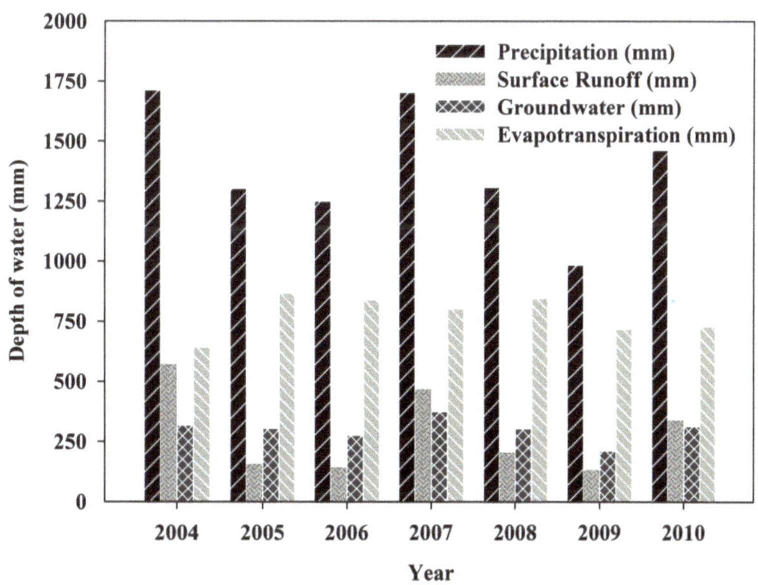

of the basin. Moreover, it may also be observed that the contribution of run-off is arbitrarily less than the groundwater flow, reflecting the presence of depressive topography near the centroid of the basin, subsequently triggering the interflow 2 (RG1). The maximum evapotranspiration of about 60 % is observed from sub-basin 1 followed by sub-basin 4 with 51 % of the aggregated precipitation. This is due to the presence of maximum agricultural class of LULC in the corresponding sub-basins.

Table 12.6 Aggregated sub-basin-wise water balance of the study area

Sub-basin	Area (Km^2)	Precipitation (mm)	Surface run-off (mm)	Groundwater (mm)	Evapotranspiration (mm)
1	787.56	645.33	61.35	78.77	390.15
2	1,872.59	295.78	74.27	72.49	112.32
3	2,194.93	764.71	222.34	231.43	223.39
4	741.25	269.98	36.33	54.50	136.60
5	1,601.75	586.17	106.14	154.69	252.21

During the HRU-wise water balance analysis, it is observed that out of 587 HRUs spread over 5 sub-basins of the study area, 184 HRUs are covered by forest, and 91 HRUs comprise the agriculture land (cultivable area). With forest cover LULC class, the average surface flow and evapotranspiration from HRUs were found to be about 23 % and 42 % of rainfall and with agricultural coverage are 14 % and 45 %, respectively. Therefore, there is a need to take up several conservation measures such as grassed waterways, strip cropping, etc. in some of the sub-basins to restrict surface run-off from agricultural areas by enhancing in-basin utilisation of water.

12.6.4 Hypothetical Variations in Precipitation and Temperature

A distributed nature of the J2000 hydrological model readily allows an assessment of the hypothetical precipitation and temperature change scenarios (Nepal et al. 2014). The global average surface temperature has increased by 0.74 ± 0.18 °C during the last 100 years (1906–2005), and it is likely to increase further by 1.1–2.9 °C for their lowest emission scenario and 2.4–6.4 °C for their highest emission scenario (IPCC 2007) and +10 % and +20 % change from the baseline precipitation (IPCC 2001). In this study, the lowest and highest emission scenarios of temperature from baseline temperature (scenario 1) and ±10 % of precipitation (scenario 2) from baseline precipitation were also analysed for run-off prediction in the context of global CC. The rise in the temperature in the modelling context means increase in maximum, minimum and average temperatures. This analysis was carried out to examine the effect of alteration in climate scenario on the resultant hydrological cycle and particularly run-off pattern.

The analysis of the scenarios indicates that the evapotranspiration was severely affected which in turn affects the run-off and water balance of the basin. The percent deviation of the run-off generated applying each scenario from the simulated run-off of the basin is presented in Table 12.7. From Table 12.7, it can be observed that the average percent deviation was found to be 3.95 and 10.38 for the lowest emission scenario and 8.61 and 22.73 for the highest emission scenario (scenario 1).

For +10 % of precipitation scenario, the percent deviation from run-off was found to be −17.99, and for −10 % of precipitation scenario, the percent deviation of 17.43 was observed. The hypothetical temperature and precipitation variations are presented in Figs. 12.15 and 12.16. From the figures it can be seen that the temperature rise of 6.4 °C, i.e. of highest emission scenario, shows the maximum deviation of 22.73 % which accounts a reduction in annual average run-off of 126.9 mm from the basin, whereas the minimum deviation of 3.95 % (temperature rise of 1.1 °C) accounts a reduction in annual average run-off of 22.0 mm from the whole basin. Moreover, the ±10 % of precipitation scenario reports a rise in annual average run-off volume of 103.4 mm and reduction of 100.7 mm from the basin. This analysis quantifies the impacts of altered temperature and atmospheric circulation on water availability for domestic, agriculture, hydropower generation and ecological environment in the basin which ultimately affects the social economy, and a superior integrated conservation measure should be employed to nullify the effect of CC.

Table 12.7 Percent deviation from simulated run-off as a consequence of hypothetical climate variation

	Scenario 1				Scenario 2	
Year	1.1 °C	2.9 °C	2.4 °C	6.4 °C	+10 %	−10 %
2003	2.15	5.68	4.70	12.55	−18.74	16.86
2004	3.17	8.38	6.93	18.27	−18.49	18.37
2005	5.08	13.41	11.11	29.23	−19.32	19.30
2006	5.15	13.53	11.21	29.30	−19.47	18.93
2007	3.55	9.52	7.84	21.70	−16.52	16.73
2008	4.19	10.96	9.09	23.84	−16.23	15.94
2009	4.54	11.79	9.80	25.69	−17.43	16.48
2010	3.79	9.80	8.16	21.23	−17.73	16.82
Average	3.95	10.38	8.61	22.73	−17.99	17.43

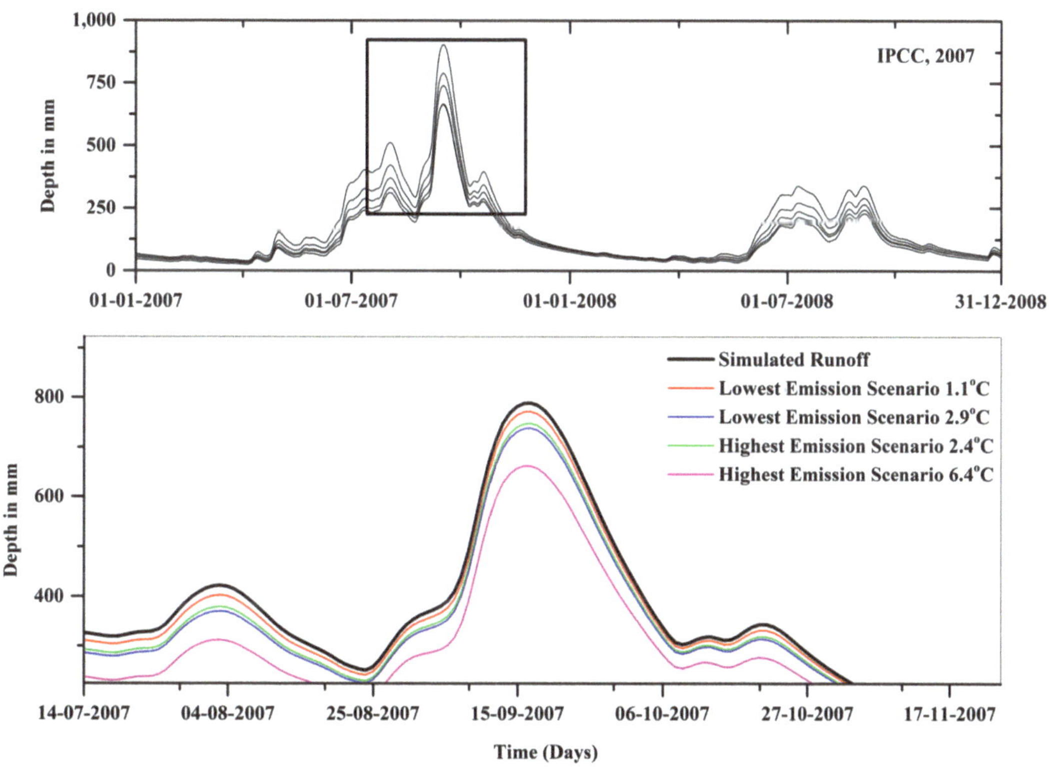

Fig. 12.15 Hypothetical temperature rise scenario of lowest emissions (1.1–2.9 °C) and highest emissions (2.4–6.4 °C) – IPCC 2007

12.7 Conclusions

In this study, a process-based J2000 hydrological model was validated on monthly and daily time steps for run-off simulation under Indian condition for the years 2003–2010 with sensitivity analysis of the model parameters. The sub-basin and HRU-wise water balance of the study area was also analysed to check the contribution of various hydrological processes. Furthermore, the effect of hypothetical variations of temperature and precipitation (under the impression of IPCC 2007) on run-off volume and their deviation from present scenario were analysed.

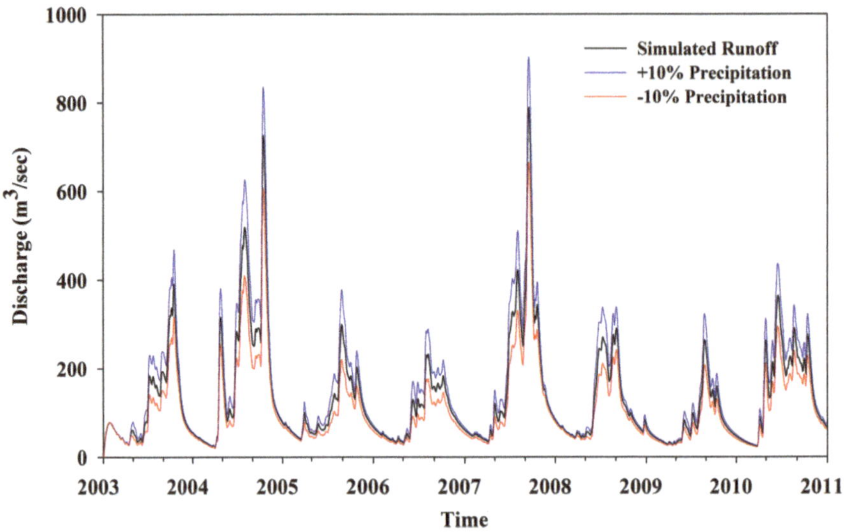

Fig. 12.16 Hypothetical variation effect of ±10 % precipitation on run-off

The following conclusions are drawn from the present study:

1. RSA method of sensitivity analysis was used to check the sensitivity of the parameters, and it was observed that the recession coefficient for overland flow (soilConcRD1) is the most sensitive parameter, followed by recession coefficient for interflow (soilConcRD2) and linear reduction coefficient for AET (soilLinRed) for the Kopili basin.
2. For daily time steps, the values of NSE, R, PBIAS and RSR were found to be 0.67, 0.85, 4.73 and 0.46 during calibration, and the values of NSE, R, PBIAS and RSR were found to be 0.85, 0.92, −1.5 and 0.29, respectively, during validation, exhibiting a good performance of the model.
3. For monthly time steps, the values of NSE, R, PBIAS and RSR were found to be 0.72, 0.88, 4.73 and 0.43 during calibration, and the values of R, NSE, PBIAS and RSR were found to be 0.94, 0.89, −1.5 and 0.26, respectively, during validation, which also reflects a very good performance by a physics-based distributed model in Northeast (NE) India.
4. Around 23 % of annual precipitation flows out of the basin as surface run-off. Therefore, there is a need to adopt several soil and water conservation measures such as grassed waterways, strip cropping, etc. to restrict surface run-off by enhancing in-basin utilisation of water.
5. The hypothetical rise in temperature scenarios indicated that the highest and lowest reduction in annual average run-off volume from the basin was 126.9 to 22 mm for both emission levels and further signifies that the evapotranspiration process might be largely affected resulting in low surface run-off and groundwater contribution. For +10 % of precipitation variation, there will be arise in the volume by 103.4 mm and a slump in volume of around 100 mm for −10 % precipitation.

The overall analysis concludes that the J2000 model is able to represent hydrological dynamics of the Kopili basin. Furthermore, the J2000 model can be considered as a promising tool for prediction of run-off and to understand the constantly changing hydrological processes of Northeast (NE) India. This analysis can provide vital information for the planning of more sustainable strategies using integrated water resource management principles to overcome the aftermath of CC within Northeast India. Similar studies should be carried out in the other parts of India to understand hydrological system dynamics and potential impact from CC.

Acknowledgements The financial assistance provided by the University Grants Commission (UGC), India, is gratefully acknowledged. Thanks are also due to the Department of Water Resource Development and Management, for providing logistical support and assistance during this study, which is a part of PhD research work at IIT Roorkee.

References

Abbaspour KC, Yang J, Maximov I, Siber R, Bogner K, Mieleitner J, ... Srinivasan R (2007) Modelling hydrology and water quality in the pre-alpine/alpine Thur watershed using SWAT. J Hydrol 333(2):413–430

ASCE Task Committee (1993) Criteria for evaluation of watershed models. J Irrig Drain Eng 119(3):429–442

Bende-Michl U, Kemnitz D, Helmschrot J, Krause P, Cresswell H, Kralisch S, Fink M, Flügel WA (2007) Supporting natural resources management in Tasmania through spatially distributed solute modelling with JAMS/J2000-S. In: MODSIM 2007 International Congress on Modelling and Simulation, Modelling and Simulation Society of Australia and New Zealand

Beven KJ (2011) Rainfall-runoff modelling: the primer. Wiley, Chichester

Blöschl G, Reszler C, Komma J (2008) A spatially distributed flash flood forecasting model. Environ Model Software 23(4):464–478

Braud I, Ayral PA, Bouvier C, Branger F, Delrieu G, Le Coz J, ... Wijbrans A (2014) Multi-scale hydrometeorological observation and modelling for flash flood understanding. Hydrol Earth Syst Sci 18:3733–3761

Bussi G, Francés F, Montoya JJ, Julien PY (2014) Distributed sediment yield modelling: importance of initial sediment conditions. Environ Model Software 58:58–70

Chu Y, Salles C, Cernesson F, Perrin JL, Tournoud MG (2008) Nutrient load modelling during floods in intermittent rivers: an operational approach. Environ Model Software 23(6):768–781

Dickinson RE (1984) Modeling evapotranspiration for three-dimensional global climate models. In: Climate processes and climate sensitivity. pp 58–72

Farr TG, Rosen PA, Caro E, Crippen R, Duren R, Hensley S, ... Alsdorf D (2007) The shuttle radar topography mission. Rev Geophys 45(2)

Fink M, Krause P, Kralisch S, Bende-Michl U, Flügel WA (2007) Development and application of the modelling system J2000-S for the EU-water framework directive. Adv Geosci 11(11):123–130

Fischer C, Kralisch S, Krause P, Fink M, Flügel WA (2009) Calibration of hydrological model parameters with the JAMS framework. In: Proceedings of the 18th world IMACS congress and MODSIM09 international congress on modelling and simulation, Cairns. pp 866–872

Fischer C, Kralisch S, Flügel WA (2012) An integrated, fast and easily useable software toolbox which allows comparative and complementary application of various parameter sensitivity analysis methods. In: Managing resources of a limited planet, sixth biennial meeting, Leipzig. International congress on environmental modelling and software

Funke R, Schumann A, Schultz G (1999) Regionalization of parameters in mesoscale hydrological models. IAHS-AISH publication. pp 171–179

Gao T, Kang S, Krause P, Cuo L, Nepal S (2012) A test of J2000 model in a glacierized catchment in the central Tibetan Plateau. Environ Earth Sci 65(6):1651–1659

Gupta HV, Sorooshian S, Yapo PO (1999) Status of automatic calibration for hydrologic models: comparison with multilevel expert calibration. J Hydrol Eng 4(2):135–143

Gupta HV, Beven KJ, Wagener T (2005) Model calibration and uncertainty estimation. In: Anderson MG (ed) Encyclopedia of hydrological sciences. Wiley, New York

Haan CT, Johnson HP, Brakensiek DL (1982) Hydrologic modelling of small watersheds. American Society of Agri. Engineers Monograph No. 5, 533 pp

Hornberger GM, Spear RC (1981) An approach to the preliminary analysis of environmental systems. J Environ Manage 12:7–18

IPCC (2001) Climate change 2001: the scientific basis. Contribution of working group I to the third assessment report of the intergovernmental panel on climate change. Weather 57(8):267–269

IPCC (2007) Climate change 2007: the physical science basis. Contribution of working group I. In: Solomon S, Qin D, Manning M, Chen Z, Marquis M, Averyt KB, Tignor M, Miller HL (eds) Fourth assessment report of the intergovernmental panel on climate change. Cambridge University Press, Cambridge, UK

Kannan N, White SM, Worrall F, Whelan MJ (2007) Hydrological modelling of a small catchment using SWAT-2000–ensuring correct flow partitioning for contaminant modelling. J Hydrol 334(1):64–72

Kralisch S, Krause P (Jul 2006) JAMS-A framework for natural resource model development and application. In: Proceedings of the iEMSs third biannual meeting. Summit on environmental modelling and software. Burlington

Kralisch S, Krause P, Fink M, Fischer C, Flügel WA (2007) Component based environmental modelling using the JAMS framework. In: MODSIM 2007 international congress on modelling and simulation. pp 812–818

Kralisch S, Zander F, Krause P (2009) Coupling the RBIS environmental information system and the JAMS modelling framework. In: Proceedings of the 18th world IMACS congress and MODSIM09 international congress on modelling and simulation. pp 902–908

Krause P (2001) Das hydrologische Modellsystem J2000: Beschreibung und Anwendung in groen Flueinzugsgebieten, Schriften des Forschungszentrum Jülich. Reihe Umwelt/Environment; Band 29

Krause P (2002) Quantifying the impact of land use changes on the water balance of large catchments using the J2000 model. Phys Chem Earth A/B/C 27(9):663–673

Krause P (2010) Technical documentation of J2000 modelling system, Friedrich Schiller University Jena. Friedrich Schiller University Jena

Krause P, Bende-Michl U, Fink M, Helmschrot J, Kralisch S, Künne A (Jul 2009) Parameter sensitivity analysis of the JAMS/J2000-S model to improve water and nutrient transport process simulation-a case study for the Duck catchment in Tasmania. In: 18th World IMACS congress and MODSIM09 international congress on modelling and simulation, modelling and simulation Society of Australia and New Zealand and International Association for Mathematics and Computers in Simulation. pp 3179–3186

Legates DR, McCabe GJ (1999) Evaluating the use of "goodness-of-fit" measures in hydrologic and hydroclimatic model validation. Water Resour Res 35(1):233–241

Ma L, Ascough JC, Ahuja LR, Shaffer MJ, Hanson JD, Rojas KW (2000) Root zone water quality model sensitivity analysis using Monte Carlo simulation. Trans ASAE 43(4):883–895

Maniak U (1997) Hydrologie undWasserwirtschaft. Springer Verlag, Berlin

McKay MD, Beckman RJ, Conover WJ (1979) Comparison of three methods for selecting values of input variables in the analysis of output from a computer code. Technometrics 21(2):239–245

Moriasi DN, Arnold JG, Van Liew MW, Bingner RL, Harmel RD, Veith TL (2007) Model evaluation guidelines for systematic quantification of accuracy in watershed simulations. Trans ASABE 50(3):885–900

Murty PS, Pandey A, Suryavanshi S (2014) Application of semi-distributed hydrological model for basin level water balance of the Ken basin of Central India. Hydrol Process 28:4119–4129. doi:10.1002/hyp.9950

Nash J, Sutcliffe JV (1970) River flow forecasting through conceptual models part I—a discussion of principles. J Hydrol 10(3):282–290

Nepal S (2012) Evaluating upstream downstream linkages of hydrological dynamics in the Himalayan Region (Doctoral dissertation, Jena, Friedrich-Schiller-Universität Jena, Diss., 2012

Nepal S, Krause P, Flügel WA, Fink M, Fischer C (2014) Understanding the hydrological system dynamics of a glaciated alpine catchment in the Himalayan region using the J2000 hydrological model. Hydrol Process 28(3):1329–1344

Pandey A, Chowdary VM, Mal BC, Billib M (2008) Runoff and sediment yield modeling from a small agricultural watershed in India using the WEPP model. J Hydrol 348(3):305–319

Rödiger T, Geyer S, Mallast U, Merz R, Krause P, Fischer C, Siebert C (2014) Multi-response calibration of a conceptual hydrological model in the semiarid catchment of Wadi al Arab, Jordan. J Hydrol 509:193–206

Singh VP, Woolhiser DA (2002) Mathematical modeling of watershed hydrology. J Hydrol Eng 7(4):270–292

Singh VP, Sharma N, Ojha CSP (eds) (2004) Brahmaputra basin water resources, vol 47, Water Science and Technology Library Series. Kluwer Academic Publishers, The Netherland, p 632

Singh J, Knapp HV, Arnold JG, Demissie M (2005) Hydrological modeling of the Iroquois river watershed using HSPF and SWAT. ISWS CR 2004–08. Illinois State Water Survey, Champaign, Available at: www.sws.uiuc.edu/pubdoc/CR/ISWSCR2004-08.pdf

Srivastava PK, Han D, Rico-Ramirez MA, Islam T (2014) Sensitivity and uncertainty analysis of mesoscale model downscaled hydro-meteorological variables for discharge prediction. Hydrol Process 28(15):4419–4432

Tolson BA, Shoemaker CA (2007) Cannonsville reservoir watershed SWAT2000 model development, calibration and validation. J Hydrol 337(1):68–86

Dheeraj Kumar Department of Water Resources Development and Management, Indian Institute of Technology Roorkee, Roorkee, Uttarakhand, India

Ashish Pandey Department of Water Resources Development and Management, Indian Institute of Technology Roorkee, Roorkee, Uttrakhand, India

Wolfgang-Albert Flügel Department of Geoinformatics, Hydrology and Modelling, Friedrich-Schiller-University Jena, Jena, Germany

Nayan Sharma Department of Water Resources Development and Management, Indian Institute of Technology Roorkee, Roorkee, Uttrakhand, India

Part V

River Training

River Management with Submerged Vanes

13

A. Jacob Odgaard

Abstract

Submerged vanes are an unobtrusive and cost-effective way for river engineers to address many problems associated with river management. The vanes are small flow-training structures designed and installed on the riverbed to modify the near-bed flow pattern and redistribute flow and sediment transport within the channel cross section. The structures are laid out so they create and maintain a flow and bed topography that is consistent with that of a stable channel creating optimum conditions for managing the river. A relatively new technology, submerged vanes are a low-impact method for restoring river banks, stabilizing or re-meandering river reaches previously modified (straightened) by humans, increasing flood flow capacity, reducing sediment deposits, and helping maintain or enhance the ecosystem in and around rivers. Following laboratory research and feedback from field installations, guidelines are now available for designs that are effective and sustainable. The experience summarized herein suggests that vanes are an attractive technology for managing streams of all sizes ranging from small creeks to large, braided rivers like Brahmaputra.

13.1 Introduction

Rivers play an important role in society. They provide water for irrigation, water supply, power generation, and many other uses. They also cause disasters primarily during floods when they inundate portions of the floodplain and destroy property and infrastructure. The sediment they transport from the watershed to the ocean often interferes with both navigation in the rivers and infrastructure along them. One of the river engineer's major tasks is to help facilitate optimum usage of this resource and at the same time provide protection against disasters. River training is the common solution. Unfortunately, many of the traditional river training strategies are expensive and sometimes even a detriment to

A.J. Odgaard (✉)
IIHR-Hydroscience and Engineering, College of Engineering, University of Iowa, Iowa, IA 52242, USA

Civil and Environmental Engineering, University of Iowa, Iowa, IA 52242, USA
e-mail: jacob-odgaard@uiowa.edu

N. Sharma (ed.), *River System Analysis and Management*, DOI 10.1007/978-981-10-1472-7_13

the environment. The submerged vane technique offers an alternative that in many river environments is less expensive and equally effective.

13.2 Theory

Submerged vanes are small flow-training structures designed to modify the near-bed flow pattern and redistribute flow and sediment transport within the channel cross section. The structures are installed at an angle of attack of typically 10–20° with the flow, and their initial height is 0.1–0.4 times the local water depth at design stage.

The vanes function by generating secondary circulation in the flow (Fig. 13.1). The circulation alters magnitude and direction of the bed shear stresses and causes a change in the distribution of velocity, depth, and sediment transport in the area affected by the vanes. As a result, the riverbed aggrades in one portion of the channel cross section and degrades in another (Fig. 13.2). A stability analysis will tell where aggradations and degradation need to take place to optimize channel stability.

Typically, vanes are installed in arrays along one side (or both sides) of a river channel long enough to create a desired flow and bed redistribution. Their advantage over traditional training structures, such as dikes and groins, is that they can produce a given redistribution of flow at less resistance to the flow and at less cost. Basic theory is described in Odgaard (2009), Odgaard and Wang (1991a), and Wang and Odgaard (1993).

Figure 13.3 is a laboratory demonstration of how vanes can redistribute flow depth in a straight alluvial channel. The channel is 2.44 m wide (at top), 0.6 m deep, and 20 m long. The bed consists of sand with a median diameter of 0.41 mm and a geometric standard deviation of 1.45. Both the water and the sediment it transports are recirculated through the channel and return circuit. The vanes are 0.8-mm-thick sheet metal plates, 7.4 cm tall (initially), and 15.2 cm long. They are installed in arrays with four vanes in each array, angled 20° toward the bank. The discharge that formed this bed topography was 0.154 m^3/s, the average flow depth was 18.2 cm, and the slope of the water surface

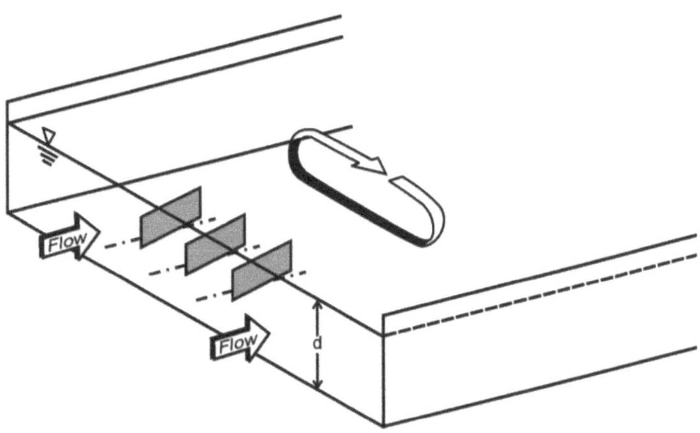

Fig. 13.1 Schematic showing circulation induced by an array of three vanes (Source: Odgaard (2009), ASCE)

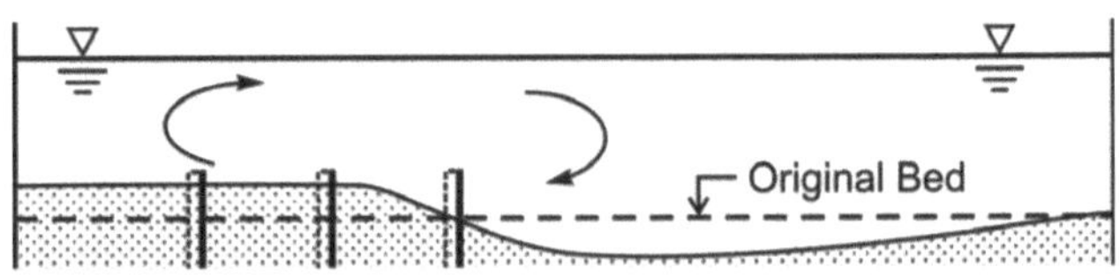

Fig. 13.2 Schematic showing change in bed profile induced by an array of three vanes (Source: Odgaard (2009), ASCE)

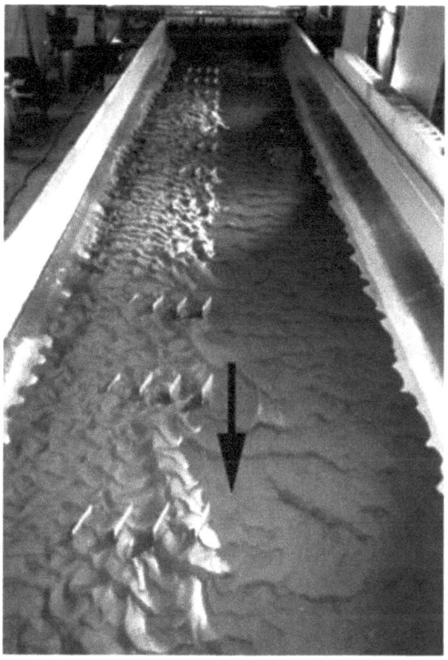

Fig. 13.3 Upstream view of nearly drained, straight channel with vanes (Source: Odgaard and Wang (1991a), ASCE)

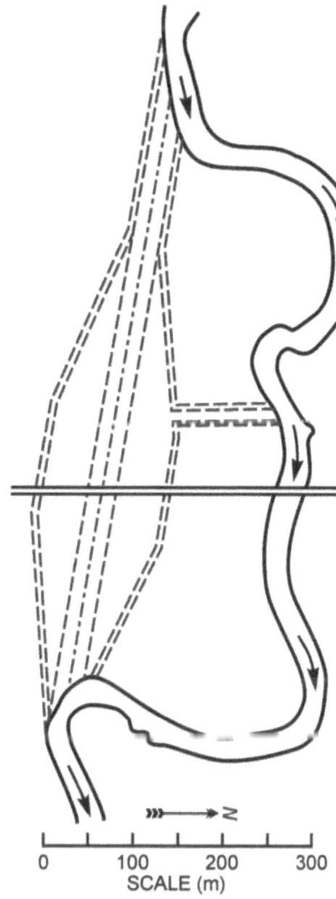

Fig. 13.4 Excavation plan for West Fork Cedar River channel straightening (Source: Odgaard and Wang (1991b), ASCE)

was 0.00064. The vanes reduced the depth near the right bank by about 50 %. This caused the depth near the left bank to increase by 20–30 %. Figure 13.3 is a view of the bed topography after draining most of the water from the flume. It shows the sediment accumulation in the vane field and the associated degradation of the channel outside the field.

13.3 Stabilization of Alignment

Many river management projects include stabilizing the alignment of a river channel or re-meandering of a reach previously straightened. In such cases, design is preceded by a channel stability analysis. Such an analysis consists of either a review of historical, sequential aerial photos of stable upstream and downstream channel reaches, or a formal perturbation analysis, or both (Odgaard and Abad 2007).

The following is a stabilization project in which a vane *system* was installed to ameliorate a channel instability problem caused by channel straightening. The river is the West Fork Cedar River, Iowa, USA (Odgaard 2009). The channel was straightened and widened at the time of construction of a new bridge (1970) to allow a 100-year flood flow to pass through. Figure 13.4 shows the excavation plan for the project. The bridge is a 150-m-long, 9-m-wide, six-span, I-beam bridge with the road surface about 5 m above the low-flow stream bed. The top width and bank-full depth of the river upstream of the excavation are 30–40 m and 1.9–2.1 m, respectively. The bed material is sand with a median particle diameter of about 0.5 mm. Annual mean flow in the river is about 14 m^3/s and bank-full flow about 100 m^3/s. The 5-year flood exceeds 200 m^3/s. By 1984, a considerable portion of the excavation upstream from the bridge had filled in and become vegetated. Figure 13.5a shows the 1984 bank line and a sandbar that subsequently

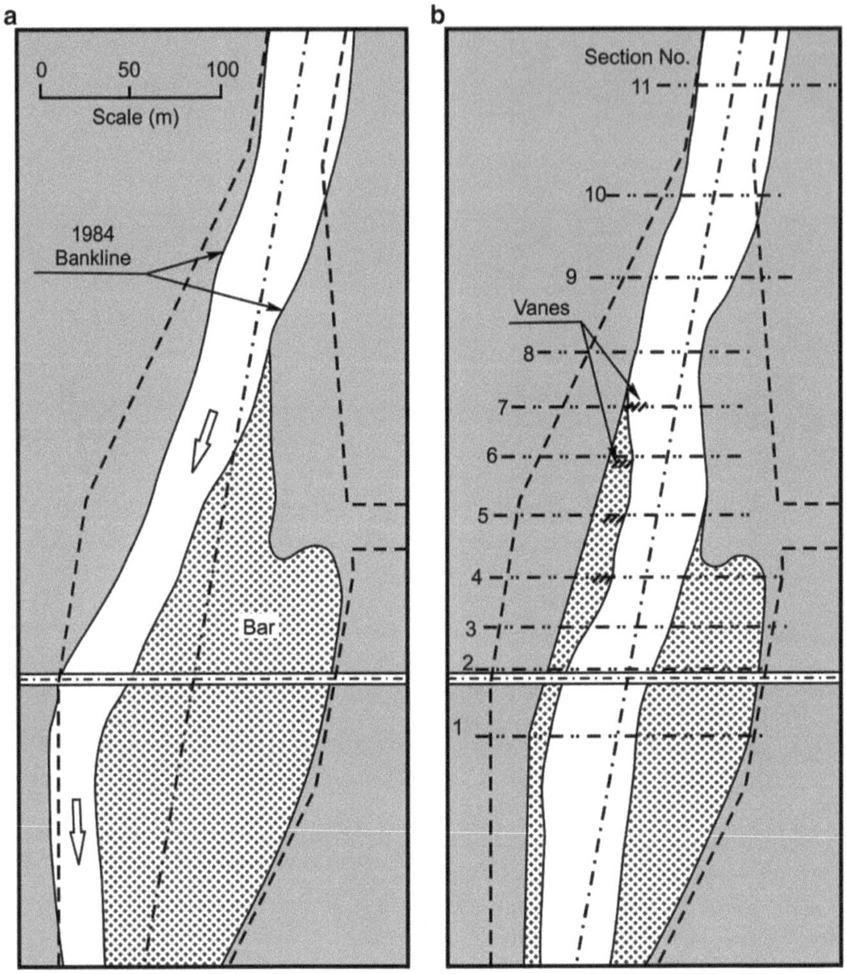

Fig. 13.5 Plan of West Fork Cedar River bridge crossing, (**a**) prior to vane installation in 1984 and (**b**) 5 years after vane installation (Source: Odgaard and Wang (1991b), ASCE)

developed along the left bank. The sandbar occupied four of the six spans, causing the flow to be thrown toward the right bridge abutment, where it undermined and eroded the bank. Annual dredging was necessary because the sandbar grew in size after each storm. It was clear from aerial photos that the bar formed as part of the river's adjustment to the 1970 channel straightening, which essentially eliminated two meanders and shortened the channel segment by 482 m (from 1189 to 707 m). The straightening resulted in a 69 % increase in the local channel slope, from 0.00049 to 0.00083. This increase in slope caused the channel reach to transition from a meandering regime toward a braided regime.

A system of 12 vanes (4 arrays with 3 vanes in each array) was installed in the summer of 1984. The layout is shown in Fig. 13.5b. Each vane consists of vertical sheet piles driven into the streambed and aligned at a 30° angle with the main channel. With this angle, the vanes are at about 20° with the 1984 mean flow direction, which is indicated by the arrows in Fig. 13.5a. Each sheet piling is 3.7 m long, and its top elevation is 0.6 m above the stream bed. The vane system was designed to cause flow depth and velocity to decrease along the right bank and increase along the centerline. As seen in Fig. 13.6, the system has accomplished this. A permanent, protective berm now is seen along the

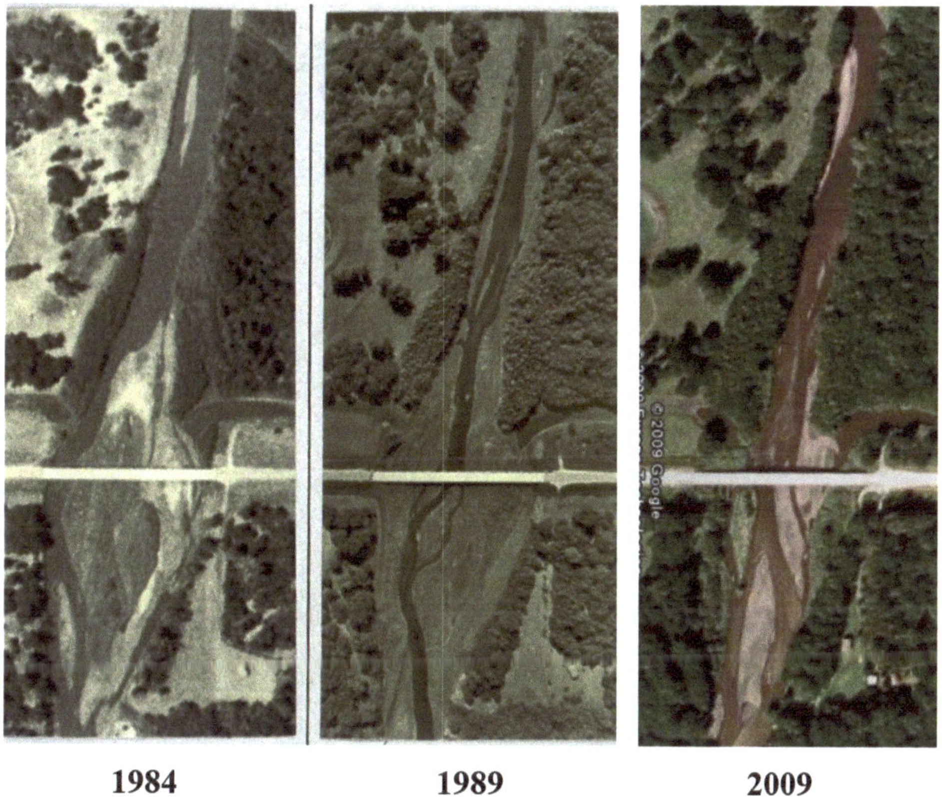

Fig. 13.6 Aerial photos of the West Fork Cedar River bridge crossing (*left*) prior to vane installation in 1984; (*middle*) in 1989, 5 years after vane installation (along right bank only); and (*right*) in 2009, 25 years after vane installation (Source: Odgaard (2009), ASCE (*left* two images) and DigitalGlobe (*right* image))

bank that was previously eroding (about 450 m). In fact, the vanes are now maintaining a cross-sectional bed profile similar to that designed when the bridge was constructed. Installation cost was $5,000 (in 1984), and, so far, maintenance has not been necessary. Nine of the 12 vanes are now permanently covered with vegetation.

13.4 Stabilization of Banks

Often, stabilization of a river alignment also brings vegetation back to the river banks and restores the bank's natural ecology. However, adjustment of alignment is not always possible. Infrastructures such as highways and bridges are often in the way, and a more localized approach is needed. In such cases, submerged vanes are also an effective measure for restoring and stabilizing the banks. Typically, the most unstable reaches of a river are the bends. In bends, the interaction between the vertical gradient of the velocity and the curvature of the flow generates a so-called secondary or spiraling flow. The secondary flow moves high-velocity, near-surface current outward and the low-velocity, near-bed current inward, thereby producing larger depths and velocities near the outer banks. The deepening of the channel diminishes the toe support of the bank and the larger velocities attack it, setting the stage for bank erosion (Fig. 13.6). As indicated in Fig. 13.7, the vanes are laid out so that the vane-generated secondary current eliminates the centrifugal-induced secondary current. The vanes stabilize the toe of the bank. The following case studies show vanes can effectively restore banks that have been destroyed by erosion (Fig. 13.8).

Figure 13.9 shows aerial views of a bend of Wapsipinicon River, USA, the left taken in 1988

Fig. 13.7 Bank erosion and secondary current in river bend (Source: Odgaard (2009), ASCE)

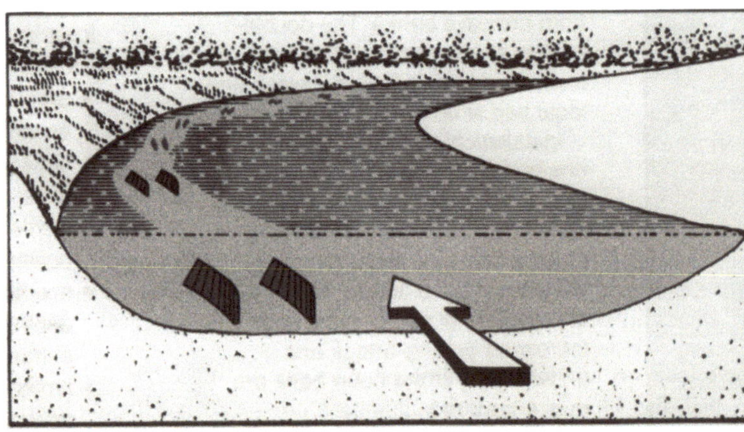

Fig. 13.8 Submerged vanes for mitigating bank erosion. Vane-induced secondary current eliminates the naturally occurring secondary current and stabilizes the bank (Source: Odgaard (2009), ASCE)

Fig. 13.9 Aerial view of Wapsipinicon River in 1988 (*left*) and in 2009 (*right*) (Courtesy of Robert DeWitt, River Engineering International (*left* photo) and DigitalGlobe (*right* photo))

and the right in 2009. Prior to 1988, the bank was eroding at a rate of 3 m per year toward a county road and was endangering a bridge structure (seen to the right in the photo). The bank height is 3.5 m and bank-full flow about 600 m^3/s. A system of 28 vanes was installed along approximately 100 m of the bend in May 1988. The vanes were fabricated of reinforced concrete and each mounted on an H-pile, which was driven approximately 4.6 m into the streambed. The vanes were oriented at approximately 20° with the direction of flow at bank-full flow. It is seen that the vane system not only restored the bank but also created a more favorable approach-flow condition for the bridge opening.

13.5 Design Example

In this example developed with data from multiple case studies, a vane system is designed to stabilize a 2-km-long river reach that has braiding tendencies. Figure 13.10 is a plan view of the reach. The objective is to stabilize the reach as a single-thread channel allowing the ecology to be restored and the area to better accommodate the needs of the nearby community. First, the alignment of the new channel is determined such that it is consistent with that of a stable meander. Alternatives 1 and 2 in Figs. 13.11 and 13.12 are two possible solutions for channel alignment. The most stable alignment is determined by a stability analysis. Using data from the field for the situation depicted in Fig. 13.10, the upstream stable reach has average width, depth, slope, and grain size of 75 m, 3 m, 0.0009, and 0.6 mm, respectively. Bank-full (channel forming) discharge is 423 m^3/s. At this discharge, the Darcy-Weisbach friction factor is $f = 0.06$ and average velocity 1.88 m/s. According to Leopold and Wolman (1957), these values of discharge and slope place this channel reach at the transition between a braided channel and a meandering channel (Fig. 13.13), close enough to the meandering regime that a single-thread channel is sustainable. A perturbation stability analysis (Odgaard 2009) shows that dominant wavelength is about 1600 m. That means alternative 1 in Fig. 13.11 is the most stable alignment for the reach. The alignment consists of three circular arcs of radii, 850 m, 1300 m, and 850 m, respectively, connected with straight-channel segments. Using a two-dimensional bend-flow model (Odgaard 1989, 2009), maximum scour depth in this alternative is estimated to be 4.3 m and maximum near-bank velocity about 2.3 m/s. The near-bank velocity is about 20 % larger than cross-sectional average velocity. Considering that the channel is a constructed channel, the banks have to be protected to allow vegetation to get established.

A system of submerged vanes is used for stabilizing the alignment, much like in the aforementioned West Fork Cedar River case. Vanes are installed along the portion of the bank with greater than average near-bank velocity. Using

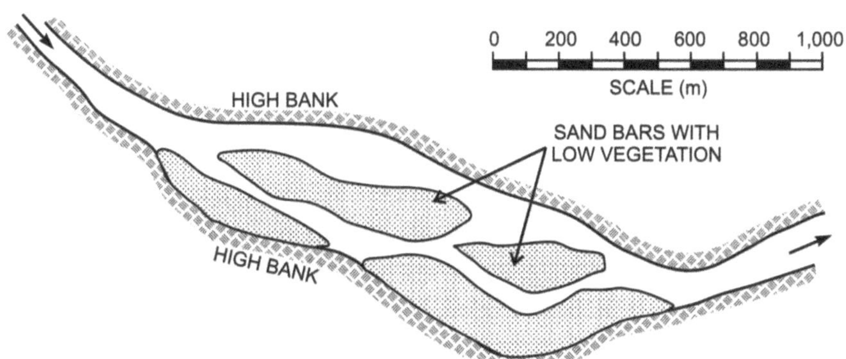

Fig. 13.10 Channel reach to be stabilized (Source: Odgaard (2009), ASCE)

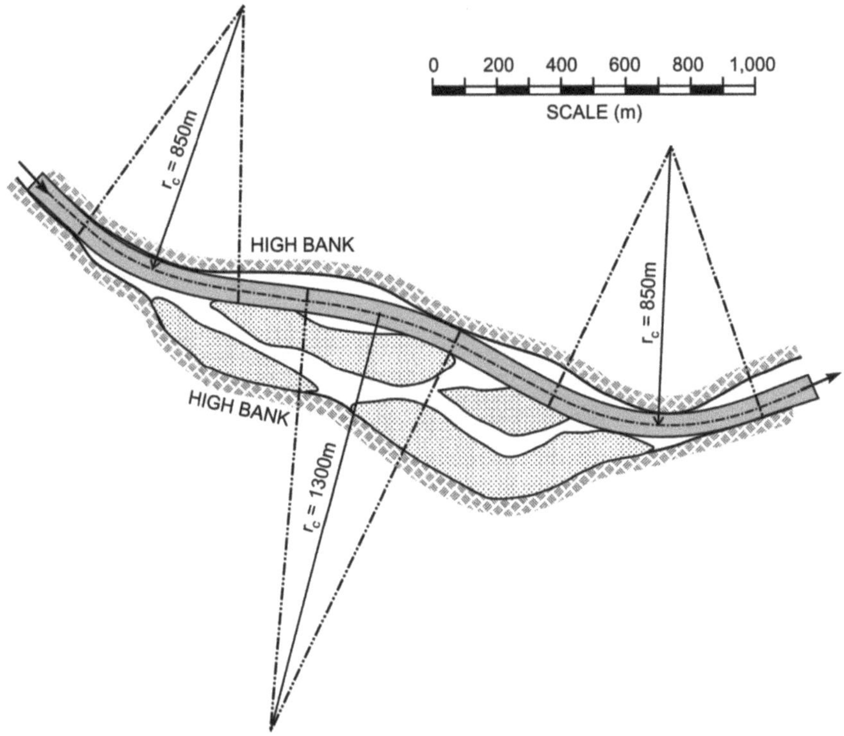

Fig. 13.11 Alternative 1 alignment through the reach (Source: Odgaard (2009), ASCE)

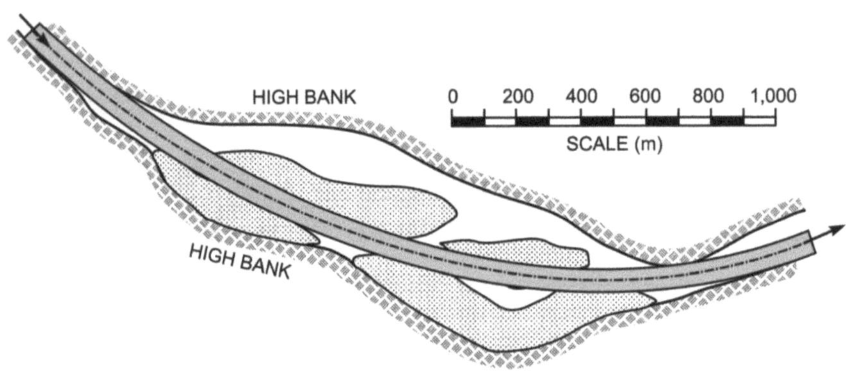

Fig. 13.12 Alternative 2 alignment through the reach (Source: Odgaard (2009), ASCE)

the design guidelines presented in Odgaard (2009), a system of 1.34-m-tall vanes placed in arrays of three and spaced longitudinally, a distance of 20–25 m will increase the bed elevation along the outer bank to approximately the elevation of the average bed. Such a system will reduce near-bank velocity to cross-sectional average velocity with no change in channel slope and will thus prevent erosion from occurring along the bank. With no erosion occurring along the bank, vegetation should be able to take hold. The first vane array of each system is installed at the point of estimated "first outer-bank erosion occurrence," which in this case is calculated to be at the halfway point between the start of the bend and bend apex. Lead-in vanes

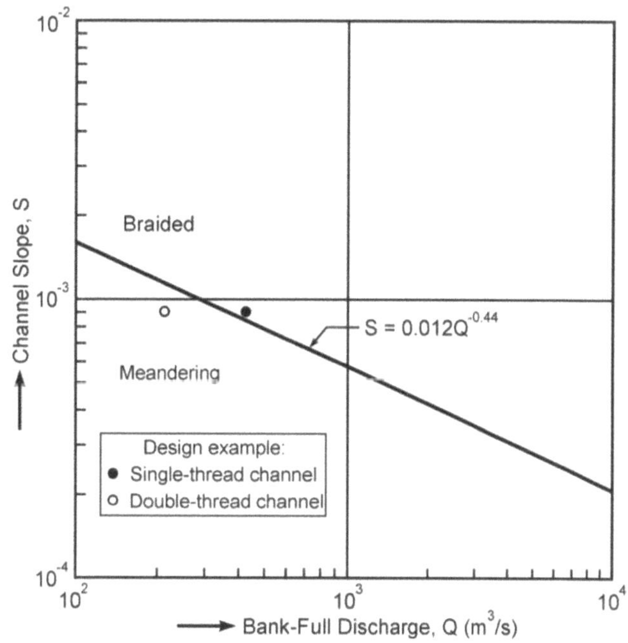

Fig. 13.13 Leopold and Wolman's threshold relation

are added at the beginning of the reach to promote the development of the first meander curve. For detailed design procedures, readers are referred to Odgaard (2009).

In general, the selection of channel alignment must be, to some extent, based on existing topography, for example, a high-bank alignment as depicted in Fig. 13.10. When the calculated wavelength differs from that fitting in naturally with the existing high-bank alignment, the selected wavelength should be adapted to fit the existing high-bank alignment as much as possible. The alignment shown is the most appropriate for the given high-bank alignment and the calculated wavelength. Had the calculated wavelength been twice that calculated above (if, say, channel width and grain size had had different values), a more appropriate channel alignment fitting with the given high-bank alignment could have been alternative 2.

13.6 Stabilization of Braided Rivers

The aforementioned example suggests that vanes can also be a useful tool for stabilizing reaches of braided rivers. By redistributing flow and sediment transport in strategic sections of a braided river, for example, at nodal points, vanes may help stabilize not only channel alignment but also flow distribution between each channel of a braided channel system. By redistributing flow among a select number of (parallel) channels, and stabilizing each channel using the aforementioned guidelines, vanes have the potential for helping reduce the lateral extent of a braided river segment. Indeed, vanes have been used in the past to successfully close off secondary branches of a river (Chabert et al. 1961). A braided reach could be narrowed by closing off the outside branches of the reach. Brahmaputra would be an obvious candidate for such an application.

Referring to the aforementioned design example, had design flow rate Q been significantly larger (and slope unchanged), the reach would be well within the braided regime; the data point in Fig. 13.13 would be further to the right. In this case, stabilizing the reach with a single-thread meandering channel would be difficult if not impossible. However, splitting the flow into two parallel channels could potentially make both become stable meandering channels. The flow split could be an even split, or it could be a split

Fig. 13.14 Stabilization by channel split

as indicated in Fig. 13.14, in which the design flow (channel forming discharge) of the upper channel is smaller than that of the lower channel. If total flow rate were 423 m^3/s as in the previous example, and the flow were split evenly between the two channels, the flow point for each channel would be that circled to the left of the threshold line in Fig. 13.13. In other words, each channel would be well within the meandering regime.

The design process would include determining sets of corresponding discharges and meander wavelengths and selecting the sets for which the meander wavelengths have the closest match to those upstream and downstream from the reach. The amplitudes of the meanders would be determined based on natural lateral constraints and on the slope of the channel immediately upstream and downstream of the reach. Submerged vanes appropriately placed at the bifurcation and at strategic points along the channel may control the flow split between the two channels.

In braided channels one of the management challenges is to limit the lateral expansion of the flow area. In some large rivers like the Kosi River in Nepal and the Brahmaputra River in India, the braiding causes the flow area to be several km wide, often leading to unpredictable shifts in the braiding pattern from year to year. These shifts can be devastating to the communities living around the river. One approach to help prevent this behavior is to identify so-called nodal points along the river, points where the lateral expansion is limited and relatively stable. Often, because of inherent flow instability, a small flow adjustment immediately downstream from such a nodal point can affect flow patterns over a large distance downstream. A limited number of submerged vanes appropriately placed at the exit from such nodal points can create and maintain a desired flow pattern for many kilometers downstream possibly all the way to the subsequent nodal point and thus help restrict the lateral expansion tendencies of the river in between the nodal points. It should be noted that submerged vanes have already proved successful in closing off secondary branches of a river channel (Chabert et al. 1961). By orienting the vanes such that they intercept and deflect sediment into the entrance region of the secondary branch, as shown schematically in Fig. 13.15, velocities decrease and a gradual channel aggradation occurs in the entrance section that eventually closes off the entrance. If such vanes are made of bamboo or other relatively inexpensive material, maintenance would be straightforward and possibly be a local community effort.

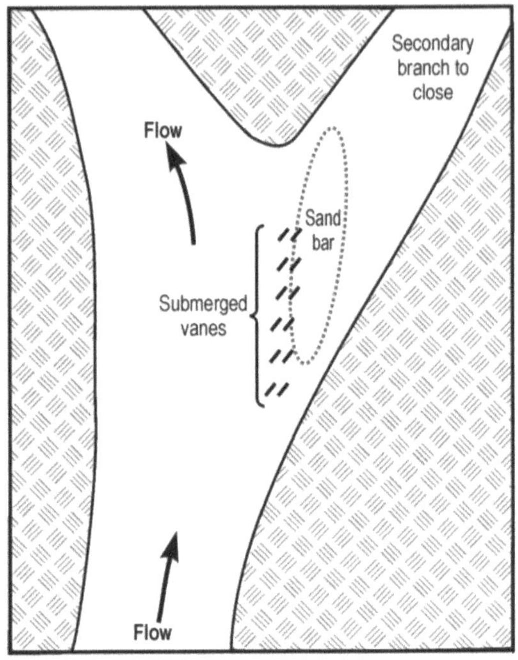

Fig. 13.15 Schematic showing how submerged vanes could help close off a secondary branch

13.7 Conclusions

The case studies described show that submerged vanes are an effective method for addressing several aspects of river management, including channel realignment (re-meandering), and the stabilization of river banks. The design process includes (1) site inspection and surveys, (2) review of sequential aerial photos, and (3) a formal perturbation stability analysis. The process typically provides dominant wavelength and amplitude of channel alignment, most stable channel cross section, flow and depth distribution throughout channel, and bank erosion patterns. Feedback from field installations, like the ones described above, and laboratory results have provided guidelines for the development of specifications for vane layouts. The specifications include vane height, length, orientation, number of vanes and array configuration, longitudinal and lateral spacing, distance to bank, location of first and last vane in system, and staging. The experience to date suggests that vanes may be effective in managing braided rivers; they may be used to close off select secondary branches and reduce the lateral extent of a braided channel system.

References

Chabert J, Remillieux M, Spitz I (1961) Application de la circulation transversal a la correction des rivieres et a la protection des prises d'eau. Proceeding of the Ninth Convention of IAHR, Dubrovnik, Yugoslavia, pp 1216–1223 (in French)

Leopold LB, Wolman MG (1957) River channel patterns – braided, meandering and straight. U.S. Geological Survey Professional Paper 282B, U.S. Geological Survey. Washington, DC, United State Government Printing Office

Odgaard AJ (1989) River-meander model. I: development. J Hydraul Eng ASCE 115(11):1433–1450

Odgaard AJ (2009) River training and sediment management with submerged vanes. ASCE Press, Reston

Odgaard AJ, Abad JD (2007) River meandering and channel stability. In: Garcia M (ed) Sedimentation engineering: processes, measurements, modeling, and practice, ASCE manuals and reports on engineering practice No. 110. ASCE, Reston

Odgaard AJ, Wang Y (1991a) Sediment management with submerged vanes. I: theory. J Hydraul Eng ASCE 117(3):267–283

Odgaard AJ, Wang Y (1991b) Sediment management with submerged vanes. II: applications. J Hydraul Eng ASCE 117(3):284–302

Wang Y, Odgaard AJ (1993) Flow control with vorticity. J Hydraul Res IAHR 31(4):549–562

A. Jacob Odgaard IIHR Hydroscience and Engineering, College of Engineering, University of Iowa, Iowa, IA, USA

Civil and Environmental Engineering, University of Iowa, Iowa, IA, USA

Behaviour and Training of River Near Bridges and Barrages: Some Case Studies

14

S.K. Mazumder

Abstract

Large numbers of bridges and barrages are being constructed across innumerable rivers in India and abroad to serve different purposes for the benefit of the people. It is important to understand the river behaviour before and after the construction of bridges and barrages for their proper planning, design, construction, operation and maintenance. Costly protective measures are often needed to train the river against uncontrolled scouring, silting, meandering, anabranching and many other problems for the safety of the structures as well as the approach and marginal/afflux embankments. Breaching of embankments, outflanking of the structures, river avulsion, etc. cause disruption of traffic, unprecedented damages and unimaginable sufferings of the people and often defeat the very purpose of the structures. In this paper, the author has discussed with figures and photographs about the morphological changes that were found to occur upstream and downstream of some bridges and a barrage and the protective measures adopted to train the rivers in the Himalayan region of India.

14.1 Introduction

Proper understanding of river behaviour in the vicinity of hydraulic structures (Mazumder 2004) like bridges and barrages is extremely important for their planning, design, construction, operation and maintenance apart from the safety of the structures. Except in their mountainous reaches, most of the rivers in the Himalayan region have wide flood plains, and the normal waterway is often restricted (Mazumder et al. 2002) due to economic considerations. Such restriction often changes the river regime and stability. The flow field which used to prevail prior to the construction of the structures also gets changed. There is afflux subjecting the channel to backwater effect upstream. Hydraulic and energy gradients are decreased in the backwater reach. As a result, stream power, proportional to the product of discharge and energy slope (QS_e), decreases and the sediment-carrying capacity of

S.K. Mazumder (✉)
Department of Civil Engineering, Delhi College of Engineering (Delhi Technology University), New Delhi, India
e-mail: somendrak64@gmail.com

N. Sharma (ed.), *River System Analysis and Management*, DOI 10.1007/978-981-10-1472-7_14

the stream is considerably reduced. This causes deposition of sediments upstream of the structures, leading to aggradations and flow instability. Immediately downstream of the structures, there is degradation due to release of water with less sediment load. Often there are flow distortion and localized erosion (Mazumder 1993a, b) due to residual kinetic energy of flow and high turbulence level of the released water. In barrages, there is diversion of flow from the main channel, and as such, there is reduction of flow (Q) in the main channel, and the sediment-carrying capacity of river is reduced since stream power (α QS_e) is reduced. Uncontrolled aggradations and degradation (MOWR 2004) often lead to serious problems of river training near these structures.

Depending upon the extent of constriction and location of the structure in the flood plain, the approaching river may often be unstable and asymmetric. Such an unstable river may shift its course and wander anywhere within the flood plain, resulting in meandering and widening of the river, localized erosion of bed and banks and a bowl-like delta formation in the vicinity of these structures (Fig. 14.1) sometimes leading to their outflanking (Fig. 14.2). Costly river training measures are often found necessary to protect

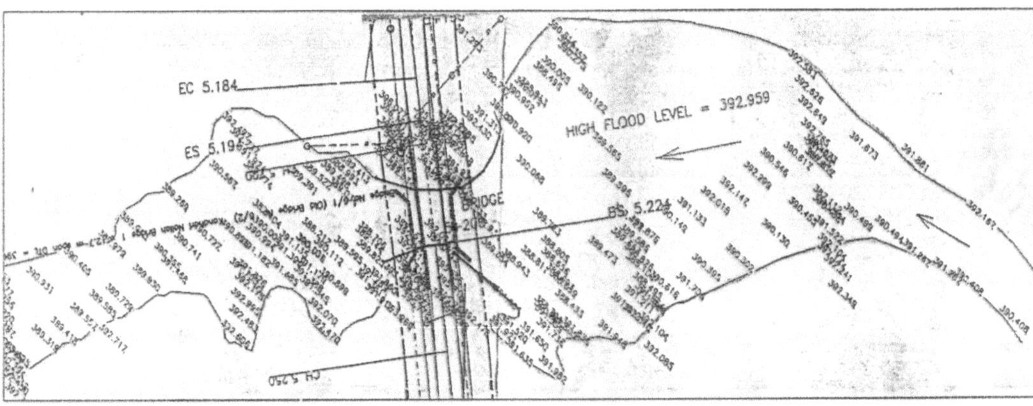

Fig. 14.1 Formation of a bowl upstream and downstream of a bridge (in a state highway in Madya Pradesh, India) due to excessive restriction of waterway. The bridge is likely to be outflanked

Fig 14.2 Outflanking of a vented causeway on the stream 'Danab Khola' in Nepal

Fig. 14.3 Disastrous flood in Malda district in West Bengal after breach of left marginal embankment upstream of Farakka Barrage in the year 1998

Fig. 14.4 Devastation brought about by river Kosi due to the change of its course on August 18, 2008, due to breach in left embankment about 12 km upstream of Kosi Barrage

afflux/approach embankments and arrest wild meandering and possible shift of the existing river course (flow avulsion) and outflanking of the structures. Often the river is found to breach the protection embankments resulting in flood damages and unprecedented sufferings of the people living nearby (Figs. 14.3 and 14.4).

If the river is in a meandering state, the process of aggradations and degradation occurs simultaneously. Islands (locally called chars/shoals) are formed upstream due to sediment deposition, and the main flow shifts away from the shoals, inducing curvature to the stream and formation of secondary current. The outer side of the curved flow undergoes constant erosion, and the eroded materials are deposited in the inner bank side, resulting in further growth of the chars/shoals. This process of erosion of outer bank and deposition on inner bank results in further increase in curvature, stronger secondary current and greater erosion of the outer bank causing migration of the meander on the outer side (Fig. 14.5) till a state of stability occurs.

One of the primary objectives of writing this paper is to discuss about the above-mentioned river behaviour with particular reference to two bridges on rivers Mahananda and Bagmati and a barrage on river Ganga (Farakka Barrage). Normal waterway has been restricted, and approach embankments have been constructed in all these structures built in wide alluvial flood plains as observed in most of the rivers in the Himalayan region.

14.2 Morphology, Stability and Planforms of Rivers

Understanding the behaviour of a river is complicated due to interrelated geomorphologic, hydrologic, hydraulic and sediment parameters. A geologist and a geomorphologist are concerned about evolution and changes in the river over a long period of time. A river engineer concerned with design of bridges and barrages and river training/improvement works usually studies the past history of the river and its morphological changes over a much shorter period, generally about 100 years or so, depending on the length of life of the structures. The interrelation between channel planforms, hydraulic and sediment parameters and relative stability of a river is illustrated in Fig. 14.6 (Schumm 1980; Leopold and Wolman 1957). It may be seen that the different planforms of a river, e.g. straight, meandering, anabranching, braided, etc., depend on the geometry, sediment load, bed slope and discharge of the river. Quantitative prediction of stream response to climatological or watershed changes is based on the fundamental relation given by Eq. (14.1) (Lane 1955):

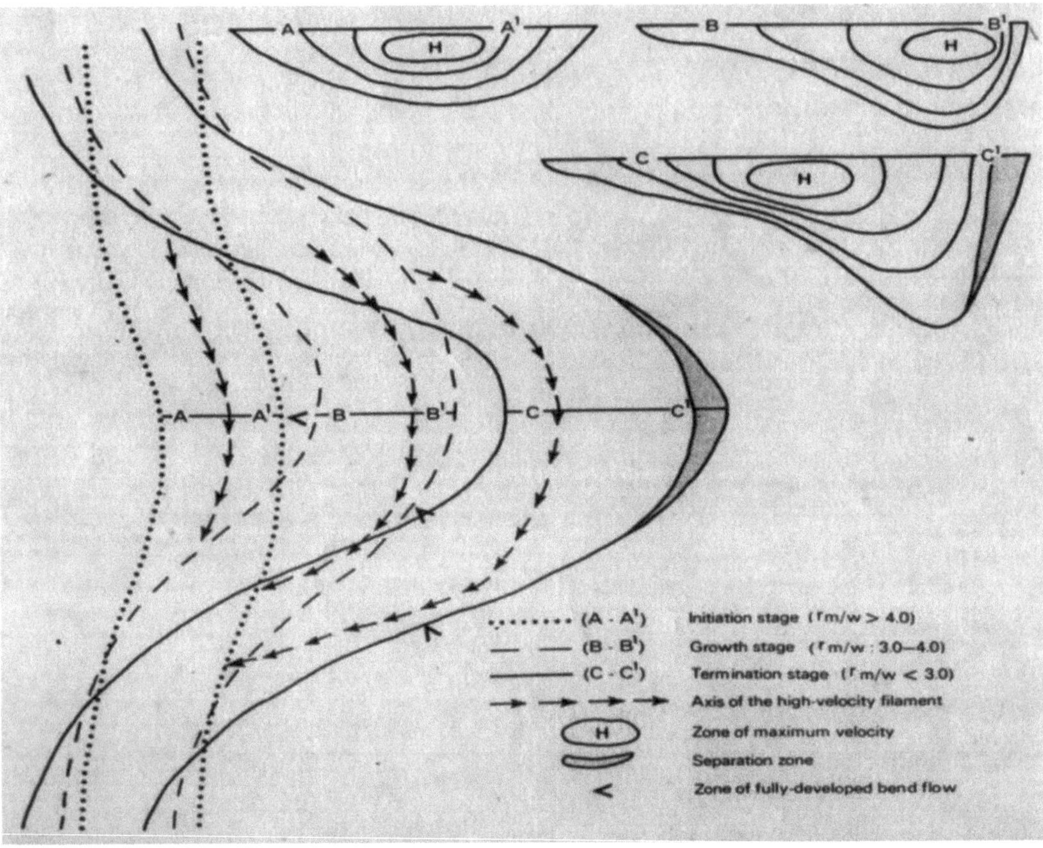

Fig. 14.5 Lateral migration of meander and development of stream cross section in a typical meandering bend with time

$$QS_e \ \alpha \ Q_s d_{50} \tag{14.1}$$

where Q is the discharge, S_e is energy slope, Q_s is sediment transport rate and d_{50} is mean sediment size. Garde (2006) used area-velocity-flow relation, Manning's equation and sediment transport equation and modified the above equation as

$$Q^{6//7} \ S_e^{7/5} \ \alpha \ Q_S \ d_{50}^{3/4} \tag{14.2}$$

Increase in sediment load due to erosion, mining operation, land slide, etc. in the catchment areas results in rise in Q_s. Since Q and d_{50} remain the same, it invariably leads to aggradations and increase in energy slope (S_e), till the stream power (QS_e) is sufficient to carry the increased sediment load 'Q_s' and the relation given by Eqs. (14.1) and (14.2) are satisfied (Garde 2004).

Farakka Barrage was constructed on river Ganga with the primary objective of flow diversion to river Bhagirathi/Hoogly (a tidal river), so that its flow (Q) and hence sediment-carrying capacity (Q_s) are increased. Hoogly River – the lifeline of Kolkata port, the cities/towns and the industrial hub on both banks of Hoogly River – was silting up very fast, resulting in deterioration of its navigability and increase in salinity, especially during lean flow period. With higher upland discharge (Q) in Bhagirathi/Hoogly (due to flow diversion from Ganga through the Feeder Canal), Q_s increased and hence the navigability of Hoogly river improved, and the salinity decreased after the construction of Farakka Barrage.

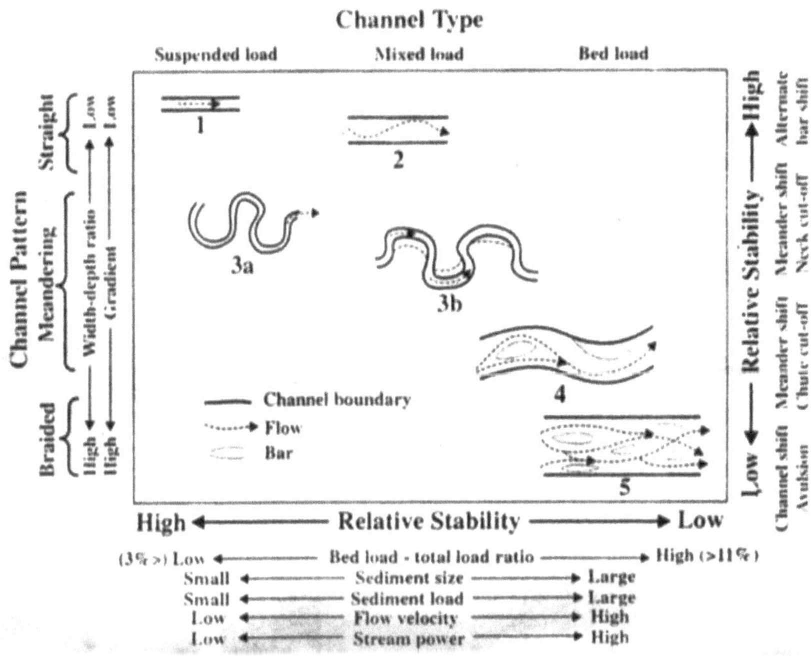

Fig. 14.6 Interrelation between channel type, hydraulic and sediment parameters and relative stability of streams

Quantitative relationships between channel planforms, bed slope (S_o) and mean annual flow (Q) are illustrated in Fig. 14.7 (Lane 1957).

A non-cohesive stream bed composed of silts and sands is predicted to meander when

$$S_o Q^{0.25} > 0.00070 \quad (14.3)$$

and braided when

$$S_o Q^{.25} > 0.0041 \quad (14.4)$$

The challenge to the river engineer is to understand the hydrologic, hydraulic and morphologic balances prevailing in the river system and their catchments and to design projects within the framework of these balances. Such an approach generally proves to be more efficient than continually trying to maintain the system against the natural tendencies of the river.

14.3 River Regime and Meandering

Aggradations and degradation in the vicinity of hydraulic structures like bridges and barrages are principally due to the loss in balance between sediment supply and transport rates. Rivers attain a stable regime over thousands of years through adjustment of its slope and cross section according to the volume of water and sediment carried over time. Commendable work on river regime has been done by Lacey (1930), Blench (1957), Chang (1988), CBIP(1989), Diplas (1990), Engelund (1973), Yalin and Silva (1999), Garde (2006), and Garde and RangaRaju (2000) for the prediction of stable river geometry based on the sediment size in bed and banks and the dominant flow carried by the river.

The major cause of change in stream characteristics can be attributed to human activities. Regardless of degree of channel stability, human activities may produce dramatic changes in the river behaviour and fluvial morphology locally and throughout the river. Man-made hydraulic structures like bridges and barrages and river training works like embankments, groynes, revetments, etc. often result in great departure from the equilibrium/regime state that existed prior to the construction of these structures.

Fig. 14.7 Interrelation between stream forms, bed slope and mean annual discharge

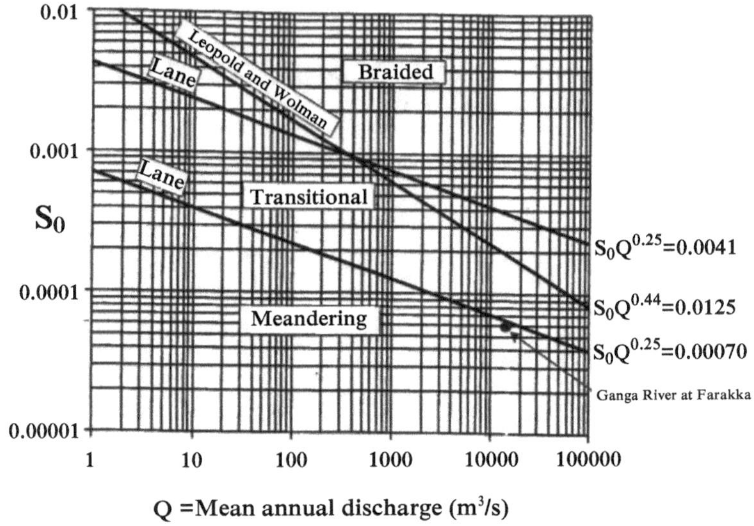

A typical straight stream is rarely stable. As shown in Fig. 14.6, streams with very small sediment load, low gradient and low velocity, low variability in flow and low aspect ratio (width-to-depth ratio) may be stable for some distances (about 10–15 km). Development of lateral instability associated with deposition and erosion on alternate riverbanks gives rise to thalweg pattern. Uncontrolled deposition and erosion ultimately give rise to meander formation as illustrated in Fig. 14.5. A lot of research work on meandering river have been carried out by river engineers like Rozovski (1957, Zimmermann and Kennedy (1978), Engelund (1973), Odgaard (1986), Yalin and Silva (1999), and Chitale (1981). Wang et al. (1992) developed a mathematical model of the meandering process to prove that the typical cross slope developed in a meander with lower bed elevation on the outer side of bend and higher bed elevation on the inner bank side (arising out of secondary current) essentially provides stability to the meandering stream.

Meanders are dynamic and found to move both laterally (at faster rate) and longitudinally (at slower rate). Lateral migration of meanders in Yellow River in China measured by Chien (1961) was found to vary from 20 to 100 m per day due to extremely high sediment load. He also observed that high lateral shifting of meander occurs downstream of constriction/control section along the river course. Lateral migration rate of river Ganga upstream of Farakka Barrage is reported to be 300–400 m per year as per Keskar Committee (1996) report.

Hickin and Nanson (1984) described the lateral migration rate (M) of a meander by the functional relation

$$M = f(\Omega, b, G, h, \tau_b) \qquad (14.5)$$

where Ω is stream power ($\tau.v$), b is a parameter expressing planform geometry of the stream, h is the height of outer bank (degree of incision) and τ_b is the erosional resistance offered by the outer concave bank undergoing erosion. Plotting the measured migration rate against relative curvature (r/w, where r is the radius of curvature and w is the stream width), Hickin concluded that the migration rate is maximum when meander stabilizes at an approximate value of $r/w = 2.5$ and obtained the relation

$$M_{2.5}(m/\text{year}) = \rho g\, QS/\tau_{b.h} \qquad (14.6)$$

where $M_{2.5}$ is the maximum rate of migration corresponding to $r/w = 2.5$. However, this relation is applicable for natural meanders without any interference from man-made hydraulic structures like bridges and barrages.

14.4 Hydraulics of Flow in the Vicinity Bridges and Barrages

Hydraulic structures like bridges and barrages often cause constriction of waterway either vertically or laterally or both. In bridges, for example, the constriction is only lateral, whereas in the case of dams and barrages, it is mostly vertical and sometimes both lateral and vertical. Depending on the degree of such restriction of waterway, the flow may be free or submerged. In free flow past a bridge and barrage, the flow gets choked with the structure acting as a control point. In choked flow, afflux is very high due to the minimum specific energy requirement, and there is hydraulic jump on the downstream side. Weirs and barrages have low solid obstructions, i.e. low crest elevation. Nowadays, barrages are generally made of low crest height with high headgates for the purpose of creating increased head and storage and for flushing out sediments deposited in the reservoir. The flow over the weir/barrage during flood may be free or submerged/drowned depending on the crest height and modular limit (Mazumder and Joshi 1981) of the structure. Depending on whether the flow is choked or not, hydraulic jump may or may not form. In case of choked flow, there is always a difference in energy level (ΔE) across the structure. If the actual loss of energy ($\Delta \acute{E}'$) within the jump is equal to the drop in energy level (ΔE), there is no residual kinetic energy of flow downstream of the structure (Mazumder 1985), and the flow is normal and uniform downstream of the structure. If the energy dissipation is inadequate/incomplete within the jump, there is a residual kinetic energy of flow which causes turbulence and non-uniformity in the flow distribution, *since a given flow with a given depth and a mean velocity can contain the excess residual kinetic energy only through non-uniformity and distortion of flow* – resulting in production of high level of turbulence (Mazumder 1993a). Mazumder and Sen (1991) found that in many of the low height barrages in India, the pre-jump Froude number of flow lies between 2 and 4. It is well established that the hydraulic jump in this region of inflow Froude's number is either undular or oscillating in nature, and the jump efficiency (as energy dissipater) is very poor. As a result, the flow downstream has high non-uniformity and is often found to swing either on the left or right bank side due to flow instability (Mazumder 1993b; Mazumder and Kumar 2001). It becomes highly turbulent causing erosion of bed and banks on the side where the turbulent *wall-jet-type flow* adheres to. Deposition of sediment occurs on the opposite bank side creating cross slope and meander formation.

In North and Northeast India, most of the rivers are found to be moving in a wide flood plain formed principally due to meandering/braiding channel formation. When a bridge or a barrage is constructed on such a wide flood plain (khadir), usually the waterway for the bridges (Mazumder 2009a) and barrages are kept limited up to Lacey's regime waterway. The khadir width is restricted (Mazumder 2008a, b) by providing approach embankments and guide bunds as shown in Fig. 14.8. Such restriction may or may not be symmetrical. As a result, there is a considerable afflux (Mazumder and Dhiman 2003) and backwater upstream of the structure, resulting in sedimentation and lateral instability of flow upstream. The main flow is often found to hug onto one of the banks causing erosion, and the eroded materials are deposited on the opposite bank resulting in meandering flow.

On the one hand, the outer bank of a meandering bend is subjected to high concentration of flow and drag; on the other hand, the critical shear stress for incipient motion of bed and bank materials is extremely low on the outer bank side principally due to the formation of secondary current (Mazumder 2010). This results in caving in of the outer bank material even with negligible drag, especially where the outer bank is made of fine alluvial soil of extremely poor shear strength.

Uncontrolled and progressive erosion of outer bank causes migration of meander as shown in Fig. 14.5. This may ultimately lead to breaching, outflanking of hydraulic structures and flow avulsion when the river shifts its course and join other low-lying rivers as observed in case of Kosi River upstream of Kosi barrage (Mazumder 1985, 2009b; Chitale 2009) in August 2008 as

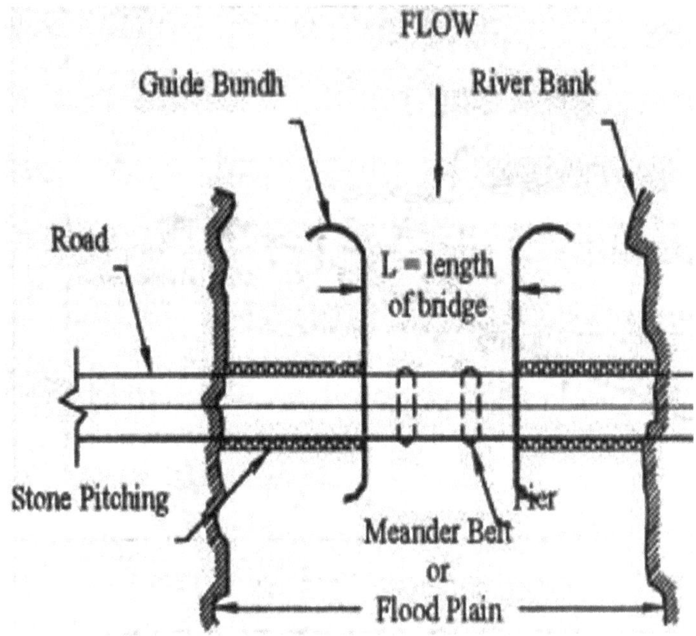

Fig. 14.8 Restriction of waterway in bridges in flood plain with guide bunds and approach embankment

predicted earlier by Chitale (2000). Similar tendency is also observed in river Ganga upstream of Farakka Barrage that is discussed afterwards under 'Case Study of River Behaviour and Training'. The approach flow is often found to separate at the head of guide bund resulting in a skew flow both upstream and downstream of the bridge and barrage, resulting in high flow concentration and scouring. As a result, the very purpose of providing guide bund is defeated sometimes. Non-uniformity (obliquity) of approach flow causes not only deep scour due to high flow concentration, it creates a large cross slope along the bridge/barrage axis resulting in stronger secondary current and greater scour. High degree of non-uniformity of flow approaching and departing the barrage is reported in the case of both Kosi and Farakka Barrages.

14.5 Training of River Near Bridges and Barrages

When damages are caused by a river prior to construction of any structure, people blame God. But the same people curse the project authorities and ask for compensation when even less damage occurs after the construction of structures like bridges and barrages on a river. Thus, river improvement and river training works should be an integral part of hydraulic structures like bridges and barrages for the safety of the structures and the people living near the riverbanks. River training measures like afflux/flood embankments, guide bunds, groynes, pitching, launching apron, etc. are often found necessary to protect the structure and to ensure that the river continues to flow along the desired course without causing any damage to the bridges and barrages. Adequate protection of the approach and flood/afflux embankments and their annual maintenance is vitally needed to prevent breaching, outflanking and possible flow avulsion. Gate operation schedule in a barrage is to be decided based on the nature of sediment deposition in the vicinity of barrage. Procedure for planning, design, construction, operation and maintenance of the different river training works are available in Indian Roads Congress (IRC), Bureau of Indian Standards (BIS) and Railways Dessign and Standards Organisation (RDSO) codes/guidelines. The choice of appropriate river training measure is site specific and requires both knowledge and experience about river behaviour before and after the construction of bridges and barrages. Some of them are briefly discussed under section 'Case Study of River Behaviour and Training'.

14.6 Case Study of River Behaviour and Training

14.6.1 Bridge on NH-31 Crossing River Mahananda

Figures 14.9, 14.10 and 14.11 illustrate the river-bank erosion upstream of Mahananda River bridge on National Highway (NH)-31. The length of the bridge is 636 m with 12 equal spans of 53 m each. It has a catchment area of about 8000 km^2 up to the bridge site. Design discharge is about 5000 cumec with 50-year return period and the corresponding High Flood Level (HFL) is 36.74 m. Due to restriction of its large flood plain extending up to 3 km or more, a big central island has been formed (Fig. 14.9) upstream of the bridge due to deposition of sediments resulting in anabranching (bifurcation) of the river and erosion on the outer banks of the branches as shown in Fig. 14.9. Costly river training measures consisting of permeable spurs and cut-offs were recommended to protect the villages and agricultural lands on either side of the banks and to prevent bank erosion and outflanking of the bridge.

Fig. 14.10 Erosion on right bank u/s of Mahananda bridge

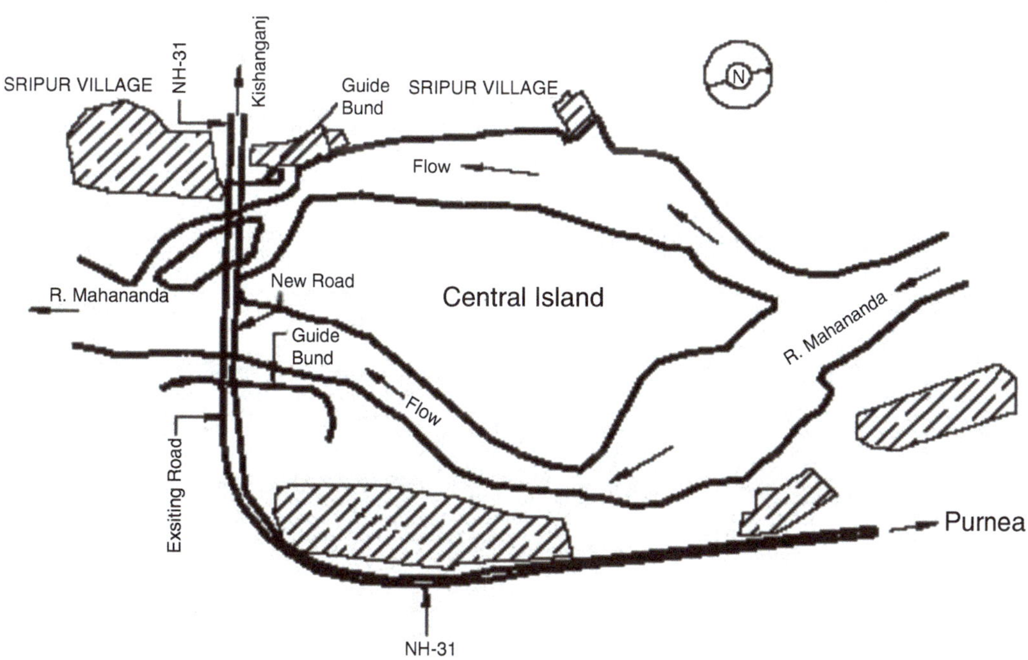

Fig. 14.9 Plan showing anabranching of river Mahananda u/s of bridge on NH-31

14.6.2 Bagmati River Bridge on NH-57

Bagmati bridge on NH-57 across river Bagmati is 100.8 m long with 4 equal spans 25.2 m each. The river is severely threatening both the national highway and the bridge. Originating from Nepal, Bagmati River has a catchment area of about 1200 km^2 up to the bridge site. Design discharge is about 1000 cumec with a return period of 50 years and the corresponding HFL is 50.81 m. Like other rivers in north Bihar, the river Bagmati carries huge amount of sediments along with water. Due to deposition of these sediments and development of sharp meandering that bends both Upstream (u/s) and downstream (d/s) as shown in Figs. 14.12 and 14.13, the conveying capacity of the river has reduced considerably over the years.

Due to high restriction of flood plain, the river spills its low height banks and spreads over a large width both upstream and downstream of the bridge during monsoon. The river is on the verge of forming natural cut-offs both upstream and downstream (Fig. 14.12) of the bridge. Training measures (Mazumder 2008a, b) consisting of artificial cut-offs, spurs and guide bunds have been recommended to straighten the river and make it flow axially under the bridge for the safety of the bridge and the approach embankment.

Fig. 14.11 Embayment of right bank of River Mahananda u/s of Mahananda Bridge on **NH-31**

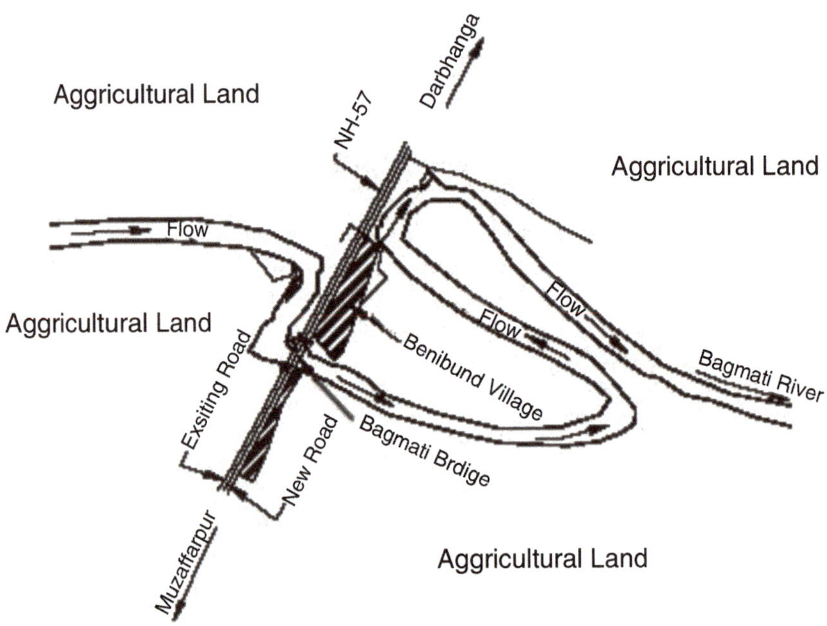

Fig. 14.12 Plan view of River Bagmati showing sharp bends u/s and d/s of bridge on NH-57

Fig. 14.13 Erosion on left bank of Bagmati River damaging the habitats on outer bank of 90° Bend u/s of Bagmati bridge

14.6.3 Farakka Barrage on River Ganga

Commissioned in 1975, a 2244 m long Farakka Barrage across river Ganga is a project of national importance. The main purpose of the project is to forcibly divert 1135 cumec of fresh upland flow of river Ganga to its tributary Bhagirathi/Hoogly as illustrated in Fig. 14.14. River Hoogly is the lifeline of Kolkata (including Kolkata port) and a large industrial complex developed on both sides of Hoogly over the years. During lean flow season, the river was getting dried up due to silting of its offtake point near Jangipur. A 38.3-km-long feeder canal has been constructed for forcible diversion of flow from Ganga to Hoogly. The barrage is designed for a flood discharge of 76,455 cumec with design afflux of 0.5 m. Further details of the barrage are available elsewhere (Mazumder 2004; Naresh Kumar et al. 2010).

With a longitudinal bed slope of 1 in 20,000 and a mean annual flow of 12,200 cumec, the river is in a meandering state near Farakka as indicated in Fig. 14.7. On an average, river Ganga carries 800 million tons of sediments (Sanyal 1980) every year up to the barrage, and it is estimated that approximately 1.3 mha.m of sediments have already been deposited upstream of the barrage. Several shoals have been formed upstream of the barrage, resulting in lateral flow instability, meandering and strong flow curvature resembling a braiding delta-like pattern. On several occasions, the marginal embankments were breached causing unprecedented flood damages (Mazumder 2000). The river has moved about 7 km inside Malda district, as shown in Fig. 14.15, wiping out thickly populated villages near the left marginal embankment. Four hundred fifty people died, and properties worth about rupees 10,000 million were damaged in the 1995 and 1998 floods alone. Twenty-seven spurs were constructed to protect the marginal embankment upstream. But the river has swallowed most of these spurs due to deep erosion near the left bank marginal embankment. At the offtake of river Pagla near Panchanandpur village, as many as ten retired embankments (Fig. 14.16) were constructed successively to prevent further migration. Any avulsion of the mighty river Ganga upstream of the barrage may cause change in its course, and it may capture rivers like Pagla, Kalindri and Mahananda. The barrage will be ineffective and it will cause colossal damage to Malda district including the National Highway

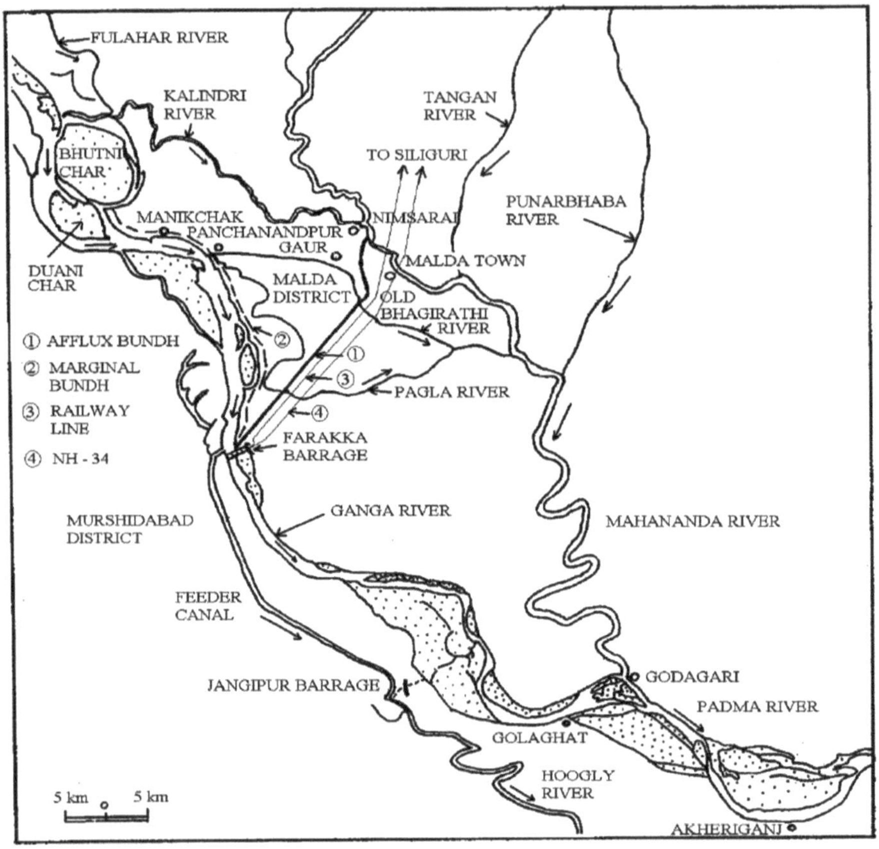

Fig. 14.14 River Ganga and its tributaries near Farakka Barrage

34 (NH-34), railway line and afflux bund protecting Malda town as shown in Fig. 14.14.

Downstream of Farakka Barrage, river Ganga has scoured the right bank in Murshidabad district of West Bengal. A typical meander is developing with Malda on the outer side of the upstream meander (left bank) and Murshidabad on the outer side of the downstream meander with Farakka Barrage at the centre acting as a nodal/fixed point. The river is threatening the existence of several towns located on right bank and loss of very fertile land. If the erosion continues further, the river may merge with Feeder canal, defeating the very purpose of the barrage. Railway line and NH-34 will be washed out. Ninety-six submergible-type boulder spurs were constructed to arrest erosion from Farakka to Jalangi, a distance of about 100 km. Several spurs and revetments have been washed out. A master plan of riverbank protection both upstream and downstream of the barrage has been drawn at a cost of about rs 927 crores (at 1995 prices) as per the recommendations of Pritam Singh and Keskar Committee appointed by the Ministry of Water Resources, Government of India.

14.7 Conclusion

Understanding river behaviour is important for proper planning, design, construction, operation and maintenance of bridges and barrages. Morphological study of river before and after the construction of these structures is essentially needed for the safety of the structure, for the approach and afflux embankments and for planning and design of appropriate river training measures. Due to high restriction of wide flood

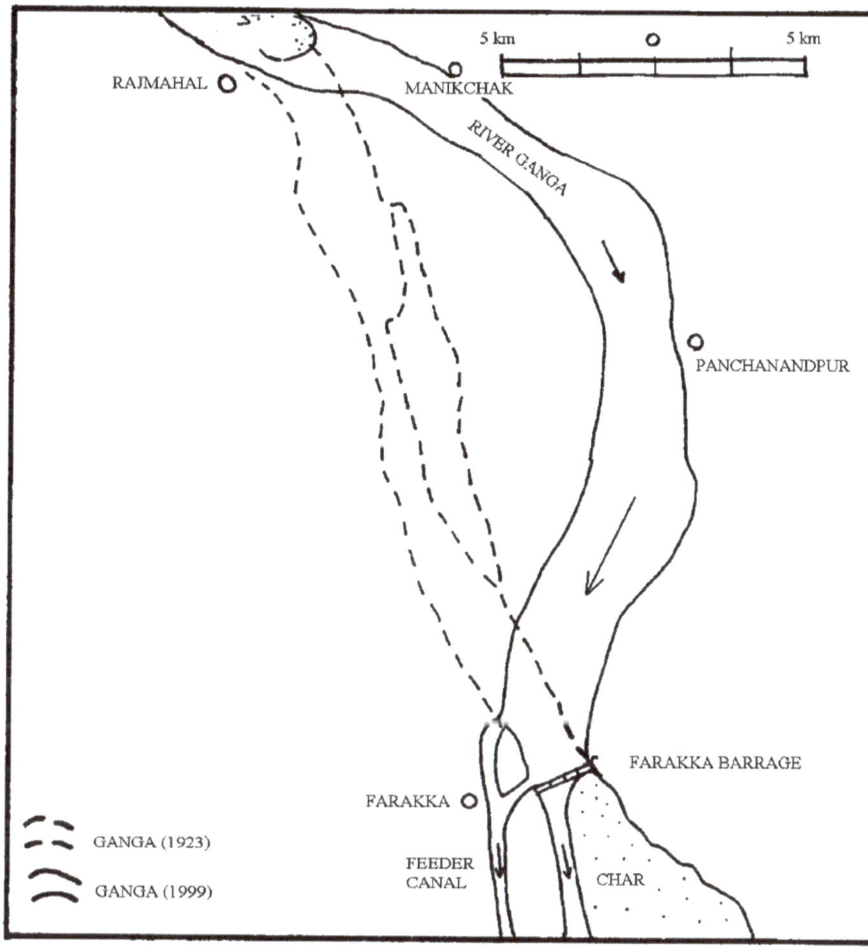

Fig. 14.15 Change in course of river Ganga over the years 1923–1999

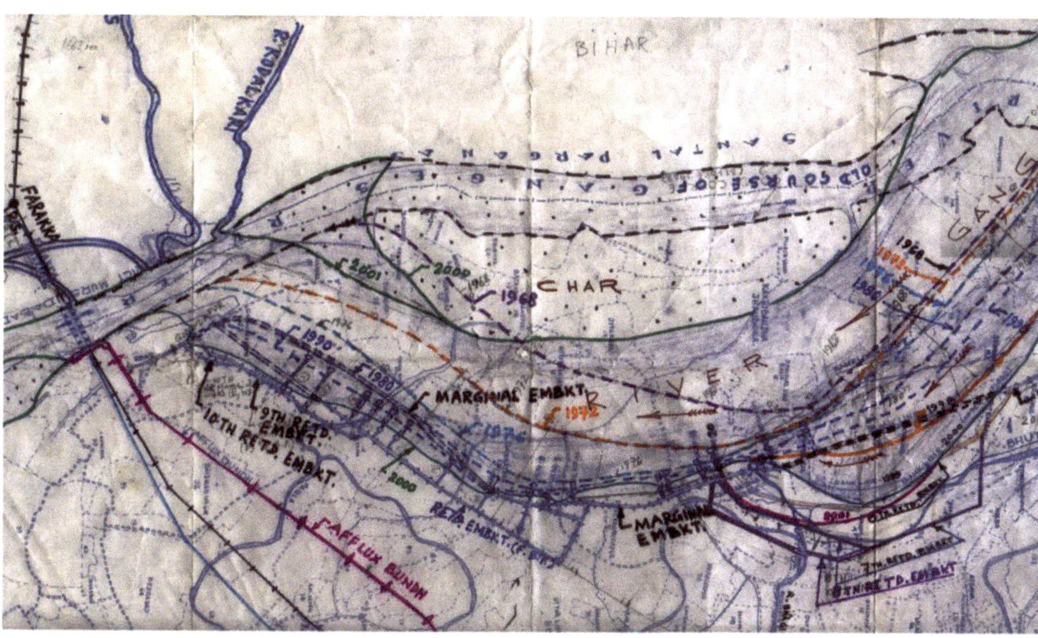

Fig. 14.16 Construction of retired embankments to prevent migration of meander towards left bank upstream of Farakka Barrage near Panchanandpur

plains of the Himalayan rivers carrying large-volume sediments during flood seasons, some of the hydraulic structures are creating unforeseen problems arising out of flow instability, meander formation, deep scour and formation of shoals. Uncontrolled erosion and deposition process in the vicinity of bridges and barrages often create serious problems of river training and threat to people living nearby. Case studies of bridges on NH-31 across river Mahananda and Bagmati bridge on NH-57 across river Bagmati have been discussed with figures and photographs. A similar case study of Farakka Barrage across river Ganga has been made, and the need of proper training measures has been stressed for the safety of the barrage and protection of life and properties both upstream and downstream of the barrage.

Acknowledgement The author wishes to thank Aquagreen Engineering Management and Intercontinental Consultants and Technocrats Pvt. Ltd. (ICT) and Scott Willson India Pvt. Ltd. (SWI) authorities for extending all facilities needed for writing the paper.

References

Blench T (1957) Regime behaviour of canals and rivers. Butterworth Scientific Publishing, London

CBIP (1989) River behavior, management and training. Central Board of Irrigation and Power, Malcha Marg, Chanakyapuri, New Delhi

Chang (1988) Fluvial processes in river engineering. Wiley Eastern, New York

Chien N (1961) The Braided stream of the Yellow River. Sci Sin 106(6):734–754

Chitale SV (1981) Shape and mobility of river meanders. In: Proceedings of the XIX Congross of IAHR, vol 2, New Delhi, pp 281–286

Chitale SV (2000) Future of the Kosi River and Kosi project. J Inst Eng (I) 81:109–114

Chitale SV (2009) Kosi disaster-caution and pre-caution. ISH J Hyd Eng. 15(1):201–207

Diplas P (1990) Characteristics of self formed straight channels. J Hydraul Eng ASCE 116(5):205–228

Engelund F (1973) On the origin of meandering and braiding in alluvial channels. J Fluid Mech 59(2): 289–302

Garde RJ (2004) Aggradation and degradation in alluvial streams, Theme paper pub in seminar on silting of rivers – problems and solutions org. by Ministry of Water Resources through Central Water Commission and CW & PRS, Pune, February 12–14. New Delhi

Garde RJ (2006) River morphology. New Age International (Pvt.Ltd), New Delhi

Garde RJ, RangaRaju KG (2000) Mechanics of sediment transport and alluvial stream problems, 3rd edn. New Age International Public Pvt. Ltd., New Delhi

Hickin EJ, Nanson GC (1984) Lateral migration of river bends. J Hydraul Eng ASCE 110(11):1957–1967

Keskar Committee Report (1996) Bank erosion problem of river Ganga-Padma in the Districts Malda and Murshidabad in West Bengal, vol I and II. Planning Commission, Government of India

Kumar N, Saran V, Sambharia RR (2010) Planning of river training works- Farakka Barrage Project-a case study. Water Energy Int CBIP J 67(2)

Lacey G (1930) Stable channel in alluvium. J Inst Civil Eng UK Paper no. 4736, 229:259–292

Lane EW (1955) The importance of fluvial morphology in hydraulic engineering. In: Proceedings of the ASCE, Paper 745, pp 1–17, July

Lane EW (1957) A study of the slope of channels formed by natural stream flowing in "Erodible material" U.S. Army Corps of Engineers, Missouri River Division, Omaha, Sediment series 9

Leopold LB, Wolman BG (1957) River channel patterns: braided, meandering and straight. USGS Proceedings of the Paper 282B

Mazumder SK (1985) Effectiveness of impermeable type Groyens in river training with particular reference to River Kosi in India. 2nd Int. Conf. On "The Hydraulics of Floods and Flood Control" org. by British Hydromechanics Research Association, Fluid Engg. Centre, Cambridge, England 24–26 September

Mazumder SK (1993a) River erosion downstream of a barrage. In: Proceedings on "River Scour" org by CBIP at Varanasi, 28–29 April

Mazumder SK (1993b) Stability of river downstream of hydraulic structures. In: Proceedings of the of VIII APD- IAHR Congress, vol II org by CW&PRS Pune, 20–23 October, pp 273–282

Mazumder SK (2000) Role of Farakka Barrage on the disastrous flood at Malda (West Bengal) in 1998. Proc. 8th ICID Intl. Drainage Workshop 'Role of Drainage and Challenges of 21st Century Vol. III, sponsored by ICID –CIID & Min. of Water Resources and org. by ICD & WAPCOS, New Delhi January 31–February 4. New Delhi

Mazumder SK (2001) Training of River Ganga near Farakka Barrage: Proc. of National conference on Hydraulics and Water Resources HYDRO –2001, org by Indian Soc. for Hydraulics (ISH) & CWPRS, Pune, December 6–7, vol 7, pp 40–50

Mazumder SK (2004) River behavior upstream and downstream of hydraulic structures. Proc. Int.Conf. on "Hydraulic Engineering Research and Practice (ICON-HERP, 2004)", Org. by Department of Civil Engineering, University of Roorkee (now IIT, Roorkee) in honour of Prof. Rangaraju, October 26–28

Mazumder SK (2008a) River behavior near bridges with restricted waterway and afflux-some case study, IRC 'Indian Highways' published by IRC, 37(10), October
Mazumder SK (2008b) Training river near a bridge with particular reference to river Bagmati Crossing NH-57, In: Proceedings of the national conference on hydraulics, water resources and environment, HYDRO-2008 held at Jaipur, December 15–16
Mazumder SK (2009a) Determination of waterway under a bridge in Himalayan Region-Some Case Studies. J IRC 70–2, July–September
Mazumder SK (2009b) Discussion on the paper by S.V. Chitale "Kosi Disaster-Caution and Pre-Caution" Published in ISH J Hydraul Eng 15(2)
Mazumder SK (2010) Critical tractive stress in a straight and curved channel with non-cohesive soil, Paper published in ISH J Hydraul Eng October 2010
Mazumder SK. Joshi LM (1981) Studies on modular limit of critical flowmeter. In: Proceedings of the XIX Congress of IAHR, New Delhi
Mazumder SK, Dhiman R (2003) Computation of afflux with particular reference to Widening of Bridges on roadway. In: Proceedings of the national conference of hydraulics and water resources, HYDRO-2003, CW&PRS, Pune, December
Mazumder SK, Kumar P (2001) Sub-critical flow behaviour in a straight expansion. ISH J Hydraul Eng 1
Mazumder SK, Sen P (1991) Design, operation and maintenance of diversion structures in alluvial rivers. J Inst Eng (India),. C.E. Division, vol 72, Part CV 3, September
Mazumder SK, Rastogi SP, Hmar R (2002) Restriction of waterway under bridges. J Indian Highw 30(11)
MOWR (2004) Proceedings of the Seminar on "silting of rivers- problems and solutions" organisation by Ministry of Water Resources, Government of India, New Delhi, February 12–13
Odgaard AJ (1986) Meander flow model 1: development. J Hydraul Eng ASCE, HY 12:1117–1136
Rozovski JL (1957) Flow of water in bends of open channels. Academy of Sciences of the Ukranian SSR. Translated in English by Prushansky, Y. Israel Programme of Scientific Translation
Sanyal N (1980) Effect of embankment of River Kosi. In: Proceedings of the international workshop on 'Alluvial Rivers' organisation by University of Roorkee, March 18–20
Schumm SA (1980) Plan form of Alluvial Rivers. In: Proceedings of the international workshop on alluvial river problems held at University of Roorkee, March 18–20
Wang P et al. (1992) Numerical prediction of transverse bed slope and bed profiles in curved alluvial streams. In: Proceedings of the APD – IAHR VIII Congress at CW&PRS, Pune, vol II, October 20–23
Wang P, Zaho Shiquiang, Fangduo Ciajinde (1994) Computation of depth averaged longitudinal velocity and boundary shear stress in channel bends. Proc. of APD-IAHR, IX Congress, vol 2, Singapore, pp 409–416
Yalin MS, Ferreira D. Silva (1999) Regime channels in cohesionless alluvium. J Hydraul Eng Res, Special Issue on Fluvial Hydraulics, IAHR, 37(6)
Zimmermann C, Kennedy JF (1978) Transverse bed slope in curved alluvial steams. In Proceedings of the Institution of Civil Engineers, London, pp 525–538

S.K. Mazumder Department of Civil Engineering, Delhi College of Engineering (Delhi Technology University), New Delhi, India

Design Development and Field Application of RCC Jack Jetty and Trail Dykes for River Training

15

Anupama Nayak, Nayan Sharma, Kerry Anne Mazurek, and Alok Kumar

Abstract

While river training structures are important tools to provide solutions to river engineering problems, conventional structures can be expensive and can create adverse environmental impacts. There is a need to develop affordable permeable river training structures. Jack jetties have been in use in the USA for various purposes, although there was no scientific methodology to support design of the structures when used for river training works. Therefore, laboratory studies have been carried out to develop this design methodology, which was verified by field-testing. Trail dykes were also tested in a similar fashion. Both types of structures show promise for use as inexpensive river engineering works.

15.1 Introduction

River training works are essential measures to train a river for protecting its banks to avoid excessive meandering, to prevent shifting its course and to maintain navigability. Protection of river banks is normally accomplished by a variety of conventional river training works including marginal embankments or levees, guide banks, guide bunds, groynes or spurs, submerged vanes, cut-offs, pitching of banks, pitched islands, sills, closing dykes and longitudinal dykes. Some of these measures are less expensive than others. Considerations in their use, besides effectiveness, include the cost of construction, environmental impacts and aesthetics.

In recent decades, the conventional river training techniques deployed for management

A. Nayak (✉)
Water Resources Department, Government of Odisha, Bhubaneswar, Odisha, India
e-mail: nayak.anupama@gmail.com

N. Sharma
Department of Water Resources Development and Management, Indian Institute of Technology Roorkee, Roorkee, Uttarakhand, India

K.A. Mazurek
Department of Civil and Geological Engineering, University of Saskatchewan, Saskatoon, Canada

A. Kumar
Water Resources Department, Government of Jharkhand, Jamshedpur, Jharkhand, India

N. Sharma (ed.), *River System Analysis and Management*, DOI 10.1007/978-981-10-1472-7_15

of stream bank erosion, flood control and fairway development for inland navigation have become increasingly expensive due to various causes. This has made conventional river training techniques practically unaffordable when thousands of kilometres of erosion-affected stream banks await remedial action. The increasing demands for bank protection works in many river reaches have focused attention on the need to develop cost-effective and eco-friendly river training measures. This chapter focuses on the development of the design and field application of jack jetty and trail dykes to be deployed as effective river training structures.

15.2 Jack Jetty

Grassel (2002) describes that the steel jack jetty system had been in extensive use in the USA for several decades for the purposes of erosion control and the maintenance of navigation channels. Grassel (2002) in her report has given a general outline of the jetty jack, which was invented by H.F. Kellner in the early 1920s. This was a permeable form of bank protection and performed the job at a lower cost than the non-permeable types of bank protection then in use. Kellner started his experiments on a small stream near his home in Topeka, Kansas. He made his first jack with three willow poles tied together at the midpoint. To keep the willow poles extended, he laced them with wire. Later, he replaced the willows with steel angles.

The structural unit of the system is called a jack and is shown in Fig. 15.1 modified by Sharma (2008). All jacks deployed in the USA followed a standard size and configuration given by Kellner. A jack was composed of three 4.88 m long, 0.1 × 0.1 × 0.0006 m steel angles. These three angles were bolted together at their midpoints. The angles were placed back to back with their longitudinal axes at right angles to each other. The angles were fastened into place and formed three sets of intersecting planes with their common point at the centre. The planes were maintained by lacing them with wire at 0.381 m intervals. Then they were linked together in a line with thick cable to form a jetty. The cables extend in a continuous line through the units and were fastened at each end of the jetty. A jack jetty field that has been

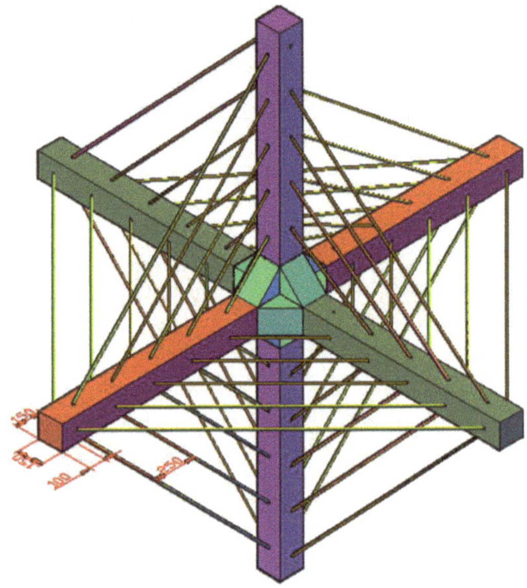

Fig. 15.1 Three member jack jetties

properly laid out traps sediment and debris during flood events and essentially builds up its own levee to confine the river channel. It is designed to conform to the existing regime of a river; a concave bank of a meandering channel is preferred for placement.

Steel jacks and jetties have been used successfully by the US Army Corps of Engineers, highway departments, railway companies and others to prevent damage to riverbanks, levees, bridge abutments and other structures. The US Bureau of Reclamation and the Corps of Engineers have used them to stabilize the channel of the Rio Grande within the floodway in the Middle Rio Grande Valley. In order to rehabilitate the Middle Rio Grande Conservancy District, structures like levees, channel improvement works and jetty fields were constructed by the Bureau of Irrigation and Drainage Systems, as part of the Rio Grande Channelization Project, covering 100 miles of the reach below Cochiti. The Espanola floodway also underwent extensive levee rehabilitation and channel straightening using these jetty fields. After the implementation of these structures, the Rio Grande was converted to a highly modified storage and water conveyance system (Crawford et al. 1993; Lagasse 1980; Najmi 2001).

Jetty fields incorporate some of the good features of walls and groynes but were also permeable, reducing the possibility of over-confining the river and causing substantial scour, such as occurs at the ends of solid groynes. Lines of jacks have also added the desirable quality of being flexible; the jacks will settle as scour occurs and conform to the bed.

The ideal operation of a jetty field may be described as follows. Lines of jacks in the flow area provide additional resistance to the water passing through the field, which in turn reduce the flow velocity. This reduces the sediment carrying capacity of the flow and sediment is deposited in the field. Vegetal growth in the deposited sediment provides additional flow resistance. Sufficient sediment is accumulated to form a new riverbank and induce the river to flow in the desired channel. Channelization causes the riverbed to scour and results in a lower water surface.

The steel jack jetty system had shown promise as a river training measure for channelization of rivers and the protection of levees. However, the performance of the jack jetty failed with reduced sediment concentration levels in a river (Grassel 2002; U.S. Army Corps of Engineers 2003). In many areas, the jetty jack fields became a non-functional eyesore that often complicated efforts towards restoration and reduction activities (a pre-emptive measure in the reduction of fire threat and/or severity by the removal of dead-and-downed vegetation). Although not a permanent structure, the jetty jacks were often entrained within depositional sediments and/or vegetation and thus defied easy removal (Grassel 2002).

In the example of the Rio Grande (US Army Corps of Engineers 2011), its riparian ecosystem continued to provide habitat for a wide variety of wildlife species, although in a much reduced and degraded state compared to its historic status. The Rio Grande was a critical travel corridor for many species, especially migratory birds. Both the degradation of the hydrologic and geomorphic character of the river and the decline in aquatic and riparian habitat threatened this diversity. Though the jetty field did result in a change in the channel alignment, the environmental issues and difficulty in removing fire hazards obligated the removal of the jetty.

Herein jack jetties are reconsidered for use as cost-effective river training measures in alluvial rivers rich in sediment. A limitation associated with the use of the jack jetty was that there was no scientific design methodology available. This chapter presents results from studies to support the development of a design methodology for the jack jetty system. Instead of steel jetty, a modified version of a jack constructed from Reinforced Cement Concrete with cables (an RCC jack) was used. This is because of increasing steel prices in developing Asian countries that make the Reinforced Cement Concrete (RCC) jack a less expensive alternative. The RCC jack jetty system could also easily be fabricated and placed on site.

15.3 Experiments, Observations and Results for RCC Jack Jetty Testing

15.3.1 Objectives

The goal of the experimental studies on the jack jetty described herein was to develop potential design indices and performance parameters to assess sediment deposition in the jetty fields (lines of jack jetty). The work was accomplished in the laboratory, with some verification of the results in a field test of a jetty system. The work summarized here is presented in more detail in Nayak (2012). The specific objectives of the experiments were as follows:

- To develop design indices and performance parameters to represent different alternative jetty field layouts to optimize performance in capturing sediment for berm formation;
- To investigate the effect of submergence level of the jetty field on sediment deposition within the field;
- To develop thresholds for achievement of various design objectives for erosion control and reclamation using a jetty field; and
- To gain an insight into the effect of various jetty field configurations on channel bed changes.

Table 15.1 Details of experimental conditions in phase I of the study

	Expt	Water depth (m)	Discharge (m^3/s)	Froude number (F_r)	D_{50} of sediment (mm)	Layout of jacks on the flume bed
SERIES-I	1	0.10	0.013	0.11	0.25	One 0.08 m model
	2	0.10	0.013	0.11	0.25	One 0.10 m model
	3	0.10	0.013	0.11	0.25	Four 0.10 m model
	4	0.10	0.013	0.11	0.25	Three 0.10 m model
	5	0.10	0.013	0.11	0.25	Two 0.10 m model
	6	0.20	0.039	0.12	0.25	Four 0.10 m model
SERIES-I	7	0.20	0.039	0.12	0.25	Three 0.10 m model
	8	0.20	0.039	0.12	0.25	Two 0.10 m model
	9	0.20	0.039	0.12	0.25	4 rows 0.10 m models, 2 in each row
	10	0.20	0.039	0.12	0.25	3 rows 0.10 m models, 2 in each row
	11	0.20	0.039	0.12	0.25	2 rows 0.10 m models, 2 in each row
	12	0.20	0.039	0.12	0.25	4 rows 0.08 m models, 2 in each row
	13	0.20	0.039	0.12	0.25	3 rows 0.08 m models, 2 in each row
	14	0.20	0.039	0.12	0.25	2 rows 0.08 m models, 2 in each row
SERIES-II	15	0.30	0.071	0.12	0.59	0.20 m model
	16	0.40	0.107	0.11	0.59	One 0.30 m model
	17	0.40	0.107	0.11	0.59	Two 0.20 m models
	18	0.40	0.107	0.11	0.59	Three 0.20 m models

15.3.2 Experimental Setup and Experiments

To study and analyse the effect of jack jetty on the flow domain and pertinent fluvial parameters, the experimental programme was divided into three phases, namely, (1) the 'micro' level laboratory experiments, (2) the 'macro' level laboratory experiments and (3) the field prototype study in the river Ganges at Nakhwa site in India. Tables 15.1 and 15.2 provide the details of the testing programme performed in the laboratory. The field experiments are discussed in a later section.

In the first phase of experiments, the effect of a jack jetty on the flow was studied. Velocity measurements were undertaken for varying depth of flow, discharge and sediment size for various model configurations (see Table 15.1). These were conducted in an over 20 m long, 1.2 m wide, 0.6 m flume with a bed of sediment, first at the Hydraulics Laboratory at Indian Institute of Technology Roorkee then at the Hydraulics Laboratory at the University of Saskatchewan. In the second, macro level phase of the study, the effect of various configurations of jack jetties on the deposition of sediment from sediment-laden flow was investigated for varying discharge, sediment concentration and flow depth (see Table 15.2). The experiments were carried out in a 0.5 m wide flume in the River Engineering Laboratory of IIT Roorkee. The goal was to find the jetty field design parameters that would result in erosion control, moderate reclaim or high reclaim for these varied conditions of flow and sediment concentrations. Experiments were performed with two types of sediments having median diameter as 0.248 and 0.59 mm. The relative densities of all sands were 2.65. The bed slope of the flume was kept constant in all phases of the experiments at 0.000133.

Table 15.2 Details of experimental conditions in phase II of the study

Expt. no.	*qs* (ppm)	D_{50} of sediment (mm)	L_r/L_s	Discharge (m^3/s)	Froude number (F_r)	Water depth (m)	Submergence ratio
1	500	0.25	0.5	0.0225	0.11	0.25	0.53
2	500	0.25	1.0	0.0225	0.11	0.25	0.53
3	500	0.25	1.5	0.0225	0.11	0.25	0.53
4	500	0.25	2.0	0.0225	0.11	0.25	0.53
5	500	0.25	0.5	0.018	0.13	0.20	0.41
6	500	0.25	1.0	0.018	0.13	0.20	0.41
7	500	0.25	1.5	0.018	0.13	0.20	0.41
8	500	0.25	2.0	0.018	0.13	0.20	0.41
9	500	0.25	0.5	0.013	0.15	0.15	0.22
10	500	0.25	1.0	0.013	0.15	0.15	0.22
11	500	0.25	1.5	0.013	0.15	0.15	0.22
12	500	0.25	2.0	0.013	0.15	0.15	0.22
13	250	0.25	0.5	0.013	0.15	0.15	0.22
14	250	0.25	1.0	0.013	0.15	0.15	0.22
15	250	0.25	1.5	0.013	0.15	0.15	0.22
16	250	0.25	2.0	0.013	0.15	0.15	0.22
17	750	0.25	0.5	0.013	0.15	0.15	0.22
18	750	0.25	1.0	0.013	0.15	0.15	0.22
19	750	0.25	1.5	0.013	0.15	0.15	0.22
20	750	0.25	2.0	0.013	0.15	0.15	0.22
21	1000	0.25	0.5	0.013	0.15	0.15	0.22
22	1000	0.25	1.0	0.013	0.15	0.15	0.22
23	1000	0.25	1.5	0.013	0.15	0.15	0.22
24	1000	0.25	2.0	0.013	0.15	0.15	0.22

15.3.3 Observations, Results and Analysis

To analyse the experimental data, terminologies need to be introduced to describe the jetty field and these are summarized below. A single or stand unit of the model is called a jack, as shown in Fig. 15.2, and when the jacks are connected together in a line with a cable, they form a jetty. When lines of jetty are laid parallel to the bank of the channel, they are called 'diversion lines', and when the jetty are projected into the river at a certain angle with the bank, they are called 'retards'. Combinations of retard and diversion lines form a jetty field. Figure 15.3 shows schematic diagram in plan of a jetty field showing retards and a diversion line, where L_r is the length of the retards and L_s is the centre-to-centre spacing of retards. Figure 15.4 shows a sketch representing the elevation of a jack in water, where h is the height of jack and H is the depth of flow. Additionally new design indices and performance parameters for the jack system were developed: the Jetty Field Density Index (JFDI), the Jetty Field Submergence Index (JFSI), the Bed Deposit Factor (BDF) and the Jetty Field Length Factor (JFLF). These are defined as follows.

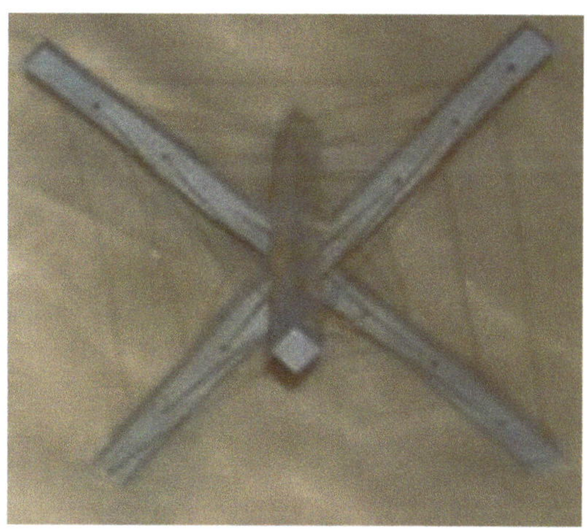

Fig. 15.2 Photograph of the jack used in the laboratory experiments

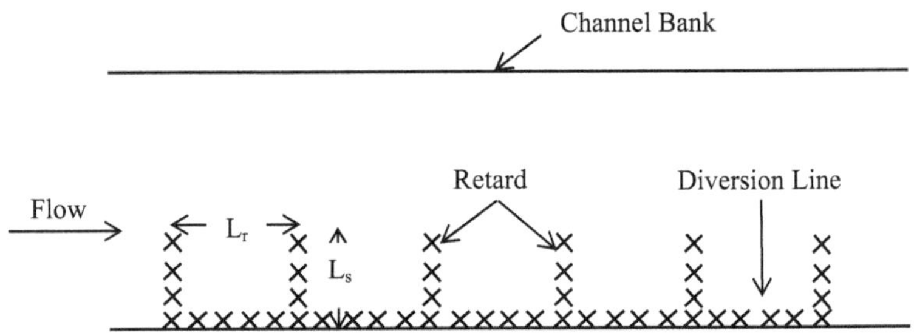

Fig. 15.3 Layout (in plan) of a jetty field showing a diversion line and retards

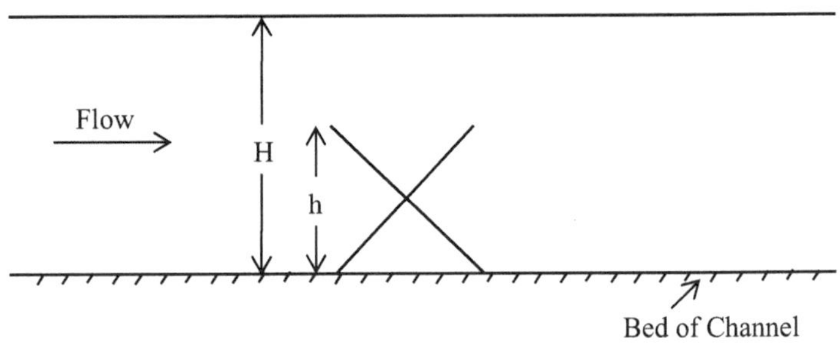

Fig. 15.4 Elevation of a jack in water

Jetty Field Density Index the length of retard to the centre-to-centre spacing of the retards or (L_T/L_S). Smaller values of the Jetty Field Density Index correspond to sparse jetty fields, and higher values represent densely configured fields.

Jetty Field Submergence Index the ratio of the depth of water above the top of the jack to the total depth of water or (H-h)/H. Smaller values of the Jetty Field Submergence Index represent smaller submergences.

Bed Deposit Factor the depth of sand deposited on the sand channel bed to the total depth of water. Larger values of the Bed Deposit Factor suggest larger amounts of sedimentation in the jetty field.

Jetty Field Length Factor the length of one compartment (the area enclosed by two adjacent retard lines) of the jetty field to the total cumulative length of jetty field compartments.

Threshold values for the Bed Deposit Factor for various combinations of the Jetty Field Density Index, Jetty Field Submergence Index and sediment concentrations of the bed load and suspended in flow were developed to aid in developing guidelines for the design of the jack jetty systems. To determine these thresholds, the jack jetty systems were tested with various Jetty Field Density Indices for four sediment concentration values of the bed load and suspended load in flow and three Jetty Field Submergence Indexes to ascertain the efficiency of the jetty field in creating sedimentation in the channel bed. Sand deposition in the jetty field was monitored and measured for all experiments and analysed in terms of Bed Deposit Factor. It was also found that the efficiency of the jetty field in providing sedimentation is enhanced at smaller submergence and for higher sediment concentrations in the flow.

For each experimental run, the Jetty Field Length Factor was plotted against the Bed Deposit factor. The trend line followed a second-order polynomial trend with high correlation coefficient. From this plot average and maximum Bed Deposit Factor values were determined. The thresholds later were discretized from these values. Based on the thresholds, the design guidelines were developed.

Table 15.3 shows the average and maximum Bed Deposit Factor developed from the bed profile plots for a suspended sediment concentration of 500 ppm for various layouts of the jetty field with varying Jetty Field Density Index for various Jetty Field Submergence Indices. Similarly Table 15.4

Table 15.3 Bed deposit factors for various jetty field submergence indices for a fixed sediment concentration in the flow of $q_s = 500$ ppm

JFSI	Average bed deposit factor		Maximum bed deposit factor	
0.53	0.05	Erosion control	0.07	Erosion control
0.53	0.07	Erosion control	0.08	Erosion control
0.53	0.08	Erosion control	0.08	Erosion control
0.53	0.08	Erosion control	0.2	Moderate reclaim
0.41	0.11	Moderate reclaim	0.15	Moderate reclaim
0.41	0.1	Erosion control	0.12	Moderate reclaim
0.41	0.1	Erosion control	0.11	Moderate reclaim
0.41	0.09	Erosion control	0.19	Moderate reclaim
0.22	0.11	Moderate reclaim	0.16	Moderate reclaim
0.22	0.11	Moderate reclaim	0.14	Moderate reclaim

Table 15.4 Threshold values for various jetty field density indices for fixed jetty field submergence index = 0.22 and varied sediment concentration

Qs	JFDI	Average bed deposit factor		Maximum bed deposit factor	
250	0.5	0.08	Erosion control	0.11	Moderate reclaim
500	0.5	0.12	Moderate reclaim	0.16	Moderate reclaim
750	0.5	0.14	Moderate reclaim	0.19	Moderate reclaim
1000	0.5	0.18	Moderate reclaim	0.23	Heavy reclaim
250	1.0	0.1	Erosion control	0.13	Moderate reclaim
500	1.0	0.1	Erosion control	0.14	Moderate reclaim
750	1.0	0.14	Moderate reclaim	0.19	Moderate reclaim
1000	1.0	0.16	Moderate reclaim	0.24	Heavy reclaim
250	1.5	0.11	Moderate reclaim	0.15	Moderate reclaim
500	1.5	0.17	Moderate reclaim	0.21	Heavy reclaim
750	1.5	0.18	Moderate reclaim	0.22	Heavy reclaim
1000	1.5	0.21	Heavy reclaim	0.25	Heavy reclaim
250	2.0	0.09	Erosion control	0.15	Moderate reclaim
500	2.0	0.12	Moderate reclaim	0.16	Moderate reclaim
750	2.0	0.18	Moderate reclaim	0.25	Heavy reclaim
1000	2.0	0.19	Moderate reclaim	0.24	Heavy reclaim

illustrates and presents the average and maximum Bed Deposit Factor evaluated from the bed profile measurements with fixed submergence and varying layout of the jetty field with varying Jetty Field Density Index and varying sediment concentrations. The results were categorized into three bed deposit levels, erosion control, moderate reclaim and heavy reclaim. If the average Bed Deposit Factor is less than 0.1, then one can expect for erosion control. If it is between 0.1 and 0.2, moderate reclaim can be expected. If the bed deposit factor is more than 0.2, then heavy reclaim might be expected.

It was evident that the efficiency of the jetty field performs better with lower submergences than higher submergences and with higher sediment concentrations in the flow as seen in Tables 15.3 and 15.4. The average Bed Deposit Factor values vary in the range of 0.05–0.08, and for medium submergences and a moderate concentration around 500 ppm, that overall vary in the range of 0.1–0.19. Similarly for low submergence and high sediment concentration, it varies between 0.2 and 0.3.

15.3.4 Design Methodology for Jetty Fields

In this section, the goal is to provide design guidelines for the Jetty Field Density Index,

Table 15.5 Guidelines for jetty field design with a fixed sediment concentration

Qs	JFSI	Requirement	Bed deposit factor	JFDI
500	0.22	Erosion control	<0.1	<0.5
		Moderate reclaim	Between 0.1 and 0.2	0.5–1
		High reclaim	>0.2	1.5–2
	0.41	Erosion control	<0.1	<0.5
		Moderate reclaim	Between 0.1 and 0.2	0.5–2
		High reclaim	>0.2	–
	0.53	Erosion control	<0.1	0.5
		Moderate reclaim	Between 0.1 and 0.2	2
		High reclaim	>0.2	–

Table 15.6 Guidelines for jetty field design with fixed JFSI

JFSI	q_s	Requirement	Bed deposit factor	JFDI
0.22	250	Erosion control	<0.1	0.5
		Moderate reclaim	Between 0.1 and 0.2	0.5–1
		High reclaim	>0.2	–
	500	Erosion control	<0.1	<0.5
		Moderate reclaim	Between 0.1 and 0.2	0.5–1
		High reclaim	>0.2	>1.5
	750	Erosion control	<0.1	<0.5
		Moderate reclaim	Between 0.1 and 0.2	0.5–1
		High reclaim	>0.2	1.5–2
	1000	Erosion control	<0.1	<0.5
		Moderate reclaim	Between 0.1 and 0.2	0.5
		High reclaim	>0.2	1–2

which would provide the designer basic information of what configuration and layout of the jetty field should be adopted for achievement of desired goals for sedimentation in the field. Table 15.5 provides information about JFDI to be adopted for various sets of experiments with fixed sediment concentration and varying JFSI. Table 15.6 provides information about JFDI to be adopted for various sets of experiments with fixed JFSI and varying sediment concentration.

15.4 Field Testing of the RCC Jack Jetty

The results from the laboratory study on the jack jetty provided guidance on potential design indices and performance parameters for jetty systems. A further field study was conducted at a site (Nakhwa) on the Ganga river to test and generally validate the effectiveness of the laboratory-based results. The 2525 km long Ganga river originates in the western Himalayas in the Indian state of Uttarakhand and flows south and east through the Gangetic Plain of North India into Bangladesh, where it empties into the Bay of Bengal. By discharge it ranks among the world's top 20 rivers and is the second largest in South Asia and the longest in India. The water depth in the Ganga varies from 8 to 10 m at low flow and 20–21 m at high flows, while the river width ranges from about 2 km to more than 10 km. It is the largest alluvial river in India, with a high degree of braiding. It was not possible to test multiple configurations of jetty field installations in the field tests because of the costs. Therefore, one jetty field configuration was tested to help validate the observations seen in the laboratory.

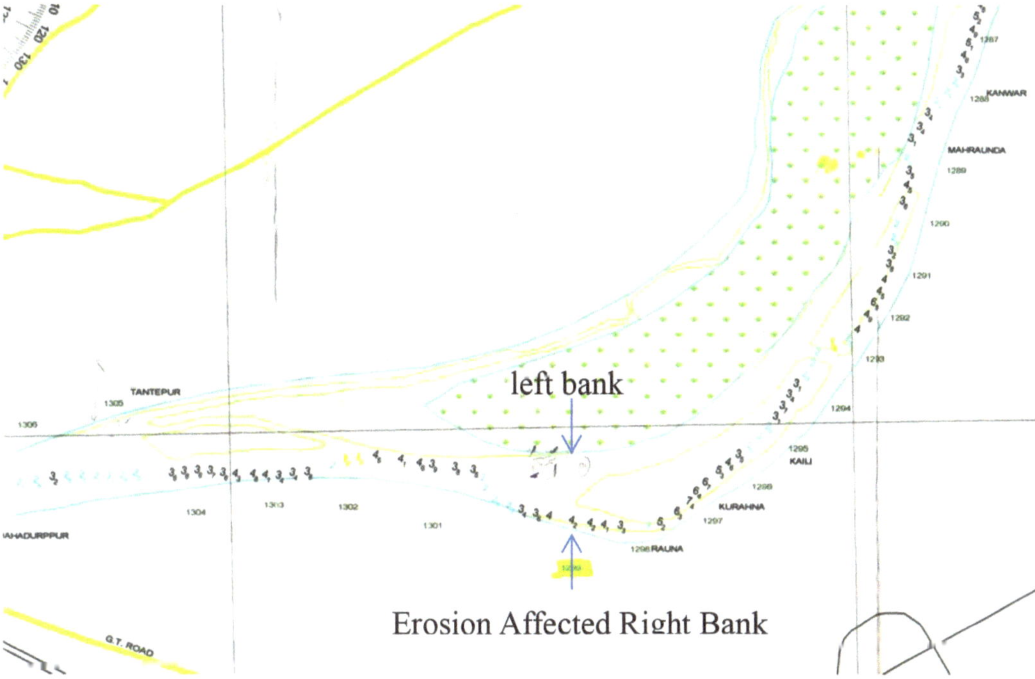

Fig. 15.5 Plan view of Nakhwa site in the Ganga River

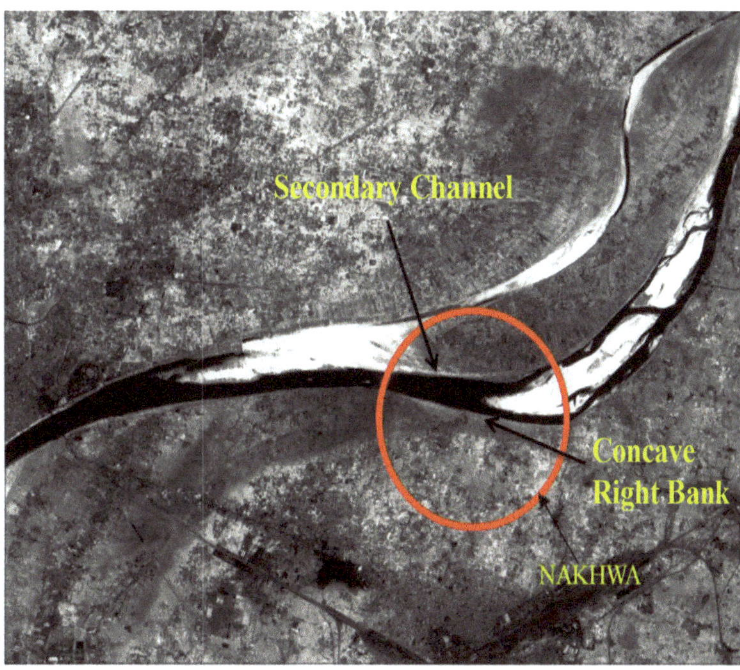

Fig. 15.6 Satellite image of the Ganga River at the Nakhwa site (Map image from IWAI 2008)

15.4.1 Study Area

The modified RCC jack jetty field was used in India for the first time in the river Ganga at Nakhwa. The test site, shown in Figs. 15.5 and 15.6, is located 11 km downstream of Varanasi (25.28°N, 82.96°E). The problem at this site was streambank erosion with bar formation causing insufficient navigation draught. The presence of the subsidiary channel on its left side diverted part of the stream flow towards the left channel decreasing the concentration of flow in the main channel providing insufficient navigation draught. The other issue at this site was a narrow channel width of 100 m along the right channel that resulted in an average water velocity as high as 1.5 m/s. Remedial action was needed to increase the depth and channel width in the right channel, which in turn would reduce the flow velocity in that channel so that conditions conducive for navigation could be developed. It was thought that these conditions could be achieved by minimizing the flow in the secondary or left channel so as to create a concentrated flow condition along the main or right channel. The flow in the wider left channel was already diminishing and the right channel had higher concentration of flow resulting from the diverted left channel flow. Thus it was thought out to be easier to address the issue by blocking the almost dead left channel and enhancing the flow in the active right channel. However, the flow is diverted towards the right channel, and then the higher flow in the right channel would result in erosion on the immediate concave right bank of the right channel. The right bank therefore needed to be provided with bank protection or other erosion control measures. The minimum and maximum water levels in 2009 and 2010 were recorded as 57.935 m and 65.310 m, respectively, with datum at 54.00 m. The minimum and maximum depths of water in 2009 were 3.9 m and 11.3 m, respectively. The suspended sediment concentration in the river was 1000 ppm.

15.4.2 Methodology

The issues at the Nakhwa site were to be tackled in two stages. In the first stage, the left channel was to be choked or partially closed by inducing sedimentation with RCC jetty screens. In the second stage, the right bank of the river was to be provided with river training measures (RCC

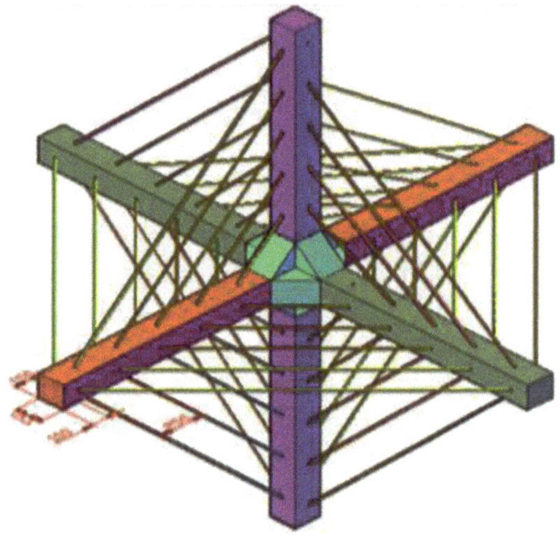

Fig. 15.7 Three dimensional picture of an RCC jack jetty used in the study site of size 3 × 0.15 × 0.15 m

jetty field) for erosion control. The jetty screen consists of continuous lines of RCC jack jetties laid together across the width of the channel (in this case, the left channel). The jack used for construction of both the RCC jetty screens and RCC jetty fields was of size 3 × 0.15 × 0.15 m (Fig. 15.7). Three RCC members of size 3 × 0.15 × 0.15 m were joined orthogonally at the centre. Holes were made every 15 cm of each member and the members were laced with iron wires through these holes. Submergence varied from 24 % to 73 % in the low and high flows, respectively. Three rows of 300 m long RCC jack jetty were positioned about 100 m downstream of the channel bifurcation. The right channel was provided with two rows of diversion lines and six retards to protect a 300 m long stretch of the river. The length of retard was kept 10 m with clear spacing between retards as 53 m providing a JFDI almost of 0.2. In the jetty screen, the lines of jetty were placed 5 m apart and the diversion line in the RCC jetty field were kept back to back. The detailed arrangement of the jetty screen and jetty field is shown in a design layout in Figs. 15.8 and 15.9.

In 2009, the jetty screens were installed in the left channel, and in 2010, a jetty field was installed in the right channel to provide bank protection.

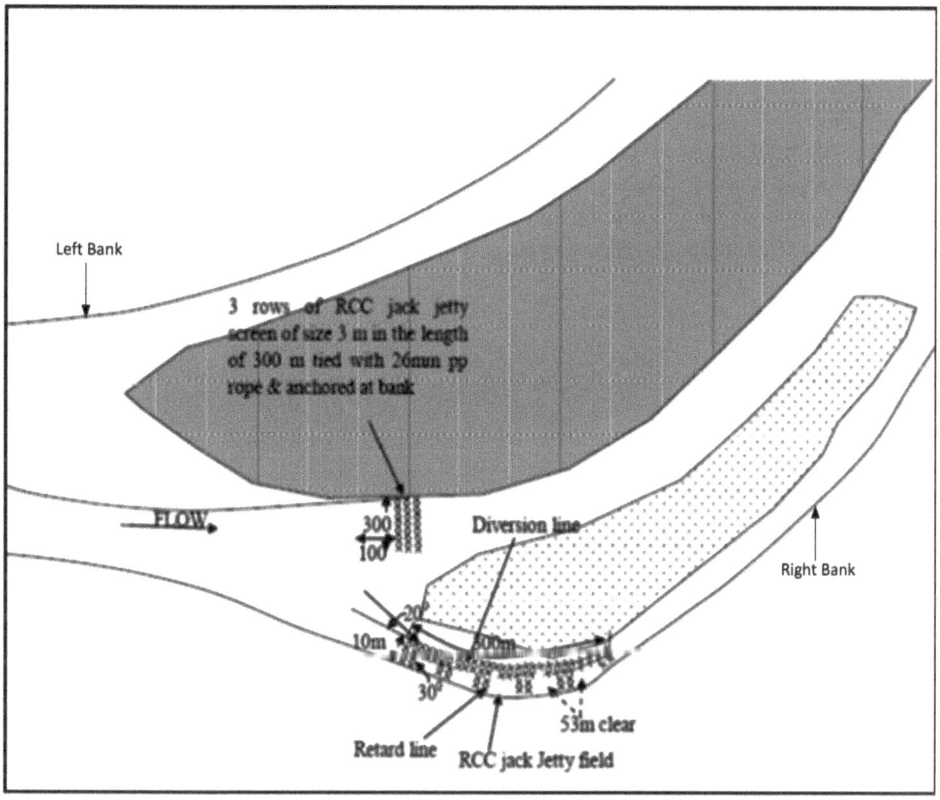

Fig. 15.8 Details of arrangement of the RCC jetty screen and jetty field at the Nakhwa site

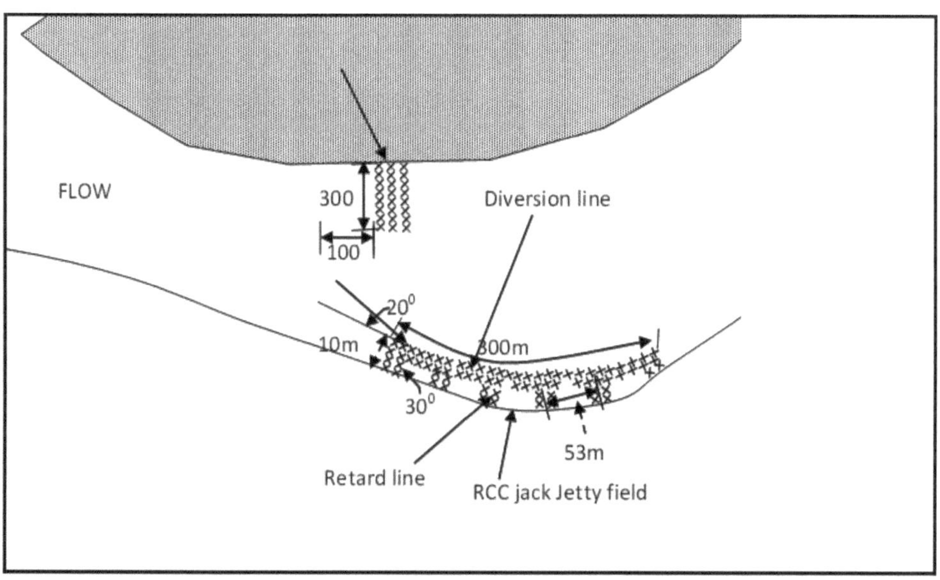

Fig. 15.9 Zoomed image of the RCC jetty screen and jetty field at the Nakhwa site

15.4.3 Results and Analysis of Field Testing

15.4.3.1 Analysis of Satellite Images

The effects of the field on the river were monitored with the help of satellite images. The morphological changes in the river were verified by comparing the pre-jack jetty installation in a PAN image of the site from 2005 (Fig. 15.10) and post-jack jetty installation in an October 2011 image (Fig. 15.11) along with GIS-based measurements. The GIS measurements were summarized in Table 15.7 which shows the width of the left channel that was earlier 246 m previous to jetty screen installation and was significantly reduced to 50 m by 2011. The width of

Fig. 15.10 Satellite image 2005 (Map image from IWAI 2008)

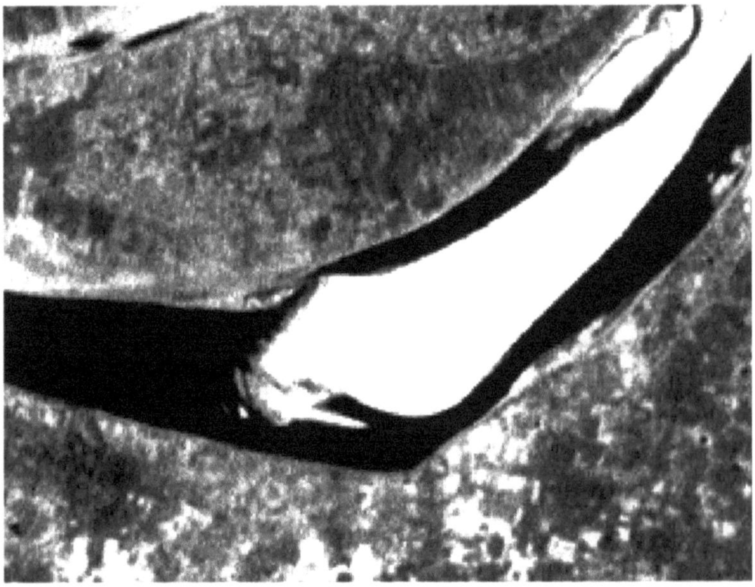

Fig. 15.11 Satellite image October 2011 (Map image from Google; Map Data: Google, Digital Maps 2011)

the right main channel widened to 341 m from 180 m. As mentioned previously, the main problem at this site was narrow channel width at the right channel resulting in very high velocity which could now be well resolved with increasing the width by almost 160 m by the installation of jetty screen and field systems making it conducive for navigation.

15.4.3.2 Analysis with Survey Data

As the second part of monitoring the field test site, a topographical survey of the study area at the Nakhwa site was conducted for pre- and post-flood season to monitor the stream bed changes post-jack jetty installation in the right channel. The main channel was delineated into 22 cross sections to quantify the bed level changes in the left channel and the main right channel post-jetty screen and field installations. After analysing the pre- and post-survey data, the stream bed changes post-jack jetty installations were calculated. The stream bed changes were calculated as the difference of pre-jetty field installation depth and post-jetty installation depth and plotted for all the 22 sections; one cross section is shown in Fig. 15.12 and details are given in Nayak (2012).

Table 15.7 GIS measurements of the width of channel

Channel location	2005 image	October 2011 image
Left channel	246 m	50 m
Right channel	180 m	341 m

The topographical survey information showed that after installation of the river training measures, sedimentation on the right bank of the right channel was arrested. The depth of the river in the middle of the channel increased that can now provide sufficient navigation draught for cargo ships. One more interesting fact to note here is that the river is more prone to erode the inner or concave bank of the river and induce sediment on the outer or convex bank of the river. The jetty field helped to induce sedimentation on the right bank. Photographs taken post-jetty installation after the flood season show a clear indication of sedimentation and channel closure (Fig. 15.13).

Analysis of the Bed Deposit Factors has been done with the field data in a similar manner as explained in the previous sections. Average and maximum Bed Deposit Factors were calculated for comparing and checking validation of the laboratory experimental results. The average BDF of the prototype study was found to be varying in between 0.1 and 0.2. Thresholds evolved with the experimental data suggested BDF in the range of 0.1–0.2 which represents moderate reclaim. The laboratory data suggested that if the sediment concentration is in the range of 1000 ppm, then jetty field layout of JFDI = 0.5 could expect moderate reclaim. For the field study, the sediment concentration of the river is in the range of 1000 ppm, and the submergence varied in between 24 % and 73 % which is equivalent to 0.24–0.73 for a Jetty Field

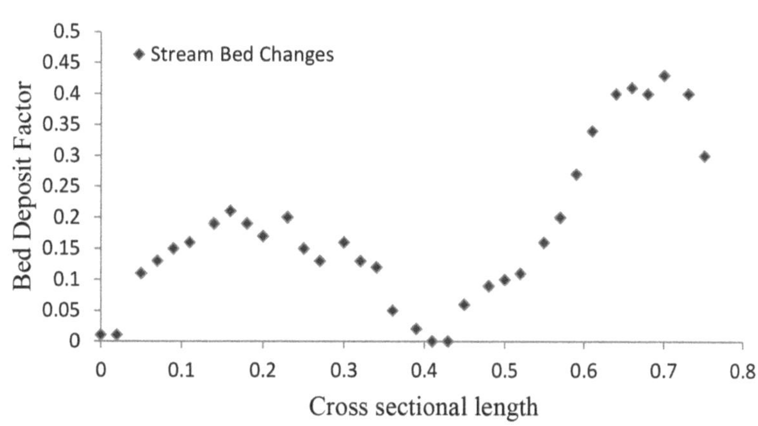

Fig. 15.12 Cross section showing the bed level changes post-jetty installation

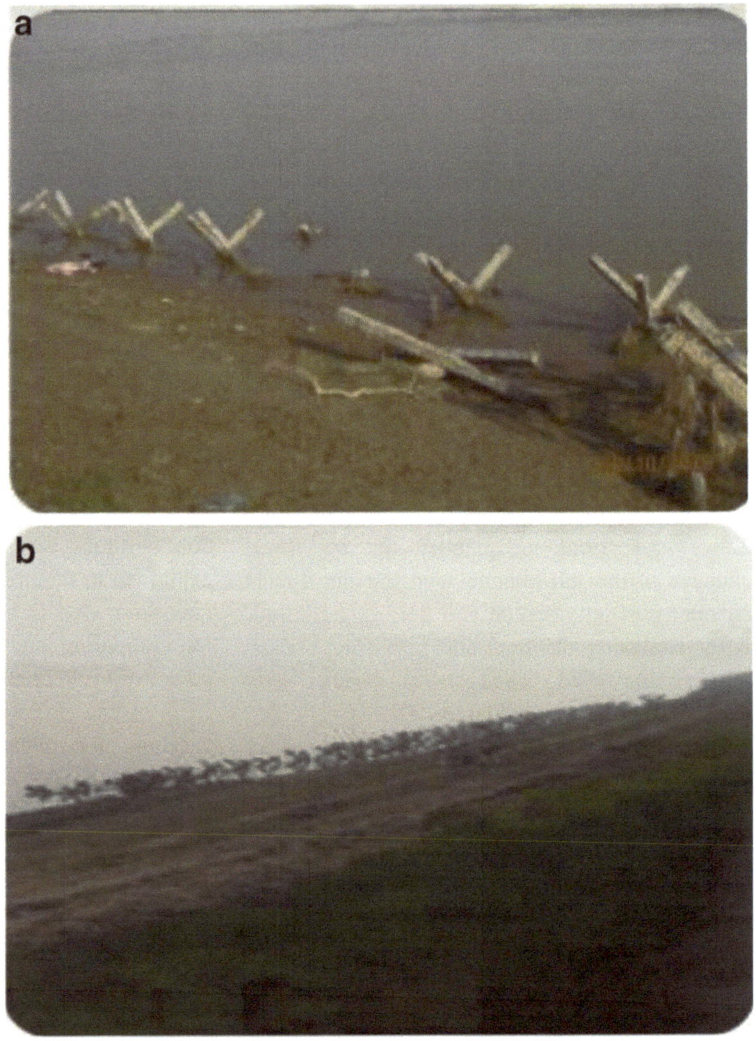

Fig. 15.13 Silting of banks with the RCC jack jetty (a) view of jacks along banks of channel (b) looking down to the channel banks

Submergence Index. In the field study, it was observed that for the lesser value of JFDI (0.2) with 1000 ppm, threshold falls in the range of 0.1–0.2.

15.5 Trail Dykes

Dykes are the structures generally placed at an angle to the flow and are mainly used to direct flow from one bend into the next bend downstream, smoothen sharp bends to a large radius of curvature providing better channel alignment, close off secondary channels and old bend ways along with concentrating flow in a restricted width within a wider channel (Petersen 1986). Based on the method and material of construction, dykes are classified as permeable (timber pile dykes) and impermeable (stone-filled/gabions) types (Mansoori 2012; Zhang and Nakagawa 2008). Besides the purposes mentioned above, such impermeable-type river training dykes are continuously providing shelter to the aquatic habitats due to the creation of reduced velocity zone between and in the vicinity of the dyke fields (dyke field pools) in many rivers like the Lower Mississippi River (Shields 1984, 1995; Shields and Milhous 1992).

Trail dykes (TD) are impermeable dykes and made up of stone gabions using manufactured

Fig. 15.14 Trail dykes made of gabions using manufactured D.T. hexagonal wire crates

DT hexagonal wire crates (Fig. 15.14). These dykes have an additional feature extending downstream from the main dyke body parallel to the channel line and work on the principle of distributing the energy of flow over a larger area. It restricts the sediment movement downstream of the dyke by reducing the generation of strong eddies there.

Trail dykes are commonly referred to as L-head dykes (ASCE Task Committee on Inland Navigation 2013b). Damage to the body can be easily and safely avoided by the addition of these structural features as these features attract most of the scour forming eddies towards it. Trail dykes can be used to enhance the spacing between dykes, to weaken the scouring eddies on the stream end of the dykes, or to increase the effects of dyke systems further downstream.

The additional section extending downstream is usually lower in elevation by 1–5 ft (Joseph 2012). The lowering of elevation to the additional dyke section of the trail dyke along with the safe provision of enhanced spacing between the successive dykes makes the dyke system more cost-effective in comparison to the other dyke systems.

Trail dykes have also beneficial impacts on habitat diversity through the creation of deep scour holes as these holes provide important shelter for fish communities during the winter low-flow season. The area behind many of the emergent dykes and especially trail dykes is providing quiet, backwater-type habitat in the open river environment of the Middle Mississippi River where backwaters are in shortages (Atwood 1995). Trail dykes were also adopted to protect the agriculture and habitat land from the channel bulging of the left bank of the Kansas River at 70 km upstream from the confluence of the Missouri and Kansas River (Blancett and Jharchow 2009). While in the case of the Lower Mississippi River near Providence, LA, trail dykes in combination with vane dykes installed as bank protection measures are performing well (ASCE Task Committee on Inland Navigation 2013a). Thus trail dykes may be treated as more cost-effective and environment friendly structures. Sedimentation around trail dykes on the Mississippi River is shown in Fig. 15.15.

15.6 Experimental Setup, Observations and Result Analysis for Trail Dykes

15.6.1 Objectives

The main objectives of the experimental study are described as follows:

Fig. 15.15 Sedimentation around trail dykes on the Mississippi River

1. To investigate the effect of different spacing, configuration and submergence ratio on the performance of trail dyke by plotting velocity contours;
2. To investigate the variation in velocity fields by the trail dyke from the non-trail dyke case;
3. To investigate the effect of different submergence ratios on the bed scour pattern in the vicinity of dykes;
4. To investigate the joint performance of trail dyke and jack jetty; and
5. To evolve a rationale design approach of trail dyke after comparing laboratory study with the field result.

15.6.2 Laboratory Flume Setup

All the experiments have been conducted out in a 0.5 m wide lab flume made of mild steel with side walls made of transparent Perspex sheet. Upstream end of the flume was associated with plastic perforated sheet acting as flow conditioner, and an adjustable tailgate at the downstream end of the flume was used to maintain flow depth in the flume. The incoming discharge of 5 L/s was controlled by using a valve gate and measured by means of V-notch weir. The areas surrounding the dykes in the flume were divided into grids of size 5 cm in the direction of flow as well as transverse to the flow. The trail dykes used in the experiments were made up of 3 mm thick Perspex sheets and indicated by T.D-A, T. D-B and T.D-C in the direction of flow. The length of each trail dyke transverse to the flow (L1) was fixed to 10 cm and varied in the direction of flow (L2) as per the requirement of experiment. The velocity at each nodal point of the grid in the plane of middle depth of dyke was measured using an Acoustic Doppler Velocimeter. For all the experiments, the initial bed surface was levelled to the horizontal. A point gauge was used to measure the bed level before and after the run of each experiment to measure the bed scour patterns.

A series of clear water experiments were conducted by changing the configuration, spacing and submergence ratios in order to investigate the effects of hydraulic parameters on the performance of the trail dyke system as per Fig. 15.16.

15.6.2.1 Experimental Setup for Spacing of Trail Dykes

Keeping the same dyke length configuration (L1:L2, 10 cm:5 cm) and submergence ratios (0.50) for each of the dykes, three experiments were conducted by varying spacing between

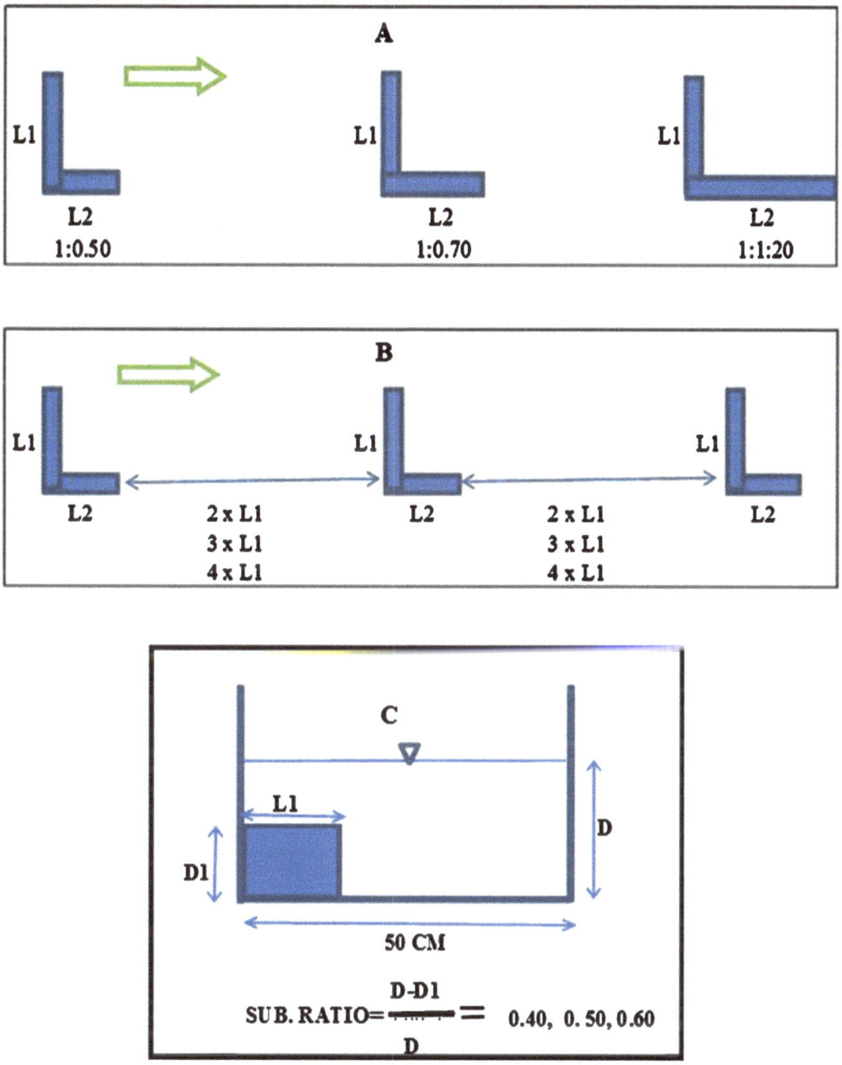

Fig. 15.16 Experimental setup conducted for configuration (**a**) spacing (**b**) and submergence ratio (**c**) in flume

dykes as two, three and four times of the effective length of dyke (L1) represented by grids 1a, 1b and 1c, respectively. Thus the spacing between dykes was 20 cm, 30 cm and 40 cm for grids 1a, 1b and 1c, respectively, and was used for comparing respective velocity contours. Grid 1d was used to compare variation of velocity for pre- and post-installation of dykes for the case whichever was found to yield the best velocity contour (Fig. 15.17).

15.6.2.2 Experimental Setup for Dyke Length Configuration of Trail Dykes

Keeping the same spacing (two times of L1) and submergence ratios (0.50) for each of the dykes, three experiments were conducted by varying dyke length configuration (L1:L2) as 1:0.5, 1:0.7 and 1:1.20 represented by grids 2a, 2b and 2c, respectively. Grid 2d was used to compare variation of velocity for pre- and post-installation

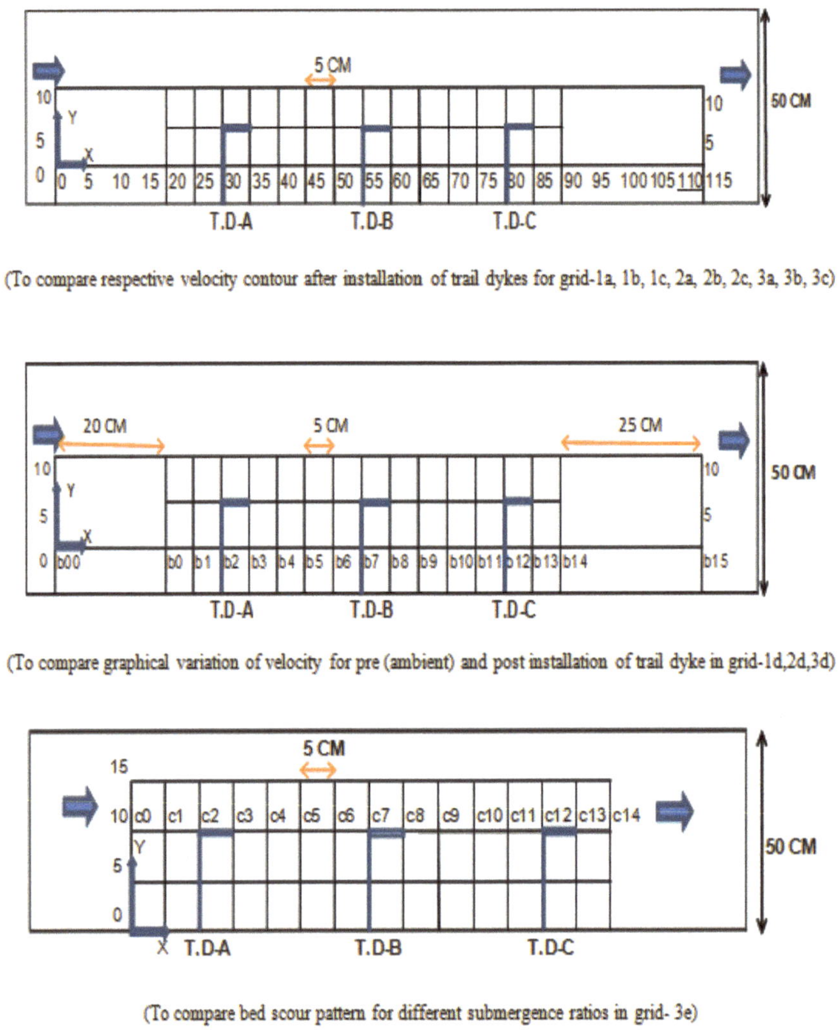

Fig. 15.17 Typical grid setup for different combinations of spacing, configuration and submergence ratio in flume

of dykes for the case whichever was found to be yielded the best velocity contour (Fig. 15.17).

15.6.2.3 Experimental Setup for Submergence Ratio of Trail Dykes

Submergence ratio is defined as the ratio of clearance depth of flow above dyke to the depth of flow in flume. Keeping the same spacing (two times of L1) and dyke length configuration (L1:L2, 10 cm:5 cm) for each of the dykes, three sets of experiments were conducted by varying submergence ratios as 0.40, 0.50 and 0.60 represented by grids 3a, 3b and 3c, respectively. Grid 3d was used to compare variation of velocity for pre- and post-installation of dykes for the case whichever was found to yield the best velocity contour (Fig. 15.17).

15.6.2.4 Experimental Setup to Access Bed Scour Pattern for Different Submergence Ratios of Trail Dykes

Bed level measurements with a point gauge were taken at each nodal point along the line L2 before the run of each experiment with submergence ratio 0.40, 0.50 and 0.60, respectively, as per the grid 3e (Fig. 15.17). It was kept with the same spacing (two times of L1) and dyke length configuration (1:0.50) for each of the dykes during all three experiments. After the installation of dykes, the experiments were carried out and again bed levels were observed after each run.

15.6.3 Observations, Results and Analysis

15.6.3.1 Effect of Spacing on the Performance of Trail Dykes

Under the same discharge condition, conceptually it may be observed from Fig. 15.18 that grid 1b with spacing three times the dyke length provides the most significant reduction of velocity throughout the grid. Grid 1c is indicating the least reduction of velocity, while in the case of grid 1a, the velocity reduces but not to the same extent as compared to grid 1b. In spite of reduced spacing between the dykes in grid 1a as compared to grid 1b, the reason for the lessened effectiveness of grid 1a may be understood in such a way that the eddies generated due to the spiral flow downstream of dyke in grid 1a are intercepting and overlapping each other due to much closer spacing so as to make the dyke unable to spread the energy of flow over a larger area resulting in lesser efficiency for the reduction of velocity. Also the increased space-to-length ratio in the case of grid 1c may lead to highly influenced turbulent eddies resulting in a lessened reduction of velocity. Graphical comparisons of the ambient velocity distribution with the velocity after installation of dykes were done for the dykes placed with spacing three times of the L1 as per the grid 1d (Fig. 15.17). It has been observed that the dykes with spacing three times of L1 shows a reliable uniform pattern of reduction of velocity throughout the grid and thus justifies the velocity contour above (Fig. 15.19).

15.6.3.2 Effect of Dyke Length Configuration on the Performance of Trail Dykes

It has been observed that under the same discharge condition, as the configuration ratio increases, the velocity reduction efficiency also increases (Fig. 15.20). So, grid 2c with series of trail dykes having configuration ratio 1:1.20 is producing better result in the reduction of velocity. Graphical comparison of the ambient velocity distribution with the velocity after installation of dykes was done for the dykes placed with dyke length configuration ratio 1:1.20 as per the grid 2d (Fig. 15.17). This comparison indicates that the experiment with configuration ratio 1:1.20 is yields consistent performance throughout the grid in reducing the velocity and thus justifies the comparisons of the velocity contours above (Fig. 15.21).

15.6.3.3 Effect of Submergence Ratio on the Performance of Trail Dykes

It has been observed from plotting the velocity contour that with the increase in submergence ratio, the effectiveness of the dyke system decreases in terms of reducing the velocity (Fig. 15.22). So, grid 3a with a series of trail dykes having submergence ratio 0.40 is indicating a better performance of the dyke systems. Graphical comparison of the ambient velocity distribution with the velocity after installation of dykes was done for the submergence ratio of 0.40 as per the grid setup 3d (Fig. 15.17). This comparison indicates that the experiment with the submergence ratio of 0.40 is yielding consistent performance throughout the grid in reducing the velocity, especially around the first dyke and thus justifies the comparison of velocity contours above (Fig. 15.23). This is generally treated as the most critical region to be protected in the field.

The sudden increase in the velocity after the installation of dykes at the points just near to the trail dyke is mainly due to the formation of strong

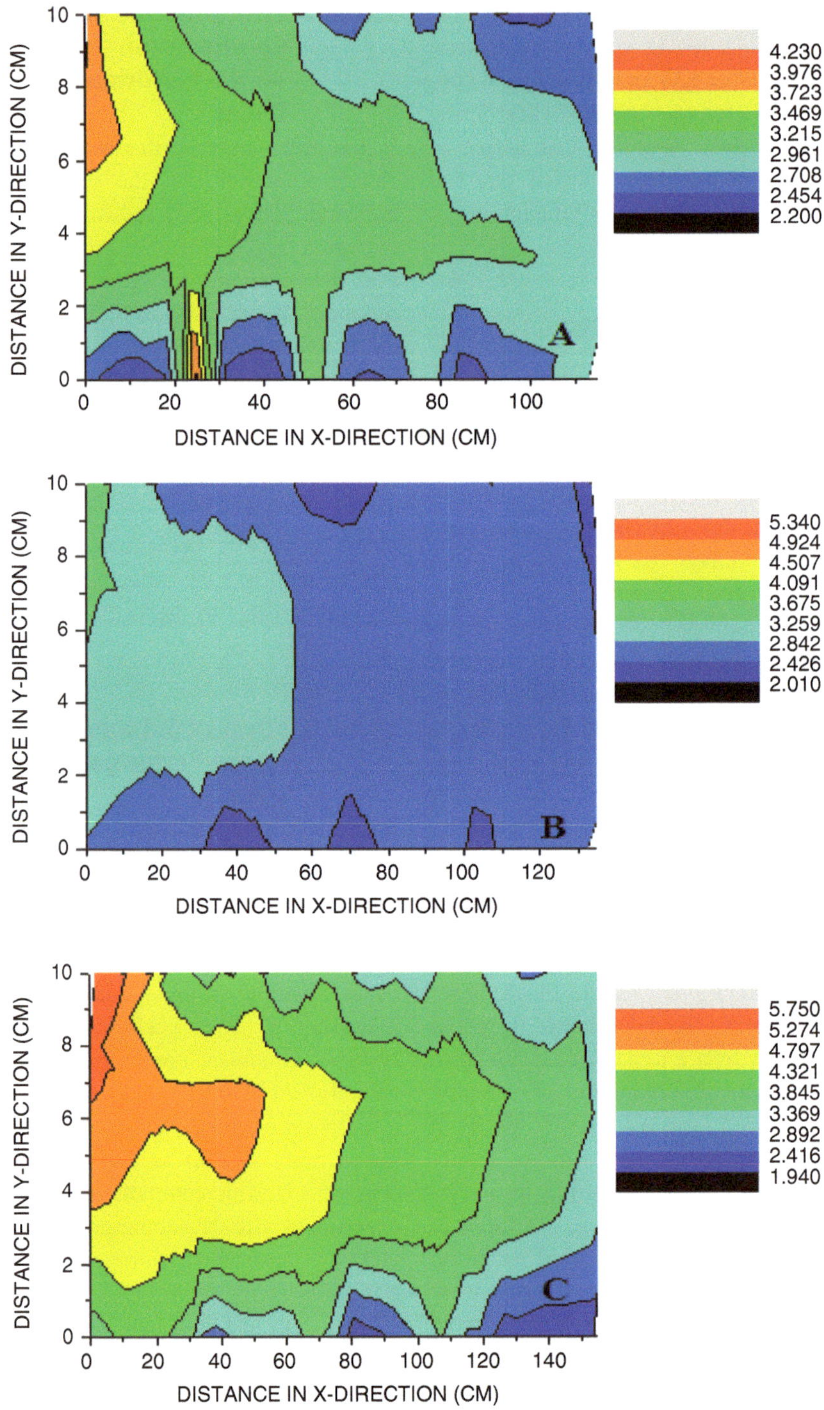

Fig. 15.18 Variation of velocity for different spacing (**a** grid 1a, **b** grid 1b, **c** grid 1c)

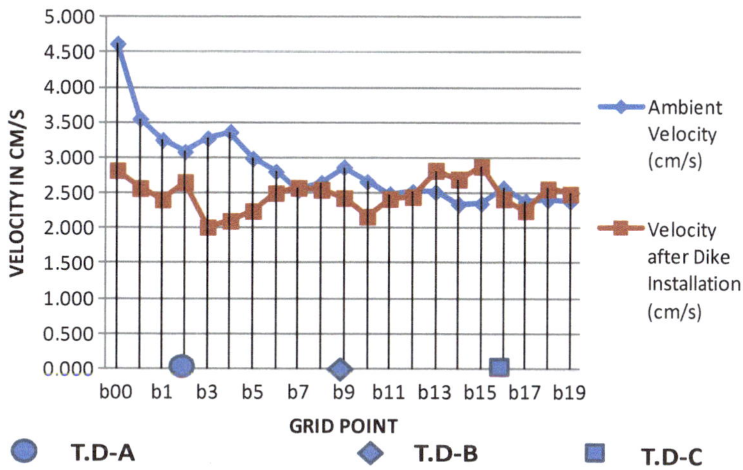

Fig. 15.19 Comparison of velocity as per grid 1d (Spacing with three times of L1)

eddies causing heavier scour mostly around the additional structural feature L2.

15.6.3.4 Effect of Submergence Ratio on the Bed Scour Pattern in the Vicinity of Trail Dykes

The experimental analysis shows that the bed deformation pattern is very complicated in the vicinity of the dykes as it depends upon many hydraulic parameters such as the dyke configuration, submergence ratios, Froude number and properties of the bed materials. However, this chapter emphasizes the effect of different submergence ratios on the bed scour pattern in the vicinity of dykes. From Fig. 15.24, which represents the variation of the difference of pre- and post-flume bed level against the respective grid points, it has been observed that with the increase in submergence ratio or decrease in the blockage to flow, the scour depth also decreases accordingly. So the experiment with submergence ratio of 0.60 shows the least scour depth in the vicinity of dykes. But as the experiment with lower submergence ratio produces the highest reduction of velocity as described in Sect. 15.6.3.3, so it is recommended to use an adequate stone apron or other protective measures to counteract the local scour in the vicinity of dykes in case the protection is to be done with dykes having lower submergence ratios.

15.6.4 Development of Design Approach for Trail Dykes

Experimental observations evolved the following design approach of the trail dyke system (Table 15.8):

In addition to the above, the following has been concluded:

- With the increase in submergence ratio or decrease in the blockage to flow, the scour depth for the trail dykes decreases accordingly; and
- Adequate stone apron or other protective measures should be provided to counteract the local scour in the vicinity of dykes with lower submergence ratios.

15.7 Effect of Combined Performance of Trail Dyke and Jack Jetty

15.7.1 Laboratory Flume Setup

After conducting the experiments for the trail dykes and result achieved, experiments with the combination of trail dykes and jack jetty were

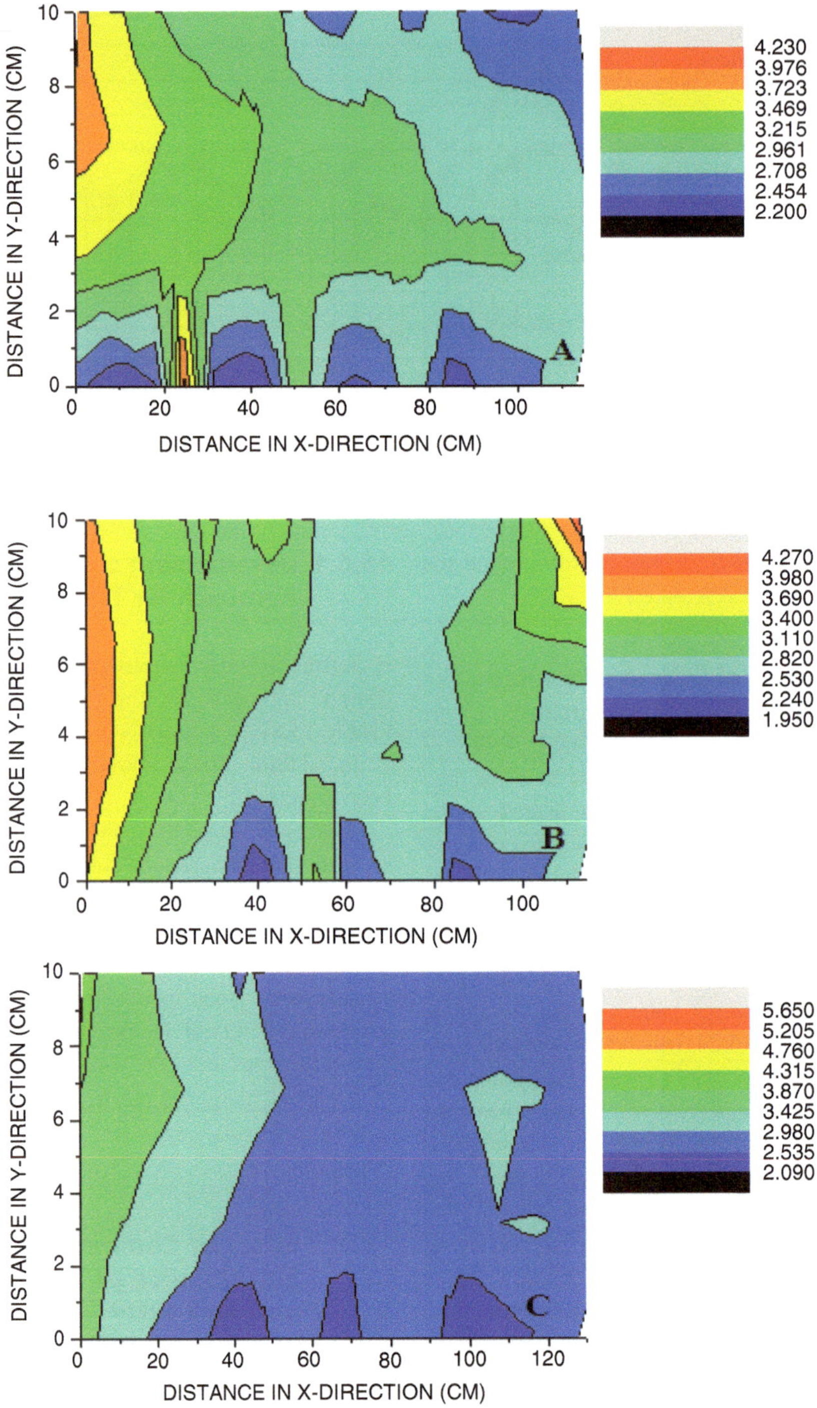

Fig. 15.20 Variation of velocity for different dyke length configuration ratio (**a** grid 2a, **b** grid 2b, **c** grid 2c)

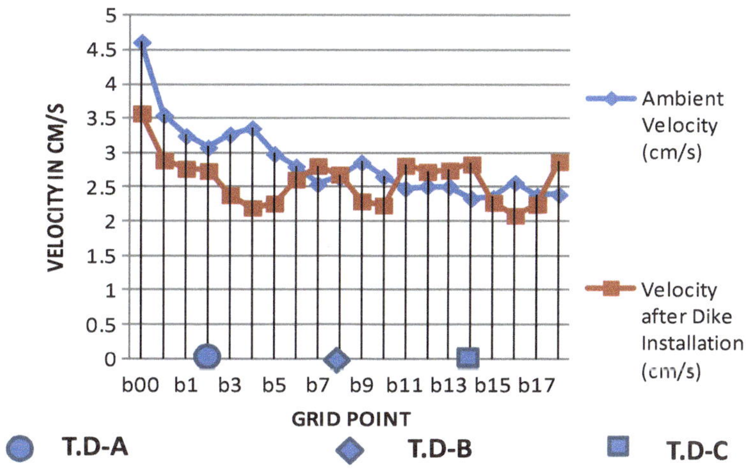

Fig. 15.21 Comparison of velocity as per grid 2d (Configuration ratio as 1:1.20)

carried out to investigate the effectiveness of structures at higher discharges. As per the result achieved in previous sections, the configuration (L1:L2) and submergence ratio for the trail dykes T.D-A, T.D-B and T.D-C were kept as 1:1.20 and 0.40, respectively. The spacing between all the dykes was taken as three times the effective length of the dyke (L1), i.e. 30 cm. One row of diversion line was laid along the line of L2 of all the dykes. One row of retard line of the jetty field was placed @1.5:1. It means that the retard line was taken after 15 cm from each end of the dyke. The first retard line u/s of the T.D-A and the last retard line d/s to T.D-C were laid out at an angle of 45 and 135, respectively, with the bank line as shown in Fig. 15.25. Clear water experiments with a discharge of 14.18 L/s have been conducted through the installed structures.

15.7.2 Results and Analysis

Graphical comparison (Fig. 15.26) of ambient velocity distribution with the velocity after installation of dykes and jetty was done as per grid 4 (Fig. 15.25). It has been observed that the significant reduction of around 22 % of velocity occurred after the installation of structures witnessing the effectiveness of the conjunctive application of trail dyke and jack jetty system.

15.8 Field Application for the Conjunctive Use of Trail Dyke and RCC Jack Jetty

15.8.1 Study Area

The Solani River, an important tributary of the River Ganga, flows in south easterly direction with a considerable discharge of around 130,000 cusecs. This river is basically having the tendency of causing significant damages to the lowland area nearby its bank by eroding the bank soil and shifting of the river course. The River catchment along with the Dhanori escape contributes runoff, which has the potential of causing flooding of approximately 9860 ha of farming land. Flooding in the river affects around 60 villages of the Haridwar and Muzaffarnagar districts in Uttarakhand state of India. Two MW solar power plants are situated near the left bank of Solani River at Bhagwanpur site (lowland area) near Roorkee in Uttarakhand state of India. The height of the channel banks varies from 3 to 4 m and the slope of the bank is 1H:1V. The Solani River also has the tendency to bulge along the left bank, which causes a threat to the power plant.

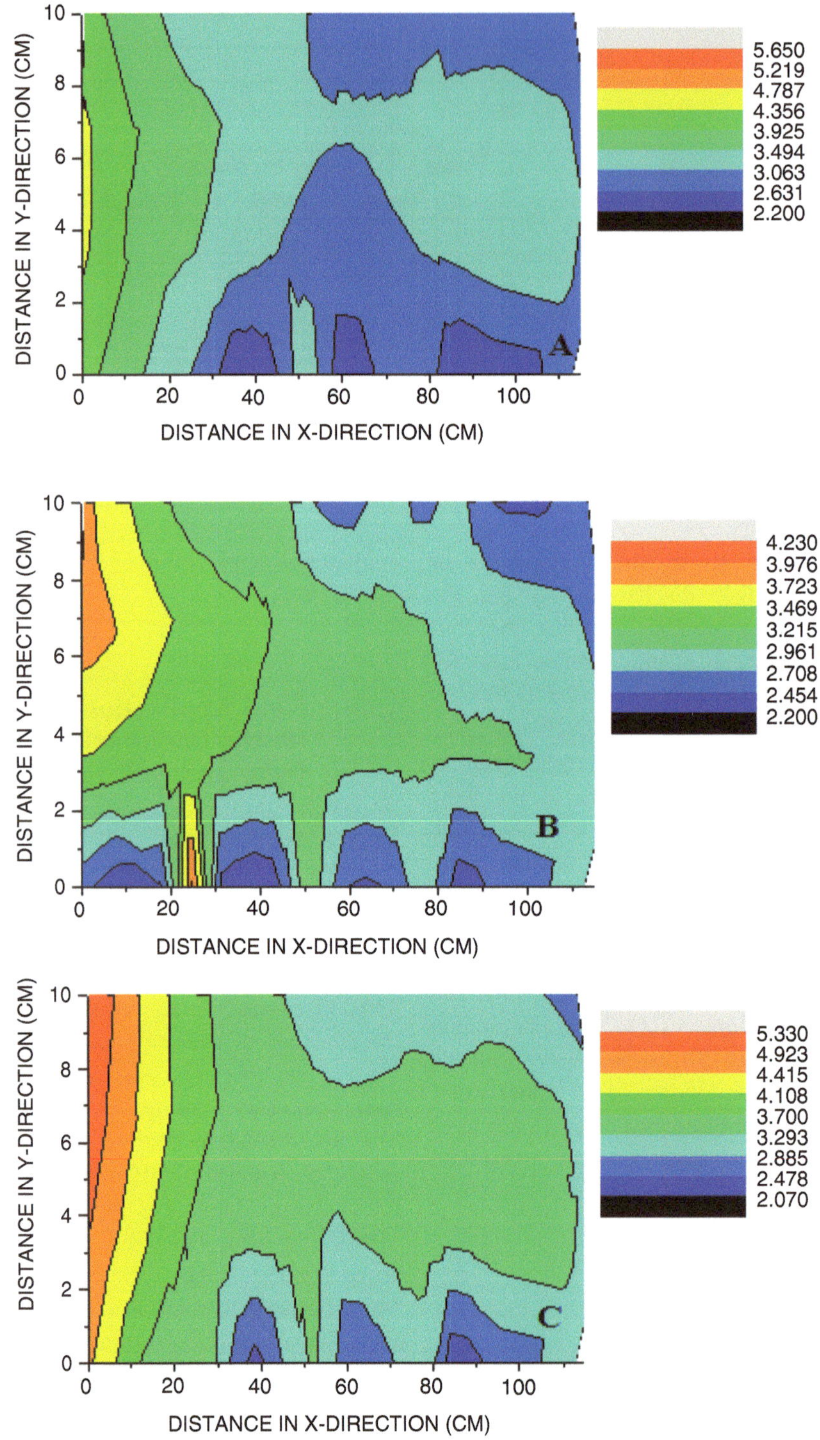

Fig. 15.22 Variation of velocity for different submergence ratios (**a** grid 3a, **b** grid 3b, **c** grid 3c)

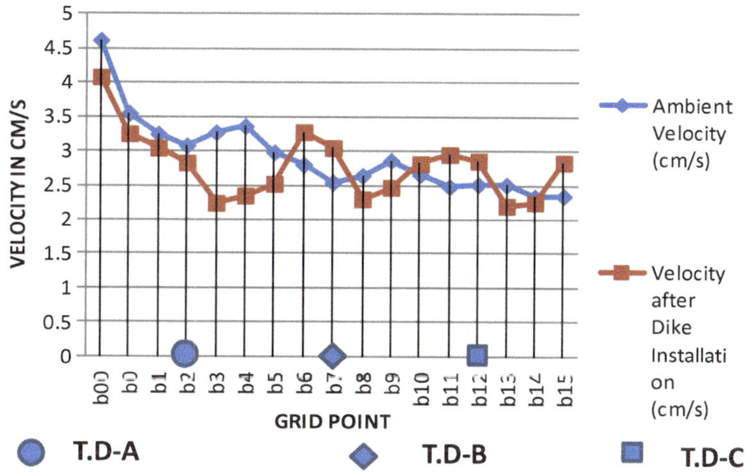

Fig. 15.23 Comparison of velocity as per grid 3d (Submergence ratio as 0.40)

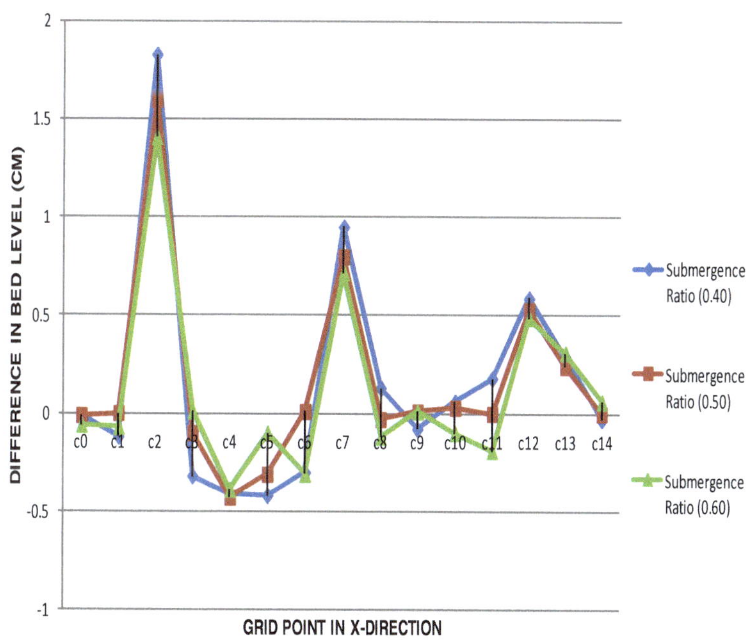

Fig. 15.24 Comparison of bed scours as per grid 3e for submergence ratio 0.40, 0.50, 0.60 (Scouring and deposition indicated by positive and negative values, respectively)

Table 15.8 Design approach of trail dyke system

Design parameter	Specification
Spacing between dykes (S)	$\leq$ Three times of the effective length of dyke (L1)
Configuration ratio (L1:L2)	$\geq$1:1.20
Submergence ratio (D-D1)/D	$\leq$0.40

15.8.2 Methodology

Before the monsoon season of the year 2012, six trail dykes (T.D-1 to T.D-6) were constructed. The implementation of trail dykes from T.D-1 to T.D-6 in the Solani River was based on reviewed literature and past experiences in the field. Keeping in view the damages that occurred to

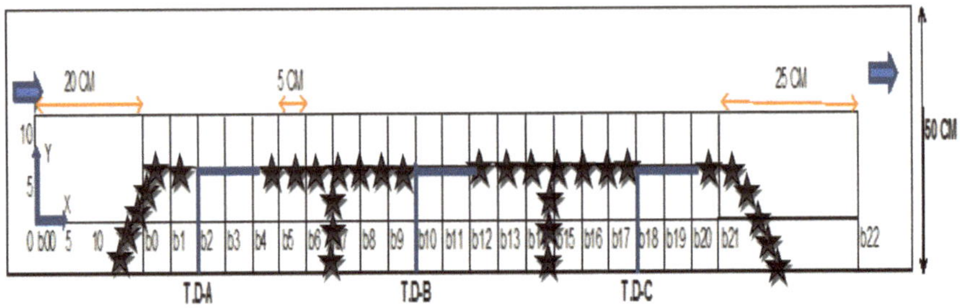

Fig. 15.25 Exp. setup for the joint performance of the trail dyke and jack jetty in flume (grid 4) (To compare pre- and post-installation velocity)

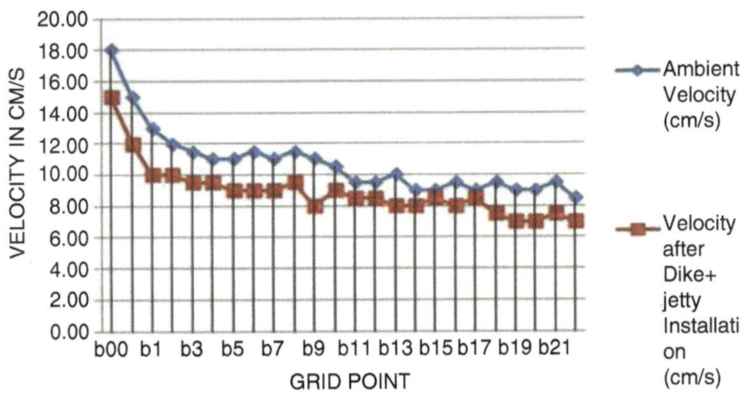

Fig. 15.26 Comparison of velocity in grid 4 (Trail dyke + jack jetty)

some portions of the bank near the area of the plant during the monsoon seasons of the years 2012 and 2013, the banks were supplemented by two additional trail dykes, one T.D-1U with L1:L2 as 15 m:7.5 m and the other T.D-1D with L1:L2 as 20 m:10 m before the monsoon season of 2014.

The length of the dyke in transverse and parallel to the direction of flow is indicated by L1 and L2. The submergence ratios of 0.40 were kept for both T.D-1U and T.D-1D. An initial high-intensity monsoon flood of the year 2014 shifted the bank about 30 m towards the plant. Due to the early occurrence of this flood, it was not possible to extend the respective lengths of L2 for T.D-1U and T.D-1D according to the laboratory findings, as their construction with boulders in deep muddy water was not possible.

A RCC jack jetty system of size 1.5 × 0.1 × 0.1 m with one row of diversion and retard line in one tier each was also installed hurriedly between the trail dyke T.D-1D and the T.D-1 especially for the purpose of restricting further shifting of river towards the plant in successive floods. However, no jetty was installed in the portion between T.D-1U and T.D-1. The purpose was to assess and compare the behaviour of trail dyke with and without the supplementation of jack jetty. The layout of bank protection works adopted in Solani River is sketched in Fig. 15.27.

15.8.3 Results and Analysis

To emphasize the effectiveness of trail dyke T.D-1U and T.D-1D and the combined performance of T.D-1D with RCC jack jetty, topographical survey was conducted before and after the monsoon season of the year 2014. From the survey result, it was found that an average deposition of around 0.75 ft up to a length of 45 m downstream of the dyke T.D-1U (trail dyke without jetty field) has occurred. Also, it has been observed that the deposition of sediment decreases as we move beyond 45 m downstream of the dyke which is basically three times of the effective length of the dyke (L1 = 15 M).

Similarly at the site of jack jetty installation between T.D-1 and T.D-1D, it has been observed

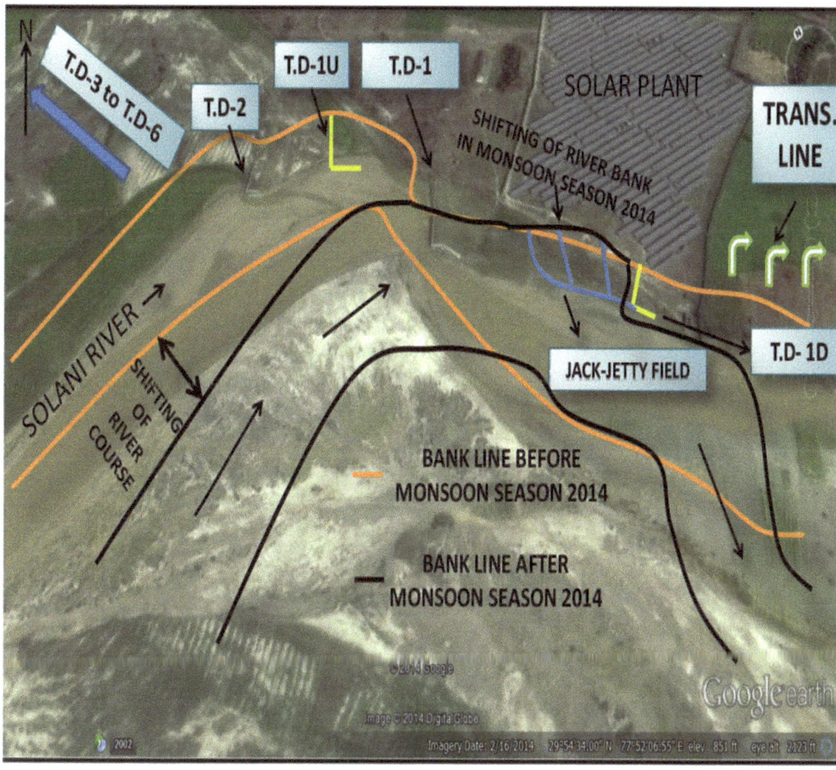

Fig. 15.27 Layout of bank protection works adopted in the Solani River

that accretion of land occurred with a significant average deposition of 1 ft mainly around the line of 10 m distance parallel to the bank. The discussed line was basically the line of diversion for jack jetty. As these jetty fields were installed hurriedly during the monsoon season, their implementation was not executed according to the design criteria available. In spite of that, the achieved result in this portion concludes that the implementations of this jack jetty with trail dyke are successfully contributing to the effective sedimentation of land.

The conjunctive field application of trail dyke T.D-1D along with RCC jack jetty system in the Solani River shifted the erosive channel currents away from the bank. In spite of the occurrence of most of the devastating flood in between T.D-1 and T.D-1D, further bank erosion was inhibited in this region. As from Figs. 15.28 and 15.29, it can be clearly observed that the repetition of high-intensity flood in the period from 31/08/2014 to 15/09/2014 could not shift the river bank further towards the plant. It was a big relief as otherwise this might be damaging the plant causing immense loss of property.

Similarly average deposition of sediments around 1.5 ft up to the length of 60 M downstream of the dyke T.D-1D has been observed. Accretion of land up to 60 M downstream of the dyke T.D-1D was measured as three times the effective length of the dyke (L1 = 20 M). It was also observed that the pre-bed levels are higher than the post-bed level as we move beyond 20 m towards the centre of the river. This has happened due to the fact that the effective length of the dyke T.D-1D was 20 m and scouring tendency of the bed generally occurs beyond this effective length due to the formation of strong turbulent eddies generated by high velocity. The installation of jetty field upstream of the trail dyke T.D-1D was also a key factor towards the significant deposition of sediment downstream to the dyke T.D-1D.

The performance of these structures also afforded protection to the transmission line of the power plant in the downstream of T.D-1D (Fig. 15.29). The protection of these transmission lines was a huge success as the supply of electricity was supposed to be stopped to the grid before the implementation of these structures.

Fig. 15.28 Implementation of RCC jack jetty to inhibit further erosion between T.D-1 and T.D-1D

Fig. 15.29 Siltation, inhibition of bank erosion between T.D-1 and T.D-1D and protection of transmission line

The significant sedimentation effect downstream to the respective trail dykes T.D-1U and T.D-1D up to three times of the effective length of the dykes (L1) justifies the result achieved for the required spacing between dykes from the laboratory study. Submergence ratios as 0.40 were kept the same for both the study. In spite of the construction of below standard configuration ratio of T.D-1U and T.D-1D as compared to the laboratory findings, the structures performed well which validates the laboratory results. For the present case of Solani River, it is recommended to extend the respective lengths of L2 for the trail dykes T.D-1U, T.D-1D and T.D-1 and construct additional trail dykes in between them to maintain the required spacing between dykes as per the design approach evolved.

15.9 Summary

This chapter dwells upon the design development of RCC jack jetty and trail dykes as cost-effective and relatively environmental friendly river training structures. Implementation of conventional river training structures like marginal embankment or levees, guide banks, groynes or spurs, cut-offs, pitching of banks, pitched islands, sills, closing dykes and longitudinal dykes is becoming unaffordable and relatively less efficient due to various factors. In the current scenario, it has become a priority for the river experts to consider not only the effectiveness of the structure but also to be deeply involved in the aspects like construction cost, environmental impact and aesthetics.

With the increase in steel prices and other construction materials, RCC jack jetty may be used as a cost-effective and workable river training structure. Behaving as permeable structures, they also produce positive impacts on the ecology. Trail dykes, acting as impermeable dykes, are emerging as more economical river training structures as compared to the other impermeable dykes like conventional spur, groynes and longitudinal dykes due to their effectiveness in working performance. Generally, the trail dykes are made submerged, so their tendency of distributing the energy of flow over a large area and restricting the generation of strong eddies in the vicinity of structure makes them effective as well as eco-friendly. Thus aquatic biota may find shelter in the vicinity of structure as aquatic habitat.

New design indices and performance parameters were evolved in this chapter which provides primary guidelines for developing design of a RCC jetty field based on the desired design objective of erosion control, moderate reclaim and heavy reclaim. Threshold values of these parameters were obtained from the analytical study of the laboratory data to categorize the desired design objectives for field condition. The field evidence and analytical results of the prototype study support the laboratory findings of RCC jetty field. This chapter also focuses on the establishment of a design approach of trail dykes based on the field application and a recent laboratory study. The design approach evolved for the trail dykes was presented in terms of design parameters such as spacing, configuration and submergence ratios.

However, there is an imperative need to undertake further studies to gain an insight into the hydraulics of complex flow mechanism involving expenditure of turbulent kinetic energy in and around jacks and trail dykes. Since the experiments for trail dykes were conducted with the clear water condition, the researchers have the future scope to repeat the experiments with the sediment-laden water for fine-tuning the results. Experiments with the lowering in elevation to the additional structural feature of the dyke (L2) towards its end may be warranted not only for the reduction of scour but also making the structure more cost-effective and environment friendly.

References

ASCE Task Committee on Inland Navigation (2013a) Chapter 5 - Dikes. Inland navigation: Channel training works. ASCE, Reston, Virgina, USA, pp 46–48

ASCE Task Committee on Inland Navigation (2013b) Chapter 10 - Environmental design. Inland navigation: Channel training works. ASCE, Reston, Virgina, USA, p 101

Atwood B (1995) Fishing the middle Mississippi, Middle Mississippi River Guide, Illinois Department of Natural Resources, USA, pp 1–25

Blancett J, Jarchow P (2009) Kansas river bank stabilization and post-project condition. ASCE, World Environment and Water Resources Congress-2009: Great Rivers, pp 3400–3410

Crawford CS, Cully AC, Leutheuser R, Sifuentes MS, White LH, Wilbur JP (1993) Middle Rio Grande ecosystem: Bosque biological management plan, US Fish and Wildlife Service, Albuquerque, New Mexico, USA, 295 pp

Grassel K (2002) Taking out the jacks: Issues of jetty jack removal in bosque and river restoration planning, Water Resource Programme, University of New Mexico, Publication No. WRP-6. Online at: https://www.unm.edu/~wrp/wrp-6.pdf

Joseph WJ (2012) Chapter 7 - River training structures and secondary channel modification, in Environmental Design Handbook, Upper Mississippi River Restoration, U.S Army Corps of Engineers, Rock Island, Illinois, USA, pp 21–22

Lagasse PF (1980) An assessment of the response of the Rio Grande to Dam construction—Cochiti to Isleta Reach. Technical report for the US Army Corps of Engineers, Albuquerque District, Albuquerque, New Mexico

Mansoori A R, Nakagawa H, Kawaike K, Zhang H, Safarzadeh A (2012) Study of the characteristics of the flow around a sequence of non-typically shaped spur dykes installed in a fluvial channel. Annuals of Disaster Prevention Research Institute, Kyoto University, 55B:453–458

Najmi YV (2001) The middle Rio Grande conservancy district and bosque management: a planning framework and guidelines for restoration projects. Professional Project, Community and Regional Planning, University of New Mexico, Albuquerque, New Mexico, USA

Nayak A (2012) Experimental study of RCC jack jetty systems for river training. Ph.D. thesis, Indian Institute of Technology, Roorkee, India

Petersen MS (1986) River engineering. Prentice-Hall, New York

Sharma N (2008) Project report on design specification for navigation channel development at six locations on the Ganga river between Varanasi & Farakka, submitted to Inland Waterways Authority of India, p 40

Shields FD (1984) Environmental guidelines for dyke fields. In: Elliot CM (ed) Proceedings of the conference rivers' 83, river meandering. ASCE, New York, pp 430–441

Shields FD (1995) Fate of lower Mississippi river habitats associated with river training Dykes. Aquat Conserv Mar Fresh Water Ecosyst 5:97–108

Shields FD, Milhous RT (1992) Sedimentation and aquatic habitat in river systems. J Hydraul Eng 118:669–687

U.S. Army Corps of Engineers (2003) Method and cost evaluation report for the Middle Rio Grande Bosque Jetty Jack Removal Evaluation Study, U.S. Army Corps of Engineers Report, Albuquerque District, Albuquerque, New Mexico, USA, 10 pp

U.S. Army Corps of Engineers (2011) Middle Rio Grande Bosque, New Mexico General Investigation Study Final Feasibility Report, U.S. Army Corps of Engineers Report, Albuquerque District, Albuquerque, New Mexico, USA

Zhang H, Nakagawa H (2008) Scour around spur dyke: recent advances and future researches. Annuals Disaster Prev Res Inst Kyoto Univ 51B:633–652

Anupama Nayak Water Resources Department, Government of Odisha, Bhubaneswar, Odisha, India

Nayan Sharma Department of Water Resources Development and Management, Indian Institute of Technology Roorkee, Roorkee, Uttarakhand, India

Kerry Anne Mazurek Department of Civil and Geological Engineering, University of Saskatchewan, Saskatoon, Canada

Alok Kumar Water Resources Department, Government of Jharkhand, Jaipur, India

Kusaha Breach Closure of Kosi River: A Case Study

16

D.P. Singh

Abstract

The eastern afflux bundh upstream of Kosi Barrage situated at Bhimnagar in Nepal near Birpur in India got breached at Kusaha (Nepal) on 18 August 2008. The Kosi floodwater gushed out to the countryside and caused large-scale devastation lower down in one district of Nepal and five districts of Bihar. A population of about 3.3 million in India (Bihar) and 100,000 in Nepal got flood inundated. They suffered extreme damages to their lives, properties and standing crops. About 500 persons lost their lives. Important infrastructures such as railways, eastern Kosi canal systems, roads, electric transmission lines, public and private tube wells, lift and surface minor irrigation schemes, public and private buildings, etc. got severely damaged. On a rough estimate, the colossal damage to properties was of the order of about Rs. 15,000 crores. Thus, it was one of the biggest tragedies due to flood fury in the recent history.

After the breach, discharge of the Kosi approaching the Barrage about 13 km downstream became almost negligible, and the entire river avulsed into a new course through various old abandoned streams (dhars) of the river.

The gigantic task of bringing the river back to the original course was taken up on war footing, and 1.73 km-wide breach was finally closed on 30 January 2009. The allied flood protection works such as spurs, bedbars, slope, apron pitching, etc. were also completed by 1 May 2009. Thus, the Kosi River was again brought back to flow along its established course through the Barrage.

The story of this nature's fury versus human grit which has been acclaimed as one of the most challenging engineering endeavours of our time has been described in this paper as a case study.

D.P. Singh (✉)
Water Resource Department, Government of Bihar,
Patna, India
e-mail: dpsingh_eng@rediffmail.com

N. Sharma (ed.), *River System Analysis and Management*, DOI 10.1007/978-981-10-1472-7_16

16.1 Introduction

Kosi River originates in Tibet at an elevation of about 5500 m from the huge Everest and Kanchenjunga glaciers. In its upstream reaches, it is made up of three tributaries, namely, the Sun Kosi, the Arun and the Tamur, which meet at Tribeni (near Barahachhetra) about 10 km upstream of Chatra in Nepal. Here, it is called Sapta Kosi. It debouches in the plains at Chatra where it is the third biggest Himalayan river next only to the Indus and Brahmaputra. The main tributaries that join the river in the plains are the Trijuga, the Bhutahi Balan, the Sugarwe, the Kamala Balan and the Bagmati. It joins the Ganga at Kursela after traversing a distance of 310 km from Chatra. The river slope in this gorge is about 2 m per km or steeper, and the average annual sediment concentration at Barahachhetra in the gorge is 0.21 % of runoff which is much more than 0.083 % for the Mahanadi at Hirakud (Orissa) and 0.112 % of this the Sutlej at Bhakra (Punjab). The river slope gradually flattens to 0.06 m per km at Kursela where it meets the Ganga from 0.2 m per km at Chatra (Nepal) where it enters the plains. Due to this topography, near Kursela the average suspended sediment concentration remains only 24 % of that of the gorge above Chatra. The remaining 76 % of the sediment load is deposited on the plains which has formed this alluvial fan of the Kosi extending over 112 km eastward and 160 km north-south. This alluvial fan is one of the biggest in the world. The Kosi carries about 107 million m^3 of suspended load per year. The bed load has been estimated to be of order of 15–20 % of the suspended load. On an average 95 % of the sediment load comes down the river during the monsoon (June–October) and only 5 % during the remaining period of the year. The gradation of sediment had been assessed to be 16 % course (0.6–0.2 mm), 28 % medium (0.2–0.075 mm), and 56 % fine (<0.075 mm).

The Kosi had been notorious for its vagrant behaviour. The devastation and destruction brought about by its floods made people call it *Sorrow of Bihar*. After deposits were formed where the river slope had become flatter for carrying sediment further downstream, the adjoining part of the cone remained lower inviting the river to shift towards the lower ground. When the flowing channel of the river in the process got silted up, the new side channels opened up and got developed. The river thus progressively shifted from one end of the fan to the other, traversing a distance of 112 km from Purnea on the east to Nirmali on the west in a period of 200 years (Fig. 16.1). The base plain on which the Kosi fan is built up has a slope from north to south and west to east.

On this plane the Kosi has built up a conical delta. The limit of this conical delta on the east and the west and the toe line along which the Kosi flows in its end reach are shown in Fig. 16.2. The river has migrated westward in the plains sticking to its original channel at Chatra where it comes out of gorge and at Kursela where it outfalls in the Ganga (Fig. 16.1). By an expert committee constituted in 1952, the flood problems of the Kosi were summarised as follows:

1. Uncontrolled floods inundating vast area of fertile land and causing heavy damage and miseries to the people
2. Heavy deposition of masses of sand on fertile land rendering the same unproductive

Though the study of the flood problem of the Kosi commenced as early as 1893, however, the project to provide immediate relief was drawn up only in 1953 which followed the usual pattern of a flood embankment project. It consisted of the eastern flood embankment, western flood embankment and a barrage at Hanumannagar (Nepal) about 45 km d/s of gorge at Chatra with afflux bundh on both sides to be tagged to high points. The embankment were constructed during 1955–1959 with some extension during 1961–1962 and the barrage during 1959–1963. The Kosi River has been very much unstable. The riverbed is progressively rising. The shifting of the course also continues within the embankments both upstream and downstream of the barrage. The river has been attacking the embankments through sudden development of braided channels. A breach under the present situation and condition of the river means jumping of the river from its present higher bed

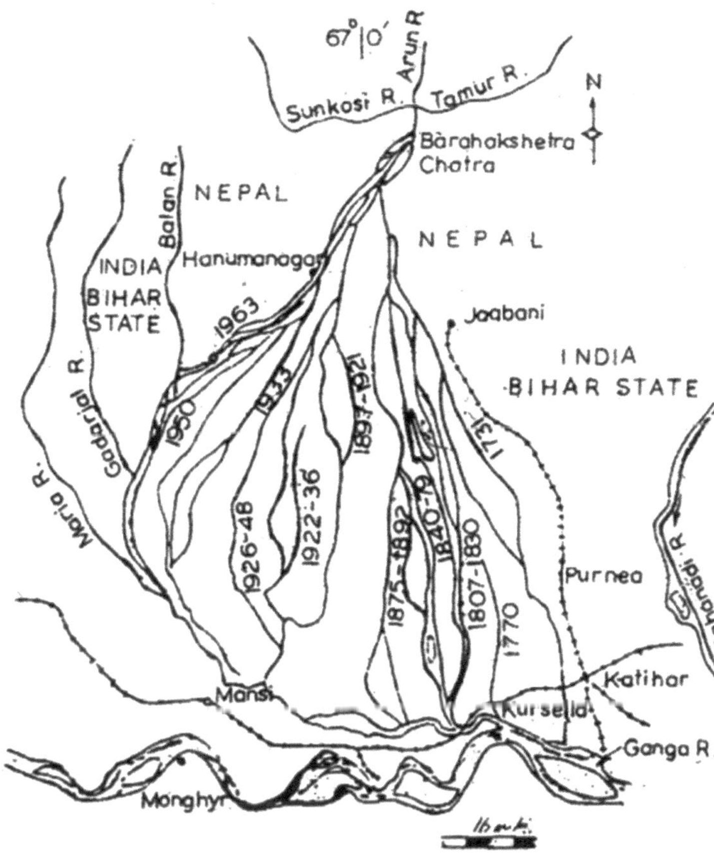

Fig. 16.1 Shifting courses of the Kosi River

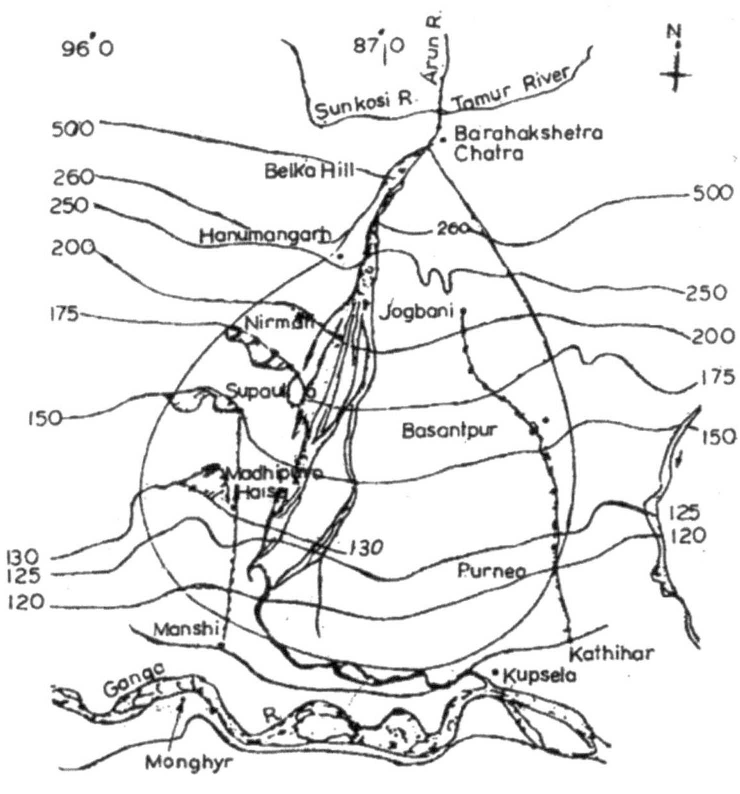

Fig. 16.2 Alluvial fan of the Kosi River

to the lower countryside causing havoc since development in its flood plains of eastern Kosi canal system has transformed it into a high agriculturally productive command area. However, since the commissioning of the Kosi project in 1963, the embankments have breached more than 18 times, Kusaha breach in 2008 being the most devastating.

16.2 Breach of Eastern Kosi Afflux Bundh at Kusaha

On 18 August 2008, the Kosi breached its eastern afflux bundh and took the shortest possible course to reach its outfall in Ganga near Kursela through some of its old abandoned dhars inundating villages and farms in its new course and displacing a large number of people in the districts of Supaul, Saharsa, Madhepura, Araria and Purnia in Bihar (India) and district of Sunsari in Nepal.

A population of about 3.4 million in India and 0.1 million in Nepal suffered extreme damages to their lives and properties. About 500 people lost their lives. Important infrastructures like railways, canals, roads, transmission lines, minor irrigation structures like lift irrigation schemes, tube wells and public and private buildings got severely damaged. In the five districts of Bihar (India), 340,000 ha of well-developed irrigation command area spread in 35 blocks and 943 villages of the eastern Kosi canals command got devastated. On a rough estimate, the colossal damage to the properties was of the order of about Rs. 15,000 crores. Thus, it was one of the worst tragedies due to flood fury in the recent history. In due course, the breach widened to 1.73 km, and the discharge of the Kosi River approaching the barrage about 13 km downstream became almost negligible. In geological terms, the entire river avulsed into a new course through various old abandoned streams (dhars) of the river. The gigantic task of bringing the river back to the original course was taken on war footing from November 2008.

16.3 Geomorphological Features of Kosi Leading to the Breach

Kosi River travels laterally over the Himalayas foreland and plains by episodic major shifts across watersheds by moving into and then out of pre-existing adjacent less actively aggrading streams. The migration of the river is unidirectional because after a channel is filled with sediments to instability, floodwater will drain preferentially into a new adjacent lower watershed or back to the next abandoned channel. Also Kosi flows through a tectonically disturbed area. It has been changing course through centuries. The fan of the Kosi Delta 180 km long and 150 km wide is one of the largest among the river fans of the world and thus is rightly called *Kosi mega-fan*. In this mega-fan, the rate of rising of catchment of Kosi is higher than the rate of undercutting of the river.

With the subsurface which is constantly unstable being a neotectonic zone, the sinking at 1 m per thousand year, the river behaves abnormally on the surface of the 'mega-fan'. Recent studies show that the 'Kosi mega-fan' is tilted westward, though the general slope of the country in the Ganga basin is eastward. The westward tilt skews the course of the Kosi and also affects the rate of undercutting by the river. Naturally, the deposit left by the river in the fan is much thicker and larger. Thus, several factors like river hydrology, sediment load, tectonics, regional slope and geomorphology control the Kosi behaviour. The breach in eastern afflux bundh of the river at Kusaha described in this paper should be examined and studied in the light of the above-mentioned adverse geomorphological conditions of the river.

16.4 Methodology Adopted for Breach Closure

The Methodology adopted for successful breach closure of Kusaha was devised by the Kosi Breach Closure Advisory Team (KBCAT) under the leadership of an eminent Water Resources Engineer, Sri N. Sanyal, Retd. Engineer-in-Chief. KBCAT

Fig. 16.3 Kosi discharge (1975–2007)

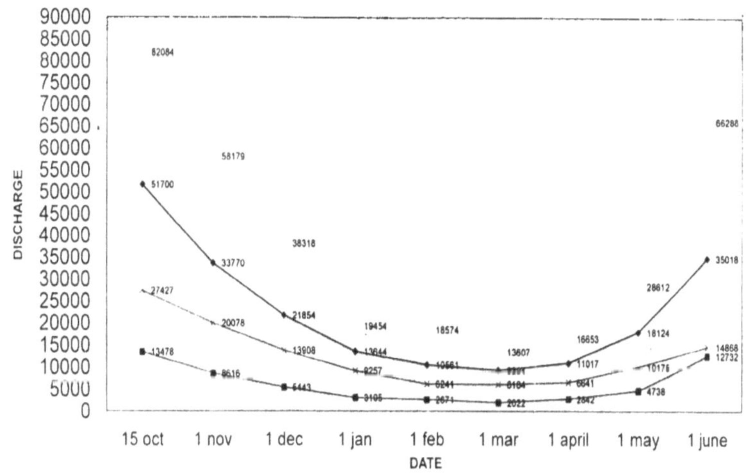

was constituted by the Govt. of Bihar on 5 September 2008 with the purpose of suggesting and devising the flood protection and allied works needed for restoration of the damaged and breached portion of the eastern Kosi afflux bundh along with corrective measures for channel improvement and smooth functioning of the Kosi Barrage to avoid such happenings in future.

The KBCAT made the first visit to the breach side on 27 September 2008 and found that the Kosi River has almost avulsed through the breach to the old abandoned *Sursar* Dhar of the river and flows through the barrage which had dwindled to almost negligible discharge. The KBCAT felt that the Kosi had a past history of changing its course and no attempt was made to redirect each time avulsed Kosi back to its abandon course resulting in the formation of a vast delta known as 'Kosi mega-fan'. But after construction of the Kosi Barrage and embankment downstream and afflux bundh upstream of the barrage, the river was jacketed in the present course and made to flow along this course for over the last 47 years. During all these years, the once flooded ravaged inland delta of the Kosi has developed into a vast irrigated agricultural command with network of canals, new township and prosperous villages. The area has rightly been rechristened as *rice bowl of North Bihar*. If Kosi were allowed to flow in the new avulsed course of post 2008 breach, the vast developed irrigated agricultural eastern Kosi canal command would suffer from flood damages year after year. In opinion of the KBCAT, the alternative to the breach closure would have been to build another Kosi Barrage project needing several thousand crores rupees, major land acquisition, rehabilitation and several years of phased construction, none of which was acceptable to the people. The KBCAT, thus, recommended the works needed for the breach closure and allied flood protection and erosion control to incorporate the following elements:

1. Redirect the river to its old established course through barrage and closing of the breach.
2. Ensure safety of the newly constructed bundh at breach site against the onslaught of river during 2009 and subsequent year flood season.
3. Revival of the silted up established Kosi channel through the barrage and making it active again.
4. Treatment of deep scour holes close to the breached bundh site with appropriate anti-erosion measures.
5. Other allied works needed to accomplish the above tasks.

The KBCAT felt that due to this breach leading to river avulsion, only a complete closure would have any reducing impact on the damage being done to the countryside. Hence, decision on the date of commencement of breach closure was very relevant. For this purpose, the KBCAT studied deeply the pattern of maximum, minimum, mean and standard deviation of discharges of the river on the specific dates: 15 October, 1 March, 1 April, 1 May and 1 June. These values are shown in the graph (Fig. 16.3). It transpired

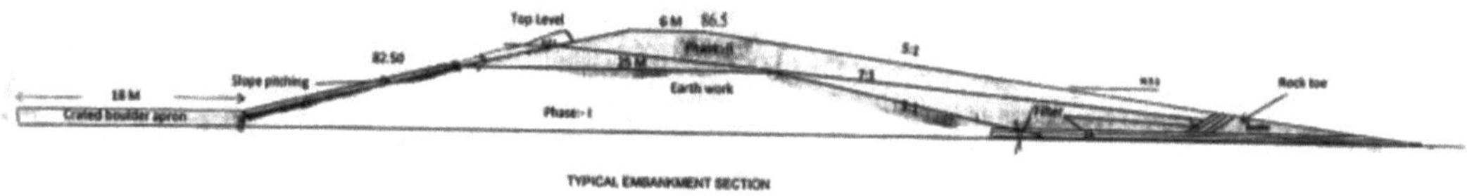

Fig. 16.4 Phase-I and phase-II section of closure bundh

from the study that from 15 October, these discharges have a trend of reduction till February. From March the trend is reversed and river discharge goes on progressively increasing up to commencement of monsoon. This was found to be on the pattern of typical glacier-fed Himalayan river. Thus, based on this observation, the KBCAT advised to keep 1 March 2009 as the most suitable date for first assault on the river for completion of the breach closure when the expected discharge in the river was expected to be the minimum. Later on the KBCAT suggested that the breach closure date may be preponed to some suitable date in January 2009 as discharge of the river did not vary much during January to April. However, co-operation of the Govt. of Nepal was required for taking off as much Kosi flow in the upstream Chatra canal as possible to make a bit easier for the final assault for breach closure and allied works of coffer dam for river diversion to the old established channel.

The KBCAT preferred preponing the closure date from March to January because breach closure of an avulsed river being a tricky operation, failure due to factors beyond control like unexpected flash flood, piping due to loose sandy formation undetected in flowing river course and human error in planning were not ruled out. It would, thus, have provided scope for second attempt for final breach closure during April 2009 if first attempt failed in January 2009.

It was found that the river just upstream of the breach had developed in a manner which might induce fresh attack on the restored closure bundh threatening the very safety of the bundh itself by inducing serious erosion under the post closure scenario. Anti-erosion works of a very dependable nature such as necessary number of repelling spurs, slope revetment and apron pitching, erection of porcupines for inducing siltation in the active channel approaching the breach along EKAB and coffer dams to divert the active channel towards west bank were planned to be constructed. Pilot channel were also planned to be excavated side by side of main breach closure work. By the end of September 2008, about 16 porcupines have already been dropped as screen to induce siltation in the eastern bank active channel.

With reduced discharge of 10,000–20,000 cusecs during January–March 2009, the formation level required for the closure bundh was estimated to be 82.50 m above MSL. It was decided to construct the closure bundh of the breach at formation level of 82.5 m with top width 25 m and side slope 1V:3H in phase I coinciding it with the completion of coffer dams upstream of the breach to be constructed with the purpose of complete diversion of lean period discharge along the western bank through pilot channel. This would facilitate the first assault for the breach closure at almost negligible river discharge through the breach. The earth, boulders, crates, sandbags, sal bullahs, etc. in sufficient quantity were stored in advance for taking up these works. It was planned to complete the phase-I works mentioned above by March 2001 so that the phase-II work of development of full section at formation level at R.L 86.50 m with top width of 6 M and R/S slope of 1 V:3H and country side slope of 1 V:5H. with appropriate apron and slope bolder pitching could be completed by May 2009. The phase-I and phase-II sections of the closure bundh is illustrated in Fig. 16.4. A battery of new spurs each 150 m long at interval of 150 m was also planned to be constructed to cover the entire length of the closure bundh. This provided additional fortification against erosion to the restored length between 11.9 and 13.6 km in addition to the entire heavily armoured eastern afflux bundh (Fig. 16.5).

The complete diversion of the avulsed river water flowing down the abandoned old dhar

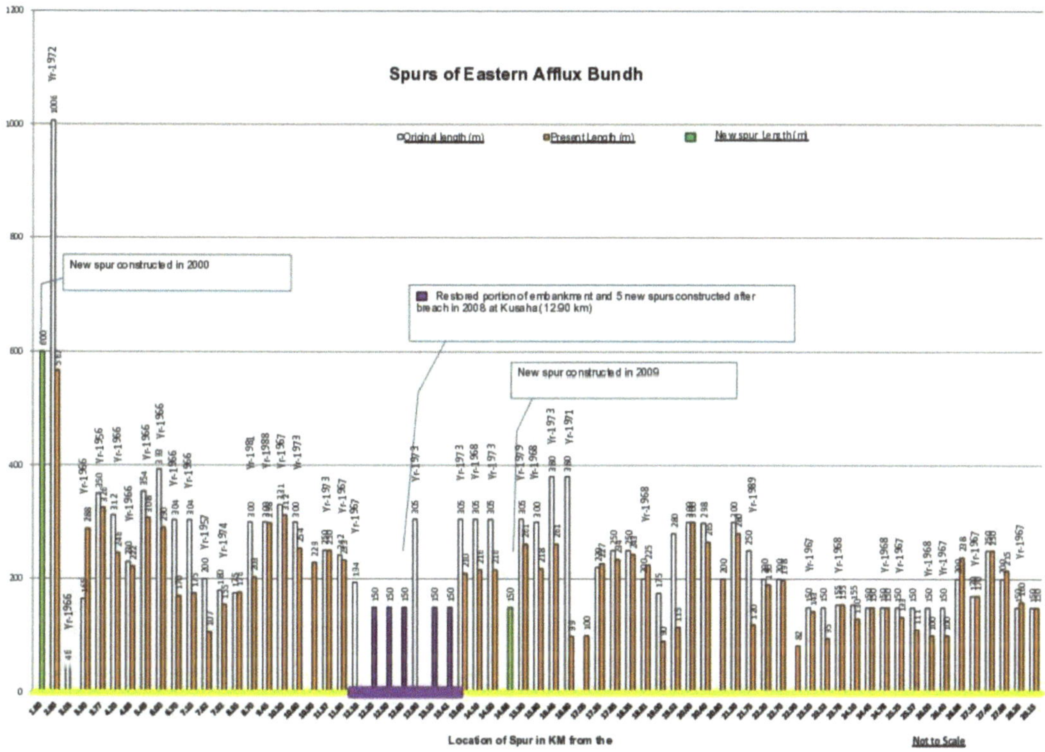

Fig. 16.5 Existing and new spurs in eastern Kosi afflux bundh

through the breach towards the established channel through the Kosi Barrage by constructing cofferdams and pilot channel and restoring the breached EKAB was a complex task. However, with the able technical guidance of KBCAT, the task was accomplished well before the target datelines. The breach was closed in a length of about 1200 m out of the total 1730 m by 5 January 2009. The phase-I work of the breach was completed by 30 January 2009. The breach was closed in full section with all allied anti-erosion and flow diversion works of phase II by 1 May 2009. The pre-breach, post-breach, progress of breach closure work and the final closure work have been illustrated in satellite imageries placed in Figs. 16.6, 16.7 and 16.8.

16.5 Remedial Measures to Check Such Breaches in the Future

A high-level expert team (HLET) constituted in Sept. 2008 and headed by the chairman of the Central Water Commission, Govt. of India, after the breach has suggested remedial measures to avoid occurrence of such breaches in Kosi embankment as described in brief hereunder.

As immediate measure, the afflux bundh and embankment should be strengthened wherever the Kosi River has come close to the embankment. Modern techniques for flood management should be adopted based on physical and mathematical model studies. The reach between 7 and 23.8 km of the afflux bundh has a big soal of about 176 km^2 in the very pond area of the Kosi Barrage. This has been declared as *Kosi Tappu Wildlife Reserve* by the Govt. of Nepal in 1976, many years after the commissioning of Kosi Barrage in 1963. The thick plantation carried out by Nepal in Tappu area had compounded the adverse effect of the shoal on straight approach of Kosi flow to the barrage apart from hindrance in round-the-clock access for the project official for flood fighting and carrying out critical anti-erosion work. The HLET has suggested removal/trimming of the

Fig. 16.6 Pre-breach and post-breach scenario of EKAB at Kusaha

shoals like Kosi Tappu. It has also suggested closure of secondary oblique channels by inducing siltation by dumping concrete porcupines. As intermediate measure, HLET has suggested stocking all requisite construction material in advance for use in the emergencies arising out during flood seasons. It has also suggested upkeep of good approach roads with proper lighting during night, procurement of surveillance boats, barge-mounted excavators, dredger and high-speed motor boats to tackle such emergent situation in future.

As long-term measure, the HLET has suggested creation of comprehensive digital database on meteorology, hydrology, river morphology, river hydraulics and sediment transport for future modelling and simulation studies and development of decision support systems. It has

Fig. 16.7 5 January 2009. The breach in a length of 1200 m was closed out of total 1700 m by the end of December 2008. A small flow in the drain of Sanjay Dhar was still flowing

Fig. 16.8 5 May 2010. The breach was finally closed and protection work was in the final stage of execution. Most of the water was diverted in pilot channel opposite the breach portion

also suggested appropriate catchment area treatment measures to effectively cut down the sediment load of the river apart from the ultimate remedy of constructing storage dams on Kosi and its tributaries in Nepal. However, till such time as these are realised, the immediate and midterm suggested measures seem the only option available to keep a check on occurrence of such breaches in the future.

16.6 Conclusion

The Kosi Project with its multipurpose benefits of flood control, irrigation and hydropower generation had developed its once flood-ravaged large delta into a vast agro-based economy. The breach of 18 August 2008 threatened to wipe out those gains. The unprecedented and gigantic works undertaken by the Govt. of Bihar after the breach in restoring the Kosi to its established course through the barrage was directed to restore the devastated eastern Kosi canal command. In the process, Kosi Barrage and its gates have been restored and modernised with SCADA system. The devastated eastern Kosi canal system is being renovated and modernised under the AIBP of Ministry of Water Resources, Govt. of India. The Govt. of Bihar has been relentless in its efforts towards attaining not just the pre-breach status with respect to the Kosi Project but a vastly improved one. In pursuance of these efforts, it has launched the Kosi Flood Recovery Project with the World Bank assistance. The project (US $ 30 million) has laid emphasis on non structural knowledge management and capacity building measures such as technical studies, modelling, flood forecasting, etc. in addition to some structural measures like road communication on embankments and training/canalization of old abandoned dhars in Kosi command.

It is now widely acknowledged that plugging of nearly 1.7 km-wide breach and allied works in a rather short time and making the Kosi River flow back to its established course through its Barrage was one of the most challenging engineering endeavours of our time.

D.P. Singh Water Resource Department, Government of Bihar, Patna, India

Part VI

Water Quality and Ecology

Preliminary Assessment and Attempt to Maintain Minimum Ecological Flows in Upper and Middle Ganga River

17

Ravindra Kumar

Abstract

The impact of releasing additional water from Tehri Dam (114 m^3/s), and subsequent releases downstream barrages at Bhimgoda, Bijnor, Ramganga feeder canal (from Kalagarh Dam on Ramganga River), and Narora Barrage (71 m^3/s) particularly during Kumbh bath festival for cultural/spiritual/ecological reasons during lean flow months (December–March) to augment river flows at Har ki Pauri (Hardwar) and at Sangam (Allahabad before confluence of Ganga with Yamuna River) at the cost of irrigation water seems to be unattractive due to negative water balance during non-monsoon period between Hardwar downstream catchment and Allahabad. The better option to raise water level at Sangam appears to be closing of lift pump canals situated between Kanpur and Raebareli (34 m^3/s) and escaping Sharda Sahayak canal water (11 m^3/s) from Bhadri escape into the Ganga River 40 km upstream of Allahabad. The other option may be construction of barrages at suitable places – Chhatnag (d/s Sangam nose, Allahabad), Kalakankar (Pratapgarh), and Bhitaura (Fatehpur) – to augment 64 m^3/s water at 75 % dependability to maintain water depth 1.2 m at Sangam. Dredging of the active channel may provide adequate depth of water for bathing and navigation.

The combined effect of low flow and discharge of polluting effluent into the Ganga has caused severe deterioration in the quality of water, sediment, and aquatic biodiversity in the river. But the Ganges' *pollution* and *low flow issues* are two entirely different things. Similarly, trade-off between food security and river ecological services has to be decided on objectives of the society. The preliminary assessment of environmental

R. Kumar (✉)
WWF-India, New Delhi, India

Advisory Group, DST-CPR, BBA (Central) University, Lucknow, Uttar Pradesh, India

State Water Resources Agency (SWaRA), Lucknow, Uttar Pradesh, India
e-mail: ravindra53@yahoo.co.in

N. Sharma (ed.), *River System Analysis and Management*, DOI 10.1007/978-981-10-1472-7_17

flows (EFs) for the Upper Ganga basin by WWF-India (2012 assessment of environmental flows for the Upper Ganga Basin, HSBC Water Programe) at Kaudiyala (rafting site 30 km u/s Rishikesh), Kachla bridge (d/s Narora), and Bithoor (u/s Kanpur Lav Kush barrage) suggests 72 %, 45 %, and 47 % MAR natural, respectively, using building block methodology (BBM). This study further needs refinement based on actual flow regimes.

17.1 Introduction

1. The impounding of water behind Tehri Dam since 2005 and cascading effect of 194 other mini to large reservoirs connected by tunnels envisaged in Uttarakhand to generate 20,768 MW hydropower, the Bhagirathi – a source river to Ganga – is anticipated to go dry in 114 km distinct stretches by 2015–2018. Similarly, the Alaknanda River which contributes about 80 % of flows to Ganga by joining at Devprayag, a barrage commissioned at Vishnuprayag, further reduces the flows in Ganga, particularly winter low flows.
2. The issues of cumulative impact of hydroelectric projects in Alaknanda-Bhagirathi basins have been widely debated and researched by institutions and civil society, to name a few, AHEC (IIT Roorkee 2011), Inter-Ministerial Group (IMG GoI 2013), and Expert Body (EB 2014) directed by the honorable Supreme Court of India and HPC (GOI 2015).
3. Figure 17.1 shows the line diagram of Upper Ganga River basin as per Central Water Commission. Here, Gomukh to Rishikesh is taken as UGS and Rishikesh to Varanasi as Middle Ganga segment (MGS).
4. About 27.3 BCM Ganges' water is already committed for irrigation through runoff-river canal systems off-taking from Hardwar (PUGC/UGC, EGC), Bijnor (MGC-I and II), Narora (LGC/PLGC), and lift pump schemes at places. Post-Tehri Dam, Uttar Pradesh, is getting 113.28 m^3/s additional water at Bhimgoda Barrage, Hardwar, for the purposes of augmenting existing Ganges Canal systems to irrigate 270,000 ha of land and drinking water needs 14 m^3/s for cities

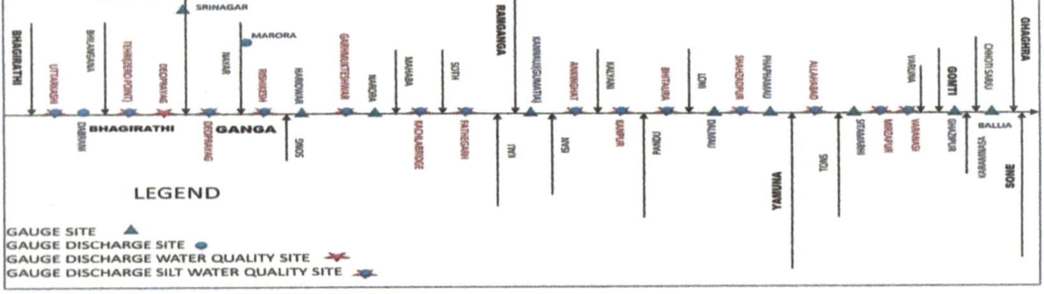

Fig. 17.1 Line diagram of Upper Ganga basin (Credit: UGBO, CWC, MOWR, India)

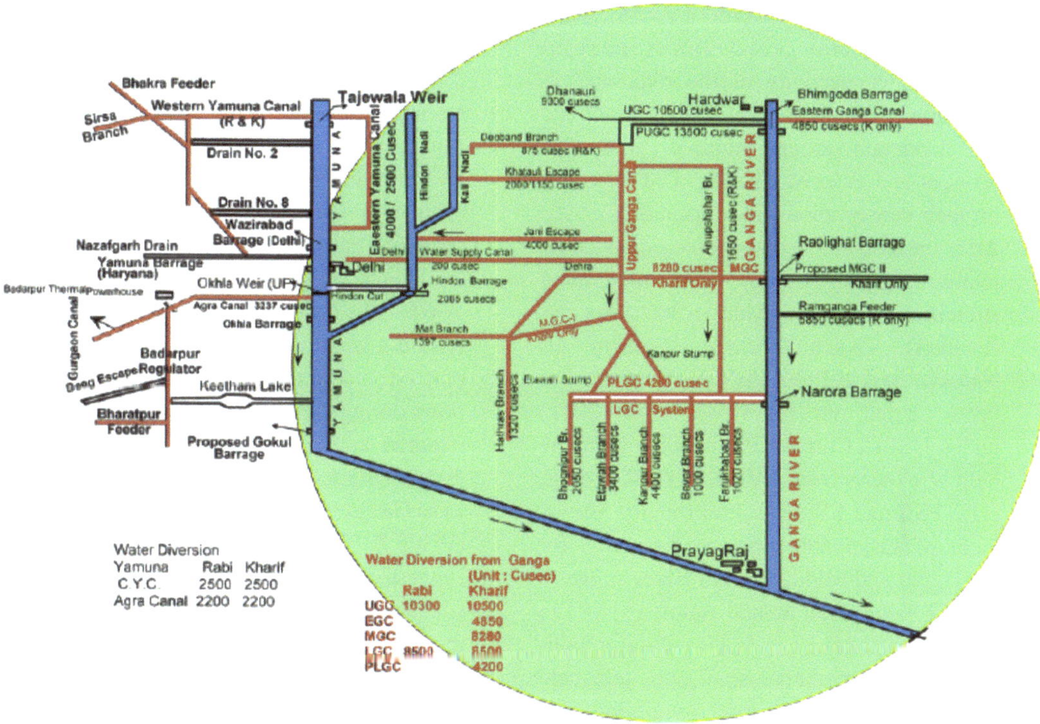

Fig. 17.2 Schematic diagram showing major ROR schemes on Ganga and Yamuna

of states New Delhi and Uttar Pradesh. The Upper Ganga Canal (UGC) which off-takes from the right bank of Ganga River from Bhimgoda Barrage has been increasingly modernized for higher capacity now for 382.32 m^3/s (2007–2008) from 297.33 m^3/s (1951–1952), 191.16 m^3/s (1938), and 185 m^3/s (1855) being its original capacity to serve agricultural, industrial, and drinking water needs in Ganga-Yamuna doab to augment Yamuna canal (by 31.15 m^3/s), Agra canal (by 31.15 m^3/s), Eastern Ganga Canal (by 28.32 m^3/s), and new Jasrana Canal (4.24 m^3/s) apart from drinking water needs of cities of Uttar Pradesh (5.67 m^3/s) and for NCR Delhi (8.49 m3/s) with provision for water losses of 4.24 m^3/s. Major runoff-river (ROR) irrigation schemes in Ganga-Yamuna doab are schematically shown in Fig. 17.2 (Plate 1.1).

5. The comptroller and auditor general of India while preparing the performance audit on "Water Pollution in India" for Central Audit Report year ending March 2010 was looking information pertaining to the following:
   - Inventory of water resources and assessment of quality of water
   - Identification and dissemination of risks of polluted water to biodiversity and human health
   - Adequacy of policy, legislation, programs, and institutions to address water pollution
   - Planning, implementation, and monitoring of programs addressing water pollution
   - Managing funds for controlling water pollution
   - Impacts of measures to control water pollution
6. The regulatory bodies like the Central Pollution Control Board (CPCB) design the river monitoring programs from a largely one-dimensional, channel-oriented perspective (i.e., to assess impacts of point sources

of pollution along the length of rivers), while ignoring the hydrological dynamics and river-floodplain connectivity of the river and its tributaries (Venkatesh et al. 2015). River and its floodplain offer multiple ecosystem services and deserve an integrated approach for their conservation and restoration.

7. An assessment of ecological flow requirements of Ganga River at critical stretches for Kaudiyala (30 km u/s Rishikesh), Kachla bridge (Badaun d/s Narora), and Bithoor (u/s of Kanpur) has been carried out by WWF-India (2012) under HSBC Climate Partnership with the help of experts from IWMI, UNESCO-IHE, IITs, PSI, and SWaRA. Preliminary assessment of sustainable flow requirements for the Upper and Middle Ganga River suggests that 72 % of mean annual runoff (MAR) in the upper and 45–47 % MAR natural in middle reaches perhaps shall be adequate enough for cultural/spiritual, livelihood, biodiversity point of view, and hydrogeological integrity of the river. These estimated flows in terms of present flow conditions post-Tehri would be still more.
8. The consortia of IITs have submitted Ganga River basin environment management plan (IITC 2015) and E-flows assessment for Ganga (IITC 2011). A three-member high-power committee (Tare 2015) has recommended E-flows during lean flow period above 60–80 % of average virgin flows and above 50 % of annual flows in the upper segment of Ganga basin as summarized in Table 17.1
9. WWF-India also made an assessment of EFs at Sangam (WWF 2013) of the order of 270–310 m^3/s. Toward a healthy Ganga, a research project (2015–2017) for cleaner waters and improved ecosystems led by IWMI is under progress, aiming to address challenges like reduced flows and the discharge of untreated toilet waste and wastewater along the Ganga and its tributaries. WWF-India has recently completed its project toward a healthy Ramganga with HSBC support. According to this finding, over 6–60 m3/s flows during the leanest flow month of May is required at Bhikia Sain to Dabri along the 600 km stretch of the Ramganga River that joins Ganga in Hardoi district of Uttar Pradesh (WWF 2015). WAPCOS on the other hand has suggested construction of three barrages at suitable site to store water within the embankment so that releases may be made at the time of requirement during Magh Mela every year and for Kumbh at 6 and 12 year cycle (WAPCOS 2011).
10. Catchment area yields and hydrology:

    Himalayan glaciers feed the Ganges. Nearly 8634 BCM of water is estimated to flow out of the Hindu Kush Himalayan region annually now said to retreat. The Ganga River has a drainage area of 1.08 million km^2 of which 0.86 million km^2 lie in India. The utilizable water resources of Ganga basin are estimated 250 BCM as surface water and 172 BCM as groundwater (CWC 2010).

    10.1. Catchment area: the catchment area of dam/barrages at Tehri, Hardwar, Bijnor, and Narora is given in Table 17.2.

Table 17.1 Summary of EFA results at seven select sites in Upper Ganga Segment (Source: Vinod Tare, IITK 2015)

	Monsoon		Non-monsoon		Lean flow period		Annual	
Location/river	A	B	A	B	A	B	A	B
Ranari and Dharasu on Bhagirathi	46	61	53	67	62	79	47	62
U/S of Devprayag on Bhagirathi	35	67	38	69	43	77	35	67
D/S of Rudraprayag on Alaknanda	40	64	46	69	48	71	42	65
D/S of Devprayag on Ganga	59	71	61	71	72	83	60	71
U/S of Rishikesh on Ganga	50	64	67	72	79	85	54	66
CWC Station Rishikesh on Ganga	53	64	71	72	83	85	56	66
D/S of Pashulok Barrage on Ganga	58	64	37	76	42	85	55	66

Monsoon, 1 June–October 20; non-monsoon, 21 October–31 May; lean period, 16 December–15 March; A, as % of average virgin flow; B, as % of 90 % dependable flow

Table 17.2 Catchment area of dam/barrages at Ganga (ID UP)

SN	River	Location	Long	Lat	Elevation, amsl	State	Catchment area in km^2 snow rain total		
1	Bhagirathi	Tehri	78.28	30.23	830	Ukh	NA	NA	7511
2	Ganga	Bhimgoda	78.10	29.57	290	Ukh	5988	17,387	23,375
3	Ganga	Rawlighat	78.02	29.22	221	U.P.	5988	22,942	28,930
4	Ganga	Narora	78.23	28.11	182	U.P.	5988	26,522	32,510

Source: Irrigation Department, Uttar Pradesh

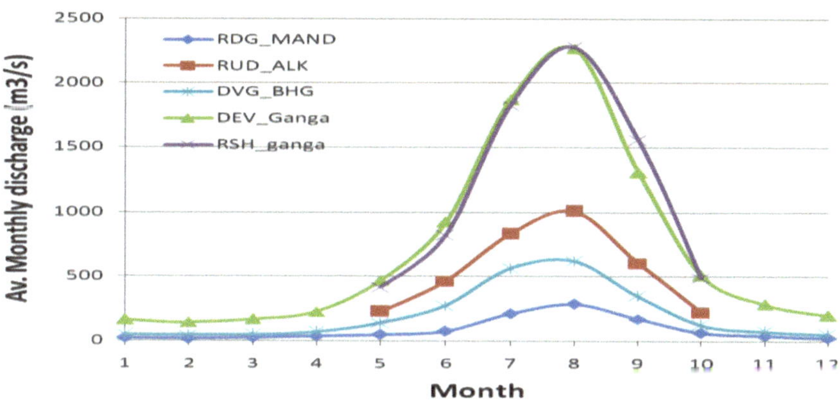

Fig. 17.3 The average monthly discharge of Mandakini, Alaknanda, Bhagirathi, and Ganga in UGS (Credit: Rajiv Sinha, IIT K)

10.2. Catchment yield: snow melt consists of 85 % of the flow at Gangotri but its contribution is reduced to 28 % at Devprayag (still in the hills of Uttarakhand). The average monthly discharge in upstream tributaries – Mandakini, Alaknanda, Bhagirathi, and Ganga at Devprayag and Rishikesh – is shown in Fig. 17.3.

10.3. Precanal and pre-Tehri Ganges natural flows

Col. P. T. Cautley (1860) measured the discharge of the Ganges River just above Hardwar (Raiwala weir), the supply of which in the month of January and December, which might be considered minimum at that point, was equal to 8000 s^3 (226.56 m^3/s) – an amount that did not differ in any great degree from that which was formerly given by Capt. Herbert about 10 years before; the lowest in the driest season of the year was measured at Hardwar 7166.1891 s^3 (202.95 m^3/s) and Garhmukteshwar 8685.2194 s^3 (245.97 m^3/s) on 1 March 1842, respectively. Maj. Abott of the Engineers Committee considered that "in abstracting 6750 s^3 (191.16 m^3/s) from the Ganges supply at Kunkul, the navigation of this river will not be injured below Kanpur."

10.4. Hydrological data analysis at Hardwar, Bijnor, and Narora barrage

Based on the 1935–1936 to 1973–1974 measured data at Bhimgoda and Raiwala, values of annual yield and time-based yield, i.e., Kharif (April to September) and Rabi (October to March) at 75 % and 50 % dependable flow as estimated by WAPCOS (1980), is reproduced in Table 17.3.

10.4.1. Flow variability in the Ganga River

The average river flows in the Ganga near Hardwar in monsoon season (July to

Table 17.3 Yield of river Ganga at different places in BCM

Sl.No	Period/time base	At Bhimgoda with dependability		At MGC Barrage with dependability		At Narora with dependability	
		75 %	50 %	75 %	50 %	75 %	50 %
1	Annual	23.17	26.70	25.48	28.67	26.72	32.01
2	Kharif (Apr to Sep)	18.98	21.13	20.02	22.45	20.93	25.70
3	Rabi (Oct to Mar)	4.19	5.57	5.46	6.22	5.79	6.31

Source: WAPCOS Study on Modernization of UGC project, Uttar Pradesh Irrigation Department

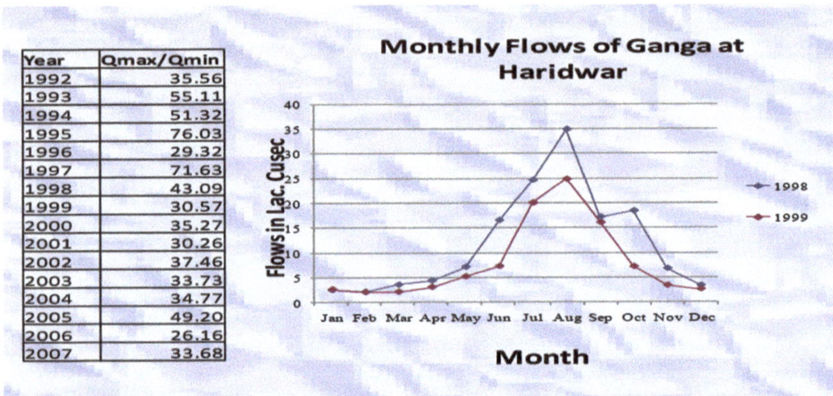

Fig. 17.4 Qmax/Qmin and monthly Ganga flows at Hardwar (Source: Author own elaboration)

September) range between 2000 and 3000 m^3/s and the lean flows are about 10 m^3/s. Average annual flow of Ganga is 23.90 BCM at Hardwar, 31.40 BCM at Narora, and 152.00 BCM at Allahabad. The discharge variability of the river systems could be accessed through the ratio between the Qmax and Qmin daily discharges to gauge the intensity of floods in the Ganga River. An idea of monthly flows in Ganga at Hardwar and Qmax/Qmin ratio pre- and post-Tehri suggests that the flood flow of the Ganga has been modified from a maximum of 76 times to lean flow in 1995 (pre-Tehri) to 26–34 times post-Tehri as shown in the table attached in Fig. 17.4. By way of comparison, it may be noted that this figure is only four for Amazon River. The Ganga River is perennial but the variation of flow is further accentuated by its diversion for various purposes, in particular, irrigation.

10.4.2. Commitments of commissioned and ongoing projects

The barrage at Bhimgoda diverts water into UGC and EGC whose authorized discharges beyond silt ejector are 297.33 m^3/s and 137.40 m^3/s, respectively. UGC is perennial canal, whereas EGC is stipulated to run from 11 June to 20 Oct. MGC barrage feeds MGC stage 1 with authorized discharge 234.46 m^3/s during

Kharif, and MGC stage II with authorized discharge 121.8 m^3/s is also proposed from left bank of the regulator of this barrage. Narora Barrage on Ganga River, LGC, with authorized discharge 240.69 m^3/s and PLGC with authorized discharge 118.93 m^3/s take off. The former canal runs both in Kharif and Rabi, while the later runs only in Kharif. LGC is also augmented from Kalagarh Dam on Ramganga. Ramganga feeder canal off-takes from Kho Barrage at Sherkot and outfalls in Ganga at Garhmukteshwar. The combined discharge capacity of all these six canals is about 1260 m^3/s (Ravindra 2011).

10.4.3. Establishing of a virgin flow series

The discharge observed at Bhimgoda Barrage in 1935–1936 to 1973–1974 is correlated with the Ganga site at Raiwala about 8 km u/s of the barrage with no major stream joining Ganga River in its stretch between Raiwala and Bhimgoda. The ten-daily utilizations of laser canal irrigation upstream of UGC and EGC have been used to get virgin flow series at Rawlighat after accounting for 10 % u/s utilization for regeneration between Raiwala and Rawlighat. Likewise, virgin flow series of Ganga River at Narora has been developed. The deference in virgin discharges at Rawlighat and Narora is attributed due to intermediate catchment contribution. After accounting for the upstream usages and the downstream commitment of LGC and PLGC and ecological need of Ganga River, water availability at Rawlighat for proposed MGC stage II has been worked out for each ten daily. The developed series of water availability in Ganga at Rawlighat for MGC stage II under construction has been arrived at. To this series the ten-daily total crop requirement is superimposed, and on its basis the failure or success years have been arrived. The project meets 75 % success criteria as it is only in 7 years out of 31 years that water requirement has fallen short in any ten-daily block. It is also emphasized here that during III ten daily of June 1989, the discharge is about 92 % of the requirement. Thus ten daily can be safely treated as success year and including it, the overall success years shall be 80 % (IDUP-2004).

11. Preliminary assessment of sustainable flow requirements for the Middle Ganga River

11.1. Water balance study at Narora: post-Tehri scenario

Water availability on 12 June 2010 at Narora u/s was 101.95 m^3/s, out of which 70.80 m^3/s released for Sangam Snan, leaves 31.15 m^3/s against LGC designed capacity at head 240.69 m^3/s. These releases from Narora including 56.64 m^3/s from Hardwar may provide temporary relief at the cost of irrigation water (about 0.16 million ha would remain un-irrigated of 12 districts in Rabi) but is an effort to solve greater problem of hydraulic integrity of the river.

Table 17.4 Water losses in Ganga River between Hardwar to Narora lean flow months

	2006			2007			2008		
Month/year	Rel.	Recd.	Losses %	Rel.	Recd.	Losses %	Rel.	Recd.	Losses %
Dec	3055	4551	−49	4949	3815	23	7212	4644	36
Jan	4388	5079	−16	5989	5393	10	9451	6828	28
Feb	4276	3299	23	5899	3551	40	10,011	4742	53
Mar	4558	4391	4	6680	5684	15	5356	4980	7
Apr	2433	2311	5	4096	3946	4	5099	4734	7
May	8139	4966	39	6254	4569	27	7611	5207	32
June	10842	7961	27	15522	8479	45	21,252	21,383	−1

	2009			2010		
Month/year	Rel.	Recd.	Losses %	Rel.	Recd.	Losses %
Dec	8764	7212	18	3898	3147	19
Jan	10,024	6996	30	5289	5238	1
Feb	7437	4742	36	2552	3066	−20
Mar	3723	3775	−1	3203	3911	−22
Apr	2895	2881	0	2736	2664	3
May	4666	4341	7	3101	2423	22
Jun	6669	4353	35	6939	3147	55

Rel sum of releases made from Bhimgoda Barrage and Kalagarh Dam through Ramganga Feeder, *Recd* water reached up stream of Narora Barrage (Source: ID U.P.)

11.2. Water losses in Ganga River course: Hardwar to Narora

Water appears as regenerated flows in river downstream (the undercurrent which percolates the gravelly bed, together with the drainage of the intermediate country), sometimes positive and in some stretches negative balance (as a consequence reduced flow is observed). Water released from Bhimgoda Barrage and Ramganga feeder in Ganga River and actual water received at Narora Barrage during lean flow months, the difference in terms of gain/loss in percent is given in Table 17.4. These losses are due to evaporation in river water and groundwater recharge by the river. The excessive groundwater pumping for agriculture in the catchment area Hardwar to Narora hardly allows to appear regeneration flows in the river during pre-monsoon period, whereas the catchment contribution of Tehri to Hardwar has been found positive throughout the year (Ravindra 2011).

11.3. Releases of water from Narora for Kumbh bath at Allahabad

70.80 m^3/s water from Narora is released since 16 December till January and 42.48 m^3/s per day in general. Impact in rise of water levels at various gauge sites of Ganga downstream of Narora due to releases made for Allahabad Kumbh bath has been studied for Narora downstream of Kachla bridge, Fatehgarh, Kannauj, Ankinghat, Bhitaura, Dalmau, Shahzadpur, Phaphamau, Sangam, and Chhatnag and shown in Fig. 17.4 for Dec 2009 to May 2010.

It is seen from Fig. 17.5 that immediate rise in water level is visible up to Fatehgarh and ripple effect after due travel time is seen further downstream, and declining trend is observed with passage of time during lean period.

It is clear from Fig. 17.6 that a release of 0.64 m water column downstream of Narora generates effective rise in water level along the river in general, 0.48 m at Kachla bridge,

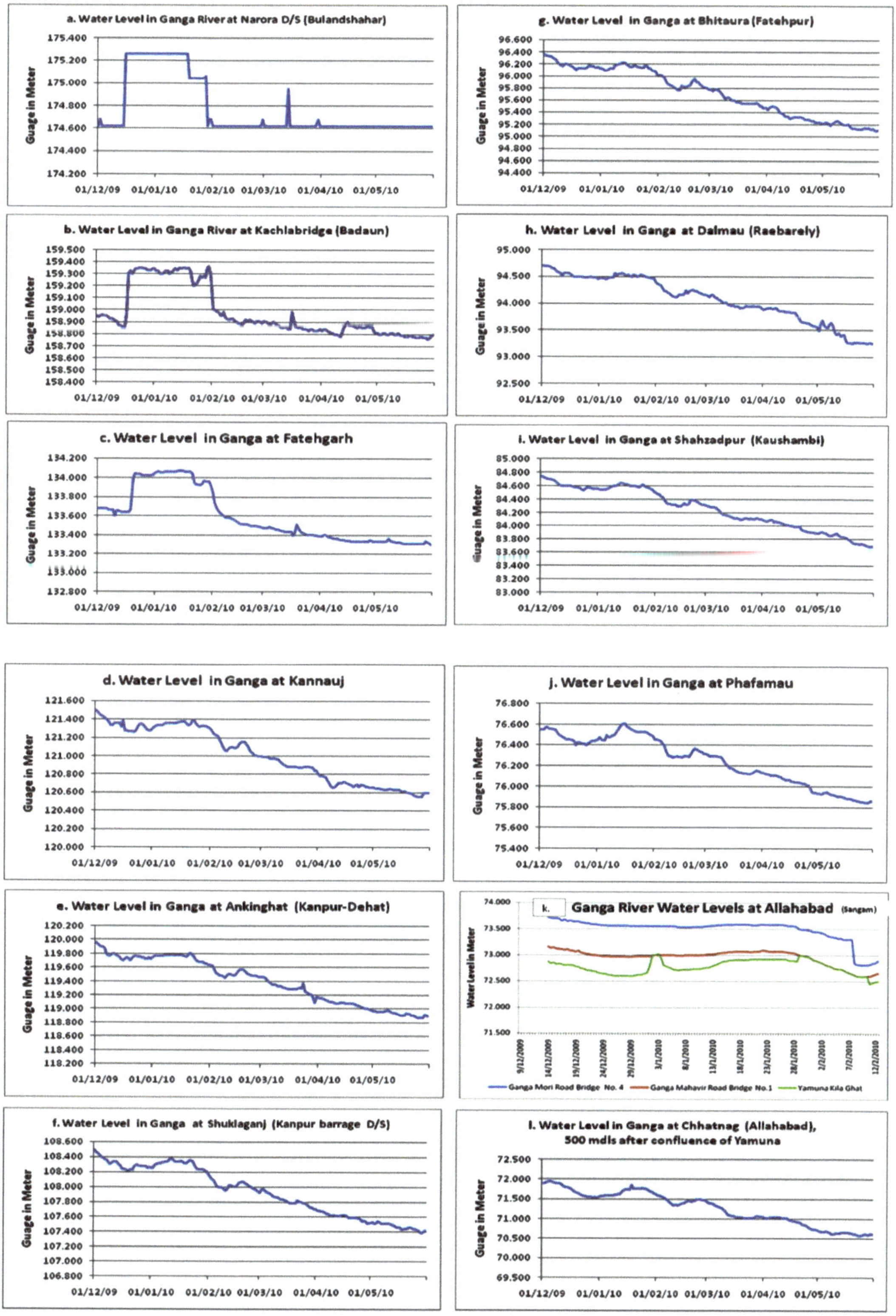

Fig. 17.5 Rise in water level downstream releases from Narora Barrage at various sites

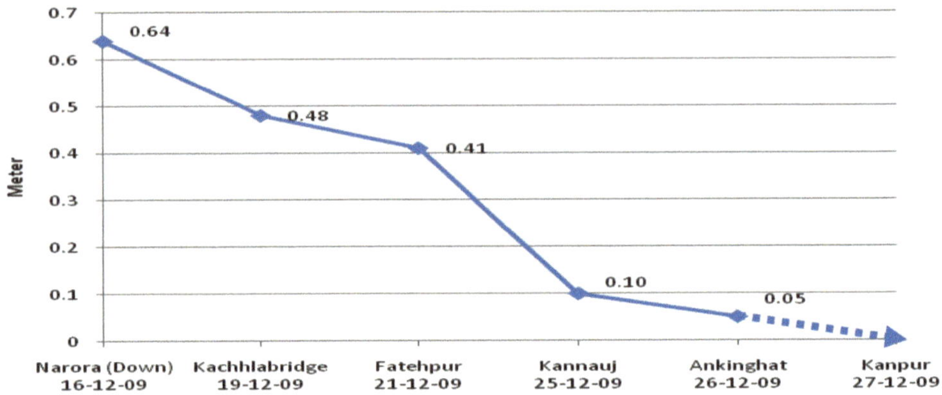

Fig. 17.6 Water depth at various monitoring sites as a result of releases downstream of Narora barrage

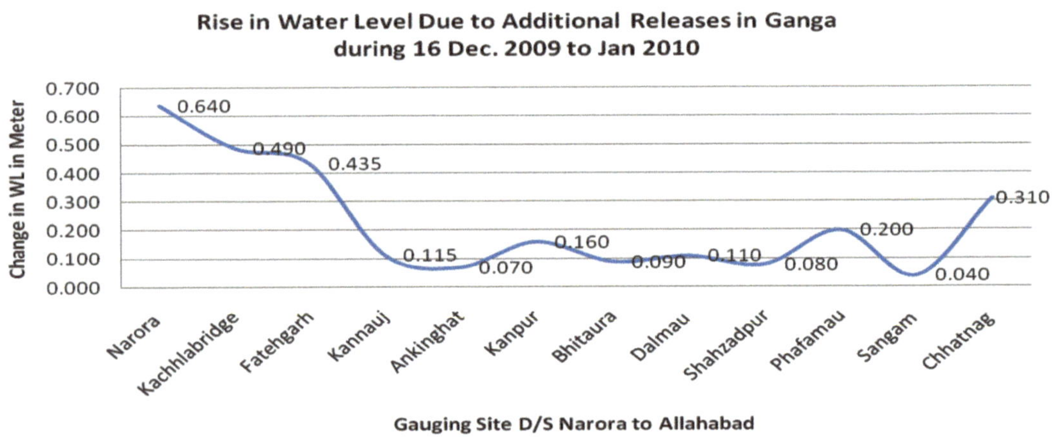

Fig. 17.7 Rise in water level due to additional releases in Ganga during 16–31 Dec 2009 (Author own elaboration)

0.41 m at Fatehgarh but of lower magnitude, simply 0.10 m at Kannauj, and 0.05 m at Ankinghat and with similar trend at Kanpur.

The presence of a barrage at Kanpur modifies rise in water level 0.16 m downstream at Shuklaganj CWC gauge site (Kanpur) that further declines to 0.09 m at Bhitaura, 0.08 at Shahzadpur with exception of 0.11 m at Dalmau (the site of lift pump canal when closed). The highest rise in water level 0.20 m at Phaphamau is resultant of closing lift pump canal Dalmau A and B and Sharda Shahayak canal water releases made from Bhadri escape into Ganga River upstream of Phaphamau. The high rise in water level at Chhatnag is due to confluence of Yamuna River with Ganga River downstream of Sangam. Monitoring of water levels at various gauge sites corresponding to releases made from Narora reveals only a small rise in water level (0.10–0.05 m) occurring due to water losses in the catchment along the course of the river downstream of Narora to Allahabad Fig. 17.7 (please see Table 17.4).

11.4. Historical discharge characteristic in Ganga at Shahzadpur, 60 km u/.

All the canal abstractions from Ganga River between Kanpur and Allahabad are situated upstream of

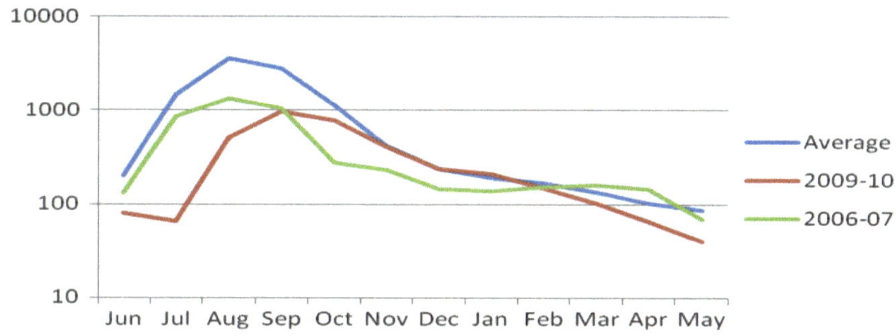

Fig. 17.8 Average flow vs. year 2006 and 2009 flows in Ganga at Shahzadpur River (Author own elaboration)

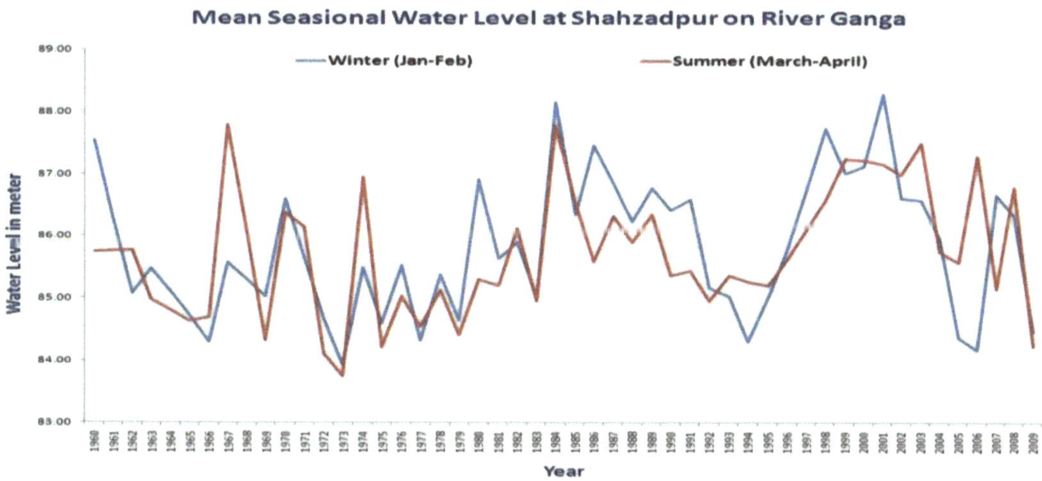

Fig. 17.9 Long-term average water level trend during winter (Jan–Feb) and summer months (Mar–Apl) at Shahzadpur, Ganga (Author own elaboration)

Shahzadpur, 60 km upstream of Allahabad, which is G-D site of CWC. Fifty-year gauge-discharge data are analyzed for (1960–2010) maximum, minimum, and average monthly flows. The monthly flows for the year 2006–2007 and 2009–2010 as compared to long-term monthly average are shown in Fig. 17.8. It is clearly visible that 2006–2007 flows are far below the average flow but after February due to occurrence of winter rainfall in upstream catchment, the river discharge is more than average during the summer months March–April and May. The current year flows 2009–2010 have been equal to average figures during lean flow months (Nov. to Feb.) but low discharges for summer months.

Therefore, in the year 2006–2007, it appears genuine to file a public interest litigation (PIL) in the High Court of Allahabad asking for more releases of water in Ganga from reservoirs but the similar demand during the year 2009–2010 does not appear to have a technical ground except cultural/religious ground. The 50-year average water level at Shahzadpur during winter (Jan–Feb) month and summer month (Mar–Apr) trend does not show a declining trend (please see Fig. 17.9).

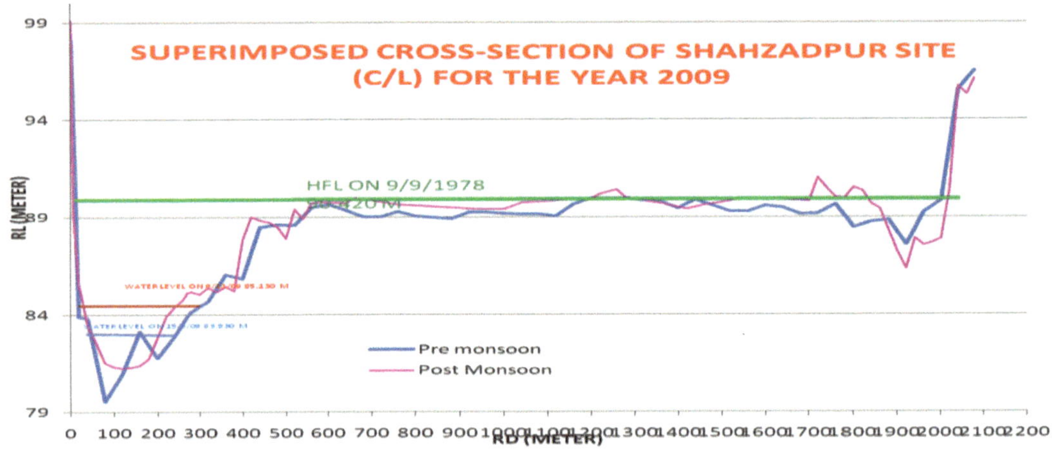

Fig. 17.10 Cross section and water level in Ganga at Shahzadpur (Source: Courtesy, Upper Ganga Basin Organisation, CWC, Lucknow)

Adequate water depth 3–4 m is available in a 200-m-wide channel at Shahzadpur as shown in Fig. 17.10. It is again stressed here that present cross section (2009) can only be completely submerged at FSL level of 9/9/1978 which corresponds to the highest flood year. Figure 17.10 also depicts changes in cross section as result of sedimentation load during post-monsoon as compared to pre-monsoon.

11.5. Homozonation of river reaches

In order to plan a detailed monitoring program to enable identification of the most sensitive zones with respect to the biodiversity homogeneous zones of stretch of Ganga has been divided in three segments:

Upper Ganga Segment (UGS): Himalayan (Gangotri to Rishikesh), 294 km, steep bed slope 1 in 70, elevation 4238–334 m

Middle Ganga Segment (MGS): Rishikesh to Varanasi, 1082 km, flat slope 1 n 5000. Elevation 334–74 m

Lower Ganga Segment (LGS): Varanasi to Ganga Sagar, 1149 km

There is a link between river cross section, habitat, fluvial geomorphology, and biodiversity with various flows. The flow requirement is based on cultural/spiritual, livelihoods, habitat, nutrients, depth, velocity for sedimentation transport, refill and pools sequencing for the maintenance year, and drought year and flood year for the month of lean and wet (i.e., January and August). Minimum flow during the driest and flash flood for the highest flood frequency was also considered. Cross sections at Kaudiyala, Kachhla bridge, and Bithoor have been rated by INRM consultant based on actual measurement of cross section. Gauge-discharge rating curves are drawn based on simulation of hydro-geological data (1995–2005) for flows using SWAT model by IWMI.

12. Discussion and summing of results

(i) Based on the requirements for a specific reach, the bare optimal flow need was considered as percentage of mean annual runoff natural. Therefore, subject to uncertainties, the workshop at Rishikesh (WWF-India 2011) concluded ecological flow results for Kaudiyala (Rishikesh), Kachla bridge, and Bithoor. Similar work was carried out regarding E-flows at Sangam

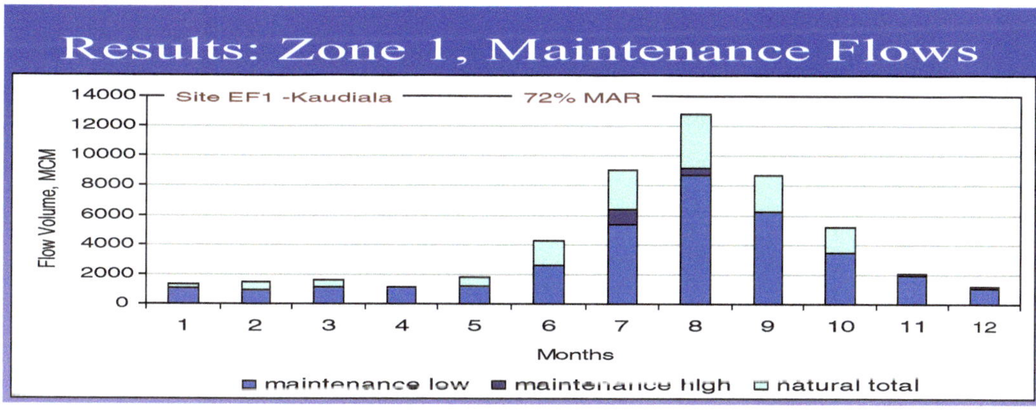

Fig. 17.11 WWF-India EF assessment in UGS (at Kaudiyala 30 km upstream of Rishikesh)

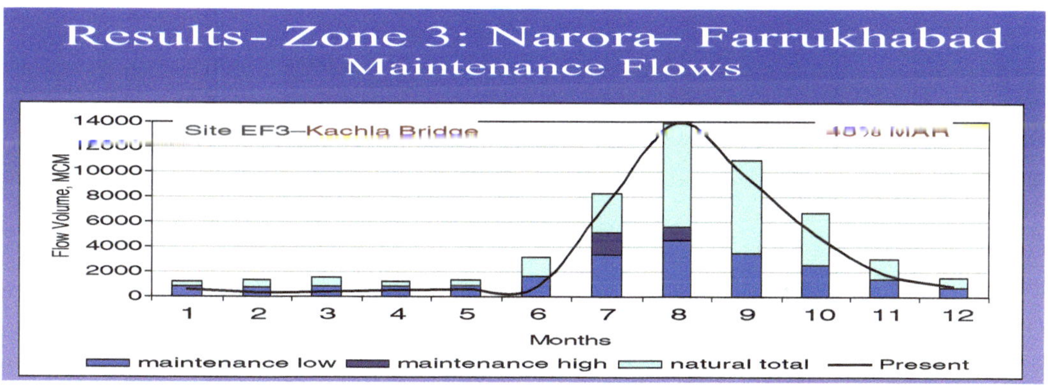

Fig. 17.12 WWF-India EF assessment in MGS at Kachhla bridge downstream of Narora Barrage

Allahabad during Kumbh-2013 by WWF-India. The EF results in the form of a graph are presented in Figs. 17.11, 17.12, 17.13, and 17.14.

(ii) Based on the recommendations of HPC, E-flows corresponding to 90 % dependable flow have been estimated by author for Devprayag and Rishikesh in UGS. The E-flows hydrograph is shown in Figs. 17.15 and 17.16.

(iii) Similar EF assessment for Ramganga river-a tributary of Ganga has been recently completed by WWF (2015), and the results is reproduced in Table 17.5.

It is observed that Katghar EF site Moradabad has historically poor flow during non-monsoon leanest month of May (pre- and postdam/barrage upstream). The virgin (natural flow) has been estimated 52 m^3/s against which present day flow is 2.8 m^3/s only. Therefore, the target EF at Katghar is estimated as 14.6 m^3/s (five times more of present day flow) for normal year and 3.1 m^3/s for drought year. The monthly distribution of EFs has been depicted in Fig. 17.17.

Improved scenario may require more flows and additional use of scenario may result in low flow requirement.

The crux of the problem is from where to get additional flows.

13. Conclusions
    1. Preliminary conclusions
        - Present conditions at the upper site are largely acceptable.

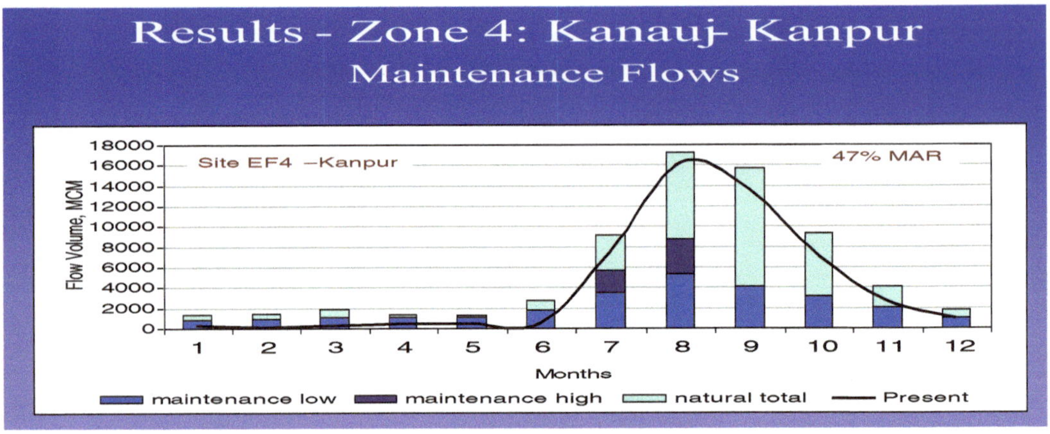

Fig. 17.13 EF assessment in MGS at Bithoor upstream of Kanpur barrage

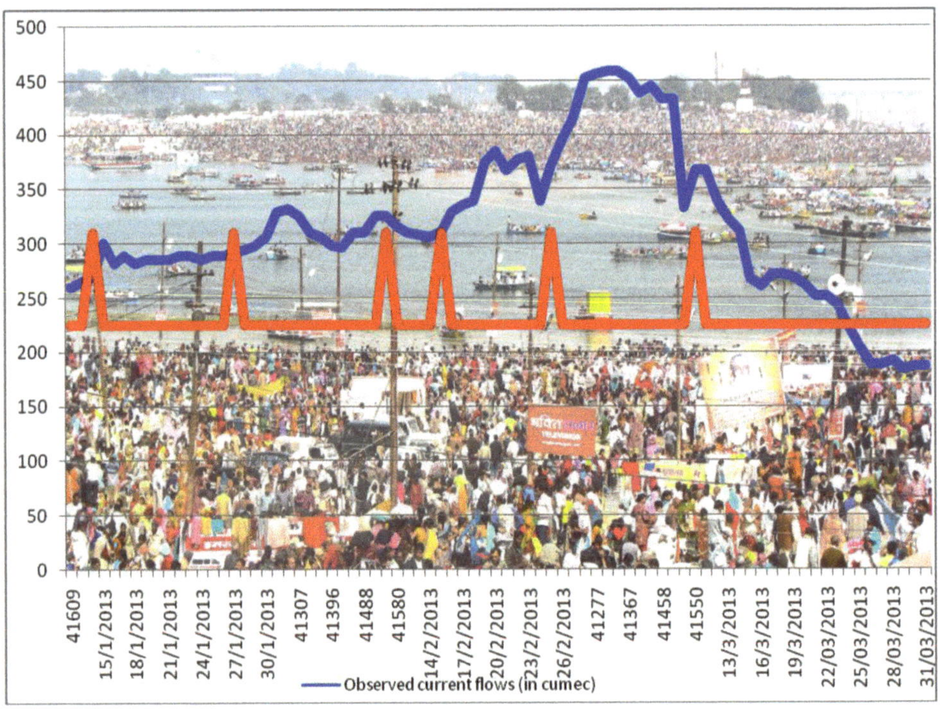

Fig. 17.14 WWF-India EF assessment at Sangam, Allahabad (MGS), and EF implementation by the Government of Uttar Pradesh during Mahakumbh, 2013

- Below Narora there are major problems in terms of water quality and quantity in the lean season.
- Conditions and motivations to improve these flows have been defined through this process.
- Water quality and ecological flows are two different aspects.
- Conventional wisdom suggests that major environmental problems that have arisen as a result of development can only be tackled through adoption of technologies for pollution control or remediation of contaminated environmental media.

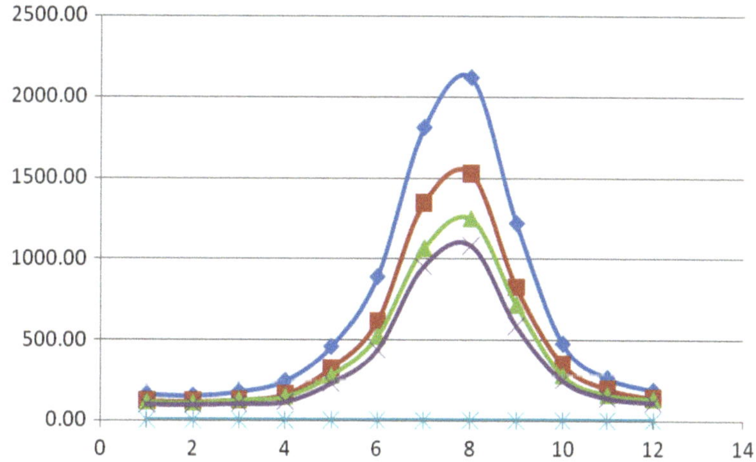

Fig. 17.15 Dev Prayag: virgin flow (*blue*), 90 % dependable flow (*red*), EF (*green*), and Min-EFR (*purple*) for Ganga River-UGS (Source: Author own elaboration)

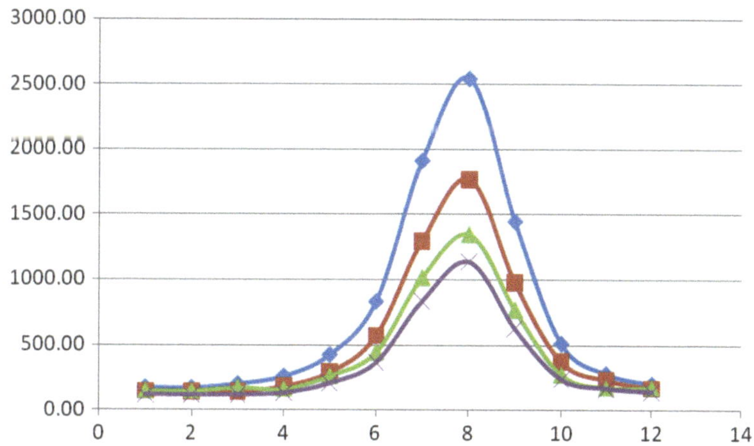

Fig. 17.16 Rishikesh: virgin flow (*blue*), 90 % dependable flow (*red*), EF (*green*), and Min-EFR (*purple*) for Ganga River (UGS) (Source: Author own elaboration)

Table 17.5 EF as % of natural flow for Ramganga River at various sites

EF site	Scenario	EF during normal year			EF during drought year		
		MN	NM	Annual	MN	NM	Annual
Bhikiasain	Target	88.4	68.8	82.8	18.3	26.5	20.6
Marchula U/S Kalagarh	Target	79.9	68.2	76.9	26.6	29.2	27.3
	Additional use	55.1	35.1	49.8	17.8	12.2	16.3
Heroli D/S	Target	42.5	37.7	40.6	14.3	16.1	15.0
	Improved	42.5	43.8	43.0	21.7	24.2	22.7
	Additional use	17.9	24.2	20.3	6.7	7.0	6.8
Agwanpur	Target	25.8	13.6	21.3	13.5	7.0	11.4
Katghar, Moradabad	Target	36.4	27.9	33.0	15.3	5.9	11.7
	Improved	42.9	36.9	40.6	18.9	11.1	15.9
	Additional use	32.6	7.6	23.2	10.8	2.6	7.7
Chaubari, Bareilly	Target	18.8	12.6	16.1	11.2	10.6	10.9
	Improved	31.5	23.7	28.0	18.8	12.6	16.1
	Additional use	11.2	8.5	10.0	5.8	7.3	6.4
Dabri, Shahjahanpur	Target	26.6	13.3	20.7	11.3	7.5	9.6
	Improved	32.0	28.6	30.4	20.0	12.7	16.6
	Additional use	14.6	11.7	13.3	8.0	5.4	6.8

Source: WWF-India Report on Ramganga EF assessment 2015 (Unpublished) based on flow data 1973–2011

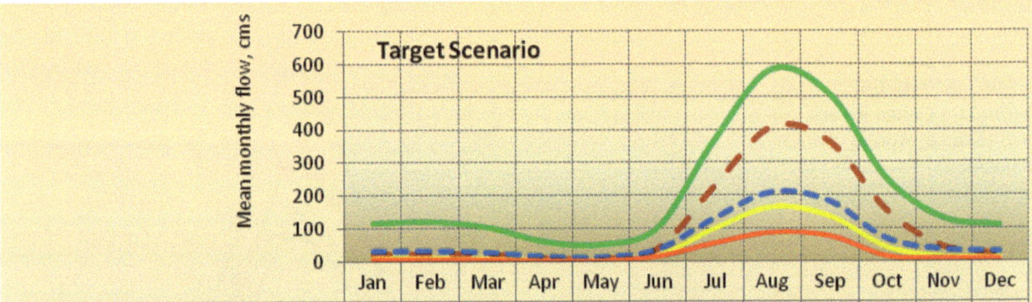

	Jan	Feb	Mar	Apr	May	Jun	Jul	Aug	Sep	Oct	Nov	Dec
Katgarh(Natural), cms	118.1	122.4	105.0	61.5	52.4	104.7	368.5	582.2	508.2	250.8	132.3	110.3
Katgarh(Unobstructed), cms	21.8	23.8	22.7	11.4	8.9	39.0	227.1	404.9	361.3	153.8	44.5	21.2
Katgarh(Regulated), cms	10.8	11.0	8.3	3.5	2.8	16.8	100.6	165.7	138.1	41.8	12.3	9.6
Eflow - Maintainance Year	32.9	34.1	29.3	17.1	14.6	37.8	132.9	210.0	183.3	69.9	36.9	30.8
Eflow - Drought Year	7.0	7.2	6.2	3.6	3.1	16.0	56.3	89.0	77.7	14.8	7.8	6.5

Fig. 17.17 EF target for Katghar site, Moradabad in Ramganga River WWF-India Report-2015 Unpublished (Credit: INRM consultant for WWF)

2. Further information requirements
   - Relationships between indicator fish, invertebrate's, etc. species and flow would allow more precise flow recommendations.
   - More precise relationships between water quality parameters and flow would have allowed the prediction of the effects of recommended flows on water quality to be assessed.
   - Additional investigation of the effects of high flows on sediment transport and channel morphology, to prevent sedimentation and maintain channel integrity.

14. Follow-up/future activities
    - Refinement of EFA
    - Finalization and dissemination of EFA report (policy and technical)
    - Cost-benefit analysis of environmental flow implementation
    - Trade-off analysis of environmental flow implementation
    - Promote implementation through policy dialogue
    - Framework for EF implementation and trade-off management

Acknowledgement The author is grateful to the State Water Resources agency, an apex water institute of the Government of Uttar Pradesh, and WWF-India Secretariat, New Delhi, for sponsoring the International Workshop at Lucknow on Assessment of Environmental Flows Requirements of Himalayan Rivers, 21–22 July 2009, organized by him and a series of follow-up workshops/meetings of experts in monitoring the Living Ganga Program conducted by WWF-India for giving an opportunity to share thoughts. The author is grateful to all those experts/consultants of WWF-India whose research works have been referred in this paper, in particular, Prof. Jay O' Keeffe of UNESCO-IHE, Netherland; Dr Vladimir Smakhtin, IWMI, Sri Lanka; Dr Luna Bharti, IWMI, Kathmandu (Nepal); Prof. Prakah Nautiyal, Garhwal University; Prof. A.K. Gosai, IIT Delhi; Dr Sandhya Rao, INRM; Prof Vinod Tare; Prof Rajiva Sinha of IIT, Kanpur; and Dr Ravi Chopra of PSI Dehradun. Special thanks to Dr Sejal Worah, program director, Living Ganga Program, WWF-India; Dr Suresh Rohila (now with CSE); Mr Suresh Babu, director, water policy and river basins; and Mr Nitin Kaushal, additional director, SWM and Wild river at WWF-India who gave him opportunity to understand the issues of EFs.

References

AHEC/IITR Report (2011) Assessment of cumulative impact of hydropower projects in Alaknanda and Bhagirathi Basins, for MoEF, GOI, New Delhi

Cautley PT (1860) Report on the Ganges Canal works from the commencement until the opening of the canal in 1854, vol I. Publisher Smith, Elder & Co., 65, Cornhill, pp 22–24

Central Water Commission (2010) Water and related statisks, WRIS, Directorate, ISO WP&PW, CWC, GOI, New Delhi

Dutta V, Kumar R, Sharma U (2015) Assessment of human-induced impacts on hydrological regime of Gomti river basin, India. Manag Environ Qual Int J 26(5):631–649

Expert Body Report (2014) "Assessment of Environmental Degradation and Impact of Hydroelectric projects during the June 2013 Disaster in Uttarakhand" Submitted to MoEF, Government of India, New Delhi

High Power Committee Report (2015) Assessment of E-flows, for MoWR, RD & RR, GOI, unpublished

IITC (2011) Environmental flows state-of-the-art with special reference to rivers in the Ganga River Basin. IIT-GRBMP thematic report – report code: 022_GBP_IIT_EFL_SOA_01_Ver 1_Dec 2011

IITC (2015) Ganga River Basin management plan – 2015: Mission 1: Aviral Dhara. Consortium of 7 IITs (IITC)

IMG (2013) Report of the inter-ministerial group on issues relating to River Ganga. submitted to Government of India, New Delhi

Irrigation Department (2004) Report of Madhya Ganga canal project-phase-11. vol 1, GoUP

Kumar R (2011) Post Tehri performance of Upper Ganga Canal System, ICID conference at Yogyarta

WAPCOS (1980) U.P. composite irrigation project, modernization of Upper Ganga Canal

WAPCOS (2011) Augmentation of water flow at Sangam, confluence of River Ganga and Yamuna, Allahabad, for Government of Uttar Pradesh

WWF-India (2012) Assessment of environmental flows for the Upper Ganga Basin, HSBC Water Programe

WWF-India (2013) Environmental flows for Kumbh 2013 at Triveni Sangam, Allahabad, HSBC Water Programe

WWF-India (2015) Assessment of environmental flows for Ramganga River Basin, HSBC HSBC water programe-rivers for life, life for rivers (under publication)

Ravindra Kumar WWF-India, New Delhi, India

Advisory Group, DST-CPR, BBA (Central) University, Lucknow, Uttar Pradesh, India

State Water Resources Agency (SWaRA), Lucknow, Uttar Pradesh, India

Part VII

Transboundary River Issues

Opportunities and Challenges in the Trans-boundary Koshi River Basin

18

Shahriar M. Wahid, Garrett Kilroy, Arun B. Shrestha, Sagar Ratna Bajracharya, and Kiran Hunzai

Abstract

The Koshi river basin is shared between China, Nepal and India and is one of the key trans-boundary river basins in the Hindu-Kush Himalayas (HKH). The basin drains an area of about 88,000 km^2 and is a river system with a high potential for investments in hydropower development as well as irrigation in the downstream areas. In addition the basin contains important ecosystems and protected areas which provide a range of biodiversity and related ecosystem services and sustain livelihoods. The basin is home to over 40 million people with agriculture as the dominant activity. However, the diverse topography, young geological formations, degree of glaciation and monsoon system make the basin particularly prone to water-related hazards like extreme flooding and landslides. Droughts are also experienced in the rain-fed tributaries of the basin during the dry season. Urbanisation and floodplain encroachment have also added pressures on the water bodies and ecosystems of the basin. Climate change will likely exacerbate these pressures with consequences for seasonal water availability, and food and energy security, highlighting the need for appropriate water management and disaster risk reduction strategies. A river basin approach, through the application of integrated water resources management (IWRM) principles, is essential to address the trans-boundary nature of many of these multifaceted

S.M. Wahid (✉) • A.B. Shrestha • S.R. Bajracharya
International Center for Integrated Mountain Development (ICIMOD), Kathmandu, Nepal
e-mail: shahriar.wahid@icimod.org

G. Kilroy
Department of Geology, School of Natural Sciences, Trinity College Dublin, Dublin, Ireland

K. Hunzai
Sustainable Livelihoods and Poverty Reduction (SLPR), International Center for Integrated Mountain Development (ICIMOD), Kathmandu, Nepal

N. Sharma (ed.), *River System Analysis and Management*, DOI 10.1007/978-981-10-1472-7_18

issues. A conceptual framework for addressing these challenges within an integrated water and land resources management perspective for the Koshi basin is presented in this paper.

18.1 Introduction

The Koshi river basin is a trans-boundary basin shared by China, India and Nepal (Fig. 18.1). The river originates on the high-altitude Tibetan Plateau and passes through eastern Nepal and northern Bihar in India before joining the Ganges. The river basin has an extreme range of altitude, from the peaks of the Himalayas to the lowland plains of India, passing through five physiographic zones, containing six different biomes and nine ecoregions. The basin encompasses a great diversity in topography, climate, vegetation, demography and culture and has a high ecological significance, serving as a vertical linkage between different habitats. It contains a rich biodiversity and is a source of valuable ecosystem services that directly sustain the lives and livelihoods of the 40 million basin inhabitants as well as contributing to the livelihoods of many more people downstream.

The basin plays a key role in the irrigation of downstream areas and has a large potential for hydropower development. However, the diverse topography, young geological formations, high degree of glaciation and strong monsoon influence make it highly prone to erosion, sedimentation and natural hazards, including glacial lake outburst floods (GLOF), landslides and debris flow, droughts and flood. These events may

Fig. 18.1 HKH trans-boundary river basins including Koshi basin

increase in magnitude and frequency in the current context of global environmental change. Increasing population, urbanisation and encroachment have added additional pressures to the basin's freshwater ecosystems. The upper part of the basin has problems associated with snowmelt, water run-off, soil erosion and land degradation, while the lower part has problems associated with waterlogging, population growth, expansion of agricultural land and urbanisation. The effects and uncertainty resulting from climate change, changes in the cryosphere, natural disasters, sedimentation and issues related to agricultural production and market access, and accessibility of ecosystem services are all key challenges for effective management of the basin.

In order to address the complex and multifaceted management challenges and promote meaningful interaction and reconciliation of the interests of the various actors and sectors and maximise benefits, such as irrigation and hydropower, while minimising adverse events, such as floods and landslides, river basin management approach offers meaningful solution. The International Centre for Integrated Mountain Development (ICIMOD), a regional knowledge development and learning centre (www.icimod.org), is developing a comprehensive understanding of the system dynamics of the basin, providing impact analysis and facilitating decision-making based on river basin management approach.

This paper discusses the challenges of water, land, environment and livelihood and presents the approach taken in the Koshi Basin Programme (KBP) to develop policy-relevant scientific, economic, social and ecological evidence-based knowledge and support for integrated, innovative and inclusive decision-making that promotes the sustainable use of trans-boundary water resources. Particular focus is given to present a conceptual framework for integrated water and land river basin management in the Koshi river basin that recognises the drivers of change and promotes wider interpretation of international river basin management.

18.2 Status, Trends and Challenges in Climate, Water, Land, Environment and Livelihood

Originating in the high-altitude Tibetan Plateau and the Himalayas, the Koshi River and its tributaries flow through Nepal's mid-hills onto the Terai lowland region of Nepal and India where a vast alluvial fan has developed over the centuries. At the confluence with the Ganges, the Koshi basin drains an area of some 87,970 km^2, approximately 32.4 % in China, 45 % in Nepal and 22.6 % in India. Figure 18.2 presents maps of both the complex nature of the basin's physical geography and the distinct population gradient ranging from the Trans-Himalayas to the floodplain (Terai).

Figure 18.2 highlights the coincidental topographic and demographic gradients extending from north to south. Topographically the basin ranges from greater than 8,000 m on the Trans-Himalayas (including Mount Everest) and some 5,000 m on the Tibetan Plateau to less than 100 m at the confluence with the Ganges.

In parallel, the population density in the Koshi basin ranges from less than five people per square kilometre on the Tibetan Plateau portion in China, increasing to 100–200 in the mid-hills of Nepal and expanding significantly to greater than 500 in Kathmandu, in some of the Terai districts of Nepal and in the northern states of Bihar in India. This topographic gradient also results in a range of geo-climatic conditions and a corresponding wide range of ecological zones. The basin includes two Ramsar Wetland Sites and several protected area, bringing a wide range of biodiversity and ecosystem services to the benefit of the local communities for their livelihoods.

The climate in the basin varies from humid tropical in the south, through subtropical and temperate, to cold and arid in the north. The climate in the southern part of the basin is strongly influenced by the South Asian monsoon, whereas to north, the Tibetan Plateau lies in a rain shadow area. In the Nepal portion, there are four distinct climatic seasons: pre-monsoon (March–May), monsoon (June–September),

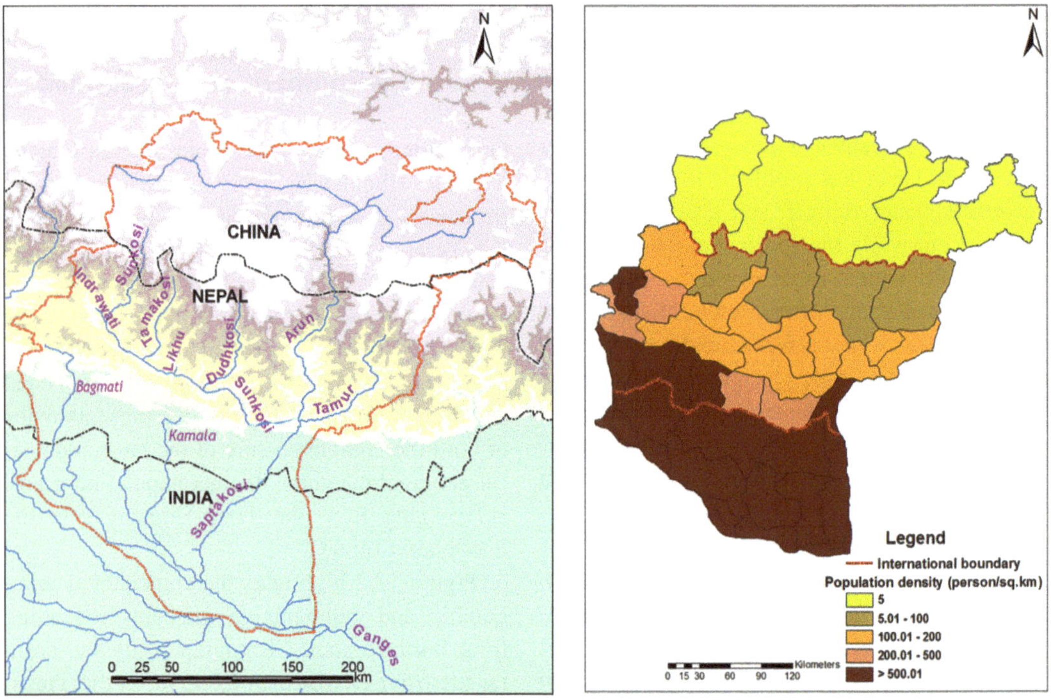

Fig. 18.2 Koshi river basin's physical geography (*left*) and population density (*right*; 2001 census data)

post-monsoon (October–November) and winter (December–February). Based on the precipitation data from Asian Precipitation - Highly-Resolved Observational Data Integration (APHRODITE) (Yatagai et al. 2012), from 1951 to 2007, there are high spatial and temporal variability of precipitation in the Koshi river basin. About 80 % of the precipitation occurred in between June and September (the monsoon season).

The northern part of the basin is also very cold and the temperature goes down to −17 °C in January. The southern part is a bit warmer. The mountains record the maximum temperature in May, while Indo-Gangetic Plains (IGP) heat up most in April. The annual rates of evapotranspiration (ET) are generally less than 1,000 mm (Subrahmanyam and Upadhyay, 1983: cited by Kattelmann 1991). However, some parts of the basin such as the Sun Koshi have extremely high potential evapotranspiration and suffer from frequent droughts and soil erosion (Fig. 18.3).

Water balance study conducted by the Koshi Basin Programme showed that mean annual precipitation is the highest (1,920 mm) in the mountains (M) and decreases in the hills (H) and Indo-Gangetic Plains (IGP). Precipitation is the lowest (914 mm) in the transmountains (TM). Annual mean actual ET shows a different trend with the minimum (231 mm) being in the TM which increases up to the IGP (660 mm). Annual mean water yield has a pattern similar to that of precipitation. Minimum water yield occurs in the TM (569 mm), whereas the maximum occurs in the mountains (1442 mm). It can be noted that the range from the maximum to minimum values of precipitation, actual ET and net water yield is very large for the mountain region. This range is the least for the TM.

There is a large temporal variation in the water balance components in the Koshi basin (Fig. 18.4). As expected, actual ET and water yield are high in the monsoon and low in the dry period of the year. Large values of Δ storage during the monsoon could be attributed to high groundwater recharge which is responsible for groundwater flow and ultimately base flow

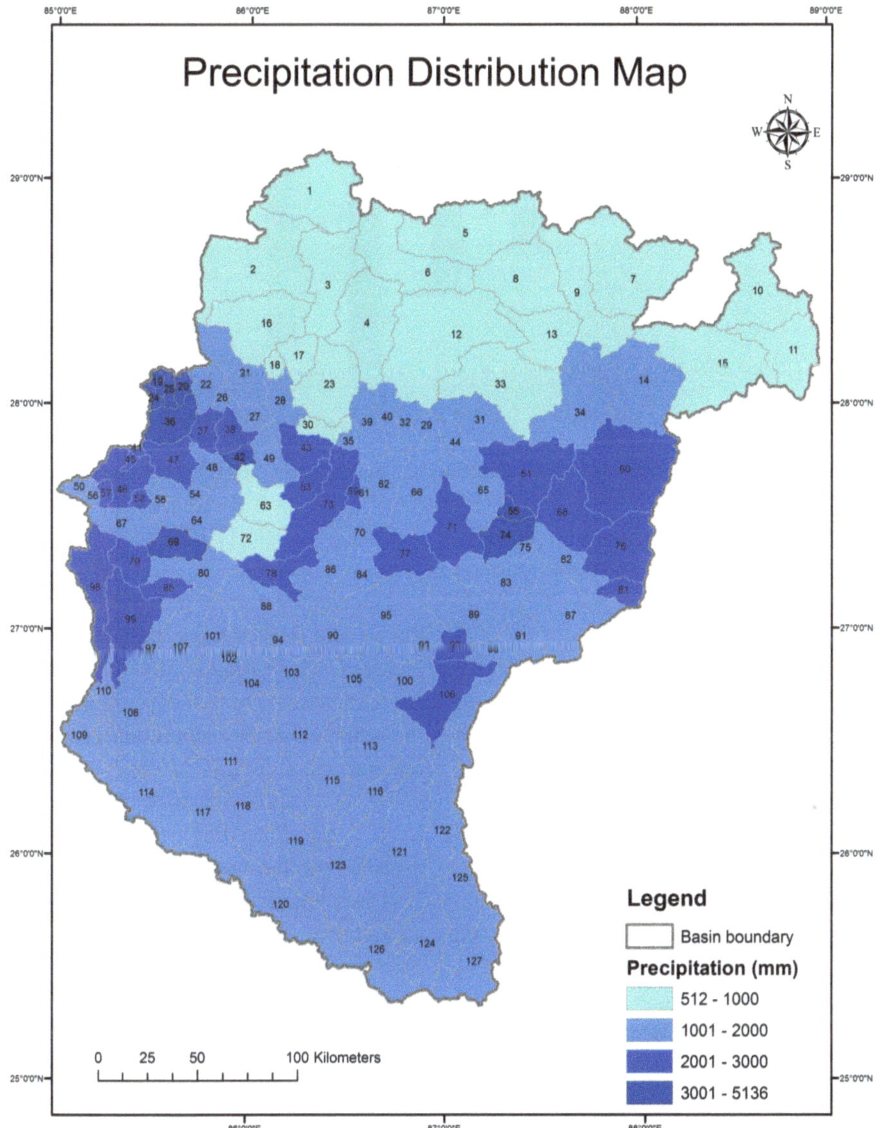

Fig. 18.3 Sub-basin wise (numbered 1–127) distribution of mean annual precipitation

during the dry period of the year. Also, the decreasing trend of Δ storage from the monsoon to the dry period is very prominent. July was found to be the wettest month with the maximum precipitation of 393 mm, and December the driest with 9 mm.

Global warming is expected to change the important hydrological processes of the headwaters of the Himalayan region (such as precipitation, evaporation and discharge) which might have serious implications in the downstream water availability (Eriksson et al. 2009). Nepal et al. (2013) carried out a scenario analysis of temperature change in the hydrological regime of the Dudh Kosi (or Koshi), one of the sub-catchments of the Koshi river basin. The results indicate that the snowmelt pattern might be adversely affected by the rise in temperature, and the river might shift from a 'melt-dominated river' to a 'snow-dominated river'. A 2 °C rise in

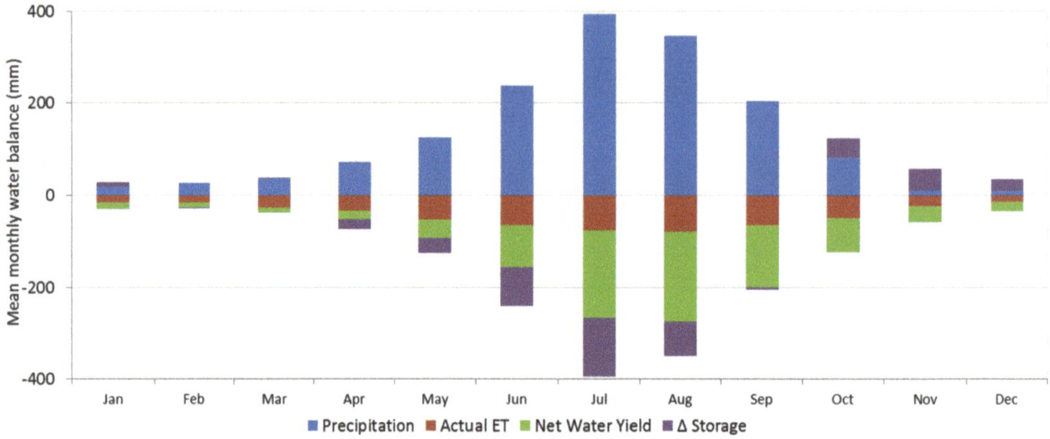

Fig. 18.4 Mean monthly water balance results from model simulation (1998–2008) for the whole Koshi Basin

temperature was projected to lead to a 31 % decrease in snowmelt by 2050 in comparison with the 1985–1997 period and a 4 °C rise in temperature to a 60 % decrease.

Bharati et al. (2012) used the Soil and Water Assessment Tool (SWAT) hydrological model to look at the potential impact of climate change projections from four General Circulation Model (GCMs) on the water balance of the Koshi basin. The analysis indicated a large temporal and spatial variability in precipitation, actual evapotranspiration and water yield in both the north and the south parts of the basin. The 2030s projections showed precipitation increases in the transmountain areas during the dry season and decreases in the other regions. The projections show a decrease in precipitation in all regions during the monsoon season. Assessment of projected future flow time series indicated an increase in the number of extreme events – both low flows and high flows. However, the authors also emphasised the high degree of uncertainty in the projected climate data. Immerzeel et al. (2012) estimated the response of glaciers in central Nepal (Langtang basin) using the climate projection data and concluded that both glacial area and glacial ice volume might be substantially decreased in the future due to increasing temperature scenarios, with glacial area decreased by as much as 32 % in 2035, 50 % in 2055 and 75 % in 2088. Projected impacts in the Koshi basin are likely to be similar.

Zhang et al. (2010) also suggest that the increasing temperature might lead to loss of glacial area in the central part of the Koshi basin and that vegetation coverage in the valleys will tend to increase, while mountain ridges show increased desertification. The loss of ice may pose a threat to indigenous species. Temperature and precipitation are important factors in the distribution of land cover types, and even small changes may result in large impacts on the ecosystems in the basin, especially in the northern part.

The climatic changes will affect the water availability, growth conditions for vegetation and the extent and impact of water-related problems and disasters and thus the basis of people's livelihoods. Populations in the middle mountains and the floodplain (Terai) region of the basin are already experiencing climate related stress such as erratic monsoon rainfall and periods of floods and droughts. Changes in the quantity, timing and intensity of precipitation are likely to affect water flow regimes and thus ecosystems and the services they provide for human well-being and livelihoods (Dixit et al. 2009). As discharge in upland sources declines, villagers will require more time, and more children will miss school, to collect water, and they will have to use water of low quality more often.

Land use and land cover are highly variable but closely aligned with altitude across the river basin. Land cover and land use are important factors for determining water usage in the basin and will influence greatly the hydrological rainfall-run-off responses in each catchment within the basin. The main land cover types are alpine meadow (shrub and grassland) to the north and forest and cultivated land to the south (Zhang et al. 2011). The major land cover classes are agricultural land including orchards (36 % of the total basin area), mainly found in the plains and mid-hills to the south; grass and shrub land (33 %), mainly found in the Tibetan Plateau to the north; and needle and broadleaved forests (20 %) in the central mountains and mid-hill areas. Glaciated areas cover around 6 % of the basin, with the largest portion in the Arun sub-basin.

Analysis by the Koshi Basin Programme showed that during 1990–2010 there was a 0.8 % decrease in forest (13,000 ha) and an 8 % increase in agricultural land (236,400 ha). The initial findings from the trend analysis suggest that the area of snow cover also decreased although the changes were variable, with a negative trend for winter and autumn and positive trend for summer and spring. However, change is often more extensive locally, and large-scale categories do not capture changes in one area compensated by changes in the opposite direction in another or the aspect of degradation, for example, reduced crown cover in forests and degraded grassland.

The Koshi basin districts are endowed with fertile land resource and water resources and potential for fisheries, hydropower and tourism. However, the incidence of poverty and low per capita income is higher in the Koshi basin districts than the corresponding national figures. In a recent study, the incidence of poverty in Koshi districts in Nepal (taken as USD 1.25 per day per person (PPP) adjusted) was found to range from 4.4 % in Kathmandu to 60.3 % in Sindhuli (WECS 2011). Chen et al. (2013) estimated that 40 % of the population in the Nepal part of the Koshi basin are below the poverty line (higher than the national average of 30 %). The Ministry of Agriculture Development estimated that 8 of 26 Nepal Koshi districts are food deficit, and the basin overall is food deficit (99,951 tonnes) (MOAD 2012). In Bihar, more than half of the basin districts have a poverty ratio above 40 %, with a range from 8 % in Madhepura to 65 % in Muzaffar (compared to 30 % for all the whole India) (Chaudhuri and Gupta 2009). In the Nepal part of the basin, more than 80 % households do not have electricity. Similarly, Bihar has the lowest per capita electricity consumption in India (only 122 kWh compared with a national average of 779 kWh).

The majority (83 %) of people in the Koshi basin districts in Nepal depend on agriculture as a source of income, and almost half (49 %) of them depend on agriculture as a primary source. Overall, 68 % of men are economically active in the Nepal part of the Koshi basin and 45 % of women (CBS 2001). Women contribute more to agricultural labour; 64 % of women are engaged in agricultural work and only 29 % of men (ADB 2012). In Bihar, women's share in nonagricultural employment (16 %) is much lower than men's (84 %). A study in 14 of the 26 districts in all the three zones in the Nepal part of the Koshi basin between 2011 and 2012 found that men worked more in agriculture, livestock raising, fishery, and other economic activities, while women contributed more to caring activities. Men were more active in seed purchase and selling agriproducts; women's roles focused more on collecting water and fuelwood. Few people had access to paid employment for 10 or more months, but the number was much higher for men than for women (0.38 persons per household compared to 0.13 for women). The 2001 and 2011 census data show that men are seven times more likely to migrate to urban areas or abroad for work or study or both (Khadka et al. 2015); the survey showed that 88 % of those migrating for employment were men, and women's contribution to remittances was correspondingly low. Households received an average value in remittances in cash and kind of USD 14 from women and USD 118 from men who had migrated for employment within the country, and USD 13 from women and USD 339 from men who had migrated abroad. Women spent an average of 5.8 h per day collecting water and fuelwood and tending livestock, and men 3.5 h.

Although agriculture is the main source of livelihoods in the basin, significant shifts are being observed in livelihood patterns, with, for example, increasing commercialisation of agriculture and replacement of subsistence by cash crops and a diversification of sources of income, including increased income from remittances of short- and long-term migrants. There are different driving forces for the livelihood shifts in rural areas. Some of the forces observed in the Koshi basin area are seasonal migration, permanent migration to urban areas or abroad, globalisation, influx of communication technology and transportation (Eele 2009). The present livelihood patterns are not only contributing to high population densities in some urban and urban-oriented areas but are also changing household dynamics by 'feminising' agricultural activities, which increases both the workload and the responsibilities of women, and creating more livelihood dependency on remittances coming into the rural areas from foreign and national employment. Globalisation and economic development are putting pressure on the rural population to participate in market-based economic activities like wage labour, tourism, off-farm production and others.

As elsewhere in the HKH region, the inhabitants of the Koshi basin face multiple hazards. In general terms, the drainage basin of the Koshi basin can be divided into three main zones: an upper erosional zone of sediment production, a middle zone of sediment transport with simultaneous erosion and deposition and a lower zone of sediment deposition. Each zone is associated with somewhat different types of hazard. Glacial lake outburst floods occur in the far upstream of the basin and can impact on the upper and middle streams, especially along the boundary between China and Nepal; debris flows are common in the upper and middle streams of the basin; landslides and gully erosion are common in the hills of the middle stream, with sediment deposition and bank cutting in the valleys; floods, riverbank erosion and sand deposition are in the middle and downstream, especially in the plains of Nepal and India; and droughts are common in the middle and downstream of the river basin (Chen et al. 2013).

18.3 Conceptual Framework to Address Challenges and Develop Opportunities

Contemporary global and regional climate and other environmental and socioeconomic changes pose immense challenges to water resources management in the Koshi river basin due to its trans-boundary nature (shared between China, Nepal and India), high spatial and temporal variation of resource endowment and upstream-downstream linkages with a high degree of interrelationship and uncertainty among water uses and users. This means that effective management of this complex basin requires a holistic approach that takes into account a wide range of natural resource management and socioeconomic issues and is designed to maintain the natural environment while ensuring provision of the ecosystem services on which the basin's inhabitants depend.

Along the vein, the Koshi Basin Programme at ICIMOD adopted a drivers-pressures-state-impact-response (DPSIR) framework (Fig. 18.5) to integrate a systems approach; generate scientific, economic, social and ecological knowledge; support decision to promote the sustainable use of trans-boundary water resources; and clarify trade-offs relating to development. The conceptual framework presented in Fig. 18.5 was first proposed by the Organisation for Economic Cooperation and Development (OECD 1994) and later adopted by numerous national and international organisations (e.g. EEA 1999). Essentially it reflects a systems analysis view of the relations between the environmental system and the human system. Drivers, such as climate change and population growth impact, may determine the status of these systems. Responses to mitigate against these impacts, including livelihood and adaptation strategies, can be developed across a number of scales from local to national to regional and from watershed to sub-basin to whole basin. The strategies can be combined with policy recommendations to develop

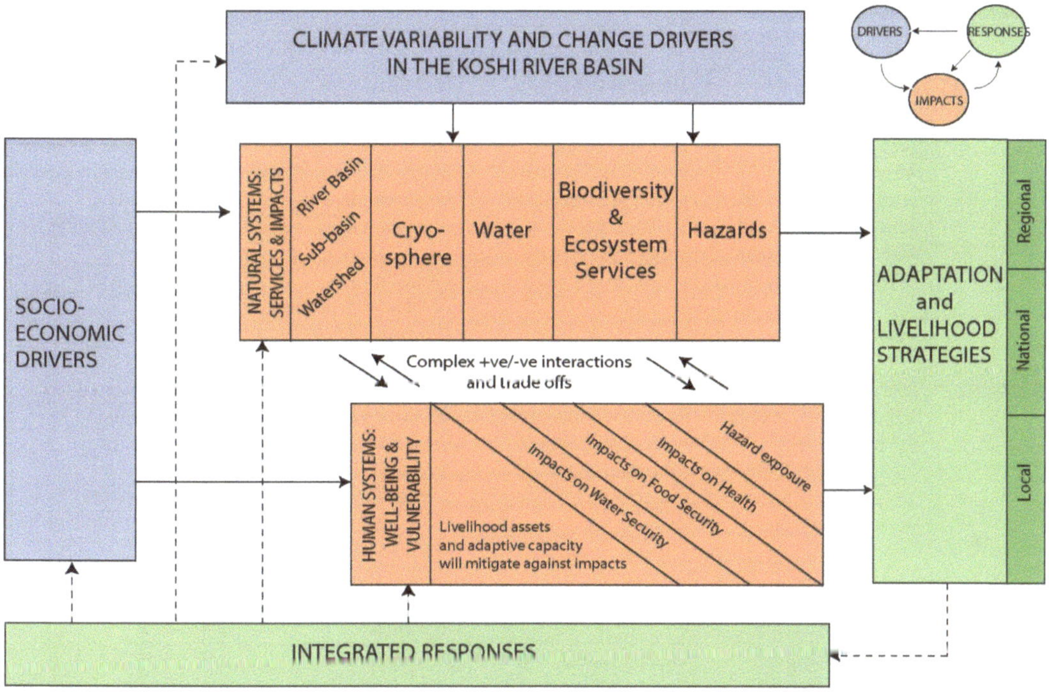

Fig. 18.5 Conceptual framework for integrated water and land river basin management in the Koshi river basin

integrated responses at appropriate scales and locations within the basin. The different components of this conceptual framework in four physiographic zones of the basin are summarised in Table 18.1.

18.4 Conclusions

This paper aimed to explore river basin management challenges and opportunities in the Koshi river basin. The challenges prevalent in the Koshi basin are multifaceted and are a function of the natural characteristics of the basin combined with socioeconomic drivers such as population increase, urbanisation and encroachment and all within a complex trans-boundary geopolitical setting. Climate change will likely exacerbate these challenges in the basin. IWRM and river basin management, and particular trans-boundary approaches, are in their infancy in the basin and have focused primarily on bilateral initiatives.

However, important steps have been made in this direction such as the Koshi River Basin Strategic Management Plan (Nepal) and the Ganga River Basin Management Plan (India). Similarly in the climate change adaptation arena, all three countries of the Koshi basin have developed national adaptation strategies, which offer much scope for integration and associated benefits this could yield. Efforts are needed to extend this work across whole basins, and Koshi basin represents a good pilot basin to trial these approaches. There are a number of opportunities for progress in this area, which have been outlined in this paper. The wider perspective of integrated water and land resources management broadens the dialogue beyond water issues to include land cover/land use change, food security, poverty reduction, water-related hazards and disaster risk reduction, conservation and biodiversity, ecosystems and ecosystems services. The conceptual framework presented in this paper offers one approach to taking this wider perspective.

Table 18.1 DPSIR framework for integrated water resource management in the Koshi river basin

	Physiographic zone			
	Trans-Himalaya	Himalaya	Mid hills	Plains (Terai)
Drivers	Climate change, livestock, rapid infrastructure development, no or sporadic presence of water and ecosystems service providers	Climate change, out migration, tourism in certain pockets, no or sporadic presence of water and ecosystems service providers	Demand for fuelwood and timber, out migration, agricultural transformation in selected pocket areas, urbanisation and infrastructure construction, limited presence of ecosystem service providers	Climate change, in migration, agricultural development, access to markets and development facilities
Pressures	Tourism development, high-altitude rangeland and livelihood	Range land and livestock management, agricultural labour, tourism, migration pattern and livelihood dynamics	Forest resources, labour, cropping pattern (crop intensity, demand for irrigation), urbanisation (drinking water, sewage and solid waste management issues), migration pattern and livelihood dynamics	Water-related disaster and livelihood dynamics. land management. cultivation and cropping system, urban expansion
State	Increased energy demand, increased water stress, decreased livestock population	Glacial retreat, decrease in productivity of pasture, fluctuation in stream flow, expansion in arable land, increased agrobiodiversity, decreasing area under traditional crops; fallow agricultural land, reduction in livestock; environmental pollution	Forest and land degradation, low agricultural productivity, water stress, increased erosion and run-off; technology use in agriculture, increase in fallow land, increase in plantation crops (horticulture and tea), stall feeding of livestock; per unit increase in fertiliser and pesticide use, capital intensive agriculture, commercial agriculture; deterioration of water quality	Flooding, land degradation, large areas of land dependent on monsoon and remain fallow with late monsoon, lowering of water table, siltation of irrigation canals; expansion of commercial agriculture, increased use of technology
Impacts	Income from livestock decreases and food insecurity increases, easy access to tourists and also outmigration, increased women's drudgery in household level activities, including agricultural production and animal husbandry	Increased remittances and household income; increased local wage rates creating inflation; increasing number of tourists putting pressure on environment, increased women's drudgery in household level activities, including agricultural production and animal husbandry; men's involvement in off-farm activities	Increasing food deficit households; higher dependency on remittances than agriculture, increase in female-headed households; increased work burden for women for managing farms and family; switch from subsistence to commercial agriculture; upstream-downstream conflicts, better market linkages; opportunities for off-farm employment	High risk involved in agricultural production; cheap labour; high pressure on natural resources base, high incidence of poverty; siltation and erosion resulting in landless farmers; higher level of gender inequality; women experience greater poverty, exclusion, workload and discrimination due to conventional gender norms and practices

(continued)

Table 18.1 (continued)

	Physiographic zone			
	Trans-Himalaya	Himalaya	Mid hills	Plains (Terai)
Responses	Livelihood diversification options to improve capacity to deal with shocks, trends and seasonality in their environment; strengthening of traditional coping mechanisms community-based conservation area management practices; women's inclusion in NRM use and decision-making		Livelihood diversification options, water conservation, recognition of upstream-downstream linkage, gender friendly technologies and linking women in value chains, gender focused case studies on impacts of water hazards on gender roles and relations, basin-wide IWRM practices to support development options	

Some topics by their unique nature will remain bilateral in focus. However, for these and wider issues, a whole basin approach will help to develop a common understanding of key issues among all jurisdictions and to develop a sound knowledge base upon which to develop policy responses to the challenges faced in the Koshi basin and explore IWRM-based development opportunities.

Acknowledgement The research paper is made possible through ICIMOD's Koshi Basin Programme (KBP), which is supported by the Australian Government through the Sustainable Development Investment Portfolio for South Asia.

References

ADB (2012) Assessment report. Technical Assistance for the Preparation of the Agricultural Development Strategy

Bharati L, Gurung P, Jayakody P (2012) Hydrologic characterization of the Koshi Basin and the impact of climate change. Hydro Nepal: J Water Energy Environ 11(1):18–22

CBS (2001) National population census 2001 – Nepal, tenth census. Central Bureau of Statistics, Government of Nepal, Kathmandu

Chaudhuri S, Gupta N (2009) Levels of living and poverty patterns: a district-wise analysis for India. Econ Pol Wkly 44:94–110

Chen NS, Hu GS, Deng W, Khanal N, Zhu YH, Han D (2013) On the water hazards in the trans-boundary Koshi basin. Nat Hazards Earth Syst Sci 13:795–808

Dixit A, Upadhya M, Dixit K, Pokhrel A, Rai DR (2009) Living with water stress in the hills of the Koshi basin, Nepal. ICIMOD, Kathmandu

EEA Report (1999) Environmental indicators: typology and overview. European Environment Agency, Copenhagen. http://reports.eea.europa.eu/TEC25/en/tech_25_text.pdf. Accessed 7 Apr 2011

Eele G (2009) Policy lessons from communities under pressure, in climate change and world food security. Glob Environ Chang 37:611–623

Eriksson J et al (2009) The changing Himalayas: impact of climate change on water resources and livelihoods in the Greater Himalayas. ICIMOD, Kathmandu

Immerzeel WW, Beek LPH, Konz M, Shrestha AB, Bierkens MFP (2012) Hydrological response to climate change in a glacierized catchment in the Himalayas. Clim Change 110:721–736

Kattelmann R (1991) Hydrologic regime of the Sapta Kosi basin, Nepal, Hydrology for the water management of Large River Basins

Khadka M, Rasul G, Bennett L, Wahid SM, Gerlitz JY (2015) Gender and social equity in climate change adaptation in the Koshi Basin: an analysis for action. In: Handbook of climate change adaptation. Springer Berlin Heidelberg, pp 1049–1076

MOAD (2012) Statistical information on Nepalese agriculture. Kathmandu, Nepal: MoAC

Nepal S, Krause P, Flügel WA, Fink M, Fischer C (2014) Understanding the hydrological system dynamics of a glaciated alpine catchment in the Himalayan region using the J2000 hydrological model. Hydrol Process 28(3):1329–1344

OECD (1994) OECD core set of indicators for environmental performance reviews: a synthesis report. Organisation for economic co-operation and development: environmental monographs, Paris

WECS (2011) Koshi River Basin management strategic plan (2011–2021)

Yatagai A, Kamiguchi K, Arakawa O, Hamada A, Yasutomi N, Kitoh A (2012) APHRODITE: Constructing a long-term daily gridded precipitation dataset for Asia based on a dense network of rain gauges. Bull Am Meteorol Soc 93(9):1401–1415

Zhang Y, Gao J, Liu L (2010) Progress of land-use and land-cover change in the Koshi Basin, central high Himalayas. In: Workshop on third pole programme, LUCC and climate adaption in Tibetan Plateau. Institute of Geographic Sciences and Natural Resources Research, CAS, Beijing, http://www.mri.scnatweb.ch/download-document

Zhang Y, Gao JG, Liu L, Nie Y, Wang Z, Yang X (2011) Land cover and climate change in Koshi River Basin, the Third Pole. AGU Fall Meet. Abstr. -1, 0649

Shahriar M. Wahid International Center for Integrated Mountain Development (ICIMOD), Kathmandu, Nepal

Sagar Ratna Bajracharya International Center for Integrated Mountain Development (ICIMOD), Kathmandu, Nepal

Garrett Kilroy Department of Geology, School of Natural Sciences, Trinity College Dublin, Dublin, Ireland

Kiran Hunzai Sustainable Livelihoods and Poverty Reduction (SLPR), International Center for Integrated Mountain Development (ICIMOD), Kathmandu, Nepal

Arun B. Shrestha International Center for Integrated Mountain Development (ICIMOD), Kathmandu, Nepal

Hydropolitics in Transboundary Water Conflict and Cooperation

19

Subash Prasad Rai, Aaron T. Wolf, Nayan Sharma, and Harinarayan Tiwari

Abstract

Water is a fundamental human need and key to economic development. Since the beginning of civilization, people have faced problems associated with river and freshwater sharing. To add on to the precarious situation, most of the freshwater rivers are transboundary rivers, i.e. they cross at least one political border, either a border within a nation or an international boundary. Water politics, commonly known as hydropolitics, are politics affected by the availability of water and water resources, which play an important role in transboundary water management. Hydropolitics relate to the ability of geopolitical institutions to manage shared water resources in a politically sustainable manner, i.e. without tensions or conflict between political entities. As the pressures of population and economic growth increase, water resources are under increasing stress. As the stress on water resources increases, the risks associated with the management of transboundary rivers increase exponentially given the hegemonic disparities of the riparians. This gives rise to risks of conflict while generating opportunities of cooperation which can be analysed with the help of risk-opportunity index developed using fuzzy synthetic evaluation technique proposed by Rai et al. (J Hydrol 519:1551–1559, 2014). It has been proposed to formulate a hydropolitical sustainability index (HypSI) keeping in view the circles of blue sustainability (blue indicates water in this chapter) which considers the social desirability, political legitimacy, economical viability, environmental sustainability and technical feasibility aspects of shared water resources.

S.P. Rai (✉) • N. Sharma • H. Tiwari
Department of Water Resources Development and Management, Indian Institute of Technology Roorkee, Roorkee, Uttarakhand, India
e-mail: subashbitsindri@gmail.com

A.T. Wolf
Department of Geosciences, CEOAS, Oregon State University, Corvallis, OR, USA

N. Sharma (ed.), *River System Analysis and Management*, DOI 10.1007/978-981-10-1472-7_19

19.1 Introduction to International Rivers

> Fierce national competition over water resources has prompted fears that water issues contain the seeds of violent conflict. – Kofi Annan

Water is one of the nature's greatest gifts to humankind. The fact that freshwater is essential to survival was realized even by early civilizations; and hence they flourished on land made productive by great rivers, be it Nile in Egypt, Indus in India and Huang Ho in China. Freshwater is required for survival of all forms of the Earth's life, including humans. The truth that we are inhabitants of the "water planet" soothes little, the reason being less than 3 % of the total water on the Earth is freshwater resources. Its distribution, too, is more often uneven, with some states suffering severe droughts every year and some having to deal with floods.

Since the beginning of civilization, people have faced problems associated with river and freshwater sharing. To add on to the precarious situation, most of the freshwater rivers are transboundary rivers, i.e. they cross at least one political border, either a border within a nation or an international boundary. According to the Transboundary Freshwater Dispute Database (TFDD), the world consists of 276 international transboundary river basins (Fig. 19.1) which cover more than 45 % of the land surface on the Earth (TFDD 2012). A total of 145 countries which constitute over 75 % of all countries have shared river basins within their boundaries, while 33 of them have more than 95 % of their territorial dominion within international river basins. The internationally shared river basins are home to over 40 % of the world's population and contribute to approximately 60 % of the global river flows (Draper 2002; Giordano and Wolf 2003; Sadoff and Grey 2005; Wolf 1999; Wolf et al. 1999).

Even more important is count of the nations that constitute certain individual basins. The Danube basin is shared by 17 countries, the highest in the world. The Zambezi, Rhine, Nile, Niger and Congo are each shared by nine or more countries, while the La Plata, Neman, Vistula, Ganges-Brahmaputra-Meghna, Tigris-Euphrates, Aral Sea, Volga, Mekong, Jordan, Tarim, Kura-Araks, Amazon and Lake Chad basins each consist territory of at least five countries (Wolf et al. 1999).

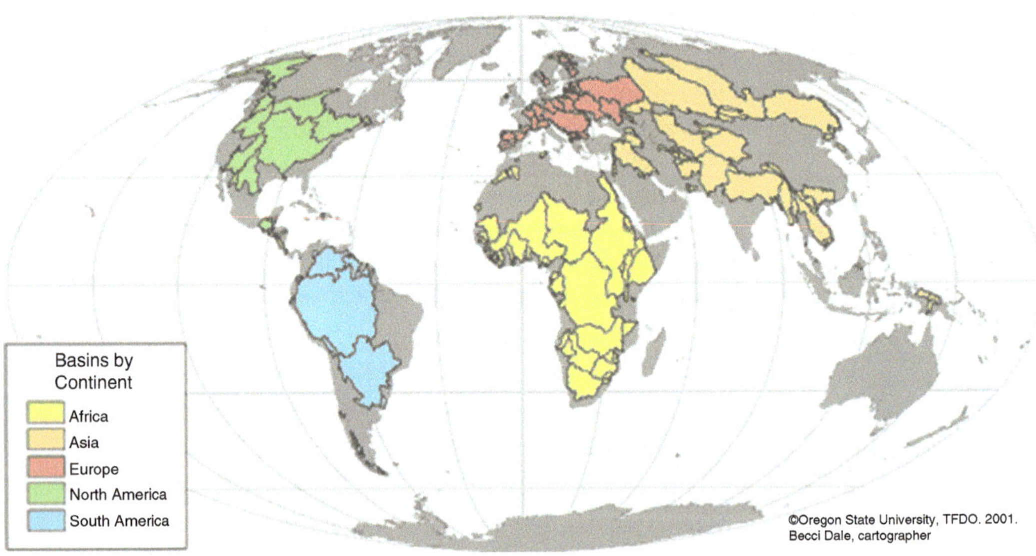

Fig. 19.1 International river basins (Source: Wolf et al. 1999, updated 2001)

19.2 Hydropolitics in Transboundary Rivers

Water makes adherence relation between riparian countries of transboundary river basins sometimes even without their consent as water is not bounded by political boundaries. As water is an essential component of our day-to-day life, riparian countries are incapable to separate from water issues. In transboundary rivers, water supply, allocation, control and use are of great consequence to survival, quality of life and economic success. The control of a nation's water resources is considered vital to the survival of a state (Daclon 2007). Water politics, sometimes called hydropolitics, is politics affected by the availability of water and water resources, a necessity for all life forms and human development. Arun P. Elhance's definition of hydropolitics is "the systematic study of conflict and cooperation between states over water resources that transcend international borders" (Elhance 1999).

The term "hydropolitics" (coined by Waterbury (1979)) can be related to the potential for conflict and violence to erupt over international waters. Hydropolitics relates to the ability of geopolitical institutions to manage shared water resources in a politically sustainable manner, i.e. without tensions or conflict between political entities. Water is a complex issue, and, as a result, it could be expected that hydropolitics will reflect this complexity. Competition for shared natural resources, particularly where limited, has caused or been additive to conflicts in the past.

With increase in population and human economic activities, the global demands of our natural resources both non-renewable and renewable have steadily increased. Water, a renewable resource, is no exception to this trend. Abstractions of water from rivers have more than tripled during the last 50 years, mainly due to extensive irrigation uses (Wada and Bierkens 2014). The socioeconomic developments of the continents have resulted in rapid drawdown of the amount of water all over the world, making it a scarce resource. And because it is finite, it is a resource of much greater importance than other resources with commodity value alone. There is no viable alternative to water. Without sufficient volumes or desired quality of water, economic development including industrial and agricultural production grinds to a halt, resulting in threats to human sufferings and societal stability. As the exponential growth of global population continues, and as changes in environment threaten the quality and quantity of water resources, the ability of countries to peacefully resolve water conflicts over transboundary international river basins will play a major role in stable and secure international relations (Jägerskog et al. 2007; Nicol et al. 2001).

The widening gap between freshwater supply and demand due to increased consumption of water will drive nation's attitudes towards rivers. The politics of freshwater in international contexts are becoming increasingly contentious. As many rivers cross political borders, the water available to one nation depends on the usage of water by the upstream nations. The issues of cross-border water sharing, use and management need greater attention as giant hydropower and irrigation projects are gradually performing key roles in defining international relations. Until there is no cooperation over shared water resources, each state can use the water in the river to the best of its advantage before it crosses political borders and becomes inaccessible.

Transboundary rivers have upstream-downstream management challenges. Consumptive use of water by an upstream country reduces the volume of water available for downstream countries. Upstream-downstream interactions can trigger conflicts or generate opportunities for cooperation. As the pressures of population and economic growth increase, water resources are under increasing stress. Management of rivers occurs in a complex political and economic context. For transboundary rivers, management is especially complex, given the multiple economic interfaces between co-riparians. The challenges of transboundary water cooperation are often exacerbated by capacity and power disparities between riparian countries.

19.2.1 Hegemonic Hydropolitics

Policy-makers and politicians have greater leverage over the direction that transboundary water interaction takes. Application of critical hydropolitics is thus useful for interpretation of the power plays that grease or block the cogs of the decision-making machinery. We argue that various riparian states are endowed with highly asymmetric capacity to use both overt and covert forms of power. As we will see, overt and covert forms of power are also commonly understood as, but not directly analogous to, "hard" and "soft" forms of power. We assert that the power asymmetries determine to a significant (not total) extent the fundamentally political distributional issue of "who gets what, when, where and why" (Lasswell 1936). The question is answered only by considering how political decisions of distribution and allocation of resources (namely, natural, political and financial) are made. Allocative politics are affected by their socioeconomic and political contexts, of course, at multiple scales (from the individual to the global) (Cascão and Zeitoun 2010).

Hydropolitics has also been strongly associated with the "water war" concept, wherein interstate armed conflicts were expected to occur in any number of "hydropolitical security complexes" such as the Tigris and Euphrates (Schulz 1995). The report of the World Economic Forum (2015) on global risks identified "water crisis" as having the greatest risk in terms of economic impact (Fig. 19.2). These risks are often greatest and increasing most rapidly in the major transboundary basins of the world. The analytical dyads of "water conflict" and "water security" are among the major forms of bias in the hydropolitics literature. The consistent association of hydropolitics with conflict or security issues has led to an impoverished debate and hindered understanding of hydropolitics as a dynamic and ongoing process involving several other key dynamics – notably society, environment and culture (Cascão and Zeitoun 2010).

Water is a very complex resource because it attains the forms of natural, social, cultural, economic and political resources. The recent decades have witnessed water more as a strategic asset owing to its economic and political nature. Power plays a significant role in influencing transboundary water relations and allocative outcomes and must therefore be incorporated into any analysis. Hydropolitics are considered to be characterized by hegemonic configurations, wherein the most powerful riparian states have an advantage over their riparian neighbours to influence the allocation of the resources. Notably, the power available to the "basin hegemon" assumes different forms – geographical, material, bargaining and ideational. The pillars of hydro-hegemony (Fig. 19.3) suggest that a hegemonic situation on transboundary waters is built on the four fields of both covert and overt forms of power.

The suggested relative measures of each field for the Eastern Nile are given in Fig. 19.4. The lengths of pillars are relative to other basin states and not quantified. These plots are based in the current political context, but as we know power and hegemonic situations are not static, the plots would certainly look different as time progresses. A quick glance at Fig. 19.4 confirms what is commonly understood, for instance, that Egypt is the basin hegemon, despite the downstream riparian position it owns. It also shows that the hegemons are usually very strong in all dimensions of power, at least in relation to their neighbours.

Given the enabling role that power plays in the allocative politics of transboundary waters, it is necessary to be able to conceptualize power in a useful manner. There is no common template that can be used to interpret all transboundary contexts. It is nonetheless

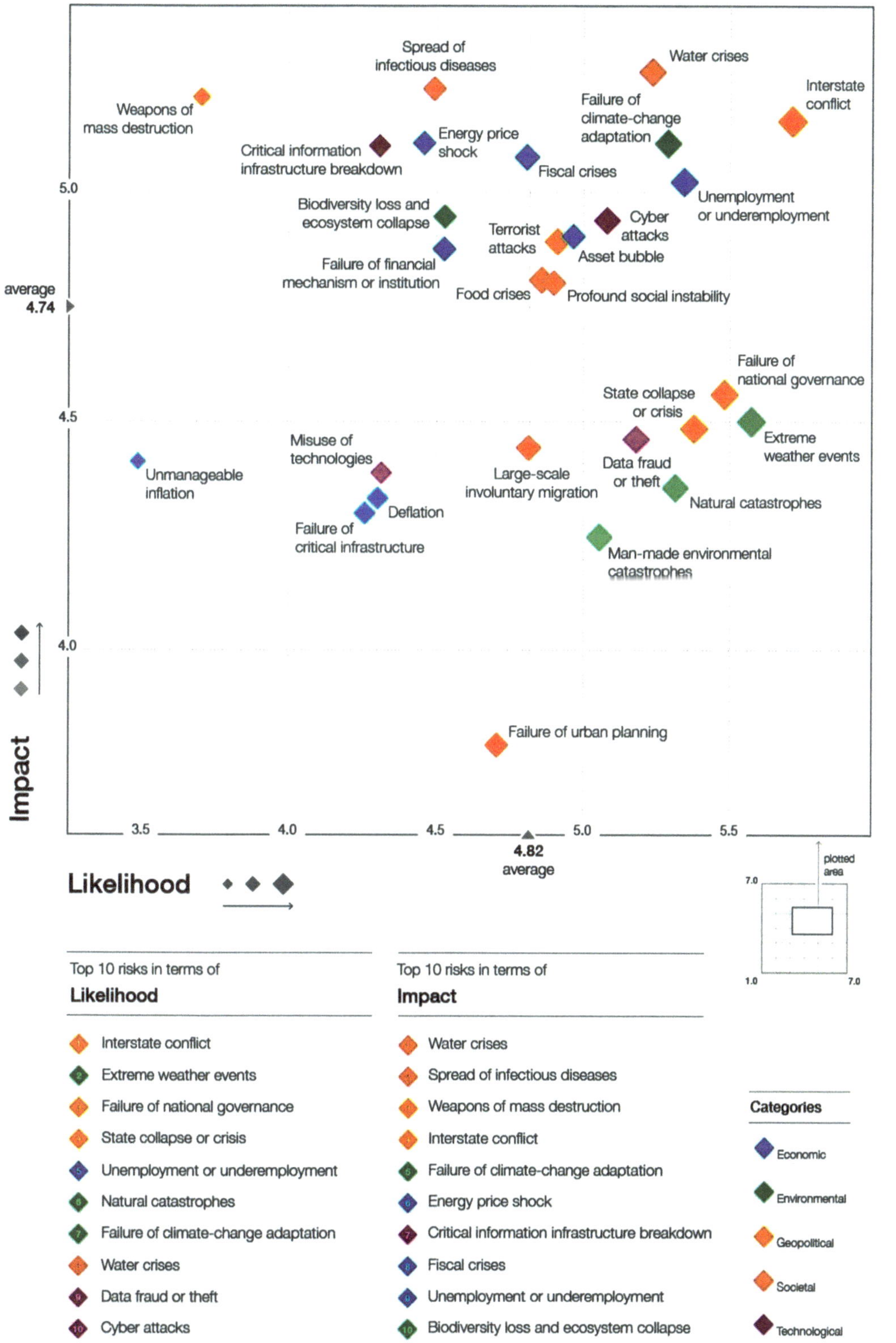

Fig. 19.2 Global risk landscapes 2015 (Source: World Economic Forum (2015))

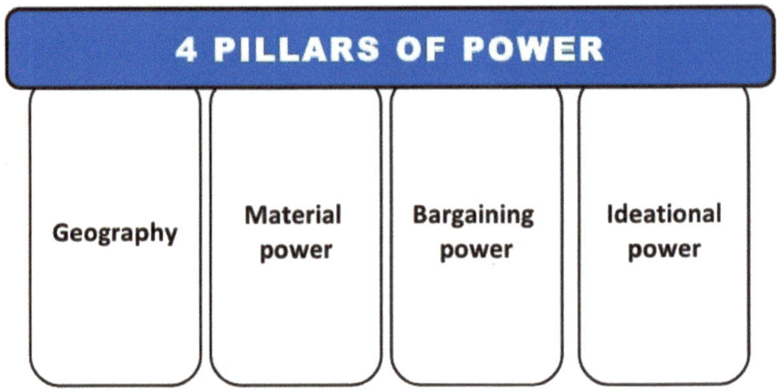

Fig. 19.3 Pillars of hydro-hegemony (Source: Zeitoun and Warner 2006)

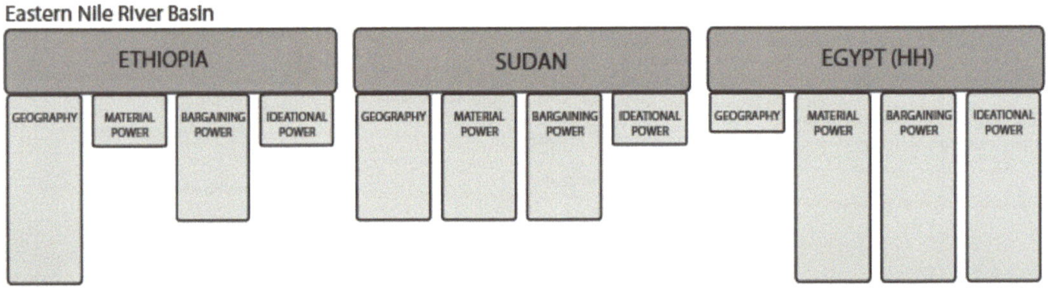

Fig. 19.4 Suggested plots of hydro-hegemonic configurations in the Eastern Nile, river basin (2009 estimates) (Source: Cascao and Zeitoun 2010)

Fig. 19.5 Range of forms of interaction over transboundary water resources (Source: Zeitoun and Warner 2006)

helpful to group contexts according to specific criteria – in terms of the character of the control exerted, for instance, as shown in Fig. 19.5. At the right-hand extreme of the continuum, control over the transboundary waters is openly competed for, sometimes in somewhat hostile political environments. At the opposite extreme is the cooperative form of interaction, based on the principle of full equality and manifested in terms of economic integration, equitable distributive politics and collective decision-making processes. The bulk of current transboundary water interaction lies conceptually between these two extremes, however – where control is shared in principle, but not in practice.

19.3 Transboundary Conflict Analysis Paradigms

"Competition" is defined as two or more entities, one or more of which perceives a goal as being blocked by another entity. If power is exerted to overcome the perceived blockage, it is referred to as "conflict". If there is coordination of behaviour among entities to realize at least some common goals, it is "cooperation". As the freshwater resource becomes scarce, the elements of the society must respond by either becoming adept at competing with each other or, in order to survive, they must learn to cooperate and develop symbiotic relationships. Human conflict over common resources can be dealt with through either competition or cooperation. Competition begets ill will, which increases competition, while, conversely, cooperation encourages better relations, thus creating an environment conducive to increased cooperation. The choice between ever-increasing conflict or cooperation in hydropolitics was discussed by Frey (1993). The tension and threat can apparently be resolved either by sharply escalating the conflict or by accepting the necessity of some form of cooperation. Dire conditions promote cooperation, but those same conditions also make severe conflict more likely (Frey 1993).

Just as natural water flow ignores international boundaries, so, too, does the evaluation of water resources transcend the analysis of any single discipline. Water, by nature, necessitates an interdisciplinary analysis. Through its physical components, we measure the quantity, quality and variability of water sources. Because we need to develop an infrastructure to harness water for human use – storage and delivery systems, for example – an engineering component should be incorporated into the analysis. Furthermore, because water can be owned, bought, sold and traded, its analysis takes on legal, economic and political aspects as well. Finally, because water is a resource that, when scarce, can induce both conflict and cooperation, water can become a subject for alternative dispute resolution (ADR) (Wolf 1995). Apart from the above, water acts as a spiritual binder affecting the emotions and psychology of people attached with it brining in the social discipline as well. Because of the properties inherent to human water needs, competition over water as a scarce resource, when it occurs, can be especially intense. Water, in short, seems to share only the most contentious characteristics with other resources, particularly in the international setting, making analysis of international water conflicts especially difficult.

There are numerous disciplines that treat water as a resource and as a subject of conflict. The approach followed by various disciplines towards conflict in general, and international water conflict in particular, is the key to resource security. The various disciplines for analysis of international water conflicts include (1) physical sciences and technology, (2) law, (3) political science, (4) economics, (5) game theory, (6) alternative dispute resolution (ADR), (7) philosophy, (8) spiritual practice and (9) sociology. There can be other disciplines too which support the analysis of international water conflicts. Delving deep into the disciplines is not of specific interest to the authors; rather it is intended to brief the practitioners about the various analysis paradigms which can be applied in the management of transboundary water resources.

19.3.1 Transboundary Water Management Through Benefit Sharing

In recent decades significant literature on conflict and cooperation along the lines of transboundary river management has developed (Bernauer 1997; Gleick 1993; Rogers 1993; Wolf 1998; Zeitoun and Warner 2006). Benefit sharing has the potential to develop as a mechanism which has the capabilities to resolve transboundary water conflicts (Dombrowsky 2009a, b; Sadoff and Grey 2002, 2005). The principal thought of the benefit-sharing mechanism is that riparians should not share the water resources; rather they share various benefits derived out of water. Through this paradigm shift, a zero-sum game of water sharing can be replaced by a positive-sum game of benefit sharing (Biswas 1999; Klaphake 2005; Sadoff and Grey 2002, 2005). Sadoff and Grey highlighted that to negotiate the management and development of international shared rivers, riparians can focus their negotiations on the allocation of water rights or on the distribution of benefits derived from the use of water (Sadoff and Grey 2005). Thus, according to this interpretation, the sharing of rights and the sharing of benefits can be understood as alternative negotiation strategies.

It is recognized that there are different types of benefits – ecological, political, social, economical, cultural and spiritual – derived from using water, and economic benefits of which have been widely available in literature (Sadoff and Grey

2002; Sengo et al. 2005; Whittington et al. 2005) and in practice (Giordano and Wolf 2003; Sadoff and Grey 2002). Sadoff and Grey (2002, 2005) proposed four categories of cooperation and benefits/costs that exist in connection to transboundary watercourses: First, benefits to the river generated through cooperative joint management of ecosystems. Second, benefits from the river derived from efficient, cooperative management and development of shared rivers. Third, cooperation on an international river will result in the reduction of costs because of the river. And finally, as international rivers can be catalytic agents, cooperation that yields benefits from the river and reduces costs because of the river can pave the way to much greater cooperation between states, even economic integration among states, generating benefits beyond the river (Sadoff and Grey 2002, 2005).

An added move to promote cooperation is by identifying the various cooperation modes or approaches that can be adopted and employing the most appropriate mode to achieve a particular goal. The optimal type of cooperation varies with hydrologic and investment opportunities and with the consequent potential benefit-sharing mechanisms. In order to facilitate optimal cooperative management in some basins, information sharing and basin-wide strategic assessments may be adequate, while in some basins, drought and flood mitigation, water storage and joint actions in river regulation would yield significant net benefits. A continuum of cooperation can be conceived from unilateral action to coordination to collaboration to joint action (Fig. 19.6).

In the absence of any water-sharing mechanism in place, an imperative need is felt to manage and harness the common pool resource of the river. It can be suggested that transboundary water conflicts may be resolved through benefit sharing. In the past, such analysis has assisted riparians to identify and explore the various trade-offs associated with different cooperative scenarios, thereby bringing clarity to the question of how they can and why they should cooperate. This should open up the concerned players for constructive interactions.

19.4 Transboundary Risks and Opportunities

Shared river basins create some level of stress among the binding societies which generate cooperative or non-cooperative responses which can reach far "beyond the river". The tensions and responses generated are bundled with various factors including – but not limited to – historic, cultural, environmental and economic which have significant impact on riparian relations. The shared bundled dynamics of transboundary river systems have the potential to either become a powerful catalyst for conflict or cooperation. Control over transboundary waters is inextricably entwined with national security, economic opportunity, society and culture. Understanding coexisting conflict and cooperation also facilitates the work of those from both groups involved in the design and execution of negotiation strategies at the multilateral level or in the development of positive-sum solutions addressing water and benefit-sharing paradigms in a balanced manner.

The quantitative work led by Wolf on the Transboundary Freshwater Dispute Database (Wolf 2002; Wolf et al. 2005) forms the basis of debate regarding cooperative events

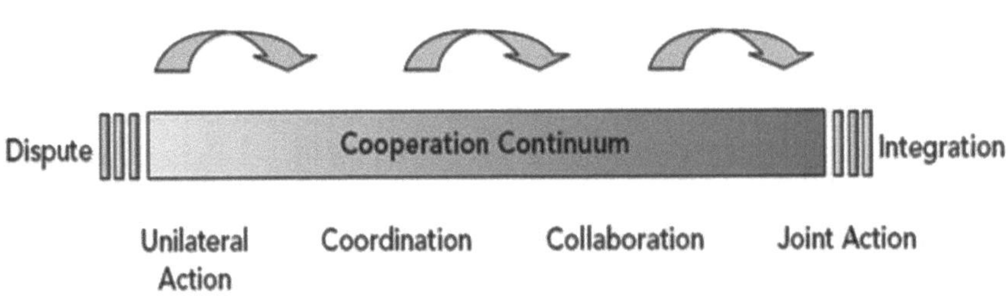

Fig. 19.6 Types of cooperation – the cooperative continuum (Sadoff and Grey 2005)

overtaking the conflict events by a big margin: 1228 cooperative events as against 507 conflict-related events. This implies that violence over water is neither economically viable, strategically rational nor hydrographically effective. The events of cooperation primarily cover water quantity, quality, economic development, hydropower and joint management. In contrast, a majority of conflict-laden events (~90 %) are related to quantity and infrastructure (Wolf 1998). Water can act as both irritant by making good relations bad and bad relations worse and a unifier by promoting cooperation along water and other allied sectors as well. International waters have the potential to act as a unifier in basins with relatively strong institutions.

19.4.1 Basins at Risk (BAR) Index

The history of transboundary rivers has a rich collection of both cooperative and conflicting international transboundary water events. International conflict and water appear with increased frequency both in policy literature as well as popular press (Elhance 1999; Gleick 1993; Homer-Dixon 1994; Hughes Butts 1997; Remans 1995; Westing 1986). The literature frequently discusses numerous indicators used to study and analyse water conflict, which includes proximity, type of government, water availability and rapid inhabitant growth. The major drawback of the existing literature was that it consisted of specific case studies from the most volatile basins which excluded various factors influencing water and international conflict.

Despite numerous case studies examining and comparing water-related conflicts, no global scale on water and international conflicts could be developed. This prompted Wolf et al. (2003) to create a global Basins at Risk (BAR) scale. Wolf and colleagues developed a 15-point "Basins at Risk (BAR) scale" which ranged from +7 to −7, including 0. +7 represented the most cooperative event, while −7 represented the most conflictive event, and 0 represented neutral or non-significant events (Wolf et al. 2003). The BAR scale developed by Wolf et al. incorporates water-specific terminologies

Table 19.1 BAR scale

BAR scale	Event description
−7	Formal declaration of war
−6	Extensive war acts causing deaths, dislocation or high strategic cost
−5	Small-scale military acts
−4	Political-military hostile actions
−3	Diplomatic-economic hostile actions
−2	Strong verbal expressions displaying hostility in interaction
−1	Mild verbal expressions displaying discord in interaction
0	Neutral or non-significant acts for the internation situation
1	Minor official exchanges, talks or policy expressions
2	Official verbal support of goals, values or regime
3	Cultural or scientific agreement or support (nonstrategic)
4	Non-military economic, technological or industrial agreement
5	Military, economic or strategic support
6	International Freshwater Treaty; major strategic alliance
7	Voluntary unification into one nation

and considerations with the International Cooperation and Conflict Scale developed by Edward Azar (1980) and Yoffe and Larson (2002). The higher BAR scale refers to higher level of cooperation, hence low conflict potential. Table 19.1 describes the 15 categories of BAR scale as developed by Wolf et al. (2003):

19.4.2 Risk-Opportunity Index (ROI)

Water is a fundamental human need and key to economic development. As the pressures of population and economic growth increase, water resources are under increasing stress. As the stress on water resources increases, the risk associated with the management of transboundary rivers increases exponentially given the hegemonic disparities of the riparians. Very recently a study was undertaken by the World Bank (Rai and Young 2015) (unpublished report) to assess the conflict-cooperation risk and opportunities in transboundary river basin in South Asian region. The study included developing a deeper understanding of the opportunities to promote cooperation in transboundary river basins and examining the significance of hegemony and development variables on the conflict-cooperation risk and opportunities in the context of South Asian transboundary rivers.

The risks and opportunities for cooperative water resources development were assessed from two perspectives: (1) the main threats to the basin's water resources from development and utilization dynamics, in terms of water, food and energy security and storage capacity of the basin, and (2) riparian interactions affected by the hegemonic disparities of the countries sharing the river basins along the lines of political, military and economic hegemony. The hegemony and development variables are explained as follows:

Hegemony Hydro-hegemony refers to water resources control strategies enabled by the exploitation of power asymmetries within a weak international institutional context. Political processes outside the water sector configure basin-wide hydropolitical relations in forms ranging from the benefits derived from cooperation under hegemonic leadership to the inequitable aspects of domination (Zeitoun and Warner 2006). Political and economic power influence how nations develop water resources in shared river basins. Larger economies can spend more on water resources infrastructure which in turn can be used to develop the water resources by enhancing storage capacity and realizing hydropower potential. Political and military dominance affects the negotiation processes. A country with greater say in negotiations and decision-making process may obtain a greater fraction of the shared water resource.

Development It describes the pressures on natural resources due to the present scenarios of water resources availability, storage capacity and food and energy security. Rapid economic development and high population growth are impacting food and water security. The bulk of freshwater is used for irrigation and its efficiency of use is low. Demand for water has escalated in recent decades, leading to increased competition for access to water between riparian countries and different water-using sectors of the economy.

The method used has a two-step approach: (i) diagnosis of issues and (ii) in-depth assessment of the identified issues, using a fuzzy risk assessment framework. A comprehensive risk-opportunity analysis has been conducted using a composite risk-opportunity index (ROI) based on four components of development (water security, food security, energy security and storage capacity) and three components of hegemony (political, military and economy).

The risk-opportunity index (ROI) for the riparian countries sharing the river basin is expressed as

$$ROI = f(D,\ H)$$

where:

ROI = risk-opportunity index
D = development
H = hegemony

High risk is associated with higher water resources stress, development pressure as well as higher hegemony disparity. This assessment recognizes that a sustainable freshwater system can only function within an integrative operational framework that combines the natural system and the management system. The different aspects related to the natural resource base and hegemonic disparities influence water resources management and thus conflict-cooperation risks and opportunities.

The risk assessment used in this analysis is based on the fuzzy synthetic evaluation (FSE) technique proposed by Rai et al. (2014). FSE combines fuzzy logic and an analytical hierarchy process (AHP). AHP is used to assign weights to individual hegemony and development parameters. Fuzzy logic is used to determine the risk and opportunities by fuzzifying the hegemonic and development variables using suitable membership functions. The hegemony and development variables are combined through a fuzzy rule base to obtain the risks and opportunities, which is then defuzzified using a scoring method.

This is the first instance when fuzzy logic (FL) has been employed in the assessment of risks and opportunities in transboundary river basins. So far fuzzy inference system (FIS) has been used in water resources allocation problems between riparians which can deal with statistical data as well as linguistic data. Water allocation problems bear fuzzy characteristics due to imprecision and uncertainty of linguistic and numerical data. Linguistic uncertainty can be treated by fuzzy processes, which take into account partial contribution of multiple parameters from each riparian point of view. A systematic computing framework employing fuzzy rules, fuzzy logic and fuzzy inference system greatly amplifies the power of human reasoning (Zadeh 1968).

19.5 Hydropolitical Vulnerability and Resilience

The concept of sustainable development was first introduced around 30 years back by the World Conservation Strategy (IUCN 1980). Sustainable development balances the exploitation of natural resources, technology development and institutional change to enhance the potential to meet human needs and aspirations, now and in the future (WCED 1987). To achieve sustainability, all the components in the system must be also in balance (Sandoval-Solis et al. 2010). Loucks (1997) defined sustainable water resources systems as "those systems designed and managed to contribute fully to the objectives of society, now and in the future, while maintaining their ecological, environmental and hydrological integrity" (Loucks 1997). Sustainability consists of three pillars: social development, environment protection and economic development.

Recently, strong emphasis has been placed on the adaptive capacity of water resources systems, which refers to measures that reduce the vulnerability of systems to actual or expected future changes. Vulnerability is the magnitude of an adverse impact on a system. In general, concepts of "resilience" and "vulnerability" as related to water resources are often assessed within the framework of "sustainability" (Blaikie et al. 1994) and relate to the ability of biophysical systems to adapt to change (Gunderson and Pritchard 2002). As the sustainability discourse has broadened to include human systems in recent years, so too has work been increasingly geared towards identifying indicators of resilience and vulnerability within this broader context (Bolte et al. 2007; Lonergan et al. 2000; Turner 2010). In parallel, dialogue on "security" has migrated from traditional issues of war and peace towards also beginning to incorporate the human-environment relationship in the relatively new field of "environmental security" (Vogel and O'Brien 2004).

"Hydropolitical resilience", then, is defined as the complex human-environmental system's ability to adapt to permutations and change within these systems; "hydropolitical vulnerability" is defined by the risk of political dispute over shared water systems. Wolf et al. (2003) suggested the following relationship between change, institutions, and hydropolitical vulnerability: "The likelihood of conflict rises as the rate of change within the basin exceeds the institutional capacity to absorb that change". It can be said that very rapid change, either on the institutional side or in the physical systems, that outpaces the institutional capacity to absorb those changes is at the root of most water conflict (Wolf 2006).

19.5.1 Hydropolitical Sustainability

As per the United Nations 2015 report, "inadequate financing and deficient information about the state of water resources, their use and management impose constraints on water resources management and its ability to help achieve sustainable development objectives" (UNESCO 2015). Water beyond the boundaries always confronts increased challenges to social

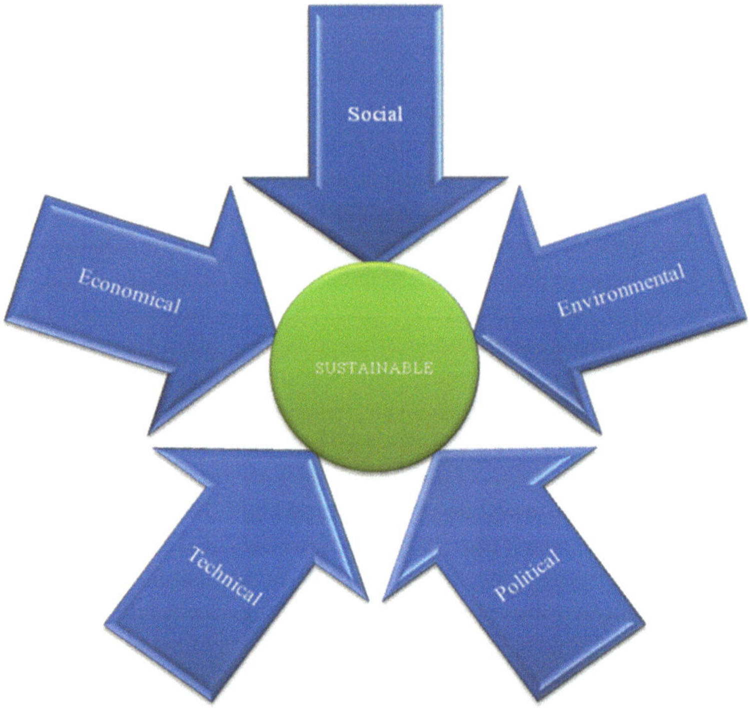

Fig. 19.7 Constituents of HypSI forming circles of blue sustainability

sustainability due to the control of political circumstances over the techno-environmental demands. "Hydropolitics" has been discussed earlier in this chapter, but in view of sustainability considering the broader viewpoint, hydropolitics can be defined as expeditiously engineered study of interactions, either conflict or cooperation, between riparians sharing transboundary water resources in the rapidly changing political and economic equations. Hydropolitics predominantly comprises two authoritative constituents: "scale" and "range of issues" (Turton and Henwood 2002). International rivers call for uncertainty and ambiguities concerning shared waters as the concerned riparians have dissimilar attitudes. The existing political scenarios point towards the need of paradigm shift in management framework which is capable to understand and delve deeply with *hydropolitical sustainability*. As proposed by the authors, hydropolitical sustainability can be defined as "the composite socio-techonomic (technical + economic) international waters system which has the ability to accommodate transpositions and modification within the political and economic context of shared water resources".

The exercise of sustainable water resources can be framed taking cue from the recent study of sustainable urban cities. City and water resources seem similar as in both the systems, human adaptation is of prime importance. Circles of Social Life is an approach which leads to engaged and collaborative exercise in making our cities, locales and systems more sustainable. As part of this overall approach, Circles of Sustainability were formulated that provides practical tools for creating sustainable cities and communities (James 2015).

In this chapter the authors propose the circles of blue sustainability (blue implies water) based on the above-mentioned concept which can be used in the formulation of hydropolitical sustainability index (HypSI). The circles of blue sustainability consider E^2SPT (environment, economics, social, political and technical) parameters. The constituent of HypSI which forms the circles of blue sustainability is shown in Fig. 19.7.

19.5.2 Components of Hydropolitical Sustainability Index

19.5.2.1 Social Desirability

Social desirability constitutes of two main interdependent features – mutual concessions and compromises. Society develops better ways to execute its aspirations to achieve its socioeconomic objectives. In terms of concession, society wishes to get the security for life as well as food. In parallel, augmentation of lifestyle plays an important role for the successful applied policy in water resources. Improvement of lifestyle alleviates poverty for the social security to individual. Individual development includes the second aspect, i.e. compromises, and it comes on the shoulder of individual responsibility (Dinar 2002). The following are the parameters which indicate social desirability:

- Life security
- Food security
- Poverty alleviation
- Individual responsibility

19.5.2.2 Environmental Sustainability

Hydro-environmental sustainability is achieved by minimizing the effect of anthropogenic actions on the water resources systems. River stability has been given precedence to promote Integrated Land and Water Resources Management (ILWRM). River health monitoring (RHM), with both quality and quantity, should be encouraged to attain hydro-environmental sustainability. Judicious utilization of both ground and surface freshwater will also stimulate the sustainability (Hanjra and Qureshi 2010). The following are the parameters which indicate environmental sustainability:

- River stability
- River health monitoring
- Conjunctive use
- Integrated Land and Water Management

19.5.2.3 Political Legitimacy

Insatiate behaviour inheres at the heart of transboundary water conflicts. It can be handled with the help of accountability, transparency and policies. Institutional capabilities should be utilized to set industrial responsibilities for the ultimate hydropolitical sustainability (Saleth and Dinar 2000). The following are the parameters which indicate political legitimacy:

- Accountability
- Transparency
- Policies
- Industrial responsibility
- Institutional capabilities

19.5.2.4 Economical Viability

To build up any eminent frame work, economic viability is the principally desired component which has to be considered. Development of market with the help of planned tariff for individual work will add financial accountability to water resources system (Johansson et al. 2002). Indigenous technology and water use optimization can input the ingredients to hydro-eco sustainability (Saleth and Dinar 2004). The following are the parameters which indicate economic viability:

- Water tariffs
- Water market
- Indigenous technology
- Treated water reuse

19.5.2.5 Technical Feasibility

In the context of hydropolitical sustainability, technical feasibility generally refers to the efficiency of the system based on their technical input and output. To define the same, quantification of separate components is required. In the technical terms, excess of water (flood) and lack of water (drought) both create a problem in the water resources system development. Quantification of these quantities can warrant the early planning and preparedness to the social security. Reliability of water supplies for municipal as well as irrigation purpose will be the major component to constitute the framework of hydropolitical sustainability (Dinar 2002; Saleth and Dinar 2000). The following are the parameters which indicate technical feasibility:

- Flood inundation mapping
- Drought mapping
- Municipal water need
- Water for irrigation – timing and reliability of the water supplies
- Land erosion and crop loss

Initially to start with the above aspects can be included in the parameterization of the circles of blue sustainability which forms the hydropolitical sustainability index. These can be modified and upgraded with more diverse parameters as research progresses in this field.

All the models of sustainability consider environmental, economical, social and political aspects. But very few consider the technical aspects as well. Technology can be used to safeguard the natural resources and promote sustainable development through cooperative mechanisms. Technical knowledge and expertise has to be promoted among water practitioners. A very small fraction of the water policy experts and practitioners involved in the decision-making process and formulation of water policies have detailed knowledge of the latest technological advancements in the management of natural freshwater reserves.

There is a strong need to develop the hydropolitical sustainability index which can respond to the environmental, economical, social, political and technical challenges posed by transboundary river systems. The international river basins at high risk are required to be studied with respect to the proposed hydropolitical sustainability index. River basins which have certain cooperative mechanisms in place need to be cross-checked with HypSI to assess the sustainability of the management practices, frameworks and mechanisms. This can open up new areas of research of practical relevance in transboundary water management and hydropolitics associated.

19.6 Conclusion

In this chapter, it has been endeavoured to delve into hydropolitics and its various related aspects. Hegemony does affect basin hydropolitics at the international arena. Political dominance affects the negotiation process and more importantly how relations between riparians flourish. The complex nature of water makes it more contentious due to the various stakeholders involved. Conflicts related to water can be analysed using numerous disciplines very much similar to the variety of human needs of water. Benefit sharing has the potential to resolve water conflicts to a greater extent as it tries to shift the focus from the physical quantities of water to the numerous benefits derived out of water. Benefit sharing has the capability to reduce the risks of conflicts and enhance opportunities of cooperation in the transboundary setup of major river basins. In the end the authors propose an index called as HypSI, i.e. hydropolitical sustainability index, to assess the hydropolitics involved in international river basins. Hydropolitical sustainability index revolves round the circles of blue sustainability (where blue implies water) which includes social desirability, environmental sustainability, political legitimacy, economical viability and technical feasibility as parameters. Further research is needed to refine circles of blue sustainability in a broader context even including more variables to create a better and deeper understanding of hydropolitical sustainability using HypSI.

References

Azar EE (1980) The conflict and peace data bank (COPDAB) project. J Confl Resolut 24(1):143–152

Bernauer T (1997) Managing international rivers. In: Young OR (ed) Global governance: drawing insights from the environmental experience. The MIT Press, Massachusetts Institute of Technology, Cambridge, pp 155–196

Biswas AK (1999) Management of international waters: opportunities and constraints. Int J Water Resour Dev 15(4):429–441

Blaikie P, Cannon T, Davis I, Wisner B (1994) At risk: natural hazards, people's vulnerability and disasters. Routledge, London

Bolte JP, Hulse DW, Gregory SV, Smith C (2007) Modeling biocomplexity–actors, landscapes and alternative futures. Environ Model Softw 22(5):570–579
Cascão AE, Zeitoun M (2010) Power, hegemony and critical hydropolitics. In: Transboundary water management: principles and practice. Earthscan, London, pp 27–42
Daclon CM (2007) Geopolitics of environment. A wider approach to the global challenges. Comunità Internazionale 62(4):681–692
Dinar S (2002) Water, security, conflict, and cooperation. SAIS Rev 22(2):229–253
Dombrowsky, I (2009a) Benefit sharing in transboundary water management through intra-water sector issue linkage. In: Conference report: On the water front: selections from the 2009 World Water Week in Stockholm, SIWI. pp 25–31
Dombrowsky I (2009b) Revisiting the potential for benefit sharing in the management of trans-boundary rivers. Water Policy 11(2):125–140
Draper SE (2002) Model water sharing agreements for the twenty-first century. ASCE Publications, Reston, p 166
Elhance AP (1999) Hydropolitics in the third world: conflict and cooperation in international river basins. US Institute of Peace Press, Washington, DC
Frey FW (1993) The political context of conflict and cooperation over international river basins. Water Int 18(1):54–68
Giordano MA, Wolf AT (2003) Sharing waters: Post-Rio international water management. Nat Res Forum 27(2):163–171
Gleick PH (1993) Water and conflict: fresh water resources and international security. Int Secur 18(1):79–112
Gunderson LH, Pritchard L (2002) Resilience and the behavior of large-scale systems. Island Press, Washington, DC
Hanjra MA, Qureshi ME (2010) Global water crisis and future food security in an era of climate change. Food Policy 35(5):365–377
Homer-Dixon T (1994) Environmental scarcities and violent conflict: evidence from cases. Int Secur 19(1):5–40
Hughes Butts K (1997) The strategic importance of water. Parameters 27:65–83
IUCN (1980) World conservation strategy: living resource conservation for sustainable development. World Conservation Union, United Nations Environment Programme, Word Wide Fund for Nature, Gland
Jägerskog A, Granit J, Risberg A, Yu W (2007) Transboundary water management as a regional public good. Financing development–an example from the Nile Basin (No. 363.61 T772t). Cambridge University Press, Cambridge
James P (2015) Urban sustainability in theory and practice: circles of sustainability. Routledge, London
Johansson RC, Tsur Y, Roe TL, Doukkali R, Dinar A (2002) Pricing irrigation water: a review of theory and practice. Water Policy 4(2):173–199
Klaphake A (2005) Economic and political benefits of transboundary water cooperation. In: Proceedings of the IHP-HWRP, the value of water–different approaches in transboundary water management. Proceedings of the International Workshop, Koblenz, pp 91–99
Lasswell HD (1936) Politics: who gets what, when, how. P. Smith, New York
Lonergan S, Gustavson K, Carter B (2000) The index of human insecurity. Aviso 6:1–11
Loucks DP (1997) Quantifying trends in system sustainability. Hydrol Sci J 42(4):513–530
Nicol A, van Steenbergen F, Sunman H, Turton A, Slaymaker T, Allan JA, de Graaf M, van Harten M (2001) Transboundary water management as an international public good. Ministry of Foreign Affairs, Sweden
Rai SP, Young W (2015) Freshwater conflict and cooperation in South Asia. World Bank, Delhi
Rai SP, Sharma N, Lohani A (2014) Risk assessment for transboundary rivers using fuzzy synthetic evaluation technique. J Hydrol 519:1551–1559
Remans W (1995) Water and war. Humantäres Völkerrecht 8(1):1–14
Rogers P (1993) The value of cooperation in resolving international river basin disputes. In Proceedings of the natural resources forum. Wiley Online Library, pp 117–131
Sadoff CW, Grey D (2002) Beyond the river: the benefits of cooperation on international rivers. Water Policy 4(5):389–403
Sadoff CW, Grey D (2005) Cooperation on international rivers: a continuum for securing and sharing benefits. Water Int 30(4):420–427
Saleth RM, Dinar A (2000) Institutional changes in global water sector: trends, patterns, and implications. Water Policy 2(3):175–199
Saleth RM, Dinar A (2004) The institutional economics of water: a cross-country analysis of institutions and performance. Edward Elgar Publishing, Cheltenham
Sandoval-Solis S, McKinney DC, Loucks DP (2010) Sustainability index for water resources planning and management. J Water Resour Plan Manag 137(5):381–390
Schulz M (1995) Turkey, Syria and Iraq: a hydropolitical security complex. In Hydropolitics: conflicts over water as a development constraint. Zed Books, London. pp 91–122
Sengo DJ, Kachapila A, van der Zaag P, Mul M, Nkomo S (2005) Valuing environmental water pulses into the Incomati estuary: key to achieving equitable and sustainable utilisation of transboundary waters. Phys Chem Earth A/B/C 30(11):648–657
TFDD (2012) International river basin register. In College of earth, ocean, and atmospheric sciences. Oregon State University, Corvallis, USA
Turner B (2010) Vulnerability and resilience: coalescing or paralleling approaches for sustainability science? Glob Environ Chang 20(4):570–576

Turton A, Henwood R (2002) Hydropolitics in the developing world: a Southern African perspective. IWMI, Pretoria

UNESCO (2015) Water for a sustainable world. The United Nations World Water Development Report. UNESCO Publishing/Earthscan, 2015

Vogel C, O'Brien K (2004) Vulnerability and global environmental change: rhetoric and reality. Aviso 13:1–8

Wada Y, Bierkens MF (2014) Sustainability of global water use: past reconstruction and future projections. Environ Res Lett 9(10):104003

Waterbury J (1979) Hydropolitics of the Nile Valley. Syracuse University Press, New York. ISBN 0815621922

Wced BC (1987) Our common future. Oxford University Press, Oxford

Westing AH (1986) Global resources and international conflict: environmental factors in strategic policy and action. Oxford University Press, Oxford

Whittington D, Wu X, Sadoff C (2005) Water resources management in the Nile basin: the economic value of cooperation. Water Policy 7(3):227–252

Wolf AT (1995) Hydropolitics along the Jordan River; scarce water and its impact on the Arab-Israeli conflict. United Nations University Press, New York

Wolf AT (1998) Conflict and cooperation along international waterways. Water Policy 1(2):251–265

Wolf AT (1999) The transboundary freshwater dispute database project. Water Int 24(2):160–163

Wolf AT (2002) Conflict prevention and resolution in water systems. In: Howe CW (ed) The management of water resources series. Edward Elgar Publishing Ltd, Cheltenham

Wolf A (2006) Hydropolitical vulnerability and resilience along international waters (five volumes: Africa, Latin America, North America, Asia, Europe). UN Environment Programme, Nairobi, p 2007

Wolf AT, Natharius JA, Danielson JJ, Ward BS, Pender JK (1999) International river basins of the world. Int J Water Resour Dev 15(4):387–427

Wolf AT, Yoffe SB, Giordano M (2003) International waters: identifying basins at risk. Water Policy 5(1): 29–60

Wolf AT, Kramer A, Carius A, Dabelko GD (2005) Managing water conflict and cooperation. In: State of the World 2005: redefining global security. The Worldwatch Institute, Washington, DC, pp 80–95

World Economic Forum (2015) Global Risk 2015. Geneva: World Economic Forum

Yoffe S, Larson K (2002) Basins at risk: water event database methodology. In: Yoffe SB (ed) Basins at risk: conflict and cooperation over international freshwater resources. Oregon State University, Corvallis, USA

Zadeh LA (1968) Fuzzy algorithms. Inf Control 12(2):94–102

Zeitoun M, Warner J (2006) Hydro-hegemony-a framework for analysis of trans-boundary water conflicts. Water Policy 8(5):435–460

Subash Prasad Rai Department of Water Resources Development and Management, Indian Institute of Technology Roorkee, Roorkee, Uttrakhand, India

Aaron T. Wolf Department of Geosciences, CEOAS, Oregon State University, Corvallis, OR, USA

Nayan Sharma Department of Water Resources Development and Management, Indian Institute of Technology Roorkee, Roorkee, Uttrakhand, India

Harinarayan Tiwari Department of Water Resources Development and Management, Indian Institute of Technology Roorkee, Roorkee, Uttrakhand, India

Part VIII

Disaster Management

Flood Disaster Management 20

R. Ranjan

Abstract

Over the past several years, the intensity and frequencies of both riverine and urban flooding are showing increasing trends. In many of the Asian countries including India, flooding due to inundation in low lying areas along river channels as well as localized unprecedented precipitation has become a common feature that has gained the status of disasters causing devastations in terms of loss of lives and huge economic losses every year. The number of people affected by riverine and urban floods is more than any other type of natural disasters. The J&K Floods in 2014, Uttarakhand Floods in 2013 and the mega flood of Chennai (Tamil Nadu) in December 2015 highlight the vulnerability status of India.

The prime reasons of flooding being the reduced carrying capacity of river channels, unusual amount and intensity of rainfall in a localized area, climate change over the past decades, poorly managed land use planning, faulty drainage design, construction & management, lack of real time warning etc. Sometimes, sudden release of water from dams along with other localized phenomena also creates flooding. The increasing ramifications of such disasters have afflicted policy planners, scientists, researchers, academicians and others concerned across the world to find out ways and means to deal with the emerging threats of this typical hydro-meteorological phenomena.

The present chapter discuss in details about the fundamental concept of the management of disasters due to floods. The flood vulnerability status in SAARC countries is also discussed along with the existing vulnerabilities of Indian states. In addition the flood management and mitigation strategies are also suggested along with do's and don'ts.

R. Ranjan (✉)
Disaster Risk Management Expert & Senior Consultant,
(NDMA), New Delhi 110029, India
e-mail: dr.rajnishranjan@gmail.com

N. Sharma (ed.), *River System Analysis and Management*, DOI 10.1007/978-981-10-1472-7_20

20.1 Introduction

Flood is a unique hydrometeorological phenomenon having widespread occurrence across the globe with varied severity and dimensions. The World Disaster Report 2014 highlights the impacts of flood with another high-profile disaster like earthquake and declares that flood incidences in the world are nearly eight times more than that of earthquakes and the number of people affected by floods is nearly 12 times more.

There are different perceptions of flood for different stakeholders. For common people, it means devastation, destructions, damage, starvation, loss of lives, damage to properties and infrastructures, etc. Those living in urban areas may treat this phenomenon as disruption in their normal functioning of society. The government machinery and policy planners treat this situation as a factor of retardation in developmental planning with additional overburden on economy and additional expenditure on rescue, relief, rehabilitation, mitigation, etc. The engineering community, especially civil engineers, considers this as a situation review of existing flood protection measures.

In some of the South Asian countries including India, there is a typical condition of flood and drought occurrence at the same time. It is therefore essential to analyze the dynamics of floods and associated phenomenon so that appropriate flood management strategies and techniques can be adopted.

The *International Commission on Irrigation and Drainage* (1995) has defined "flood" as "relatively high flow in a river markedly higher than the usual". The *World Meteorological Organization (WMO)/UNESCO International Glossary of Hydrology* (*WMO 1974*) defined flood as "rise, usually brief in the water level of a stream or water body to a peak from which the water level recedes at a slower rate" and "relatively high flows as measured by stage height or discharge."

20.1.1 Types of Floods

Floods have been classified in various ways depending upon the nature, severity, and sources of inundation:

Riverine flooding is the main flooding type that occurs due to various reasons but primarily due to heavy precipitation or glacial melt with resultant runoff. The increased discharge in river channels with decreasing carrying capacities leads to overflow causing inundation in the adjoining low-lying areas.

Flash flooding is an unprecedented situation that occurs in hilly regions and sloping lands where torrential heavy precipitation, thunderstorm, or cloud burst commonly occurred without any prior warning. This sometimes cause huge loss of lives and damage to properties.

Urban flooding occurs in regions where developmental planning has not been in tune with the geomorphological, ecological, and environmental set up, which results in the increased vulnerability in urban regions. Many urban agglomerations in India are suffering from the problem of flooding even after moderate rainfall. The situation aggravates when rainwater mix up with drain water causing many additional problems including spread of epidemics.

Coastal flooding occurs due to a number of reasons like cyclones and associated storm surge, high tides, tsunami, etc., wherein the low-lying areas in coastal tracts are inundated, as a result of which losses occur on a larger scale. In addition, salinity increases in the coastal groundwater and wells.

Glacial Lake Outburst Flood (*GLOF*) occurs in the downstream of glacial regions, where glaciers holding large quantities of water suddenly release them due to melting of ice jam. Glacial outburst is one of the prime reasons of flash floods in some of the Himalayan rivers.

Cloud Burst Flood is the manifestation of climate change and hydrological imbalance that primarily occurs as sudden heavy rainfall. Cyclonic circulations in monsoon may also lead to cloud burst.

Cyclone and storm surge flooding mainly occurs in coastal areas due to rainstorms associated with low-pressure systems. Movement of cyclonic storms in quick succession leads to severe flooding, especially in low-lying coastal areas.

20.2 Flood Vulnerability in SAARC Countries

The menace of flood in some of the SAARC countries like Bangladesh, India, Sri Lanka, and Pakistan is a recurrent phenomenon that directly affects their populace and economy and indirectly affects neighboring countries having trade relations with them. Despite adoption of several structural and nonstructural measures, flood causes huge loss of lives, damage to properties and infrastructures, and loss of economy and livelihoods. Every year about thousands of people lose their lives due to this annual hydrometeorological phenomenon with millions get homeless.

Even though several measures are being adopted in almost all the affecting nations to minimize the impact of floods, it's still a matter of concern that damages due to floods are showing an increasing trend. The impacts are so severe that during the entire period of flood, the only companions with graved communities are despair, hope, anger, stress, and agony besides their disrupted life activities.

The worst repercussion of this phenomenon is damages to infrastructures like bridges, residential buildings, road networks, electricity transmission networks, telecommunication lines, etc. For the agricultural sector, it comes as an anathema causing huge damage to standing crops, vegetation, and overall agricultural productivity.

The flood also create disasters within a disaster, when the outbreak of epidemics and other waterborne diseases sets in affecting communities, which are already crippled in a disastrous situation.

It's a matter of great concern that floods are now affecting areas in the entire South Asian regions which were otherwise considered flood-safe zone. This is showing an alarming trend in countries like India, Bangladesh, Pakistan, and Sri Lanka. The Climatologists, Hydro-geomorphologists, and other domain experts are citing this phenomenon as possibly arising due to the impact of global warming and climate change. Besides, the vast expansion of industrial and urban agglomerations and increasing developmental activities are also causing threats to the situation (Table 20.1).

The entire South Asian countries except Maldives and Sri Lanka receive most of the precipitation from the southwestern monsoon which, in general, prevails from June to September and contributes about 70–80 % of rainfall in the region. On the other hand countries like Maldives and Sri Lanka receive mostly North Eastern Monsoon. Occasionally, climatic phenomenon like western disturbance also causes rainfall in some parts of Sri Lanka, India, Pakistan, etc.

20.2.1 Afghanistan

Flood is the most common hydrometeorological disaster in Afghanistan, mainly affecting foothill regions of mountains and alluvial plains of rivers in the northern provinces. The main river of Afghanistan is *Amu Daria*, which is connected with branches having their sources lying in the mountains. During spring and early summer, these rivers are fed by snow melts, which increase discharge volume in rivers, causing

Table 20.1 Decadal flood damages and deaths in South Asia (2005–2015)

Sr. No.	Year	No. of occurrence	Total deaths	Affected	Injured	Homeless	Total affected	Total damage (000 $)
1.	2005	41	3281	36,889,173	584	268,525	37,158,282	6,220,000
2.	2006	36	1915	3,808,350	890	4,013,615	7,822,855	3,409,000
3.	2007	37	4362	53,071,942	342	3480	53,075,764	841,719
4.	2008	21	1886	13,509,075	97	2,435,818	15,944,990	248,029
5.	2009	19	1807	7,201,224	166	60,000	7,261,390	2,514,000
6.	2010	21	3236	25,653,988	3181	400,000	26,057,169	11,754,000
7.	2011	16	1358	19,213,054	948	1,069,973	20,283,975	4,657,000
8.	2012	20	1217	15,153,391	3239	2445	15,159,075	2,746,200
9.	2013	15	7139	5,015,385	5437	4314	5,025,136	2,860,000
10.	2014	17	1871	11,373,645	1433	660,000	12,035,078	18,475,000
11.	2015	13	406	319,800	333	1000	321,133	100,000
Total		256	28,478	191,209,027	16,650	8,919,170	200,144,847	53,824,948

(Source – emdat.be)

floods in low-lying areas. The heavy rainfall or avalanches in the region also sometimes cause flood, especially flash floods, e.g., flash floods in Faryab and Baghlan province on May 8, 2015. The Avalanche-induced floods occurred in provinces like Panjshir, Nangarhar, Laghman, Kapisa, Parwan, Nuristan, and the capital Kabul in February 25–27, 2015; flash floods in Jawzjan province and Badakhshan province in April to June 2014; etc. (SAARC Disaster Knowledge Network Portal) (Table 20.2).

(Courtsey-theguardian.com)

(Courtsey-en.trend.az)

Floods in Afganistan

20.2.2 Bangladesh

Geographically, the People's Republic of Bangladesh is located at the confluence of Padma (Ganges), Jamuna (Brahmaputra), and Meghna rivers and their tributaries. On account of being located in the downstream of these three rivers, the country is extremely vulnerable to

Table 20.2 Decadal flood damages and deaths in Afghanistan (2005–2015)

Disaster subtype	Events count	Total deaths	Total affected	Total damage ('000 US$)
Riverine flood	30	1116	226,005	20,000
Flash flood	13	797	195,431	3000

(Source – emdat.be)

Table 20.3 Decadal flood damages and deaths in Bangladesh (2005–2015)

Disaster subtype	Events count	Total deaths	Total affected	Total damage ('000 US$)
Riverine flood	15	1544	24,908,777	274,000
Flash flood	4	42	1,795,559	0

(Source – emdat .be)

flooding. Some of the historic floods in the years 1987, 1988 and 1998 had created great disasters in the country, leaving behind the trials of devastations and destructions. In addition, floods have also been affecting the country due to localized reasons like collapse of embankments and tidal effects in the coastal regions.

There are two types of flood occuring in Bangladesh – *annual* (*moderate*) flood known as *barsha*, which does not cause significant damage of communities, whereas high-intensity flood *Banna* creates destruction and devastation, inundating about 35 % of the areas. Some of the recent examples of flood are that due to heavy rainfall in the southeastern district of Cox's Bazar and neighboring Bandarban in June 2015 and flooding due to collapse of embankments of river Kholpetua in southwest district of Satkhira in March 2015, which caused substantial losses in the country (Table 20.3).

20.2.3 Bhutan

On account of unique geological and geographical location, Bhutan is not much prone to annual floods. The country is located in the high mountainous regions with deep, narrow gorges through which rivers flow in a confined channel. This unique topographic characteristic has made the country less vulnerable to annual floods except occassional flash floods somewhere in the lower reaches due to heavy precipitation. This sometimes carries boulders, pebbles, and cobbles along with a high flow of waters which causes substantial loss of lives and damage to properties and infrastructures of downstream cities and communities. In May 2009 heavy precipitation due to the impact of cyclone *Aila* exceeded the water bearing capacity of rivers *Punakha* and *Wangdue* affecting almost all the districts.

There are other threats of flood in Bhutan known as *Glacial Lake Outburst Floods* (*GLOF*). Due to the impact of climate change and other localized hydrometeorological factors, the Himalayan glaciers are shrinking rapidly. The retreating glaciers increase the volumes of water in the existing lakes created by glacial erosion, as a result of which lakes burst with consequent flooding occurs in the foothill regions. As per the study conducted by the Department of Geology and Mines, Bhutan, in collaboration with ICIMOD, there are 2674 glacial lakes existing in Bhutan, out of which about 562 are associated with glaciers. Twenty-five of them are considered as potentially dangerous lakes that could pose a GLOF threat in the near future (SADKN Portal).

20.2.4 Nepal

The mountainous country of Nepal is considered among the most vulnerable flood regions in South Asia. The rivers like Narayani, Kosi, Karnali, Mahakali, etc., are perennial in nature. They are fed by waters coming from snow clad mountains. During monsoon period, all the rivers receive excess water in addition to glacial melts. This causes flooding especially in Terai regions

Table 20.4 Decadal flood damages and deaths in Nepal (2005–2015)

Disaster subtype	Events count	Total deaths	Total affected	Total damage ('000 US$)
Riverine flood	16	923	12,048,61	63,429
Others	1	294	184,894	0
Flash flood	1	119	4320	0

(Source –emdat.be)

Table 20.5 Flood damage in Maldives (2005–2015)

Disaster subtype	Events count	Total deaths	Total affected	Total damage ('000 US$)
Coastal flood	1	0	1649	0

(Source – emdat.be)

of Nepal. Some of the devastating floods occurred in 1978 (Tinao basin), 1980 (Tadi basin), 1987 (Sun Koshi basin), and in August 2008, Kosi floods, that caused devastation in eastern Nepal affecting about 200,000 people. One recent flood also occurred in Taplejung district in June 2015 (Table 20.4).

20.2.5 Maldives

The group of 142 inhabited islands (source-SADKN) is particularly vulnerable to localized floods of lesser severity. These include rainfall-induced flooding, tidal flooding in low-lying coastal areas due to high tides, flooding due to storm surge during cyclones, etc. The long-term perspective also predicts climate change with the resultant sea level rise as a potential threat of flooding especially in low-lying coastal areas.

Some of the remarkable flooding in Maldives occurred in 1987 and 2007 affecting more than 2000 people in the country (source-The International Disaster Database) (Table 20.5).

20.2.6 Pakistan

Pakistan is among those South Asian countries, where flood is considered a major annual hydro-meteorological phenomenon. There are several types of flood occurring due to spatial and temporal variations across the country causing disasters. The most common among them are riverine flood due to the monsoonal impact, mostly from July to September. Due to this impact, the majority of the river basins, especially Punjab and Sind provinces, are inundated causing severe floods. The snow melt in the upper reaches aggravates the situation. Flash flood also occurs in northern provinces of Pakistan, which causes great devastations in the region. During monsoon, cities like Islamabad, Lahore, Peshawar, Karachi, etc., are severely exposed to urban flooding. In addition, provinces like Sindh and Balochistan are particularly affected by coastal flooding due to cyclonic impact and resultant storm surge. The Makran coastal regions have gained notoriety of having frequently exposed to coastal flooding.

For the last 114 years, flood hazards have claimed about 12,156 lives with a total affected population of about 5.70 crores in the region. Flood alone has caused total damages of more than 19 billion dollars. Some of the remarkable floods in recent past includes flash flood in the Pakhtunkhwa province in northwest Pakistan (June, 2015), Balochistan province due to heavy torrential rainfall (June, 2015) (SAARC Workshop Report 2012) (Table 20.6).

20.2.7 Sri Lanka

Flood in Sri Lanka is a commonly occurring phenomenon, causing destructions and devastations more than other disasters in the country. Physiographically, northeastern and southwestern parts are particularly vulnerable to floods. Flooding is mostly triggered by both the branches

Table 20.6 Decadal flood damages and deaths in Pakistan (2005–2015)

Disaster subtype	Events count	Total deaths	Total affected	Total damage ('000 US$)
Flash flood	11	2278	20,447,593	9,827,118
Riverine flood	27	3157	22,305,462	8,633,000

(Source –emdat.be)

Table 20.7 Decadal flood damages and deaths in Sri Lanka (2005 2015)

Disaster subtype	Events count	Total deaths	Total affected	Total damage ('000 US$)
Flash flood	7	65	1,387,775	50
Riverine flood	15	408	4,442,332	609,200

(Source –emdat.be)

of monsoonal precipitation – southwest and northeast monsoon. The former strikes and inundates districts like Kegalle, Ratnapura, Kalutara, Colombo, Gampaha, and Galle, whereas the latter causes extensive flooding in districts like Ampara, Trincomalee, etc. (Table 20.7)

20.3 Flood Vulnerability in India

In the entire Indian subcontinent, the flood gains the status of disaster, when normal channels of rivers are breached or flow in excess of their carrying capacity. Flood is a recurrent phenomenon in India that normally starts with the onset of monsoon; however, due to localized hydrometeorological aberrations, there has been specific flooding in localized areas. The United Nations International Strategy for Disaster Reduction (UNISDR) Report 2015 observed that out of the average annual loss of 9.8 billion USD in India, about 7.4 billion USD are accounted by the damage caused by floods.

Rashtriya Barh Ayog (RBA) constituted in India to assess the flood situation in the country has listed the flowing situations for flooding:

- Streams flowing in excess of the transporting capacity
- Backing up of water in tributaries
- Heavy rainfall
- Ice jams or landslides blocking stream courses
- Heavy localized rainfall
- Cyclones and typhoons

Out of the total geographical area of 329 million hectares, about 40 million hectares is liable to floods in India as estimated by the RBA in 1980. Subsequently 11th five-year plan working group has compiled the area liable to flood as 45.64 million hectare. It is estimated that about 25 out of 36 states and union territories are flood prone in the country. The areas stretching north to south from the extrapeninsular regions to the tip of the peninsula and from extreme desert regions of the west to the east coastal regions and northeastern regions are all prone to floods in varying magnitude and nature.

The following table shows annual and decadal damage status in India due to floods (Table 20.8).

There are about 22 major river basins in India, out of which four major river basins are typically known as flood-prone basins:

1. Brahmaputra and Barak basin
2. Ganga basin
3. North West River basins
4. Central India and Deccan river basins

The Brahmaputra basin covering northeastern states, northern part of West Bengal, and Sikkim is affected by severe and recurrent floods. The entire catchment area of this basin receives heavy rainfall from June to September. The frequently occurring earthquakes and landslides in hills upset the flow regime of rivers causing imbalance in flow dynamics. In addition, spilling of rivers, drainage congestion and tendency of some of the rivers to change courses also cause flooding. In Assam and Tripura, flooding primarily occurs due to inundation by spilling of Brahmaputra and tributaries as well as bank erosion along the Brahmaputra.

Table 20.8 A damage due to floods/heavy rains during 2000–2012 in India

India											
Statement showing damage due to floods/heavy rains during 2000–2012											
				Damage to crops		Damage to houses					
Sr. no.	Year	Area affected in m.ha.	Population affected in million	Area in m. ha.	Value in Rs. Cr.	Nos.	Value in Rs. Cr.	Cattle lost nos.	Human live lost nos.	Damage to public utilities in Rs. crore	Total damages crops, houses and public utilities in Rs. crores (col.6+8+11)
1	2	3	4	5	6	7	8	9	10	11	12
1	2001	6.175	26.463	3.964	688.481	716,187	816.474	32,704	1444	5604.461	7109.416
2	2002	7.090	26.323	2.194	913.092	762,492	599.368	21,533	1001	1062.083	2574.543
3	2003	6.120	43.201	4.268	7307.230	775,379	756.481	15,161	2166	3262.154	11,325.866
4	2004	5.314	43.725	2.888	778.694	1,664,388	879.601	134,106	1813	1656.090	3314.385
5	2005	12.562	22.925	12.299	2370.923	715,749	380.531	119,674	1455	4688.219	7439.672
6	2006	1.096	25.224	1.822	2850.668	1,497,428	3,636.848	266,945	1431	13,303.926	19,790.922
7	2007	7.145	41.402	8.795	3121.532	3280233	2,113.108	89,337	3389	8,049.037	13,283.677
8	2008	3.427	29.910	3.186	3401.563	1,566,809	1,141.891	101,780	2876	5046.481	9589.935
9	2009	3.844	29.537	3.592	4232.609	1,235,628	10,809.795	63,383	1513	17,509.353	32,551.758
10	2010	2.624	18.297	4.994	5887.380	293,830	875.952	39,706	1582	12,757.253	19,520.586
11	2011	1.895	15.973	2.718	1393.847	1,152,518	410.475	35,982	1761	6053.570	7857.892
12	2012	2.141	14.689	1.950	1534.108	174,526	240.572	31,558	933	9169.968	10,944.648
Total		**62.433**	**341.669**	**57.67**	**34,486.127**	**13,835,174**	**22,669.096**	**951,878**	**21,374**	**88,173.595**	**14,5315.3**

In the *Ganga Basin*, flood problems are mostly confined to the northern bank of the Ganga. This recurrent phenomena occurs mostly due to spilling over banks and change in river courses; however, erosion problems are confined to a few places. The problem increases from west to east and from south to north.

In *North West River basin*, covering Punjab and Haryana, inadequate surface drainage along with high rainfall causes flooding and waterlogging over vast areas.

The *Central India and the Deccan basin* covers all southern states where most of the rivers have well-defined and stable courses, but inadequate capacity in lower reaches and delta regions causes flooding. In addition, urban flooding also occurs due to heavy rainfall. The deltas of Mahanadi, Godavari, and Krishna periodically face flooding in the wake of cyclonic storm.

The varied climatic and rainfall patterns in different parts of the country create typical situation in a way that while some of the regions are

Table 20.9 Areas liable to floods in India

Sl. no	State	Geographical area (Mha)	Area liable to floods (Mha)	Percentage of areas liable to flood	Area protected (Mha) as considered by RBA
1	Andhra Pradesh	27.51	1.39	5.05	0.700
2	Assam	07.84	03.15	40.18	1.305
3	Bihar	17.39	04.26	24.50	1.566
4	Gujarat	19.60	01.39	07.09	0.362
5	Haryana	04.42	02.35	53.17	1.095
6	Himachal Pradesh	05.57	00.23	04.13	–
7	Jammu and Kashmir	22.22	00.08	00.36	0.012
8	Karnataka	19.18	00.02	00.10	0.001
9	Kerala	03.89	00.87	22.37	0.011
10	Madhya Pradesh	44.34	00.26	00.59	–
11	Maharashtra	30.77	00.23	00.75	0.001
12	Manipur	02.23	00.08	03.59	0.073
13	Meghalaya	02.24	00.02	00.89	0.075
14	Orissa	15.57	1.40	08.99	0.351
15	Punjab	05.04	3.70	73.41	2.407
16	Rajasthan	34.22	3.26	09.53	0.016
17	Tamil Nadu	13.01	00.45	03.46	0.029
18	Tripura	01.05	00.33	31.43	0.009
19	Uttar Pradesh	29.44	07.34	24.93	0.739
20	West Bengal	08.88	02.65	29.84	1.001
21	Delhi	00.15	00.05	33.33	0.023
22	Pondicherry	00.05	00.01	20.00	Neg.
	Total		**33.52**		**9.776**
	Say		34 Mha		10.00 Mha

Areas liable to floods in India (Source – MoWR, GoI)

suffering from floods due to excess of waters, at the same time, other regions might be affected due to drought conditions (Table 20.9).

The above table indicates the widespread occurrence of flood across the country, but states like Uttar Pradesh, Assam, Bihar, West Bengal, Odisha, etc., are typically more vulnerable to perennial flooding. The brief account of flood vulnerability in some states are discussed below.

20.3.1 Assam

The Rashtriya Barh Ayog (RBA) has declared the state of Assam as the most flood-vulnerable state in the country. Out of the total geographical area of 7.84 million hectares, about 3.15 million hectare land (40.18 %) is prone to floods. This makes 9.4 % of the total flood-prone areas of India (source-ASDMA). In addition to annual flooding, the state is also affected by flash floods during monsoon (Flood hazard Atlas of Assam).

Assam is situated in the middle of the two major river basins – *Brahmaputra* and *Barak*. Due to typical geo-climatic condition, there are high risks of floods in these basins. Besides, there are a number of natural and anthropogenic reasons of flood occurrences in the state. The heavy monsoonal precipitation combined with a unique physiographic setup makes the region highly susceptible to flood hazards. The situation gets aggravated when the narrow valley of the Brahmaputra compounded with steep gradient

Table 20.10 Vulnerability status of Assam

State (total distt. flood-prone districts)	Area prone to flood as assessed by RBA (Lakh hectare)	Most vulnerable flood districts	Vulnerable flood districts	Flood-vulnerable districts (as per GFCC report 2006) Distts.	Percentage area flooded
Assam (total districts, 27; flood-prone districts, 27)	31.50	1. Nalbari	1. Barpeta	1. Dhubri	51.47
		2. Morigaon	2. Sibsagar	2. Lakhimpur	35.04
		3. Darrang	3. Jorhat	3. Morigaon	27.69
		4. Lakhimpur	4. Udalguri	4. Dhemaji	26.77
		5. Dhemaji {source – Assam Flood Hazard Atlas}	5. Nowgong	5. Barpeta	25.04
			6. Goalpara	6. Jorhat	23.76
			7. Kamrup (rural)	7. Nalbari	15.13
			8. Bongaigaon	8. Sibsagar	17.22
			9. Dhubri	9. Goalpara	17.71
			10. Dibrugarh		
			11. Sonitpur		
			12. Golaghat		
			13. Tinsukia		
			14. Karimganj		
			15. Hailakandi		
			16. Cachar		
			17. Kamrup (metro)		
			18. Kokrajhar		
			19. Baska		
			20. Chirang		
			21. Karbi Anglong		
			22. North Cachar {source – Assam Flood Hazard Atlas, NRSC}		

of channels create the situation of drainage congestion and consequent flooding (Table 20.10).

20.3.2 Uttar Pradesh

Flood is a commonly occurring disaster in Uttar Pradesh that is also a recurrent phenomenon every year. Due to typical hydrometeorological and geomorphological conditions, eastern, central, and Terai regions of the state are more vulnerable to flood.

As per RBA estimation, 73.36 lakh hectares of the total geographical area are prone to flood, that are mostly concentrated in eastern Uttar Pradesh. Out of the total 76 districts in the state, 34 are flood prone (Table 20.11).

There are about eight prime rivers crossing the state that create devastations, when they are in full spate- Ganges, Yamuna, Ramganga, Gomati, Sharda, Ghaghara, Rapti, and Gandak. The annual precipitation due to southwest monsoon is the main flood-inducing factor in the region. About 80 % of the total rainfall (60–190 cm) received by the state is in the form of SW monsoonal precipitation. The pattern of flood occurrence follows the pattern of precipitation that increases from west to east and from south to north. As per the rough

NDRF ON A MISSION DURING ASSAM FLOOD-2012

Table 20.11 Vulnerability status of Uttar Pradesh

				Flood-vulnerable districts (as per GFCC report 2006)	
(Total distt. v/s flood-prone districts)	Area prone to flood as assessed by RBA (Lakh hectare)	Most vulnerable flood districts	Vulnerable flood districts	Distts.	Percentage area flooded
(Total districts, 76; flood-prone districts, 34)	73.36	1. Pilibhit	1. Gazipur	1. Mirzapur	20.77
		2. Lakhimpur Kheri	2. Unnao	2. Sidharthnagar	20.72
		3. Sitapur	3. Bulandshahar	3. Basti	17.97
		4. Bahraich	4. Lucknow	4. Ballia	15.23
		5. Barabanki	5. Bareilly	5. Farrukhabad	16.50
		6. Gonda	6. Bijnor	6. Gorakhpur	19.71
		7. Faizabad	7. Banda		
		8. Ambedkar Nagar	8. Saharanpur		
		9. Basti	9. Muzaffarnagar		
		10. Sant kabir Nagar	10. Shamli		
		11. Azamgarh	11. Gautambuddha Nagar {source – UPSDMA}		
		12. Mau			
		13. Ballia			
		14. Deoria			
		15. Kushinagar			
		16. Gorakhpur			
		17. Siddharth Nagar			
		18. Badaun			
		19. Farrukhabad			
		20. Kasganj			
		21. Balrampur			
		22. Shravasti			
		23. Maharajganj {source – UPSDMA}			

estimate, the average annual loss to crops, houses, and livestock in the state is to the tune of about INR 2000 crores {source-UP Govt. Report}.

20.3.3 Bihar

The state of Bihar is known as the *theater of natural disasters* especially flood on account of its recurrent nature every year with resultant loss of lives and substantial damages. As per the estimation done by the Water Resources Department, Govt. of Bihar, about 56 % of the total geographical area of the state is affected by flood and permanent waterlogging (Taal).

Bihar is geographically divided by river Ganges into two main regions – North Bihar and South Bihar. The former is fed by rivers flowing from the Himalayas and covering the entire North Bihar. Rivers like Ghaghra, Gandak, Burhi Gandak, Bagmati, Kamla, Kosi, Mahananda, and Adhwara have their catchment extending from Nepal to Bihar. Mostly being of glacial origin, these rivers are perennial in nature, thus maintaining an optimal level of flow round the year. During monsoonal precipitation, these rivers get an additional volume of water leading to large-scale flooding situations in the entire stretch of North Bihar. As per the estimation, 73.63% area of North Bihar is considered as flood prone {5}.

The river Ganges on its left bank receives tributaries like Ghaghra, Gandak, Kosi, etc. Since they are mostly of Himalayan origin, they carry along with them a large amount of silts, which are deposited all along their channels in the Indo-Gangetic plains. Heavy siltation combined with anthropogenic obstructions lead to overtopping of excess river water, causing widespread inundation and consequent flooding in the catchment area. The situation gets aggravated by further inundation through rivulets existing in the interfluves of all these tributaries. Due to prolonged siltations, some of these rivers have acquired characteristics of changing their courses, thereby destructing habitations and agricultural cultivation in the adjoining locations. The river Kosi is one such river, which has gained the notoriety of changing course drastically. During the last two centuries, this river has moved nearly 110 km westward, devastating the entire area that came in her path. Being very destructive in nature, this river is termed as *River of Sorrows*.

The geographical and geological setup in the south of river Ganges (South Bihar) is somewhat different. The region having a total geographical area of 44,000 km^2 is drained by rivers like Karmanasa, Sone, Punpun, Kiul, Badua, Chandan, etc. They are mostly rain fed. During monsoon period, the surplus water in these rivers get accumulated along the southern tracts of the natural levee of river Ganges, causing flood. The wetland region, lying south of river Ganges known as *Taals*, are also inundated by these surplus waters.

There are 28 flood-prone districts in Bihar, out of which 15 are considered as the most vulnerable and 13 are less vulnerable (Table 20.12).

Table 20.12 Flood-vulnerable districts of Bihar

State (total distt.; Flood-prone districts)	Area prone to flood as assessed by RBA (Lakh hectare)	Most Vulnerable flood districts	Vulnerable flood districts	Flood-vulnerable districts (as per GFCC report 2006) Distts.	Percentage area flooded
Bihar (total districts, 38; flood-prone districts, 28)	42.60	1. Darbhanga	1. Araria	1. Sheohar	45.40
		2. East Champaran	2. Bhojpur	2. Sitamarhi	39.63
		3. Katihar	3. Buxar	3. Darbhanga	38.69
		4. Khagaria	4. Gopalganj	4. Gopalganj	36.94
		5. Madhepura	5. Kishanganj	5. Saharsa	35.38
		6. Madhubani	6. Lakhisarai	6. Muzaffarpur	30.61
		7. Bhagalpur	7. Nalanda	7. Supaul	22.61
		8. Saharsa	8. Patna	8. Madhubani	20.53
		9. Samstipur	9. Purnia	9. Katihar	19.88
		10. Sheohar	10. Saran	10. Samastipur	19.66
		11. Sitamarhi	11. Sheikhpura	11. Bhagalpur	17.77
		12. Muzaffarpur	12. Siwan	12. Vaishali	17.53
		13. Supaul	13. West Champaran {source –Bihar Govt.}	13. East Champaran	16.94
		14. Vaishali		14. Purnea	15.69
		15. Begusarai {source –Bihar Govt.}		15. Araria	15.51

The historical Kosi flood in 2008 is considered as the worst disastrous flood ever occurred in the country. On August 18 2008, an unprecedented flood situation occurred when the river Kosi breached eastern afflux embankment near village Kusaha in Nepal. This led to the formation of new scattered channel east of the old channel releasing about 1.66 lakh cusecs of water spreading across the channel causing devastations mostly in districts like Saharsa, Madhepura, Purnia, Supaul, Forbesganj, Araria, Katihar, etc. Over 3.3 million people affected due to sudden onset of this disaster with over a million rendered homeless. The loss of about 3000 km^2 of fertile land was reported. About 412 g panchayats under five districts were severely affected due to this natural calamity. A total of 3,40,742 houses collapsed with more than a thousand animals died. All the basic amenities were severely disrupted in the affected districts. The railway tracks submerged, the electricity lines were disrupted, and many roads and communication networks were damaged in this catastrophe. The state government immediately started disaster response operations with the help of central government.

20.3.4 Punjab and Haryana

In the alluvial plain of Punjab and Haryana, the main reason of flooding is waterlogging and drainage congestion. In Punjab, the entire alluvial plains are basically old floodplains of rivers. The alluvial plains of Ravi Beas, Satluj, Ghaggar, and Markanda rivers along with other rivulets constitute about 10 % of the state geographical area. Rivers like Ghaggar and Markanda sometimes inundate the adjoining low-lying areas after their courses are choked and obstructed especially during the monsoon period.

The drainage system in parts of Jind, Rohtak, Hisar, and Gurgaon districts of Haryana are either poorly developed or damaged causing severe flooding (Table 20.13).

Table 20.13 Flood-prone districts of Punjab

State (total distt.v/s; flood-prone districts)	Area prone to flood as assessed by RBA (Lakh hectare)	Most vulnerable/ vulnerable flood districts
(Total districts, **20**; flood-prone districts, **11**)	37.00	1. Ropar
		2. SBS nagar
		3. Jalandhar
		4. Ludhiana
		5. Moga
		6. Ferozpur
		7. Fazilka
		8. Kapurthala
		9. Taran Taran
		10. Gurdaspur
		11. Amritsar {source –Punjab Govt. memo no. 11/12/14-5DM1/5157 dated 3-4-14}

The Punjab and Haryana plain together accounts for about 45 % of the total flood loss incurred by the country. There are various factors leading to higher degree of vulnerability to floods in both the states. In Punjab, almost 80 % of the total annual rainfall is concentrated over a short period of 3 months, which coincides with the cropping seasons. In addition, the increased developmental practices in the region have resulted to more encroachments in the floodplains, causing enhanced vulnerability of flood. The natural geomorphic structure of Haryana is like a saucer-shaped depression, along the linear axis of Delhi-Rohtak and Hisar-Sirsa. Due to heavy precipitation and poor drainage network, sometimes the entire area gets flooded, e.g., Rohtak flood (1995). In addition, some of the recently emerged urban agglomerations are facing the problems of drainage congestion and localised urban flooding due to inadequate or faulty drainage system.

Rohtak Flood 1995 (courtsey –indiatimes.com)

20.3.5 Rajasthan

The state of Rajasthan is typically known as the drought-prone region of India with generally little or scanty rainfall. There are about 13 river basins and sub-basins in the state out of which basins of three rivers – Chambal, Banas, and Luni are particularly prone to flooding due to reasons like excess rainfall in the catchment areas, sudden release of water from dams, breach in the embankment, decreased water bearing capacity of dams, etc. Besides these, the change in rainfall pattern has also increased the risk of flash floods. In 2006, the Barmer flood occurred due to unprecedented heavy rainfall in the region. Districts like Barmer, Jalore, Sirohi, Pali, Chittorgarh, Kota, Bharatpur, Bundi, etc., are among the flood-prone districts in the state.

Barmer Flood 2006

20.3.6 Odisha

On account of typical geomorphological and hydrometeorological setup combined with maritime impacts, the state of Odisha is vulnerable to flood in several ways. Primarily, flooding occurs due to heavy monsoonal precipitation, as about 80 % of the total annual rainfall is received over a short period of 3 months. Because of the heavy siltation in rivers like Mahanadi, Brahmani, and Baitarani. The surplus water overtops the main channels and inundate adjoining areas. Sometimes, embankments are also breached due to heavy pressure of flood water. Since these rivers

Table 20.14 Flood-prone districts of Odisha

State (total distt.; flood-prone districts)	Area prone to flood as assessed by RBA (Lakh hectare)	Most Vulnerable flood districts	Vulnerable flood districts	Flood-vulnerable districts (as per GFCC Report 2006) Distts.	Percentage area flooded
Odisha (total districts, 30; flood-prone districts, 24)	14.00	1. Balasore	1. Angul	1. Cuttack	19.55
		2. Bhadrak	2. Bargarh	2. Jagatsinghpur	31.70
		3. Cuttack	3. Balangir	3. Kendrapara	28.86
		4. Jagatsinghpur	4. Deogarh	4. Bhadrak	23.83
		5. Jajpur	5. Gajapati	5. Puri	22.01
		6. Kendrapara	6. Ganjam	6. Jajpur	37.32
		7. Khurda	7. Kalahandi		
		8. Nayagarh	8. Kandhamal		
		9. Puri {source – Govt. of Odisha}	9. Keonjhar		
			10. Koraput		
			11. Mayurbhanj		
			12. Nabarangpur		
			13. Rayagada		
			14. Sambalpur		
			15. Sonepur (Subarnapur) {source – Govt. of Odisha}		

have common delta, they create a disastrous situation, while in the high spate. The situation gets aggravated when flood synchronizes with the condition of high tide.

The linear coastal tract of Odisha is also vulnerable to storm surge, especially during cyclone season. This situation is often accompanied by heavy precipitation that resulted in flooding in the lower reaches of the coastal districts. The deforested catchment areas and offshore bars choke the river mouths and obstruct the free flow of waters into the sea. In 1960, the region was hard hit by unprecedented flood. In super cyclone of 1999, many of the low-lying coastal districts were completely inundated. The high population density along with poor socioeconomic condition has led to increased encroachment in the entire stretches of floodplain, which ultimately increased the vulnerability of floods in the state. Out of the total 30 districts, 24 are flood prone, including nine most vulnerable districts as depicted in Table 20.14. For the last 10 years, floods in the state have caused great havoc with

Ganjam Flood, Odisha (Source-NDRF)

resultant loss of lives, properties, and infrastructures. The most severe flood that occurred in 2006, 2007, 2008, and 2011 had caused huge devastation.

20.3.7 Gujarat

On account of the maritime impact and typical geomorphic setup, the state of Gujarat receives heavy precipitation, which often results in the inundation of low-lying areas and consequent flooding. In addition, there are other factors responsible for flooding in the state. In 1979, a flood in Morbi city occurred due to dam break in which about 12,000 people died. In August 2006 Surat city and South and Central Gujarat got affected by flood wherein about 250 people died of drowning and leptospirosis.

20.3.8 West Bengal

The state of West Bengal has the total geographical areas of 88.75 lakh hectares out of which about 37.6 lh of land is identified as flood prone spreading over 111 blocks. The state is typically vulnerable to three main types of flooding – flash floods, riverine floods, and coastal floods. The flash floods occur due to the impact of heavy torrential rainfall supplemented with cloud burst, storm surge, and cyclonic impact. In urban agglomerations, the condition gets aggravated due to poor and choked drainage system coupled with poor urban planning. Riverine flooding occurs when the surplus volume of water due to high precipitation added in the river channels, particularly when they are in full spate. Most of the rivers flowing in the state are having their origin in Nepal, Bhutan, and Sikkim (India), and they flow downstream to meet either

in the sea or neighboring country of Bangladesh before meeting to the ultimate destination. The rivers like Teesta, Torsa, Jaldhaka, Raidak (I&II) flows through the districts of Jalpaiguri and Cooch Behar. They cause intense flooding in these districts when there is high precipitation in the upper reaches.

The river Mahananda when gets high discharge in the upper catchment area, causes flooding in Uttar and Dakshin Dinajpur. The district of Malda is flooded due to high discharge of rivers Fulhar-Mahananda-Ganga. The river basins of Bhagirathi-Hooghly create flooding primarily due to excessive discharge in these rivers on account of high precipitation in the catchment area associated with drainage congestion and decreased carrying capacity of river channels. The most affected regions are from Jangipur (Murshidabad) to Kalna (Bardwan). The South Bengal in general is flood prone having threats of riverine flooding and tidal flooding. The situation acquires an alarming position when tidal bore is also at peak during flooding in rivers. The districts like Darjeeling (N Bengal), Bankura, and Purulia (S Bengal) are comparatively free from flood threats. Out of the total 20 districts in the state, 15 are considered as flood prone (Table 20.15).

Table 20.15 Flood-vulnerable districts of West Bengal

State (total distt.; Flood-prone districts)	Area prone to flood as assessed by RBA (Lakh hectare)	Vulnerable flood districts	Flood-vulnerable districts (as per GFCC report 2006)	
			Distts.	Percentage area flooded
West Bengal (total districts, 20; flood-prone districts, 15)	26.50	1. Jalpaiguri	1. Murshidabad	19.92
		2. Cooch Behar	2. Birbhum	15.55
		3. Uttar Dinajpur	3. Nadia	17.60
		4. Dakshin Dinajpur		
		5. Malda		
		6. Murshidabad		
		7. Nadia		
		8. North 24 Parganas		
		9. Birbhum		
		10. Burdwan		
		11. Hooghly		
		12. Howrah		
		13. Paschim Medinipur		
		14. Purba Medinipur		
		15. South 24 Parganas {source govt. of West Bengal}		

20.3.9 Andhra Pradesh

Traditionally Andhra Pradesh is affected by multifaceted problems of flooding. While rivers like the Godavari and the Krishna along with other smaller rivers inundate surrounding areas due to excessive inflow in their channel beyond the carrying capacity, storm surge along the coastal belt also results in flooding. In addition, the drainage congestion mostly in delta regions is also responsible for flood problems in the state. In Telangana region (now a state), Khammam district is vulnerable to monsoonal flooding that occurs due to NE monsoon. The coastal belt from Nizampatnam to Machilipatnam is affected by storm surge flooding mostly due to cyclonic impact. The delta regions of Godavari and Krishna that are the most fertile regions of Andhra Pradesh experience problems of recurrent flood and drainage congestion.

Besides monsoonal and storm surge flooding, the riverine districts are also vulnerable to floods occuring due to peak discharges. The dams located along various rivers release excess water in the downstream when they are in high spate. To cite some examples, peak discharge of river Krishna at Vijaywada from NS Dam in the years 1989, 1990, 1991, 1998, and 2009 caused flood in the districts of Guntur, Krishna, Nalagonda, Kurnool, and Mahaboobnagar, and peak discharge of river Godavari at Perur affected districts of Khammam and East and West Godavari in the years 1986, 1990, 1992, 1994, 2004, 2005, and 2006. Peak discharge of river Penna at Nellore affected Nellore district in 1988, 1991, and 2001. Similarly, peak discharge at river Vamsadhara at Gotta barrage (I & CAD) badly affected Srikakulam in 1990 and 1992. Peak discharge at river Nagavalli also affected Srikakulam and Vijayanagaram in 1990, 1991, 1992, 1994, and 1996 (source-KSNDMC)

20.3.10 Flooding in Highland States

Flooding in the highlands or extrapeninsular regions of India is unique in terms of nature and magnitude considering the unique geomorphological topography. They are also known as upstream flooding that mostly occurred in the form of flash floods, cloud burst, or glacial lake outburst floods. They occur in a small, localized, and upper parts of the basin. The states mostly affected by this hydrometeorological phenomenon are Uttarakhand, Himachal Pradesh, J&K, Sikkim, etc.

20.3.10.1 Uttarakhand (Flood 2013)

On June 16 and 17, 2013, the state of Uttarakhand received more than 340 mm of rainfall, which was 375 % more than the normal benchmark of 65.9 mm rainfall during a normal monsoon. The sudden occurrence of cloud burst near Kedarnath temple and flooding in rivers like *Bhagirathi*, *Mandakini*, *Ashiganga*, *Kali* caused flash floods. This created one of the greatest disasters in the Indian history. Four out of 13 districts were heavily affected by floods, which created huge devastations in terms of loss of lives, properties, and infrastructures. At least 5000 people reported to have been killed in the deluge that inflicted heavy damage especially in the Kedarnath valley. Many thousands of pilgrims of "Char-Dham" Yatra were also affected. The unprecedented cloud burst and subsequent flash floods caused huge loss of lives and properties particularly in *Rudraprayag*, *Pithoragarh*, *Uttarkashi*, and *Chamoli* districts.

Damaged Houses in Uttarakhand Flash Floods

Damaged Roads in Uttarakhand Flash Floods

20.3.10.2 Jammu and Kashmir (Flood 2014)

The Jammu and Kashmir state has a very peculiar geography and climate. Most of the valley regions of the state are fed by rivers like Jhelum, Indus, and Chenab. The low lying areas of the Kashmir Valley, especially Srinagar, along with parts of Jammu, are prone to floods that occur due to heavy rainfall in upper catchment areas. Heavy rain followed by flash flood in September 2014 caused great devastation in the valley that claimed at least 280 lives and stranded hundreds of thousands of residents. The flood was unprecedented in nature, where the most part of the southern district has received very high rainfall. The weekly total rainfall for most of the stations during the period Sept 2 to Sept 8, 2014, was more than 200 mm. This is very high for a terrain like Jammu and Kashmir. The table below shows spatial distribution of weekly rainfall for select India Meteorological Department stations in J&K.

Rainfall status during 2 weeks in August to Sept. 2014

Actual (mm)	Normal (mm)	Departure (percent)	Category
43.2	27.9		E

August 28 to Sept. 3 2014

Actual (mm)	Normal (mm)	Departure (percent)	Category
267.7	30.0	792	E

Sept.4 to Sept.10, 2014 (Source –IMD/Frontline)

The primary cause was incessant rainfall that occurred due to western disturbance on September 3rd 2014 and continued for 4 days. The melting of snow also added the severity of floods. The districts like *Anantnag*, *Baramulla*, *Doda*, *Jammu*, *Kulgam*, *Pulwama*, *Ramban*, *Reasi*, *Sopian*, and *Udhampur* received extremely heavy rainfall which inundated many low lying areas of the state. Huge devastation occurred in these districts in the form of loss of houses, livestock, crops, and livelihoods. Many cities were badly affected. The capital city Srinagar was marooned due to flood for many days. Many localities turned into lakes with collapsed houses and communication networks.

Geomorphologically, it is said that the huge devastation occurred due to the absence of floodplain along the river Jhelum which could have accommodated excess of runoff averting floods in the valley. The massive urbanization in the urban localities of J&K along river Jhelum led to the destruction of floodplains almost all along the river, which added the severity of floods in the valley.

After the disaster, the National Disaster Management Authority took control of the situation and started disaster response at different locations in the valley. National Disaster Response Force (NDRF) battalions pressed into action along with Indian Armed forces to provide rescue and relief. The armed forces (including Army, Navy, and Air Force) and National Disaster Response Force (NDRF) launched one of the biggest search and rescue operations in the state. Operation "Megh Rahat" and operation "Sahayata" involved all the three wings of armed forces and NDRF which rescued people from inundated areas and provided them food, water, medicines, and shelters. By September 19, 2014, over 2.37 people were rescued by the joint efforts of these forces. The army deployed about 30,000 troops for search, rescue and relief operations.

The following table gives an account of flood occurrence in India from 2005 to 2014 (Table 20.16).

20.4 Flood Forecasting and Early Warning System

The forecasting and early warning is an integral and important component of flood risk management, which enables authorities and communities to take appropriate preparedness measures for

Table 20.16 Flood occurrence in different states of India (2005–2014)

No.	Year	Date of occurrence (dd/mm)	Location (city/village, state)	No. of Population affected	No. of Population Lost	Estimated value of Damage (in 000 US$)
1.	2005	27/05	Nagaland	12	14	
2.		28/06	Gujarat	405,000	239	2,300,000
3.		09/07	Maharashtra		25	
4.		26/06	Himachal Pradesh	5000	6	23,000
5.		07/07	Assam, Arunachal Pradesh, Uttar Pradesh, Uttaranchal	1,908,000	70	–
6.		02/07	Madhya Pradesh	49,000	62	–
7.		10/07	Andhra Pradesh	10,000	12	–
8.		24/07	Gujarat, Madhya Pradesh, Maharashtra, Goa, Odisha, Karnataka, Himachal P, J&K	20,000,055	1200	3,330,000
9.		26/08	Uttar Pradesh	800,000	27	
10.		23/07	Andhra Pradesh	100,000	126	
11.		16/09	Himachal Pradesh, Uttar Pradesh, Uttaranchal	2504	23	
12.		14/09	Chhattisgarh, Orissa, Maharashtra, Madhya Pradesh	550,000	89	420,000
13.		21/10	West Bengal, Orissa	2,250,000	19	117,000
14.		22/09	Gujarat		15	
15.		23/10	Tamil Nadu	2,000,000	162	
16.		02/12	Tamil Nadu	200,000	30	
17.		05/07	Himachal Pradesh, Punjab, J&K, Haryana	2000	10	
18.	2006	15/04	Andhra Pradesh	–	20	
19.		31/05	Assam, Tripura	504,000	21	
20.		25/05	Kerala	10,800	32	
21.		31/05	Maharashtra, Gujarat		41	
22.		24/06	U.P.	300,000	130	
23.		04/07	Gujarat	8400	24	
24.		03/07	Odisha		33	
25.		24/07	J&K	800	15	
26.		28/07	Andhra Pradesh, Gujarat, Maharashtra, Chhattisgarh, Rajasthan, Madhya Pradesh, Odisha, Karnataka,	4,000,065	350	3,390,000
27.		13/08	Gujarat	50,000		
28.		18/08	Rajasthan	20,000	135	
29.		31/08	J&K	15,000	19	
30.		29/08	Uttar Pradesh	100,000	42	
31.		26/10	Tamil Nadu, Andhra Pradesh	225,000	47	
32.		01/08	Odisha, Andhra Pradesh, Chhattisgarh	2,000,000	185	
33.		26/10	Tamil Nadu, Andhra Pradesh	225,000	47	
34.		01/08	Odisha, Andhra Pradesh, Chhattisgarh	2,000,000	185	
35.		09/03	Madhya P, Maharashtra, Rajasthan	113	61	
36.		01/10	Madhya Pradesh		39	
37.	2007	18/06	Assam	200,000	15	
38.		01/07	Gujarat, Rajasthan, Madhya Pradesh	63,000	225	
39.		01/07	Maharashtra	10,000	62	
40.		30/06	Chhattisgarh	50,000	29	
41.		22/06	Andhra Pradesh, Kerala, Karnataka, Maharashtra	200,000	127	

(continued)

Table 20.16 (continued)

No.	Year	Date of occurrence (dd/mm)	Location (city/village, state)	No. of Population affected	No. of Population Lost	Estimated value of Damage (in 000 US$)
42.		21/07	Tripura			
43.		03/07	U.P., Bihar, Assam, Odisha, West Bengal	18,700,000	1103	
44.		12/08	Himachal Pradesh	15,000	76	
45.		08/08	Gujarat		16	
46.		22/09	Odisha, W Bengal	7,200,000	80	275,000
47.		16/09	Andhra P, Karnataka	20,000	94	
48.		27/10	Andhra P, Tamil Nadu	50,000	29	
49.		03/08	Odisha	500,000	15	
50.		12/07	Assam, Arunachal P, Meghalaya, Manipur	11,100,000	96	
51.		16/07	Kerala,	35,000	44	101,151
52.		11/02	Rajasthan, U.P., Madhya Pradesh, Kashmir	8	40	
53.	2008	20/03	Tamil Nadu, Karnataka	10,278	37	2000
54.		11/06	West Bengal, Odisha, Assam, Bihar, Gujarat, Goa, Haryana, Kerala, Karnataka, Maharashtra, Madhya Pradesh, Punjab, Odisha, Rajasthan, U. P., Tamil Nadu, West Bengal, Arunachala, Uttarakhand, Jharkhand	7,900,000	1063	123,000
55.		05/07	Assam, Bihar, Gujarat, Goa, Haryana, Jharkhand, Kerala, Karnataka, Maharashtra, Madhya Pradesh, Punjab, Odisha, Rajasthan, Uttarakhand, U.P., Tamil Nadu, West Bengal	50,000		
56.		05/09	Andhra Pradesh		74	
57.		14/09	Odisha, U.P., Himachal P,	2,400,000	173	
58.		30/08	Bihar, Assam	2,600,000	47	20,000
59.		27/11	Tamil Nadu	803,740	54	
60.		20/07	Assam, U.P., Bihar	225,333	142	
61.	2009	29/07	Delhi		11	
62.		00/07	Bihar, Odisha, W Bengal, Assam, Kerala, Gujarat, Karnataka,	1,886,000	992	220,000
63.		25/09	Karnataka, Maharashtra, Andhra Pradesh	4,100,000	355	2,150,000
65.		09/10	Meghalaya		20	
66.		03/11	Tamil Nadu	08	70	64,000
67.		26/08	Bihar,		52	
68.	2010	05/07	Haryana, Punjab, Kerala, Assam	400,000	53	447,000
69.		18/05	Andhra Pradesh	50,000	27	
70.		13/07	New Delhi		11	
73.		05/09	Assam	30,000		
74.		18/09	Uttarakhand, Bihar, Uttar Pradesh	3,267,183	200	1,680,000
75.		09/09	Punjab, Haryana, Uttar Pradesh	12,500		
76.		15/11	Tamil Nadu		203	22,000
77.	2011	10/08	West Bengal	700,000	47	275,000
78.		23/09	Odisha	3,443,989	239	930,000
79.		15/08	Eastern Assam, Western Meghalaya,	1000	07	

(continued)

Table 20.16 (continued)

No.	Year	Date of occurrence (dd/mm)	Location (city/village, state)	No. of Population affected	No. of Population Lost	Estimated value of Damage (in 000 US$)
80.		15/08	Assam, Bihar, Uttar Pradesh	5,549,080	204	
81.		23/07	Uttar Pradesh, Rajasthan	200,000	19	
82.		05/09	Odisha	2,100,000	42	432,000
83.		15/06	Uttar Pradesh, Uttarakhand		50	20,000
84.	2012	26/06	Assam, Arunachal Pradesh	2,200,000	120	
85.		21/08	Rajasthan,		37	
86.		19/09	Assam, Sikkim, Arunachal Pradesh	260	21	98,000
87.		21/08	Himachal Pradesh	9460	26	16,000
88.	2013	12/06	Uttarakhand, Himachal Pradesh, Uttar Pradesh, Bihar, Karnataka, Kerala, Gujarat, West Bengal	504,473	6054	1,100,000
89.		09/07	Uttar Pradesh	500,000	174	
90.		23/06	Assam	2,000,000	80	
91.		22/08	Madhya Pradesh, Uttar Pradesh, Assam	40,000	73	
92.		21/10	Odisha, Andhra Pradesh, Bengal	375,000	72	260,000
93.	2014	26/06	Assam,	18,500	27	
94.		16/07	Uttarakhand, Himachal Pradesh		26	
95.		03/08	Odisha	179,000	35	
96.		00/09	Jammu	275,000	298	16,000,000
97.		11/08	Uttar Pradesh	4000	31	
98.		24/09	Assam, Meghalaya	650,000	95	163,000
99.		16/08	Uttarakhand		27	
100.		09/08	Odisha	3,600,000	47	100,000

(Source –CRED/emdat.be)

impending flood situations in order to reduce flood damages and loss of lives.

Flood forecasting a non structural mitigation measures, which are considered complimentary to structural mitigation measures. There are primarily three types of forecasts prevalent in India and many South Asian countries – *Stage forecast*, *Inundation forecast and Inflow forecast.* While the *stage forecast or level forecast* gives information about water level in rivers, *inundation forecast* gives estimation about areas likely to be inundated or submerged during the high flood situation. The *inflow forecast* provides information about the amount of discharge in the river.

The flood forecasting system in India was established in the year 1958 in a scientific manner, when the *Flood Forecasting Unit* (*FFU*) was set up by the erstwhile Central Water and Power Commission (CW&PC) for generating and issuing flood warning at river Yamuna in Delhi. Since then, the system has been gradually expanded to other parts of the country with more advanced and scientific technology.

20.4.1 Stages of Flood Forecasting

There are four main stages of flood forecasting and warning in India

4.1.1 Real-time data collection
4.1.2 Transmission of data
4.1.3 Data processing and preparation of forecasts and warning
4.1.4 Dissemination of flood forecasts and warning

20.4.1.1 Real-Time Data Collection

The collection of real-time meteorological (rainfall) and hydrological (gauge discharge) data is the prime requisite for generating flood forecasts and warnings. While, at the central level, India Meteorological Department (IMD) and Central Water Commission (CWC) receive meteorological and hydrological data respectively through their base stations located across the country, a few states like Bihar, Karnataka, Andhra Pradesh, Tamil Nadu, etc., have established their own automatic weather stations generating and utilizing data for flood forecasting and early warning besides getting information from few national organizations.

IMD through its 10 flood meteorological offices receive daily rainfall data from more than 7000 centralized and state rain gauge stations in addition to forecasts of heavy rainfall and quantitative precipitation forecasts for various river basins. All these data are supplied to the respective flood forecasting centers of CWC.

20.4.1.2 Transmission of Data

The data generated at various hydrological and hydrometrological stations are transmitted to flood forecasting stations through different means of communication systems like VHF/HF wireless sets, telephones, V-Sat, Internet, etc.

There are 544 wireless stations owned by the Central Water Commission meant for near real-time data transmission. The high-frequency (HF) wireless sets are having 3–30 MHz frequency used for long-distance transmission of data, whereas very high-frequency (VHF) wireless sets with frequency range 30–3000 MHz are used for short-distance communication.

In addition, about 445 stations of CWC are equipped with (or somewhere under installation) satellite-based "telemetry stations" for generating automatic data communications (CWC Annual Report 2013).

During the Xth five-year plan, CWC established sensor-based telemetry stations at 223 locations across the country in basins of river Krishna, Godavari, Mahanadi, Chambal, Damodar, Yamuna, and R. Brahmaputra. Further, 222 telemetry stations were established during 11th five-year plan. It is proposed to establish another 600 stations during XII plan, which covers the entire country. All the sensors located at these stations will transmit data to earth stations located at Jaipur (Rajasthan) and Burla (Odisha) through INSAT or Kalpanasat satellites. All these received data are further transmitted through V-Sat to different modeling centers (CWC Annual Report 2013).

The transmission of data takes place twice to thrice in a day, depending upon flood situation. In case of extreme flooding, the transmission frequencies are even increased to hourly basis.

20.4.1.3 Data Processing and Preparation of Forecasts and Warnings

After receiving hydrological and hydrometeorological data from ground stations, the various modeling centers of CWC undertake data processing by using advanced software.

Primarily the precipitation and discharge data are utilized for forecast generation on a real-time basis. For example, *inflow forecasts* are generated by using rainfall runoff correlation through Windows-based hydrodynamic simulation modeling software MIKE-11. The Central Water Commission has developed a site-specific model for different locations under various divisions by using MIKE-11.

The flood forecast model for Srinagar (J&K) is currently under formulation based on MIKE-11. Similarly, the development of flood forecast models for rivers like Sankosh, Godavari, Brahmaputra, Jhelum, Alaknanda, and Yamuna basin has been taken up by the Central Water Commission (CWC).

20.4.1.4 Dissemination of Flood Forecasts and Warning

The final flood forecast information generated through various processes are disseminated to user agencies for issuing early warning and undertaking appropriate flood preparedness measures. The information is also circulated in print and electronic media for dissemination to local populace. Some of the user agencies like National Disaster Management Authority (NDMA), National Disaster Response Force (NDRF), state governments, railways and roadways authorities, defense forces, etc., are regular recipients of flood forecast information.

20.4.2 Agencies Involved in Flood Forecasting and Early Warning

20.4.2.1 Central Water Commission

The *Central Water Commission* is the technical agency dealing with water resources and flood management working under the *Ministry of Water Resources, River Development and Ganga Rejuvenation, Government of India.* The agency is entrusted with the general responsibilities of initiating, coordinating and furthering in consultation of the state government concerned, schemes for control, conservation and utilization of water resources throughout the country, for the purpose of flood control, irrigation, navigation, drinking water supply and water power development" (Central Water Commission, Ministry of Water Resources, River development & Ganga Rejuvenation, Govt. of India).

A separate *River Management Wing* has been established in CWC specifically for undertaking all activities related to flood management in the country. The wing is headed by Member (River Management) having an ex-officio status of Additional Secretary to the Govt. of India. There are various directorates in this wing headed by respective directors.

The role of this wing in the management of flood is primarily centered on collection, compilation, collation, and analysis of hydrological and hydrometeorological data and sharing of flood forecast analysis and early warning information to concerned state government and other user agencies except Ganga and Brahmaputra river basins for which two separate organizations have been created. The other functions of river management wing are as under:

- Formulation and dissemination of flood forecasts for all flood-prone rivers
- Providing support to state governments in technical matters of river and flood managements
- River morphology studies, appraisal of flood management schemes
- Provide advisory support to all user agencies including National Disaster Management Authority (NDMA)

The systematic approach of flood forecasting and early warning was initiated by CWC in 1969 with the establishment of *flood forecasting* and *warning organization* (Annual Report 2013). With the advancement of new technology, the flood forecasting network of CWC got strengthened. At present, there are 175 flood forecasting stations working under the *National Flood Forecasting* and *Warning Network* of CWC, out of which 28 stations are working for inflow forecasting and 147 are for level forecasting as listed in the table below.

River system-wise distribution of flood forecasting state

Sr. No.	River system	Types of forecasting station		Total
		Level forecasting	Inflow forecasting	
1	Ganga system	77	10	87
2	Brahmaputra system	27	–	27
3	Barak system	5	–	5
4	East flowing river system	8	1	9
5	West flowing river system	9	6	15
6	Southern system	3	1	4
7	Mahanadi	14	4	18
8	Godavari	3	6	9
9	Krishna	1	–	1

CWC Report (2013)

State wise distribution of flood forecasting station

Sr. no.	River system	Types of forecasting station: Level forecasting	Inflow forecasting	Total
1	Andhra Pradesh	9	7	16
2	Assam	24	–	24
3	Bihar	32	–	32
4	Chhattisgarh	1	–	1
5	Gujarat	6	5	11
6	Haryana	–	1	1
7	Jharkhand	1	4	5
8	Karnataka	1	3	4
9	Madhya Pradesh	2	1	3
10	Maharashtra	7	2	9
11	Odisha	11	1	12
12	Tripura	2	–	2
13	Uttarakhand	3	–	3
14	Uttar Pradesh	34	1	35
15	West Bengal	11	3	14
16	Dadra Nagar Heavily	1	–	1
17	NCT of Delhi	2	–	2
Total		147	28	175

CWC Report (2013)

The river management wing of CWC operates with the help of regional offices each headed by a Chief Engineer. At present, 14 circle offices and 25 divisions of CWC are actively engaged in flood forecasting activities.

20.4.2.2 India Meteorological Department (IMD)

India Meteorological Department (IMD) is an apex organization in the Government of India actively involved in weather forecasting and early warning in association with the Central Water Commission (CWC). Established in 1875, the organization is a leading meteorological service provider in the country with specialization in hydrometeorology, seismology, and allied subjects (India Meteorological Department, Government of India).

IMD supports CWC in formulating flood forecasts of different river basins by way of providing various meteorological inputs like *Quantitative Precipitation Forecast* (*QPF*) for 24 h and *weather condition* and *the rainfall probability forecast warning* for 24–48 h period. There are 10 flood meteorological offices located in Agra, Ahmadabad, Asansol, Bhubaneswar, Delhi, Guwahati, Hyderabad, Jalpaiguri, Lucknow, and Patna that provide region-specific weather information to flood control rooms and forecasting stations.

In addition, IMD also has the specialization in short-range weather forecasting by using Doppler Weather Radar (DWR). The DWR data are useful in numerical weather prediction models for better estimation of rainfall. The velocity and spectrum width data of DWR can provide weather forecast information in detail. The functioning of Doppler Radar is based on the principle of Doppler's effects. In case of relative motion between the source of the electromagnetic waves and the target, the waves reflected from the target has the change in frequency as compared to the transmitted waves. This change in frequency is called 'Doppler Shift' that is directly proportional to the relative velocity between target and the source of the electromagnetic waves. The DWR is operational round the clock in auto acquisition mode. The range of surveillance is 500 km; however, for the purpose of structural analysis, the range is 300 km. In India, there are six DWR installed along the east coastal regions viz. Chennai, Karaikal, Machilipatnam, Visakhapatnam, Paradip and Kolkata. One at Gopalpur is likely to be installed shortly. A Doppler Radar is installed in Mumbai in the west coastal region with three more at Kochi, Goa and Bhuj likely to be installed shortly. There are 11 DWRs installed in the inland locations in India at Agartala, Bhopal, Hyderabad, Jaipur, Lucknow, Mohanbari, Nagpur, New Delhi, Patiala, Patna and Srinagar. The DWR can be useful in the following observations:

- Amount and rate of rainfall
- Cyclone intensity and wind speed
- Direction and speed of the movement of thunderstorms, tornadoes and cyclones
- Expected storm surge height, potential destruction, etc

There are various IMD centers of prime importance operational through Delhi headquarter that provides support in weather forecasting and early warning-

(A) *National Weather Forecasting Center (NWFC)* is an integrated and automated wing of IMD that provides services in all types of weather forecasting as well as advisory support to all its regional and state offices. The center is based in IMD headquarter, New Delhi, with backup support server at IMD Pune. There are various under mentioned cells of NWFC, which generate weather forecast information by using GIS-enabled modeling software.
   - General forecasting cell
   - Aviation cell
   - Public weather service and multi-hazard monitoring cell
   - Cyclone warning and marine cell
   - Radar and satellite application cell
   - Nowcasting cell
   - Numerical weather prediction cell
   - Hydrometeorology cell
   - Agrometeorology cell
   - Climatology cell

   The products are used by different agencies working on different disasters.

20.4.2.3 National Centre for Medium-Range Weather Forecasting

National Centre for Medium Range Weather Forecasting is engaged in medium-range weather forecast and climate modeling through research, development, and application of advanced technology. The center uses various forecasting models like NGFS, NCUM, NGEFS, and VSDB for generating wind forecast, rain forecast, MSLP, meteogram, trajectory, dust forecast, temperature change forecast, etc. for India as well as Africa, Afro-Asia, Southern Ocean, and Antarctica (ncmrwf.gov.in). The products are used in operational forecasting by organizations like IMD, Indian forces, Snow and Avalanche Study Establishment (SASE), Bhabha Atomic Research Centre (BARC), Indian Institute of Tropical Metrology (IITM), Pune, INCOIS, etc.

20.4.2.4 State Agencies

In addition to CWC and IMD, there are few agencies in some of the states which are involved in weather forecasting and early warning through automatic weather stations (AWS). The information generated through these agencies is useful in the prediction of flood occurrence in the concerned river basins. At present, the states of Andhra Pradesh, Karnataka, Bihar, Tamil Nadu, Odisha, Rajasthan, etc., have established their automatic weather stations either on application or experimental basis; the details about few of them like Karnataka, Andhra Pradesh, and Bihar are as under.

(A) *Karnataka State Natural Disaster Monitoring Centre*

The government of Karnataka has established *Automatic Weather Stations* through Karnataka State Natural Disaster Monitoring Centre based in Bangalore. The automated and timely weather alerts and forecasts are helping farming communities as well as policy planners, including disaster managers to a great extent. The most significant feature of the system is automated generation of weather-related data and instant dissemination to user communities through electronic communication media.

The scope is focused on the real-time weather monitoring, data analysis, vulnerability mapping, risk assessment, report generation, and disseminating information to users (source-KSNDMC).

In the current AWS system, a dense network of GPRS-enabled solar-powered telemetric rain gauge (TRG) stations covering all the 5625 g panchayats and telemetric weather stations at all the 747 hoblies (cluster of few gram panchayats) have been designed and installed in Karnataka. Thus there is a rainfall monitoring station at every 250 km^2 in the state (source-KSNDMC).

The organization, with the help of TRGs and advanced software, generates data on rainfall, dry spells, aridity anomaly, agriculture sowing status, crop condition status of the major reservoir levels, etc. The accumulated reports are generated automatically at KSNDMC and disseminated through e-mails, SMSs, etc., to the concerned government officers and community dwellers.

An interactive help desk *Varuna Mitra* has also been established by KSNDMC to disseminate weather-related information, forecasts, and advisories to farming communities and the general populace. An AWS-based project on *urban flood monitoring and management* has been undertaken by KSNDMC for the city of Bangalore. The aim is to provide alerts and early warning to *Bangalore Municipal Authority* (*BMA*) for onward dissemination to citizens of Bangalore.

In the current system, automated data on the intensity and amount of rainfall is collected at every 15 min interval. Forecast information is generated at KSNDMC by using such data that are sent to concerned government officials of Bruhat Bangalore Mahanagara Palike (BBMP) through SMSs and e-mails.

During the situation of high rainfall, a *High-Intensity Rainfall Alert* (*HIRA*) is automatically generated and sent to these officials through the same media. Hourly rainfall maps are also generated, which are useful in rainfall variability analysis (source-KSNDMC)

(B) *Andhra Pradesh State Development Planning Society*

The state of Andhra Pradesh has established *Integrated Flood and Cyclone Warning System* by using automatic weather stations (AWS). The system is developed and implemented *by Andhra Pradesh State Development Planning Society* (*APSDPS*) based in Hyderabad.

The real-time flood warning system is being implemented for 24 river systems, including rivers like Pennar, Krishna, Godavari, Nagavali, and Vamsadhara covering a total area of 6.85 lakh square kms. The APSDPS, with the help of automatic weather stations, capture spatial and temporal rainfall data that are used for forecasting rainfall in advance for next 48 h using weather forecasting models. The forecasted rainfall data is then converted into runoff forecast by using hydrological models. This gives an estimate about the potential water levels in the rivers at certain time interval. For this purpose, the *MIKE-11* hydrodynamic model is used. The obtained data is finally loaded in the *MIKE GIS* software to generate potential *inundation map* of the area 48 h ahead of the event. This may enable local administration, policy planners, and local communities to respond appropriately to the emerging situations. The various models used for generating forecast data are *hydrological models* (for runoff estimation), *hydrodynamic models* (flood forecasting), and *rainfall atmospheric model* (precipitation estimation).

The system is providing support to the state government in getting real-time flood forecast information for the next 24 h, runoff estimation, rainfall forecast, potential areas likely to be inundated during high rainfall, etc. (source-APSDPS).

(C) *Flood Management Information System (FMIS)*

The Water Resources Department, Government of Bihar has established a network of flood forecasting and inundation modeling and information system with the help of World Bank known as the *Flood Management Information System (FMIS)*. The main objective of this system includes long-term objectives of developing and implementing a comprehensive information system to support policy planners and administrators of flood-prone areas. The other objectives are to develop flood hazard characterization and operational flood management information products, updated flood control manuals, etc. The system is based on technologies of GIS, remote sensing, rainfall forecast modeling, etc. FMIS may be extremely useful in the state in terms of providing rainfall forecasts; flood forecasting and inundation predictions; flood hazard zonation mapping; hazard, vulnerability, and risk analysis; etc. The system is being installed in a phase-wise manner in the state covering most vulnerable and vulnerable flood districts like East Champaran, Muzaffarpur, Begusarai, Samastipur, Darbhanga, Sitamarhi, Sheohar, Madhubani, Supaul, Saharsa (first phase) and Patna, Bhagalpur, and Munger districts (second phase).

20.5 Flood: Hazard and Risk Management

While defining flood, two main concept emerges out to understand – firstly flood is the phenomenon of water, streams, and rivers in a particular area and secondly no flooding may occur if the water is easily and quickly drained out or managed. Hence the concept of flood as a hazard should also incorporate these two points. The dimension of flood has different connotation for different countries in the world. The annual flood in the river Nile in Egypt is considered as the gift of God, which brings life and prosperity in the form of water and fertile silts for the people of Sahara deserts; the similar natural phenomenon brings a calamity somewhere in other parts of the world including India.

It is significant that regions where flooding gain the status of calamity are those where society or community gets affected. Flooding in uninhabited regions does not affect the society and therefore human concerns are less, whereas the same phenomenon that occurred in populated regions seriously affects the society and ultimately gains the status of disaster.

Flood as a disaster has long been a matter of concern for policy planners in India as it brings benefits as well as losses for communities. The annual floods in the Indo-Gangetic plains maintain the fertility of soil by depositing silt containing different minerals carried out from different parts of the mountains. They bring additional water for irrigation, thereby contributing to enhanced fertility in the region, whereas the same flood in larger dimension or in unprecedented situation seriously affects communities living in the floodplain or low-lying areas. In other parts of the country, the flooding occurs due to heavy torrential rainfall combined with poor drainage, that causes disasters in the society resulting in the loss of lives populations and infrastructures. The recent examples are Uttarakhand flood (2013) and J&K flood (2014).

In order to prevent this hydro-meteorological phenomenon attaining the status of disasters, there is a need to adopt a comprehensive, integrated, and scientific approach. Flood disaster management approach is essentially required in India because it affects the normal functioning of societies or communities, but above all, the optimal utilization of land and water resources is of vital importance to bring prosperity in the country.

On the account of unique and varied geo-climatic condition right from the extra peninsula in the north to peninsular tip in the south and from Arunachal Pradesh in the east to the extreme western Thar Desert, the nature and scope of flood risk varies greatly; therefore the risk management strategies and disaster management plan during the flood disaster must address all the topographical, geographical, and climatic conditions to effectively combat the potential threats.

The "National Water Policy" has suggested that there should be a master plan for flood control and management, for each flood-prone basin. It is essential to promote watershed management practices through water management, soil conservation, catchment area development, etc. to reduce the intensity of floods.

The systematic approach of flood management and mitigation in India at the policy level was started by the Government of India in the year 1954 after the unprecedented floods in different parts of the country. A policy statement by the Ministry of Planning, Irrigation, and Power was placed before the parliament under two separate categories – "floods in India (problems and remedies)" and "the floods in the country." The objective was to suggest a comprehensive framework for the management of flood disasters in the country. Since then, various committees have been constituted from time to time to suggest recommendations, strategies, and policies on various flood management and mitigation issues.

Table 20.17 Agencies constituted for flood management (1954–2004)

Sl. No.	Committee/working groups/task forces	Year	Highlights/recommendations of the report
1.	Policy statement	1954 (3rd Sept)	Submitted by union ministry of planning, irrigation and power before the parliament
			Highlighted two main statements – "floods in India (problems and remedies)" and "the floods in the country"
			Objective was to get rid the country of the menace of floods by containing and managing them
2.	Supplementary statement	1956 (27th July	Came after the policy statement in a less modified form
			Emphasized to "curb and confine the floods and do all the possible to save people from the harm and devastation they bring"
			Absolute immunity from flood damage was not physically possible, even in the distant future
3.	High level committee on floods	1957 (Dec.)	Absolute or permanent immunity from flood damage is not physically attainable by knowing the methods of flood control
			Measures like floodplain zoning, flood forecasting, and warning should be given due importance
			Flood control schemes should fit in other water related plans
			Future multipurpose project should consider inclusion of flood control aspects
			Effects of embankments on river regimes should be considered in project proposals
			Priority for soil conservation work relating to flood control
			Priority for watershed management
4.	Policy statement	1958	While substantial diminution of flood related distress is possible, immunity against flood is impracticable
5.	Minister's committee on flood control	1964 (February)	To review the national flood control policy outlined in 1954
			Recommended more attention to nonphysical measures like flood warning and forecasting, floodplain zoning, flood insurance, etc.
6.	Working group on flood control for the 5-year plans		Suggested appropriate strategies to formulate proposals, including mobilization of resources for each 5-year plan and recommended measures required for effective flood management programs in the country
7.	Rashtriya Barh Ayog (RBA)	1980 (March)	A total of 207 recommendations given covering the entire gamut of flood problem in the country
			Data collection on long-term performance and their impact
			Legislation and enforcement by states to prevent unauthorized river bed cultivation and encroachments into drains, etc.

(continued)

Table 20.17 (continued)

Sl. No.	Committee/working groups/task forces	Year	Highlights/recommendations of the report
			Separate reporting of flood damage for protected and unprotected areas and areas situated between embankments
			Legislation for management of floodplains
			A comprehensive, dynamic, and flexible approach to the problem of floods
			Priority for measures to modify the susceptibility of life and property to flood damage
			Priorities for the completion of continuing schemes
			Provision of adequate funds for maintenance
			Intensifying studies on sedimentation of reservoirs
			Forming a national council for mitigating the effect of the disaster
8.	Pritam Singh Committee Report	1980	To examine the erosion problem in West Bengal on both river Ganga upstream and downstream of Farakka barrage
			Suggested priority areas for undertaking anti erosion measures
9.	National Water Policy	1987	Basin-wise master plan for each flood-prone basin
			Sound watershed management and catchment area treatment
			Adequate flood cushion should be provided
			Emphasis on flood forecasting and floodplain zoning to minimize flood damage
10.	Report of the committee on flood management in northeastern states – **Naresh Chandra committee**	1988	Committee constituted after the great flood of Brahmaputra valley in 1987
			Recommendations of the RBA should be implemented by the state governments
			Anti-erosion works are costly and can be justified only when protection is provided for vital installations
			For drainage improvement, the adequacy of existing sluices and drainage channels should be checked in a timely manner
11.	Committee on flood management in Bihar, West Bengal, Uttar Pradesh and Orissa	1988	Committee constituted after the severe flood of Bihar, West Bengal, U.P., etc., in 1987
			Embankment should continue as cost-effective and quick measure
			Early completion of projects should be ensured
			Construction of raised platform for government or acquired land

(continued)

Table 20.17 (continued)

Sl. No.	Committee/working groups/task forces	Year	Highlights/recommendations of the report
			Ensuring adequate waterways to ease out drainage congestion
			Operation of existing reservoirs, keeping flood moderation in mind
			Floodplain zoning implementation
			Setting up of Tal development authority
			More funds for unfinished schemes in the Sundarbans of West Bengal
12.	Recommendations of the regional task forces	1996	Committee constituted after the monsoonal floods of Rajasthan and Haryana in 1996
			Govt. of India (GoI) had constituted 5 regional task forces:
			Eastern Region task force
			Northeastern Region task force
			Northern Region task force
			Northwestern Region task force
			Southern Region task force
			Main thrust of task forces was to implement the main recommendations of RBA, preparing a catalogue of embankments existing in various river systems
			10% of the annual outlay to be earmarked for flood control structure
			Studies and reviews on major reservoirs and operation/rule curves
13.	Expert committee for bank erosion problem of river Ganga-Padma in districts Malda and Murshidabad (West Bengal) – **G R Keskar Committee**	1996	Short-term measures for left bank upstream of Farakka barrage in Malda district, construction of two long spurs at Farakka barrage, repair/ restoration of existing protection works
			Short-term measures for right bank downstream of Farakka barrage in Murshidabad district – construction spurs near Bindugram, repair/ restoration of existing protection works
			Long-term measures include monitoring of performance of two long spurs upstream of Farakka
14.	National commission for integrated water resources development plan	1999	To shift strategy toward efficient management of floodplains, flood proofing, flood forecasting, etc.
			Performance of embankments has to be evaluated and suitable changes be made in design, construction, and maintenance for better results
			Network of flood forecasting and warning to be extended to remaining flood-prone areas
15.	Expert group for flood Management in U.P. and Bihar – **G N Murthy committee**	1999	Need for building the realistic data bank on hydrology, topography, geology, morphology, etc.
			Closure of gaps in the embankment
			Construction of storage reservoirs and watershed management

(continued)

Table 20.17 (continued)

Sl. No.	Committee/working groups/task forces	Year	Highlights/recommendations of the report
16.	Working group on flood control program for the 10th five-year plan – **R Rangachari working group**	2001	Suggested on the future strategy of FM
			International dimension of flood management
			Review of the implementation of RBA recommendations
17.	Report of the committee on Silting of Rivers in India – **Dr. B K Mittal committee**	2002	Studied the problem of silting in Indian Rivers and feasibility of desiltation
			Recommended catchment afforestation, right practice of land use, catchment area treatment, etc.
			Construction of suitable hydraulic structures
			Embankments along the aggrading rivers
			Selective dredging may be undertaken
18.	National Water Policy	2002	Basin-wise master plan for flood control and management
			Adequate flood cushions in reservoir projects
			More emphasis on nonstructural measures
			Strict regulations of settlements
			Flood forecasting activities to be modernized
19.	Expert committee to review the implementation of the recommendation of Rashtriya Barh Ayog – **R Rangachari committee**	2003	Flood damage assessment is not done realistically or on a scientific basis as per RBA recommendations. This needs corrective steps
			Lack of representative, scientific, and credible post-project performance evaluation
			Unabated and unplanned intrusion into the floodplains and river beds has now reached alarming dimensions
			Most of the recommendations of RBA have not been implemented
			Interstate issues in multi state river basins are a very important matter waiting to be effectively addressed
20.	Committee for the identification of critical "Anti Erosion schemes of Ganga basin states for inclusion in CSS to be implemented during 10th plan" – **C B Vashistha committee**	2003	Recommendations came after the committee visited river Ganga in Uttar Pradesh, Bihar, and West Bengal for assessment of the problem and gave its recommendations
21.	Technical group on flood control and erosion problem of North Bengal- **M K Sharma committee**	2004 (July)	Rivers like Teesta, Torsa, Raidak, and Mahananda, draining the North Bengal along with their several tributaries causes serious flood erosion problems in the region
			Design flood estimation may be done in accordance with the subzonal report of CWC (Manual of Estimation of Design Flood 1961)
			River training/activation of channel may be attempted
			Maintenance of embankments during the pre monsoon and monsoon period may be undertaken
			Comprehensive plan for flood management for the North Bengal may be prepared
			Materials such as bamboos, branches of trees, river shingle, etc., can be utilized for inducting siltation thus diverting the river flows and preventing bank erosion

(continued)

Table 20.17 (continued)

Sl. No.	Committee/working groups/task forces	Year	Highlights/recommendations of the report
22.	Task force on flood management/erosion control	2004	Looked into the recurring problem of floods in Assam and neighboring states like Bihar, West Bengal, and Eastern Uttar Pradesh
			Flood control schemes should be funded through the centrally sponsored scheme in the ratio of 90 % central and 10 % state
			Total investment of plan/flood management may be increased
			Funds in the state sector are earmarked as additional central assistance for the maintenance of embankments
			A revolving fund of Rs. 50 Crores is available to MoWR to take up emergent flood management schemes
			Funding the flood control component of the reservoir projects
			Authority in the northeastern region with all the statutory powers be set up
			Sikkim and North Bengal River Management Board be established
			Ganga flood control commission be strengthened by the addition of the post of member (works), etc.

The list of all the committees/working group constituted since 1954 is as under (Table 20.17).

Though several good policy initiatives were taken by the government of India to minimize the menace of floods in the country, the one which is of greater significance was the constitution of *Rashtriya Barh Ayog* (RBA). In 1976, the government constituted RBA under the chairmanship of Mr. Jaisukhlal Hathi, the then governor of Punjab/Haryana, to look into the contemporary situation of floods in the country and to carry out in-depth study of the long-term flood management approach. This was the most comprehensive study of flood carried out by any committee since independence. It looked into various flood problems and submitted its report in 1980 with 207 recommendations covering almost all the aspects of flood management. However, the implementation of recommendations by different states remained mostly under the slow pace. In order to review the progress of implementation of RBA, few committees were further constituted including an expert committee setup in 2001 by the Ministry of Water Resources under the chairmanship of Sri R. Rangachari. The committee reviewed each of the recommendations of RBA and opined that its implementation has been slow which requires more attention by the concerned stakeholders. The committee further suggested 40 out of 207 recommendations to be taken up on priority basis.

In many of the recommendations listed in previous pages by different committees/commissions, it was strongly felt that practically it is not possible to protect all areas against high flood and structural mitigation measures alone cannot be considered as appropriate for flood management, there should be a combination of both structural and nonstructural measures on flood mitigation, so that overall losses could be minimized. The flood management measures can be classified into the following categories and subheads as under.

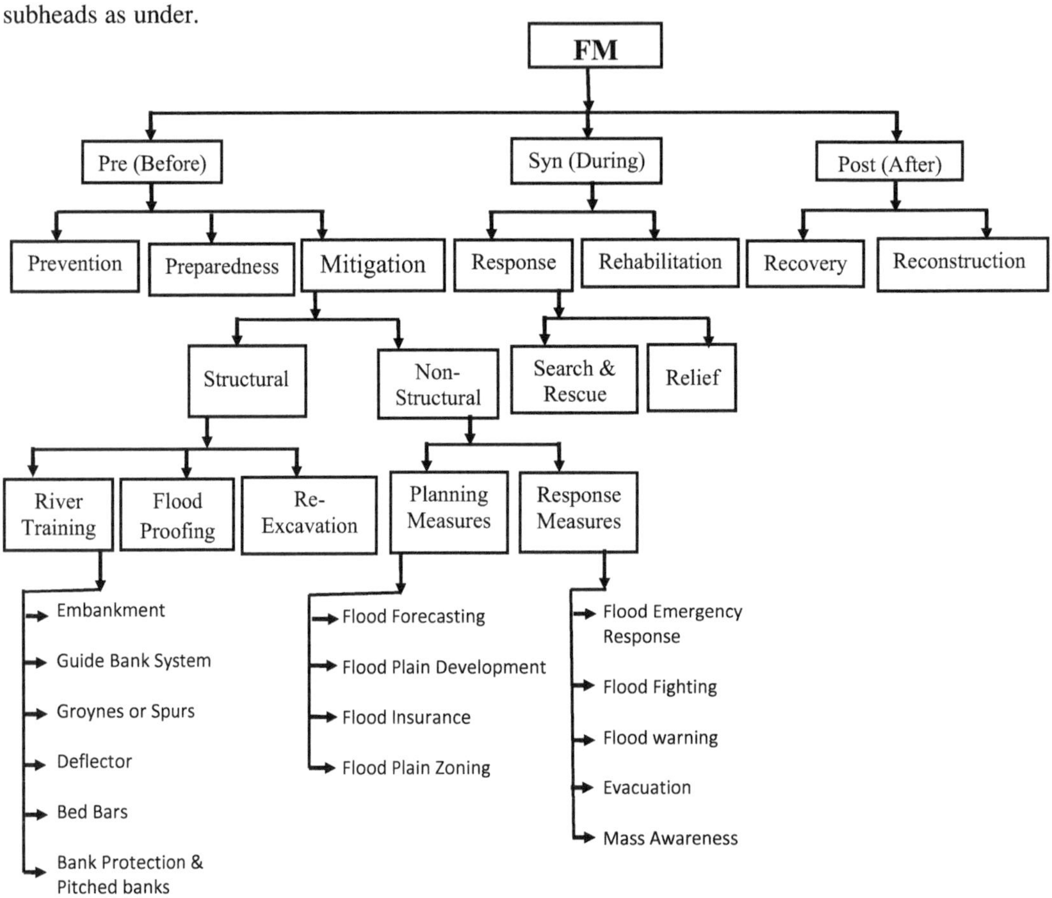

20.5.1 Flood Prevention, Preparedness, and Mitigation

Flood is a natural phenomenon of rivers and channels at the time of high discharge that cannot be prevented or checked through human interventions; however, its impact in the form of disasters may be prevented to a greater extent through better response planning, preparedness and mitigation approach. Disaster prevention, preparedness, and mitigation are such measures through which potential disaster threats can be minimized or even prevented (Ghosh 1997).

20.5.1.1

There are various preventive measures, which, if adopted in a right manner, at the right time, and at the right place, can avert the threats of potential disaster situations. The flood prevention measures can be the combination of both structural and nonstructural measures as mentioned below:

(A) *Rainfall Runoff Balance* – A complete balance of rainfall-runoff can be the best preventive measure for flood hazards. Due to the occurrence of high-intensity rainfall, resultant runoff becomes heavier and faster, which decreases the carrying capacity of rivers with resultant flooding in the adjoining areas. A large-scale catchment area treatment, including *Afforestation* and *reforestation*, may delay the surface runoff to reach river channels, thus minimizing the risk of flood disaster. The forest and vegetation cover can be helpful in a number of ways which includes delaying surplus water to reach river channels. They can promote infiltration of rainwater, thus preventing threats of flood as well as saturating groundwater. Forest and vegetation cover can also reduce soil erosion, thus preventing silts to be deposited in the river beds, thus increasing the water carrying capacity of rivers.

(B) *Smooth flow of discharge through river channels* – Developing smooth flow of discharge through river channels may be a good option for flood prevention, provided the socio-economic viability is ensured. This includes making meandering course of rivers straighter and linear, so that flood discharge may move downstream more rapidly. This was successfully experimented in 1933–1936 in river Mississippi (USA) near Greensville and on the river Missouri (USA) in Sioux city.

(C) *Detention of surplus water* – The flood storage reservoirs and detention basins are the structural measures of flood prevention that reduce the volume of water to reach river channels at the time of high discharge thus averting threats of disasters. The construction of reservoirs and detention basins helps in retaining surplus waters during peak flow, thus reducing threats of floods in downstream areas. In addition, the stored water may also be used for the purpose of irrigation and drinking water. The Tennessee Valley project in the USA and Damodar Valley project in India were undertaken for multipurpose utility, the prime being flood control (Sinha and Sinha).

(D) *Diversion of excess water to deficit regions* – At the time of high discharge, there is a need to divert flood water in low-lying areas, depressions, or secondary channels so that excess water may be used for other purposes and can be transferred to deficit areas. For example, Ghaggar Diversion Scheme in Rajasthan had used to divert about 340 cumecs of water discharge to a low-lying depression before entering Rajasthan.

20.5.1.2 Flood Preparedness

The preparedness for flood is an integral and most crucial element of the flood management plan which aims to minimize the adverse impact of flood hazard through effective precautionary measures and to ensure effective response at the time of disaster. The primary responsibility of flood preparedness in India lies with the state and district administration to make preparedness plans based on risk assessment and vulnerability analysis of floods. The following preparedness measures should be undertaken in any flood management program:

- State Authorities and the district administration should regularly check the early warning notices issued by the India Meteorological Department (IMD) and flood forecasting information issued by the Central Water Commission and other agencies.
- State governments or State Disaster Management Authorities must take prohibitive action for the blocking of natural drainage channels and sluices through appropriate policies and laws and also to improve their capacities and construct new channels and sluices to ensure the flow of excess rainwater in the area.
- In flood-prone areas, buildings constructed of earth, weak foundations, and water-soluble materials may collapse and endanger human lives and properties. Such structures may be checked in advance, so that precautionary measures can be taken in advance.
- During excessive rainfall, especially in hilly areas, physical damage in the form of landslides may occur. In addition, some structures may also get damaged due to high-intensity flowing waters. Such areas may be demarcated with the adoption of suitable mitigation strategies.
- Flood preparedness programs should be conducted at the community level, which may comprise of awareness campaigns, mock drills, NGO coordination, resource management, etc.

- District administration in the state should review the existing danger levels and warning levels in their districts, and if required, these level marks should be updated.
- Pre-monsoon inspection of rail tracks, roadways, canal networks, and drainage networks should be conducted periodically.
- In order to prevent outbreaks of epidemics and viral infections, well-coordinated medical preparedness strategies must be adopted well in advance, which may include stocking of emergency medical equipments and medicines and availability of medico and paramedic staffs.

20.5.1.3 Flood Mitigation

The flood mitigation measures are broadly categorized into *Structural* and *Non-structural* based on the kind of interventions. In structural mitigation measures, the emphasis is toward "preventing flood from affecting the society or population," whereas nonstructural mitigation measures aim to "keep people away from the flood."

A. Structural Mitigation This involves the process of constructing structures along the rivers or areas which are annually or perennially affected by floods in order to make rivers flow in a guided manner so as not to cause much damage during high discharge. Some of the prominent structural measures are as under:

1. *River Training Works* are types of engineering interventions applied on rivers to regulate and control the flow to river channels and river bed configurations, smooth navigation, control of sedimentation, etc., thereby contributing to flood protection, prevention, and mitigation in the catchment area. This is considered as one of the effective structural measures of flood management. The construction of dykes along the river Ganges for protecting adjoining cities from flood threats was an earlier attempt in the country to adopt river training works. This is considered as one of the successful techniques of preventing sedimentation and river bank erosion in perennially flood affected areas in Assam and Bihar.

There are various types of river training works that are considered helpful in flood management. They are *Bed Bar, Bank Protection and Pitched Bank, Guide Bank System, Groins or Spurs, Deflectors*, etc.

Transverse structures	Longitudinal structures
Check dams, Spurs, Sills, Screen dams, Porcupines, etc.	Embankments, Levees, Guide banks, Revetments, etc.

The river training structures are broadly classified into two types based on the alignment of structures with respect to rivers – *transverse structures* and *longitudinal structures*.

Transverse Structures

(i) *Groins or Spurs* are structures constructed in such a way that one part of it is projected toward the river course. They are also known as Spur Spur Dykes, Transverse Dykes etc. This is primarily useful for providing a directional flow to rivers and preventing river bank erosion. The "backward" sides of spurs are zones of moderate to slow flow, which promote siltation in between two spurs thereby creating natural banks along the river.

(ii) *Sill* is a transverse structure constructed on the river bed across the river to reduce downward erosion. This can be of different shapes and materials depending upon the utility of the structure and availability of materials

(iii) *Check dams* are constructed across rivers to stabilize the flow of river channels. These structures decrease the morphological gradient of the torrent bed and reduce the water velocity during a flood event by increasing the time of concentration of the hydrological basins and reducing the flood peak and solid transportation capacity of the river (Shrestha et al. 2012).

(iv) *Sediment retention structures* are created across rivers to filter debris, boulders, and other sediments reaching downstream through flood waters. Whereas the structure like *beam dam* is constructed to retain sediments and silts, *screen dams* are used for filtering materials like wood, tree trunks, branches, etc.

(v) *Porcupine structures* are useful for retarding the flow of water during peak discharge, thus helping to reduce river banks and bed erosion. They are designed mainly by timber or bamboo in such a way that they are protruded in different directions; however concrete structures are also in use. The common shapes are tetrahedral and prismatic (Shrestha et al. 2012).

Groins on river Narayani, Nepal (Source – northstarnepal.com)

Longitudinal Structures These are technically designed structures constructed along the river course for providing protection against river bank erosion, inundation in low-lying areas, and preventing rivers from meandering its course. They are of the following types:

(vi) *Embankment or Levee* is an earthen longitudinal protection structure, constructed along the course of river channels in such a way as to protect the area behind it from the overflow of flood waters. In India, since 1954, about 35,200 km of embankments have been constructed by March 2011 (XII plan report, GoI). Some of the rivers like the Ganges, Gandak, Damodar, Mahanadi, Godavari, Krishna, and Cauvery are known to have been protected by embankments on a larger scale.

The highlands or high ridges along the rivers are considered as suitable for the construction of embankment provided the soil and ground conditions are suitable. They should also be maintained and protected after construction to avoid any further disasters. In case an embankment is likely to be collapsed or likelihood of flood is more, a *loop bund* is constructed behind the embankment to provide a second line of protection.

(vii) *Bed bar* is a submerged longitudinal structure which divides the flow horizontally in two parts. The flow above the bed bar follows the weir flow, whereas the flow below the top level is obstructed by the bar and diverted toward the nose.

(viii) *Bank protection structures* are artificially constructed surface on river banks or slopes, designed to absorb energy of water waves and to protect them against erosion. Revetment is one of the most common structures of river bank protection wherein artificial slopes or surfaces are created along the bank. This can be done through "rip rap" which is a kind of arrangement of loose rocks or boulders to make the revetment structure.

(ix) *Guide banks*, as the name denotes, are embankment-like structures constructed on rivers to guide the flow of rivers or in other words to provide a directional and controlled flow of rivers. The guide bank system works on the principle that a flood partially controlled and directed by the groins should confine in a directional flow to ensure its safe passage without destroying the river banks or other structures.

2. *Flood proofing* is the long-term measure to mitigate the effects of flooding through modification of building and other infrastructures in the immediate surroundings of flood-

vulnerable locations in order to minimize the damage due to flood.

B. Non-structural Mitigation This involves planned activities to mitigate the adverse impacts of flooding without undertaking any structural changes/modifications. Unlike structural measures, this aims to adopt strategies for keeping people away from the flood vulnerability. Some of the common non-structural measures are as under:

(i) *Flood forecasting* and *warning* is the most common and important mitigation measure being adopted in India. This has already been discussed in the previous pages of this chapter.
(ii) *Floodplain management* is one of the major thrust areas of government's flood management programs, wherein more emphasis is on the developmental activities in floodplain areas, to make them encroachment-free to minimize flood damages. The objectives of such programs are:
   - To reduce future potential damage in floodplain and adjoining locations
   - To develop strategies for maximum utilization of floodplains for developmental activities during non-flood peak periods
   - To regulate and control floodplain through regulations, by-laws, building codes, policies, etc.

20.5.2 Flood Response and Rehabilitation

Flood response measures are activities undertaken at the time of flood occurrence or at the time when flood threats are imminent. The response measures undertaken by responder teams/organizations are usually planned, organized, coordinated, and effective at all levels, so that the affected communities get adequate relief at the time of calamity. In India, flood response measures are undertaken at different levels depending upon the severity of floods. In order to have a better management and control over the situation, it is essential to identify the scale of response required and the role of various responders right from the national level to the district and the village level. At the time of flood emergency, institutionalization of the system had to be properly managed at various levels. The response activities should not be confined to a single organization either by the government body or the private agencies, but this should be made as multi-organizational with close coordination of multiple stakeholders.

The role of communities as first responders is now an established fact. After any flood disaster, the community responders primarily extend their support with all the available resources; thus any response plan must take into consideration the role of community responders. The local volunteers must be trained by National Disaster Response Force (NDRF) and district/state training institutions in various skills of flood response in each district with basic training in search and rescue, medical first aid, CPR, etc. The community volunteers thus trained can assist in planning and setting up emergency shelters, distributing reliefs, identifying missing people, and addressing the need of education, health care, water supply and sanitation, food, etc. In addition, community level organizations like NGOs, Self Help Groups, community-based organizations, National Cadet Corps (NCC), National Services Scheme (NSS), Nehru Yuva Kendra Sangathan (NYKS), women's group, civil defense, etc., volunteer their services during flood and other disasters. Thus, there is a need that various task forces may also be constituted through active involvements of these stakeholders for inculcating a culture of preparedness to respond at the time of flood. The community-based preparedness and response planning coordination are required among various agencies as under:

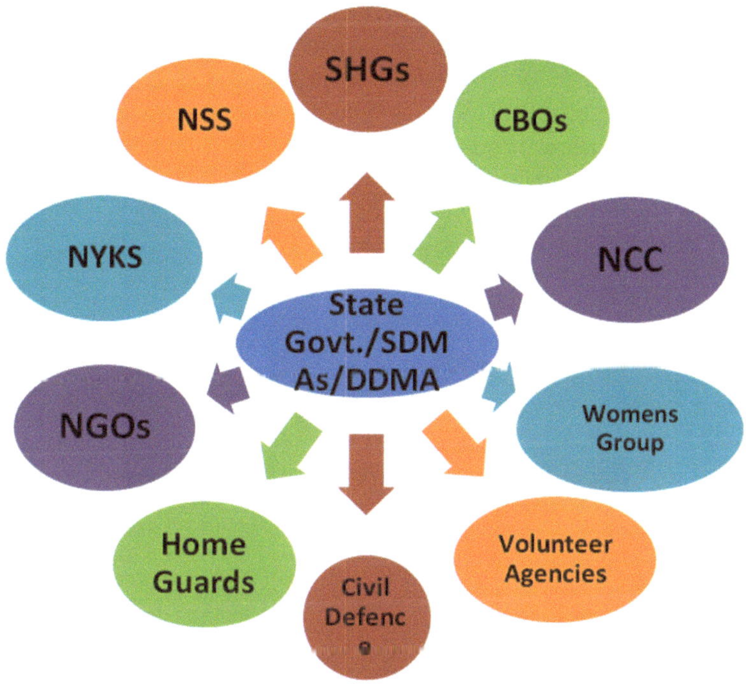

The flood response comprises a sequence of activities undertaken by a group of experts or skilled people or even unskilled community dwellers to carry out search and rescue of victims, mobilization of equipments, resources, and services. A successful response planning requires immediate planning, mobilization of resources, quick activation of essential functions etc.

During the situation of all the disasters, including floods, the response planning is strategically designed based on the severity of disasters and the ability of response authorities to deal with the situation. Accordingly, disasters are categorized as **L0**, **L1**, **L2**, **and L3** type (source-ASDMA).

Response severity level	Characteristics
L0	Normal times
	Stage to carry out prevention, preparedness, and mitigation activities
	Stage of research, documentation, and monitoring and planning
	State authority should focus on training and capacity building intervention
L1	Disaster severity is at the district level, which may be managed by the district authority at the district level
	State and central authority must be in readiness to provide assistance if required
L2	Disaster severity has expanded to more than one
	State intervention may be required
	Assistance may be provided from the state headquarters
	Central authority must be in readiness to provide assistance if required
L3	Disaster severity is large which is beyond the capacity of district and state administration to provide adequate response
	Central government intervention may be required
	Central assistance required (courtesy – asdma web site)

(continued)

Immediately after receiving early warning information, the evacuation is required to be undertaken in a systemic manner. The successful evacuation planning for population and livestock is considered as the only means to save them from flood disasters. Evacuation planning for a larger group of population becomes a difficult task, which required effective *Incident Response System* to be in place. Responsibilities are to be fixed for each stakeholder in the form of *Standard Operating Procedures (SOPs)*. The successful evacuation depends on a continuous dialogue with the following stakeholders:

The response activities should be carried out with preliminary estimation of the flood situation by studying the flood level data and inundation data received from satellites. This may help policy planners to undertake focused response activities at the right location at right time. While carrying out response activities, effective media management also plays a greater role. The state government may utilize different types of media, viz., print, electronic, and other social media to disseminate early warning of flood occurrence in any locality. This may help authorities to undertake smooth response operations. In addition, last mile connectivity is also ensured with the help of local administration and community-based organizations. Thus the effective response planning is much required in order to have a better culture of prevention, preparedness, and mitigation.

20.6 Role of Geographical Information System (GIS) and Remote Sensing (RS) in Flood Risk Management

The Mitigation of flood hazards can be successful only when the detailed knowledge is obtained about the expected frequency, character, and magnitude of hazardous events in an area as well as the vulnerabilities of the people, buildings, infrastructure, and economic activities in a potentially dangerous area. Remotely sensed imagery and the Geographical Information System (GIS) can be very effective in identifying the spatial component of flood for its better management. Remote Sensing offers a synoptic view of spatial distribution and dynamics of hydrological phenomena such as flood and River Erosion. They are used to measure and monitor the extent of flooded areas and provide a quantifiable estimate of the land area and infrastructure affected by flooding and erosion.

GIS may be defined in different ways, but the most commonly used definition is that provided by Burrough (1986) generally known as "tool box definition." He defined GIS as a *powerful set of tools that enables collection, storage, retrieval, analysis and presentation of geographically referenced information.* Remote Sensing (RS) on the other hand is defined as *the science of acquiring information about the earth's surface without actually being in physical contact with it.* The transfer of information is done using electromagnetic radiation with the aid of sensors. The Remote Sensing plays greater role in the development of GIS, both as a source of technology and as a source of data. Together with RS and modeling, GIS provides a wide range of applications in agriculture, geology, natural disaster management, hydrology, weather monitoring, business and service planning, government, logistics, and transportation and environmental management.

The role of this technology is vital in flood management as the information on the predicted flood extent is required by the government, the public, and the emergency departments in order to facilitate early preparations and plans well in advance before the actual flood event. Early preparations and planning result in the effective and efficient response, thus minimizing and/or mitigating the after flood effects. There has been widespread development in the use of hydrological models with a flood prediction component. These models are in most cases, either loosely or tightly coupled with GIS and remotely sensed data. Most of these models require different types of data input such as land cover, land use, river discharge rate, rainfall amount, surface roughness, Digital Elevation Models (DEM), and size of drainage basin, among others. In this case RS techniques can be used to obtain spatial and temporal information needed for parameterization of the distributed hydrological models. The general idea is that RS and GIS provide spatial and temporal data input required by the distributed hydrological models in order to simulate runoffs and thus floods. Remote sensing (RS) data in some studies have also been utilized to calibrate and improve the performance of distributed hydrological models. The REmote Sensing techniques provide an option of accessing information from otherwise physically inaccessible areas. GIS tools have been imbedded in the hydrological models to facilitate in data analysis, querying, and presentation of information in a more simplified way; thus they form a critical part of the distributed hydrological models used for flood prediction.

One of the key stages in flood management is the identification of areas with potential flood risk that is the product of flood hazard and vulnerability. Mapping of flood risk areas is not only important for the location of potentially hazardous zones but also for government, nongovernmental organizations (NGOs), and other planners to get an idea of where priority should be given while allocating resources. Evacuation agencies, insurance companies, and relief providers also require knowledge of spatial extent of inundated areas. This could be information about roads that may or may not be passable, worst affected areas, and areas suitable for camping during flood periods.

The Remote Sensing and GIS techniques have been proved resourceful at different stages of flood management. For example, areas of potential flood risk were able to utilize the overlaying function of a GIS to combine land cover maps with the flood-predicted zones. The resultant maps provide simplified information on the flood hazard (depth, velocity, direction of flow), elements at risk, their exposure, and vulnerability. In addition, flood hazard, vulnerability, and risk maps were drawn showing areas of low or high flood risk.

The GIS and modeling approaches, in particular, have been used in investigating the possible effects of land use changes during flood. Land use scenarios and their possible impacts in the generation of runoffs and consequently flood management may be useful. This may also be helpful in developing policy guidelines and recommendations for urban planning, land use planning, as well as settlements and types of buildings. In this way, flood impacts can be prevented or even mitigated.

The adequate knowledge of damage inflicted by flood is essentially required by the authorities and insurance companies in order to draw policy for compensation as well as to have an estimate of the cost of reconstruction. GIS has a function of overlaying layers and through this function, layers on inundated areas can be overlaid with land use maps, land cover layers, and infrastructure layers, among others. Remote sensing tool can be used for obtaining images before, during, and after flooding. These images are thereafter processed and analyzed in order to obtain information of land cover, buildings, roads, schools, and other infrastructures of the area under normal hydrological conditions (before flooding), inundated areas and flood extent (during flooding), and flood effects, deposits, and debris (after flooding). When the comparison of these images together with a pre-flood data is carried out, the extent of flood damage can be estimated.

In spite of the great potential that RS and GIS offer in flood management, their use has been limited to some extent. The presence of cloud covers during flood periods has been reported as the major challenge in the use of optical remote sensing in flood management. Using Synthetic Aperture Radar (SAR) is a better option since radar pulse has a higher penetration power to overcome the problem of cloud cover; however, its use, especially in developing countries, has been constrained by its high prices as well as limited coverage. One of the most pressing challenges of remote sensing technology is limited availability of imageries in time and space, seasonal variations, technical limitations, and above all the problem of low temporal resolution. With reference to the problem of temporal resolution, most radar images take some time before and also after the flood and in most cases the flood peak may not be captured. In other words, there is a time delay between the actual time the flood occurs and the time when satellite images are taken. Most of the current radar satellites have a long revisit time that can be up to 35 days. There are other challenges too in the application of GIS technology in natural disaster mitigation. These include high cost of digitization and raw data collection, the intrinsic complexity of predictive models, lack of appropriate raw data, inadequacy of hardware technology to handle large spatial data sets, and difficulty in GIS to manage historical data necessary for some natural hazard assessments. However, the technology has a great prospect in the country that may create revolution in the field of flood management.

20.7 Do's and Don'ts

Time	Phase	Do's	Don'ts
Before	Prevention	Design your houses as structurally flood proof	Don't encroach river channels and floodplain with illegal constructions
		The height of the plinth should be raised above the historic flood level	Don't cut trees & forests in the flood-prone areas. This may loosen soil and may cause river erosion and consequent flooding
		Always keep river channels and floodplain obstruction-free	Stop Deforestation, promote reforestation and afforestation
		Consider producing flood resistant crops	Don't cover topsoil layer with concretes/ metallic surfaces
		Promote/encourage afforestation/ reforestation in areas of flood vulnerability	Don't encourage river aggradation or degradation
		Keep some areas open for percolating excess runoff	Don't promote activities which may weaken the strength of embankment
		Maintain the depth of river channel to accommodate excess water at the time of flood	Don't compromise with the quality of materials for the construction of river training works
		Consider making strong embankments in the flood prone areas. Embankment should be aligned on the high ridges or natural banks of the river, where land and soil is suitable	Don't release debris, concrete materials, or solid wastes in river channels which may obstruct the flow-causing floods
		If you are a policy planner/engineer, consider applying river training works like construction of groins, deflectors, and bed bars to confine and regulate the flow of river channel	
		Debris or concrete or solid wastes should be disposed off suitably without disturbing river ecosystem	
	Preparedness	Get information about the safe evacuation routes of your community to reach the nearest shelter	Don't keep these items preserved for longer duration. Replace them after 6 months with fresh items
		Keep an "emergency kit" ready. The kit may contain:	Don't give importance to rumors
		Torch with spare batteries	Don't be in panic
		Candles and match boxes	
		Bottle of fresh water (pref. sealed bottled water)	
		Bottle of kerosene oil	
		Portable radio with spare batteries	
		Dry foods	
		Umbrella and **rain coats**	
		Gum boots	
		Salt and sugar	
		Keep your "first aid and medication kit" ready. The kit may contain the following items:	
		Essential life savings medicines	
		Packets of ORS or electrical powder	
		Adhesive bandages of various sizes (20 nos.0	
		5 × 10 in. sterile dressings	
		Gauze roller bandage (2 nos.)	

(continued)

Time	Phase	Do's	Don'ts
		Triangular bandage with safety pins (2nos.)	
		Sterile gauze pads 4 × 4 in. and **3 × 3 in. (1 pkt each)**	
		Antiseptic wipes	
		Hand sanitizer	
		Latex gloves (**disposable**)	
		Adhesive tapes (**25 × 5 mm**)	
		Non adhesive absorbent pad (**7.5 ×10 c. m**)- **4 nos**.	
		Saline solution (**30 ml**)	
		Wound dressing with bandage (**1**)	
		Ambu bags	
		Thermometer	
		Chlorine tablets, **etc**.	
		Always keep a watch on the expiry dates of these items. Replace them with fresh items after a certain period	
		Keep a waterproof bag ready with all your essential documents, photo ID cards, valuables, personal papers, etc.	
		Make an indoor plan of your house indicating which items would be raised or managed if water enters in your house	
		Be a regular listener/viewer of local news from radio/TV for warning and advice	
		Gather regular updates on flood warnings from local authorities like gram panchayats, BDO/Tehsildar office, weather stations, etc.	
During	Leaving houses	Ensure all your personal essential items are packed in waterproof bags	Don't forget to take personal kit, emergency kit, and first aid medication kit with you
		Intimate your neighbors/friends/local volunteers about locations you are shifting to	Don't keep yourself and your family empty stomach. Take light foods at regular interval
		Display your contact number on your house wall, so that you may be contacted in case of emergency	
		Raise all wooden furnitures, carpets, clothings, and other valuables onto the top of the roof under shed or at higher places to avoid contacts with flood water	
		Turn off all electricity/power switches and main power supply. Unplug all electrical gadgets	
		Insert sandbags in toilet bowls and all drainage inlets/outlets to prevent backflow of sewerage/floodwaters	
		Lock your home cautiously. Move through known evacuation routes	
		Always spray disinfectant, bleaching powder, etc., to keep your surroundings free of infection	

(continued)

Time	Phase	Do's	Don'ts
		Extend helping hands to government officials/volunteers distributing relief materials	
	Health care	Always use boiled water for drinking purpose	
		Keep all eatables covered	
		Don't take heavy meals	
		If exposed to diarrhea, ORS solution, rice water, etc., should be taken on periodic intervals	
	Emergency shelter	Immediately rush to higher places preferable at railway tracks or highways	
After	Returning home	Update yourself with the local situation through radio/TV	Prevents children to play or roam in flood water
		Check and repair all electrical appliances before using them	Don't enter receded water of unknown depth
		Eat only fresh foods or sealed foods	
		Before getting supplied pure water, use boiled water by your own	
		Snake bites are common during and after floods. Be careful	
		Always keep antivenom with you	

References

Annual Report (2013) Central water commission, ministry of water resources, river development & ganga rejuvenation. Govt. of India

Flood hazard Atlas of Assam, NRSC-DMS, Hyderabad http://asdma.gov.in/project_flood.html

Ghosh SN (1997) Flood control and drainage engineering 92nd Edition 0; Oxford & IBH Publishers

Manual of Estimation of Design Flood (1961) CWP&C, India

Report of SAARC Workshop on Flood Risk Management in South Asia (2012) SAARc disaster management center & NDMA, Pakistan

Shrestha AB et al (2012)

Sinha SC, Sinha SB Flood problems and management in Bihar

Web References

Andhra Pradesh State Development Planning Society. http://www.apsdps.ap.gov.in/

Assam State Disaster Management Authority Portal. www.asdma.nic.in

Central Water Commission, Ministry of Water Resources, River development & Ganga Rejuvenation, Government of India. http://www.cwc.gov.in

Flood Forecast Information System, Central Water Commission. http://www.india-water.gov.in/ffs

Flood Management Information System, Water Resources Department, Govt. of Bihar. http://fmis.bih.nic.in/

Govt. of Bihar web page. www.bihar.nic.in, http://disastermgmt.bih.nic.in/

India Meteorological Department, Govt. of India http://www.imd.gov.in

Karnataka State Natural Disaster Monitoring Center. http://dmc.kar.nic.in/

Maldives Disaster Knowledge Network. http://saarc-sadkn.org

SAARC Disaster Knowledge Network Portal. www.sadkn.org

The International Disaster Database. http://emdat.be/

R. Ranjan Disaster Risk Management Expert & Senior Consultant, (NDMA), New Delhi, India

Kolibri 360 GmbH
Ferdinand-Jühlke-Straße 7
99095 Erfurt, Deutschland
produktsicherheit@kolibri360.de